MISGUIDED

MISGUIDED

Where Misinformation Starts,

How It Spreads,

and What to Do About It

MATTHEW FACCIANI

Columbia University Press

New York

Columbia University Press
Publishers Since 1893
New York   Chichester, West Sussex

Library of Congress Cataloging-in-Publication Data
Names: Facciani, Matthew author
Title: Misguided : where misinformation starts, how it spreads,
and what to do about it / Matthew Facciani.
Description: New York : Columbia University Press, [2025] |
Includes bibliographical references and index.
Identifiers: LCCN 2024059409 | ISBN 9780231205047 hardback |
ISBN 9780231205054 trade paperback | ISBN 9780231555814 ebook
Subjects: LCSH: Misinformation | Disinformation | Communication—Social aspects
Classification: LCC HM1231 .F33 2025 | DDC 001.9—dc23/eng/20250331

Cover design: Elliott S. Cairns

GPSR Authorized Representative: Easy Access System Europe,
Mustamäe tee 50, 10621 Tallinn, Estonia, gpsr.requests@easproject.com

CONTENTS

ACKNOWLEDGMENTS

I would like to thank my friends, family, and colleagues for their unwavering assistance and encouragement as I wrote *Misguided*. I am especially grateful for my wonderful partner, Jasmine, whose invaluable help and support were indispensable in bringing this book to life.

MISGUIDED

INTRODUCTION

I magine you have an old friend named John. You went to high school with him and remember him being a friendly and curious person. You haven't spoken to him in a while, but you keep connected through social media and occasionally see pictures of his dog and his travels as they pop up in your newsfeed. One day you notice John has shared a link to a video that claims the Earth is flat. This surprises you because you don't remember John being into conspiracies. You click on his profile to delve deeper and see, scattered among travel and dog pictures, various links to articles and videos claiming the Earth is flat. As you scroll through his posts, you are surprised to see that he recently attended a Flat Earth Society meeting. You scratch your head. You wonder how your friend can believe something like this that is so clearly untrue.

John, being the curious person that he is, may have started out watching a few conspiracy videos online—whether seeking them out himself or engaging with them at the suggestion of social media algorithms. John always liked to think of himself as a contrarian and felt a sense of pride when he believed something that others did not. This personality quirk further led him to seek out other conspiracy theory videos and online discussion groups. He found out that a local Flat Earth Society group was having a meetup near him, and he decided to check it out. At the meeting, he met several friendly people and had a very positive experience. These positive social connections made him want to also attend a conference for flat-earthers that was just a few hours' drive away. These social components contributed to John's Flat Earth beliefs as well as his desire to share Flat Earth content online.

John is fictional, but his story is not far-fetched at all. In fact, a 2018 documentary titled *Behind the Curve* shares the stories of various people who believe the Earth is flat.[1] We might think this is an absurd belief, but the documentary illustrates how powerful social connection is in supporting such ideas. There are Flat Earth online communities and in-person conferences, which make it easy for flat-earthers to befriend those who already share their beliefs or to connect with or attract others who may come to share their beliefs with time and community influence. When you think about how important it is to maintain our social connections, it becomes clearer why someone might be motivated to protect a belief that connects them with people who provide friendship and a sense of belonging. Additionally, if that person takes pride in being an independent thinker, then they may be further motivated to believe something different from what other people believe. The film resonated with me because it also discusses how we are all vulnerable to social biases that influence our beliefs. Believing the Earth is flat may be an extreme example, but we probably all believe something untrue because people around us endorse it. This may be something innocuous, like believing a particular sports team is the best, or it could be something more serious and insidious, like believing an unproven supplement cures a life-threatening illness.

Daniel Clark, the director of *Behind the Curve*, was asked what lessons he hoped people would learn from his film. He answered: "My dream would be that when people watch it, they take Flat Eartherism as an analogy to something they believe in because it's so easy to demonize another group or another person for something they think, but you're kind of just as guilty if you do that."[2]

This book is about the social and psychological forces that guide us into believing false information. I hope it will help increase our understanding of why people believe falsehoods and will offer insight that helps us recognize times when we may be vulnerable to false and misleading information. It is easy to identify when someone else's beliefs are incorrect, but it is much harder to recognize our own biases. By learning about the social and psychological factors that impact how we evaluate information, we are better equipped to navigate the rapidly evolving information landscape.

THE MISINFORMATION PROBLEM

Since you picked up this book, I assume you agree that misinformation is a problem. We are not alone—70 percent of participants in a massive nineteen-country Pew Research Center poll believed the spread of false information online is a "major threat."[3] When focusing only on the United States, the poll found that

84 percent of U.S. adults agreed that made-up news and information is a "very big" or "moderately big" problem today.[4] This ubiquitous concern surrounding the spread of false information likely contributed to the selection of *misinformation* and similar terms like *post-truth* as Word of the Year by *Collins Dictionary*, Dictionary.com and *Oxford Dictionary*.[5] While the 2016 U.S. presidential election sparked a broad interest in fake news and misinformation, these problems are hardly new.

When I use the word *misinformation*, I am broadly referring to false or misleading information. I will dive into the details of how we define and study misinformation in chapter 1. We are seeing more and more people concerned about misinformation, and this trend will likely continue in the future. The advent of the internet provides us with greater access to evidence-based knowledge, but it also exposes us to more falsehoods.

Though there has certainly been more conversation surrounding misinformation in recent years, is the problem really worse now, or are we simply more aware of it? Political scientists Nancy Rosenblum and Russell Muirhead detail a concerning trend in political discourse in their book *A Lot of People Are Saying: The New Conspiracism and the Assault on Democracy*.[6] They describe how classic conspiracies focused on finding evidence that was believed to be hidden from the public. However, a new type of conspiracism has arisen where claims are asserted and believed because they are "true enough." I agree this kind of rhetoric is unfortunately commonplace and a definite threat to democratic norms and institutions. There is some evidence that fact-based arguments have been declining in books since the 1980s and that a growing proportion of emotion-based arguments has taken their place.[7] Additionally, news headlines have grown more emotionally evocative and negative in the past few decades.[8] However, I do not see strong evidence that people no longer care about the truth. People have always had biases, but the advent of the internet and social media makes it much easier for misinformation to spread. For most of human history, it was very difficult to broadcast a wild and false bit of information to a large group of people. You usually needed an existing audience or access to a spot on radio or television. Now all you need is access to the internet and a social media account.

It can be frustrating to see people not only believe but also share things that we know are false. I know I certainly have felt this way many times while browsing social media. We might think that believing and sharing false information is an irrational act, but it's important to remember there are perfectly rational reasons to do so. Inaccurate beliefs may be irrational from a logical or epistemological perspective; however, holding those same off-base beliefs may have enormous benefits from a social perspective.[9] We will look at into these social factors throughout the book.

My main goal in writing *Misguided* is to describe the psychological and sociological processes that lead to the belief in and spread of false information. Those who fall for misinformation are often regular, everyday people who have been misguided. I hope to both describe why this happens and suggest some evidence-based solutions. Because the social sciences can help us understand these processes, they can also help us counter them. We can learn about techniques to help others who have been misled and also to prevent ourselves from being misled.

Understanding Identities Is Crucial

One of the main social forces I discuss in this book is the concept of identity. Identities are sets of meanings that define the social categories or groups to which we belong.[10] We can have hundreds of identities that guide us as we navigate different situations, and we enact a different role in each identity. For example, we may be a parent, so our parent identity motivates us to take care of our children. We may also have a sports fan identity, and this identity motivates us to watch and cheer on our favorite teams. Our identities vary in their personal importance. The more important an identity is to us, the more likely it is to influence our beliefs and behavior.

As we'll learn throughout the book, our identities offer great opportunities for connecting with others and navigating our social worlds. They can provide meaning and are a major source of self-esteem, which stems from acting in ways that are consistent with our identities—that maintain consistency between the values of our identities and our environment. Take the parent identity as an example. Taking care of our children is a positive outcome that affirms our identity. In many cases, our actions can be neutral: We watch sports on TV to affirm our sports fan identity and carry out work tasks to affirm our worker identity. However, we also may be more likely to believe something is untrue if it affirms an important identity of ours. We can think back to our fictional friend John and how his contrarian identity may motivate him to believe the Earth is flat to affirm his identity as someone who thinks outside the box. That identity is also bolstered by positive social feedback from meeting friends who support that belief and engaging with a community of like-minded thinkers.

I've always been intrigued by belief systems and the identities that support them. Long before I read the academic literature on identity theory, I was curious about how certain worldviews can influence what a person believes. It's easier to think about single identities influencing a single belief, but the reality is that our selves are composed of many identities that are constantly and dynamically interacting with one another. We will learn more about how the complexity of our identity maps can shape our belief systems later in the book. On a personal level, I've held many identities and beliefs that have shifted over time or have been contradictory. I've never really felt like I fit into a neatly organized identity box, so this has forced me to be more aware of my own identities. For example, I've been both

- religious and nonreligious,
- politically liberal and politically conservative,
- a "bad" high school student and a "good" college student,
- a nerd who loves books and an athlete who loves sports, and
- a quantitative neuroscientist and a sociologist.

Many people have had contrasting identities like I have. Reflecting on my own identities definitely fueled my interest in the topic of how identities influence what we believe. However, my journey into studying misinformation and ultimately writing this book can be traced back to my first semester of college when I was introduced to psychology for the first time.

MY JOURNEY INTO STUDYING MISINFORMATION AND WRITING THIS BOOK

In my first semester of college, I was still an "exploratory" major, and I didn't know what path I wanted to take. I knew I had an interest in psychology, but I was uncertain if I wanted to major in it. It wasn't until I read *The Evolving Self* by psychologist Mihaly Csikszentmihalyi during my first semester that I realized I wanted to study social science.[11] Csikszentmihalyi focuses on the importance of learning "what we are made of, what motives drive us, and what goals we dream of" in order to build a foundation for a "stable, meaningful future." He examines how both genetics and cultural forces have shaped humanity and how learning more about these domains will ultimately improve society. In retrospect, this book was a great advertisement for studying psychology.

Csikszentmihalyi studied positive psychology, which broadly encompasses how to best promote human flourishing and happiness. I found reading *The*

Evolving Self rather uplifting, as the takeaway for me was that working toward learning more about ourselves will improve our lives as well as the lives of those around us. I hope my book contributes to that same conclusion. Of course, learning about our psychology is not always a positive experience. Discovering why humans harm others, promote extremism, and spread lies requires acknowledging the darker social and psychological forces that guide our behavior. This quote from Csikszentmihalyi's book struck me as a first-year college student and continues to resonate with me today: "Poets make much of the majestic eagle soaring freely among the snowy peaks. But the eyes of the eagle are generally focused on the ground, searching for rodents lurking in the shadows. The lives of much of humanity could be summed up in similar terms."[12]

This high-flying eagle has access to unimaginable scenery on the horizon but often misses those incredible sights because its eyes are focused on the ground. Similarly, the complex human brain is free to investigate the wonder of the universe while jumping from one independent thought to another. Yet we can get stuck in certain thinking patterns and refuse to process information in a way that conflicts with our worldview. We worry about things that we may realize are irrational on some level, but our minds continually gravitate toward them like an overpowered magnet. We often pursue goals in search of potential validation from others, even at the expense of our own happiness and fulfillment. The world of social science constantly tries to understand these complex issues that explain our behavior.

After reading *The Evolving Self* in the fall of 2007, I knew I wanted to study social science. I changed my major to psychology and minored in sociology. I sought out research projects and internships throughout my years as an undergraduate to focus my interests. During this time, I spent one summer as an intern at a hospital in Pittsburgh, Pennsylvania, where I administered neuropsychological tests to inpatients and outpatients who had traumatic brain injuries. This fueled my interest in neuroscience, and I sought out another internship the following summer at Dartmouth College, where I could get direct experience with neuroimaging research. Working in the neuroimaging lab at Dartmouth College allowed me to see a picture of my brain for the first time. It was incredible to view the object that creates my experience of reality with my own eyes. As a twenty-year-old, I still felt rather immortal, but after seeing my brain, I remember being particularly focused on my mortality. I remember riding my bike back to my apartment afterward and constantly thinking that the object within my helmet must be protected at all costs because it is everything I am.

I became fascinated with the brain and the idea that we could use neuroimaging tools to predict what a person is seeing, hearing, or thinking about. Fascinating research in this area was coming out of the University of South Carolina, where I began my PhD in neuroscience immediately after finishing my undergraduate degree. I conducted hundreds of brain scans as we researched how neural activity predicts the type of picture a subject sees or the kind of sound they hear. For example, we could predict whether someone heard something positive, like a crowd cheering, or something negative, like someone crying, by simply analyzing their brain activity. This work, published in the journal *Human Brain Mapping*, would end up being my lone neuroscience publication.[13]

As I was working on my PhD in neuroscience, I found that I particularly enjoy doing scientific outreach and sharing what I have learned about the brain with others. For example, I regularly gave a talk about the neuroscience of religious belief that people seemed to enjoy. I also volunteered to share my expertise when it seemed relevant to policy. Being active in my community and learning more about politics made me much more aware of larger social problems, such as the reality that many public policies are based on false or misleading information. These activities made me want to learn more about how people and society operate on a broader level instead of focusing on patterns of neural activity. I didn't know it at the time, but providing scientific testimony in 2014 would play a major role in my ultimately changing academic fields so I could study misinformation directly.

It was a sunny day in January in South Carolina, but my suit kept me plenty warm as I walked to the Gressette Building behind the South Carolina State House. I was about to tell a group of legislators that their bill was "foundationally flawed" during a public hearing. I was nervous, but I also felt reassured by the scientific citations I had supporting my case. The bill confidently claimed fetuses felt pain at twenty weeks, but there was not sufficient scientific evidence to back this up.[14] Not only was this claim unproven, but also abortions performed after twenty weeks were exceedingly rare and usually occurred because women lacked financial resources and struggled to find a provider.[15] The bill was using a flawed argument to remove reproductive rights from women who were already in dire situations.[16] Of course, this bill was politically motivated, and the facts didn't matter as much. But as a naive twenty-five-year-old PhD student, I didn't realize the scope of this. I gave my testimony, in which I described exactly how and why the bill was not supported by science. I even printed out copies for each legislator of the scientific articles that described the neuroscience behind why fetuses cannot feel pain at twenty weeks. It didn't matter. Abortion is such a politically and emotionally charged issue that evidence is not the way to go about changing minds or hearts. There was barely any discussion about the science at all.

One of the legislators even told me that my testimony was "very impressive" but then asked me if I had ever seen the documentary *The Silent Scream*. He was equating my recent articles from peer-reviewed scientific journals with a dramatized film from 1984 that was full of false information.[17] The same legislator then asked an abortion doctor who testified after me if a baby ever "ran away" during an abortion. These were the people writing our laws (and this subcommittee that was debating the abortion bill was all male). Despite the bill being based on a scientific argument about pain receptivity, the debate surrounding the bill largely focused on feelings, anecdotes, and false information instead of scientific evidence. Some of my local legislators did find my testimony useful and cited the research I shared during the debates on the House floor. While the bill was delayed, it eventually passed in 2015. The attack on reproductive rights continued as *Roe v. Wade* was overturned in 2022.

I already knew that people had cognitive and ideological biases, of course, but I wasn't aware of how scientific evidence could be completely dismissed in policymaking. Admittedly, abortion is an extremely polarizing topic, and there are examples of evidence-based policy. However, my experience of having scientific articles from top medical journals countered by agenda-driven documentaries from decades ago made me want to investigate the broader issue of misinformation in society. I certainly didn't understand the depth or breadth of this problem at the time, and I definitely did not have a good idea of what could be done about it. I continued to work on my PhD in cognitive neuroscience while also remaining active with various organizations in hopes of pushing for more evidence-based policy. In fact, I strongly considered going into science policy after my PhD. During my time in graduate school, I became much more active with science communication and began giving talks about my research to audiences around the country. I learned more about how strongly people hold onto their beliefs, even when they are false. This conviction in their beliefs results in large part because those beliefs are attached to an important identity of theirs. This is particularly salient in politics, as we saw in my experience providing scientific testimony. As I paid more attention to politics and policy, I realized our country is inundated with misinformation and misunderstandings of scientific evidence. It is not only ubiquitous but also built into the structure of our society and frequently wielded as a political tool. I wanted to channel the frustration I felt about misinformation into action. Since there was already a great sociology program at the university I was attending for my PhD, in 2015 I switched from my neuroscience program to a sociology program in order to study political bias and misinformation directly. Specifically, I studied the social psychology of identities, networks, and beliefs while also learning about the macrolevel sociological forces.

My awareness of the widespread nature of misinformation only increased as the 2016 election campaigns raged on. There was so much discussion about fake news that it elevated my academic interest into the mainstream, and these discussions only became more popular during President Donald Trump's administration. My interest in misinformation became particularly relevant during the COVID-19 pandemic in 2020. We were all fighting against this pandemic, but we were also fighting against misinformation at the same time. Fake news in particular and misinformation in general have become relevant issues in everyone's lives.

I finished my PhD in sociology in 2020 and completed a dissertation on political bias and misinformation. Graduating during the COVID-19 pandemic made me realize this was a crucial time to share what I'd learned in book form. I offer a unique perspective as I write this book because I am both a researcher who has scoured the academic literature on this topic and someone who has communicated science to a variety of audiences through my outreach and advocacy work. After completing my PhD, I spent time as a postdoctoral researcher at Vanderbilt University and the University of Notre Dame, where I further examined the social science of misinformation. At Vanderbilt, I focused on studying political and health misinformation, and at Notre Dame, I focused on developing media and information literacy tools. In this book, I summarize much of what I have learned from reading the scientific literature on misinformation while focusing on the perspective of identities and personal networks.

Identities and the networks that facilitate them can be wonderful tools that allow us to connect with others and make meaning from our chaotic worlds. However, they can also bias us toward believing things that are not true. Unlike the eagle, we should aim to view the majestic scenery our consciousness allows instead of focusing so much on rodents in the shadows. I hope that learning about identities and other social processes will help us do that.

MOVING FORWARD—INSIDE *MISGUIDED*

Misinformation continues to be a huge problem for our society, but this is not an inevitability. There are ways we can fight against it. In *Misguided*, I provide simple techniques to change the misinformed perspectives many people have about science-based issues. I share what these techniques are and how they work, and I give concrete examples of how to put them into practice. In chapter 1, I explain how we define and study misinformation as well as what some of its significant consequences are. After that, in chapter 2, I consider identities and the ways they influence us. Understanding the ins and outs of identity processes is vital for understanding our susceptibility to misinformation.

I wrote this book during the COVID-19 pandemic, and that certainly was a salient example of misinformation spread. In chapter 3, I use the COVID-19 pandemic as an important case study, in which I describe what went wrong. I also provide evidence-based strategies for having more productive conversations with people who disagree with us. As we will see later, having productive conversations can reduce polarization and also help us better equip ourselves to reject falsehoods that we may be susceptible to. Covering productive dialogue at the end of chapter 3 moves us into how the people we associate with—or our social networks—impact our beliefs in chapter 4. Identities are social processes and are influenced by others, so it is important to understand how our personal networks shape our identities and, conversely, how our identities shape our personal networks. In chapter 4, I cover both offline and online connections, as social media play a massive role in how we view the world. I then use the antivaccine movement as a case study in chapter 5 in order to investigate more specifically how social media and personal networks promote the spread of misinformation. I end the chapter with a discussion of the importance of rebuilding trust to counter the spread of misinformation.

In chapter 6, I describe how media and information literacy can protect us against falling prey to misinformation. Improving literacy is important, but I also describe the challenges and limitations of such an approach and conclude the chapter by integrating what we have learned about social processes with media literacy. In chapter 7, I highlight some of the main takeaways from the book and look forward to some of the most pressing issues (such as artificial intelligence) that will affect how we tackle misinformation in the future.

CHAPTER 1

THE SCOPE AND CONSEQUENCES OF MISINFORMATION

FAKE NEWS IS NOT NEW

The topic of misinformation has received a lot of attention in recent years, but false and misleading information has been common throughout history across many different domains. The fabrication of a fake country called Poyais for fraudulent purposes is one of the most striking examples of this misinformation. The existence of countries is pretty indisputable. They consist of land and people—they have a physical location on Earth. However, people once believed that a fake country was not only real but so real that they were happy to collectively invest huge sums of money in this made-up place. It may sound impossible, but this is precisely what Gregor MacGregor of Scotland got people to do in the early 1800s.[1] MacGregor began his career in the British Army and gained fame after leading several successful missions in Central America. In 1820, he bought a large piece of land along the eastern coast of modern-day Honduras and Nicaragua. This land was uninhabited, and its terrain was not suitable for growing crops or maintaining livestock. Over the next couple of years, MacGregor advertised that this piece of land—objectively unfit for settlement—was a thriving new country named Poyais.

Opting to maintain his residence in London rather than in this famed new land, MacGregor sold investment opportunities in his fictional country. By his accounts, Poyais was a utopia with plenty of fertile ground for crops to flourish, fish just begging to be caught and sold, and natives eager to befriend the British. MacGregor also claimed that there was a burgeoning new capital city that already had houses, a bank, and even an opera house. To top it all off, he declared that the rivers were full of gold. All of these falsehoods were documented in a

355-page guidebook that described the purported country in detail. MacGregor forged official documents, currency, and even the country's national anthem. None of it was true or real. MacGregor fabricated all of it, down to the national anthem and the opera house. Londoners and Scots eagerly purchased land in Poyais at seemingly great prices. It is estimated that MacGregor sold investments in his fake country worth over $1 billion (measured in modern-day dollars).

When investors from England finally arrived in Poyais in late 1822, they found a barren land, despite the promises of an idyllic paradise. Harsh conditions and disease killed most of the initial 250 explorers; only about 50 survived and could return home. These survivors thought there must have been a mistake, but MacGregor was able to keep his con going for a few more months. Eventually, too many travelers returned and reported his lies, and by the end of 1823, he was charged with fraud. Ever the con man, he blamed his managers for the false promises, fled the country, and eventually avoided jail time.[2]

How did so many people fall for such blatantly false information, risk (or lose) their lives, and invest heavily in a fake country? There were a few social and psychological ingredients that made this con easier to digest. The political turmoil in Central America in the 1800s increased the plausibility that a new country had been established. News traveled slowly in the 1800s, giving MacGregor plenty of time to spread his version of the truth before anyone could fact-check him.

MacGregor targeted Londoners, who shared his place of residence, and Scots, who shared his nationality—an important identity—thereby increasing familiarity and trust. We are more likely to trust and cooperate with those who belong to the same group we do.[3] The reputation and social capital MacGregor had built via his public military career further solidified the apparent legitimacy of his claims and presumed trustworthiness. Our social connections to other people can massively influence how we evaluate the legitimacy of the information they provide.

Additionally, investments in real Central American countries were booming in London at the time. MacGregor emphasized that the land was selling fast and potential investors needed to buy before it was too late, thereby introducing social pressure as well as the idea of scarcity. Scarcity, the idea that we will miss out if we do not act quickly, is one of the six principles of persuasion, according to psychologist Robert Cialdini.[4] All of these conditions created a perfect storm that allowed MacGregor's misinformation to thrive.

Poyais is just one of the many examples of misinformation throughout history. False information was first spread through word of mouth and then through printed matter, but the speed and modes of communication have certainly changed significantly since the 1800s. Innovations in communication, including radio and television, accelerated our communication and broadened its audience.

This expedited the exchange of information to incredible levels, regardless of whether that information was true or false. While current technology allows misinformation to be spread even more quickly and more widely, the essence of the story is the same. Those components of our psychology that help us navigate our social world also make us vulnerable to false information. Later in this chapter, we will cover how misinformation is linked to a wide variety of issues, including violence, public health disasters, and economic waste. Before covering the scope and impact of modern-day misinformation, it is important to explicitly define the various terms that describe false information in our society.

DEFINING MISINFORMATION

Misinformation is any inaccurate, false, or misleading information.[5] Importantly, it includes both intentionally and unintentionally false information. Beliefs based on inaccurate information are known as *misperceptions*.[6] For example, if someone believes that walking backward cures the common cold, then that would be a misperception, since it is a belief that is not supported by evidence. There is also a crucial difference between being misinformed and being uninformed. An *uninformed* person does not have strong prior beliefs about the issue in question. Conversely, a *misinformed* person has a strong conviction in an inaccurate belief, and this usually predicts that they hold other, related misperceptions. Misinformed people are much less likely to change their minds when presented with counterevidence compared to those who are simply uninformed.[7] Other misinformation research focuses on whether the source of the information is credible or misleading rather than on the specific content.[8] If a blogger regularly posts fake claims (e.g., walking backward can cure the common cold), then their blog can be considered a source of misinformation.[9] We will cover more about evaluating the credibility of sources later in the book.

A common example that contributes—sometimes unwittingly—to the existence of misinformation is the journalist who fails to verify the claims in their story and then later finds out these claims are inaccurate. An infamous case of misinformation occurred in 1912 when several major newspapers falsely reported that the *Titanic* did not sink. The newspapers relied on electric telegraph messages that claimed the ship was safe, but clearly they got their wires crossed.[10] Whether reviewing unverified telegraph messages from the 1900s or unverified claims from social media today, journalists should do their due diligence before breaking a story.

While we often think about misinformation in terms of news and media, there are many other forms of misinformation that we encounter more casually. Any type of information that is not true—even unverified gossip—should be considered misinformation. Researchers often use misinformation as an umbrella term to capture the various types of false information; however, other terms refer to specific types of misinformation. *Disinformation* is *intentionally* false or misleading information. Its use is often motivated by a desire to change public perception. For example, individuals are motivated to share false or misleading information to win supporters, followers, etc. *Propaganda*, on the other hand, is a form of disinformation used in an attempt to change beliefs through false, misleading, or sensationalist information. Propaganda leaflets were commonly distributed by air during World War I. Aircraft would literally dump thousands of leaflets behind enemy lines in an attempt to change the attitudes of the troops.[11] It's hard to quantify what impact, if any, these pamphlets had in terms of changing beliefs, but they provide a vivid example of a disinformation technique. In the next section, we will see how these coordinated disinformation campaigns continue today with more sophisticated technology.

Other types of misinformation can exist on a spectrum of veracity. For example, *hyperpartisan news* is information that is not necessarily false but that is extremely biased or one-sided.[12] It is found on any number of media platforms, whether in print or on air. This is particularly salient across different news channels that have a partisan viewpoint. You probably have seen how many cable news shows will demonize one side of the political aisle and focus more on conflict than substance. Comedian Jon Stewart called out cable news for leaning into the conflict-over-substance coverage of politics (i.e., partisan hackery) way back in 2004 when he was a guest on CNN's *Crossfire*.[13] It wasn't always this way! Research has shown that columns expressing outrage against political opponents were quite rare in U.S. newspapers in the 1950s. However, by 2009 the average newspaper contained almost six examples of outrage. Such emotional rhetoric has also become much more common across radio and television,[14] as this trend of polarization and sensationalism continues to accelerate. An analysis of American newspapers, magazines, and television programs from 2000 to 2012 found that uncivil discourse and mentions of polarization significantly increased during this time.[15]

Programs on CNN and MSNBC are much more likely to invite guests who donate to liberal politicians, while Fox News is much more likely to invite guests who donate to conservative politicians.[16] Additionally, a 2021 study found that both left- and right-leaning media use more positive sentiments for their own political group and more negative sentiments for their opposing political group.[17]

An analysis of media coverage during the 2008 U.S. presidential election compared how different news channels portrayed each candidate. MSNBC was much less likely to use a negative tone while covering Barack Obama, and Fox News was much less likely to use a negative tone while covering John McCain, further demonstrating the partisan bias of news outlets.[18] Changing sentiments is just one element of partisan bias in the news; more serious examples of hyperpartisan news include omitting key facts and cherry-picking data.[19] Finally, hyperpartisan politicians receive up to four times as much media coverage as their more moderate and bipartisan counterparts.[20] Whether it is appealing to the audience's political beliefs or focusing on the most extreme, hyperpartisan news plays a large role in polluting the information ecosystem.

Yellow journalism is similar to hyperpartisan news but does not have to be political, as it includes anything that is poorly researched and often has the goal of gaining attention. Headlines that are colloquially known today as clickbait fall into the category of yellow journalism, but this type of misinformation was first described over a century ago. The term is thought to have originated in the 1890s when it was used to critique the sensationalism found in the articles of two major competing newspapers.[21] Today we see constant examples of yellow journalism with tabloid or blog post titles (often in the form of ads) with claims that "THIS WEIRD TRICK WILL LET YOU EAT JUNK FOOD AND NOT GAIN ANY WEIGHT" or "THE SECRET TO BECOMING RICH THAT BANKS DON'T WANT YOU TO KNOW." Many of these clickbait titles can be quite absurd, but some are less extreme. They may claim to have secret knowledge about finance or wellness, but once you dive into the details, they lack substance. There is certainly a range of sensationalism, but overall such claims lack rigorous evidence to back them up.

Finally, there is also *fake news*, which has become a popular term in politics over the past few years. While some politicians use it to refer to news that is critical of them,[22] the definition found in recent psychological literature is "news content published on the internet that aesthetically resembles actual legitimate mainstream news content, but that is fabricated or extremely inaccurate."[23] One of the most egregious contemporary examples of fake news is the article titled "Jimmy's World" that Janet Cooke wrote for *The Washington Post* in 1980. In the article, she shared the harrowing story about an eight-year-old named Jimmy who was already a heroin addict and aspired to sell drugs in the most dangerous parts of Washington, D.C. The story gained immense attention, and Cooke even won a Pulitzer Prize for it. The city's mayor asked the police to try to locate Jimmy, but suspicion arose when they investigated the details of Cooke's story and could not locate the boy. The story was eventually found to be completely made up. Cooke admitted to the fabrication and gave back her award.[24] As we'll see later, the advent of the internet makes fake news much easier to generate and spread.

While these terms have different sources and intentions, they all fall under the broad umbrella of misinformation. Sometimes misinformation is clear-cut—for example, when a ship actually sank or a story is fabricated. At other times, it becomes more challenging to determine what is true or false—for example, when faced with complex political events and/or statements. For these situations, we can rely on a consensus of experts and/or independent fact-checking organizations. The key word is *consensus*, as any individual group may get something wrong. Studies have found that fact-checking organizations like Fact-Checker, PolitiFact, and Snopes generally agree on statements they deem to be outright true or false.[25] Furthermore, a politically balanced group of people rate news headlines similarly to a group of professional fact-checkers.[26] However, evaluating the truthfulness of statements becomes more challenging when they contain a mix of true and false elements. The ambiguous language used by politicians can add a layer to subjectivity in evaluating whether what they say is mostly true or partly false.[27] Statements that involve more than one claim also cause challenges for fact-checkers.[28] This is a limitation of misinformation research: We tend to rely on stories and statements that are clearly true or false because it is more challenging to study information that has elements of both truth and fiction.[29] It is important to be transparent about how and why we categorize something as false.[30]

Another challenge arises when information is evolving and incomplete. Such information, which is not quite misinformation or true information, has been categorized as *midinformation*.[31] During the Middle Ages, before the development of germ theory, medical experts believed the theory of miasma, or "bad air," to be true. They thought that diseases such as the plague were spread by poisonous air instead of by direct contact with individuals.[32] A modern example occurred at the start of the COVID-19 pandemic, when information about the virus was rapidly evolving. Eventually, we learned more about the virus with the completion of scientific studies, but misinformation persisted even after substantial levels of evidence were made public. In chapter 3, we will examine the misinformation around COVID-19 in detail.

Discussing misinformation can be challenging when information is rapidly changing, and the truth is not yet known, but there are techniques that can help us get closer to the truth even as our understanding evolves. While scientific knowledge is not set in stone, that does not mean we lack methods to evaluate the validity of scientific claims.[33] It's also important to note how the definitions and conceptualizations we have for misinformation are limited, keeping in mind that researchers need to draw the line somewhere in order to research the topic effectively. As I'll argue throughout this book, the narratives based on our identities are crucial for understanding how we process information. Understanding how our identity biases impact our processing of information more broadly will ultimately be more helpful than constantly trying to label each piece of information as true or false.[34] I do not want to dictate what people think or to demand censorship. Instead, my main goal in this book (and beyond) is to share techniques that help us recognize our own biases and more carefully evaluate information.

Examples of misinformation are found throughout history, and it has been a concern for quite some time. Technological advances have made both true and false information easier to access. Unfortunately, this doesn't always mean people will seek out accurate information. For decades, social scientists have known that people possess a strong motivation to find information that confirms their prior beliefs, even if these beliefs are incorrect.[35] Psychologists call this tendency confirmation bias, and we will examine it further in the next chapter. Carl Sagan was aware of this problem, and in 1995, he shared an ominous warning about what the future would hold if trust in science and truth continued to decline:

> I have a foreboding of an America in my children's or grandchildren's time—when the United States is a service and information economy; when nearly all the manufacturing industries have slipped away to other countries; when awesome technological powers are in the hands of a very few, and no one representing the public interest can even grasp the issues; when the people have lost the ability to set their own agendas or knowledgeably question those in authority; when, clutching our crystals and nervously consulting our horoscopes, our critical faculties in decline, unable to distinguish between what feels good and what's true, we slide, almost without noticing, back into superstition and darkness.[36]

Sagan's nightmarish prediction highlighted social problems that still resonate decades after his text was published. When reflecting on our current time period, I have to wonder if we have slid further into darkness in the thirty years since Sagan wrote this. While I've noted that plenty of examples of misinformation are found throughout history, there are likely plenty more to come. In the next section, I'll outline the current scope of misinformation. Modern technology gives us an immense amount of evidence-based information at our fingertips, but it also provides us with plenty of hoaxes, conspiracy theories, and pseudoscience. Comparing the total amount of false information from decades ago to that today would be extremely difficult, so it's hard to say whether people believe more misinformation today or whether it is merely more easily accessible to the masses. What we do know is that misinformation is still a major concern, and we can directly observe the impact false information has on modern society.

MEASURING MODERN-DAY MISINFORMATION

The bad news is that false information is a growing and ubiquitous problem. The good news is that many people are focused on trying to solve this problem before it gets worse. Researchers study the scope of misinformation through several methods, including surveys, experiments, content analyses, and social network analyses. I'll briefly describe how researchers use these tools to measure the scope of modern-day misinformation.

Surveys are a common tool that allows researchers to measure people's attitudes about a subject by simply asking them questions and recording their answers. If you collect data from members of a certain population, such as Americans, you can also infer how that population feels about the topic. Survey researchers need to try to collect information from as large and as representative a sample as possible if they want to have confidence that their results accurately reflect the feelings of a population. They also need to eliminate the misleading wording of survey questions that may cause people to provide biased answers. In addition, some people may not answer the survey honestly or accurately, whether intentionally or unintentionally. For example, the amount of time people spend looking at their digital devices has received significant media attention. However, studies that use objective measures of screen time (such as an app that records total use) have found that people are generally pretty bad at estimating their screen time. Thus, some of the conclusions based on these surveys may not be accurate.[37] While surveys are a vital tool for understanding public opinion, they also have limitations that should be considered.

Surveys can inform us what the general sentiment a population is toward misinformation. Of the American adults surveyed in 2019 by the Pew Research Center, 68 percent agreed that made-up news and information "greatly impacts Americans' confidence in government institutions," and 54 percent said that it is having a "major impact on our confidence in each other."[38] A 2023 KFF poll found that 83 percent of Americans believe false and inaccurate information is a "major problem."[39] These results demonstrate that a significant number of Americans are concerned about the impact of misinformation.

Surveys also can tell us where people are getting their information. Social media are frequent sources of misinformation, a situation we revisit throughout this book. A 2024 survey found that nearly seven in ten Americans use Facebook and over 80 percent use YouTube.[40] Additionally, half of U.S. adults use social media for news at least sometimes.[41] During the 2016 U.S. presidential election campaign, fake news on social media was actually "liked" and shared more often than real news.[42] While social media continue to grow in popularity, research has shown that Americans overall consume five times more television news than online news.[43] Despite declining ratings in recent years, local television is still the most popular news source for Americans.[44] Local news stations are accurate and efficient sources of information for many Americans, so their decrease in popularity is certainly a concern.[45]

In addition, surveys allow researchers to measure how many people actually believe the misinformation they have been exposed to. Research from the United States and UK demonstrates that people across various demographics, including age, gender, and political identity, believe conspiracy theories.[46] Other research has found that one-third to one-half of Americans believe at least one conspiracy theory.[47] An example of these conspiracy theories is the belief that the vapor trails from airplanes are actually chemicals meant to harm us and that this process is directed by government officials. You may find this particular theory far-fetched, but as we will learn throughout the book, we all are vulnerable to holding beliefs that are not evidence-based. Surveys allow us to look at other variables (such as political identity and trust in institutions) that predict why someone may hold a particular belief.

Experiments are another tool researchers can use to study misinformation. Instead of just asking people for their thoughts about misinformation, researchers can study whether different variables change behavior or attitudes. One particular benefit of experiments is that they allow researchers to discover information that they might not be uncover through surveys. For example, a 2021 experiment evaluated the impact of fake news on unconscious performance.[48] The study split the participants into three experimental groups where they read (1) a fake news story claiming finger-tapping speed is associated with intelligence, (2) a fake news story claiming finger-tapping speed is associated with criminal activity, or (3) an article about statistics that does not mention tapping speed. After reading one of the three texts, participants engaged in a finger-tapping exercise that tested how quickly they could tap repeatedly tap a computer key. The participants were also asked to estimate their own finger-tapping speed.

Results showed no significant differences in the estimations of finger-tapping speed across all three conditions. In other words, the participants did not believe they had been influenced by reading those stories about finger tapping. It's possible they might not have even believed they were true! However, those who read the fake news article about the correlation between finger tapping and intelligence did tap significantly faster. The participants didn't think the information they read would influence their behavior. However, when actually recording their behavior, the experimenters found that fake news did impact them on an unconscious level. This experiment shows that misinformation may influence people even if they do not believe it. Experiments can offer explanations of causal relationships between variables and uncover different unconscious processes related to behavior.

While experiments require real participants to engage in different tasks, content analyses evaluate and summarize real-world data. They provide researchers with the ability to measure the scope of misinformation in different contexts and have been especially popular for measuring online misinformation due to the accessibility and potential reach of the internet. An early example of this is a study conducted in 2002 that analyzed internet search results regarding skin conditions. The researchers found skin product sites are significantly more common than objective, high-quality educational sites. The researchers concluded their paper with a warning about how easy it is to find misinformation on the internet.[49] Two decades later the concern about online misinformation has only grown with the wider adoption of the internet. In 2005, it was estimated that about 1 billion people used the internet, but just seventeen years later, in 2022, internet use had skyrocketed to 5.3 billion people, or two-thirds of the global population.[50]

Social media are popular sources of misinformation, and there are plenty of studies analyzing the content spread of these websites. Researchers can track and analyze social media data to evaluate how people are discussing major global events. For example, a 2014 study identified and analyzed the top ten rumors that were being spread on Twitter regarding Ebola, including its ability to be transmitted through the air. The researchers also found this misinformation regarding airborne transmission was most common during early October 2014.[51] Being able to pinpoint popular false information offers could help health officials create targeted campaigns to combat them.

Finally, social network analyses are tools used to examine how information can spread through our personal networks. Networks are quite broad and are characterized by anything from brief online exchanges to longer interactions with our closest friends. A network analyses can include evaluating the small group of people we associate with most often. It can also include using complex computations to study social media connections on a website like X (formerly

Twitter). This technique can provide insights into how false information spreads and how common it is within a group of people.

While network analyses have tremendous utility, it is important to be aware of the limitations and specific parameters of each study before drawing conclusions that are too broad. For example, there has been research showing that a false story can spread on the internet much faster than a legitimate news story.[52] However, this research focuses on a very specific type of information: rumors. Alberto Acerbi reviewed this research in his book *Cultural Evolution in the Digital Age* and concluded: "What the study says is that rumors that were subsequently debunked 'diffused significantly farther, faster, deeper, and more broadly' than rumors that were subsequently confirmed."[53] The spread of completely made-up news may not be our greatest concern if we are worried about false information influencing people's perspectives. In a 2020 study, researchers from MIT, Microsoft, the University of Pennsylvania, and Harmony Labs investigated the scope of misinformation. They concluded that "public misinformedness and polarization are more likely to lie in the content of ordinary news or the avoidance of news altogether as they are in overt fakery."[54] This is consistent with a 2024 review of misinformation research in which the authors noted that completely fake news is often concentrated in a narrow fringe of social media users.[55]

Content and social network analyses are useful tools for measuring the scope of misinformation, but it is still a monumental task to quantify the total amount of misinformation with great precision. When different studies have different parameters and definitions of misinformation, it can be very difficult to directly compare their results. Determining exactly how much false information is out there is difficult, but we can be certain that it has a very real impact on our society.

THE REAL IMPACT OF FAKE NEWS

Violence

I was just starting to write this book when a misinformation-fueled event grabbed the nation's attention: the insurrection attempt at the U.S. Capitol in 2021. President Donald Trump soundly lost the presidential election to Joe Biden in both the popular vote (by over 7 million votes) and the Electoral College (306 to 232).[56] Donald Trump repeatedly claimed that the 2020 election was stolen from him, despite presenting no credible evidence. Even after he and his campaign lost over fifty court cases contesting the election results and after a number of election recounts had virtually no impact on the vote totals,[57] Trump and many of his supporters still claimed the election was stolen. A survey conducted shortly after Biden officially won the election found that over 65 percent of Republicans believed Trump actually won the election. Additionally, about 40 percent of Republicans believed that Biden would be an illegitimate president even if Trump conceded.[58]

Again, none of Trump's extraordinary claims were proven in court, yet allegations of the election being rife with fraud spread all over social media.[59] Cable news giant Fox News also regularly shared unfounded skepticism over the election results with its millions of viewers.[60] Sadly, this anger fueled by misinformation culminated in an attack on the Capitol when Congress was certifying the final electoral votes on January 6, 2021. Thousands of angry Trump supporters, believing the misinformation regarding the election being stolen, stormed the building. At least seven people lost their lives in connection with the riots, according to a bipartisan Senate report.[61] An undetermined number of rioters were injured, and over fifty police officers suffered serious injuries.[62] More than one thousand people were arrested on charges related to the January 6 riot, with more than half of them pleading guilty to their crimes.[63] This harrowing event was driven by widespread misinformation and disinformation about the 2020 election, much of which was propagated by Donald Trump and his supporters, further motivating the rioters.

I watched the attack on the Capitol take place in real time via numerous news outlets. It was surreal in many ways. It was stunning for me, as an American, to watch rioters overtake the Capitol and all it stands for with ease. The building is a symbol of democracy—not just for Americans but also for people all over the world. It was truly horrible to see this take place. As a scientist, I also thought about how much false information led to this event. I've studied polarization and false information for years and have been worried an event like this could happen. However, it was still shocking to watch. The Capitol attack solidified my belief that we must study the forces behind misinformation (as well as disinformation). Understanding why people believe misinformation and how it spreads increases our ability to combat it before it is used to incite violence.

False information spread via social media has been linked to violence around the globe.[64] One harrowing example involved the widespread violence toward Myanmar's Muslim minority group, the Rohingya. Misinformation and hateful speech directed toward this group were largely spread through Facebook. Ultimately, in 2018 the UN Human Rights Council concluded that Facebook did have a determining role in the violence.[65] It's also important to note that cell phone use skyrocketed in Myanmar shortly before these attacks occurred, with almost all of them having the Facebook app installed by default.[66]

In Latin America, false claims that doctors received financial incentives for admitting more COVID-19 patients to intensive care facilities were spread on social media. This led to mistrust of Colombian doctors and even several cases of reported violence.[67] In India, rumors and lies about immigrants abducting children were spread through the messaging app WhatsApp. These rumors incited violence against two innocent immigrants who were murdered by an angry mob.[68]

These stories of violence across different countries are powerful examples of the dangers of false information. According to a 2019 FBI report, investigations of fifty-two terrorists who acted alone found that 46 percent of them either discussed or consumed information regarding conspiracy theories.[69] This is consistent with other research showing that people are more likely to support acts of violent extremism if they also have higher levels of conspiratorial thinking.[70] More broadly, research has shown that exposure to false information is also linked to lower overall trust in the media.[71] Distrust in institutions can make it easier to embrace misinformation, which creates a vicious feedback loop. We will explore the role of trust in more detail later in the book, as it will continue to be a central factor in combating the spread of false information.

Health

Public health is commonly influenced by false information and will be a central area of discussion throughout this book. I wrote this book during the COVID-19 pandemic, which became a fertile breeding ground for misinformation. During this time, we saw a great deal of misinformation regarding the virus and potential treatments.[72] Unfortunately, because the virus was politicized, we saw Republicans take COVID-19 far less seriously, which led to more deaths from the virus along partisan lines.[73] Once a health issue becomes political, it also becomes vulnerable to evolving into a vessel for misinformation.

Another example is reproductive health in the United States. As I mentioned in the introduction to this book, one of my first and most noteworthy experiences with political misinformation came during the scientific testimony I provided in opposition to a bill that falsely claimed fetuses can feel pain at twenty weeks. Misinformation about reproductive health continues to be a major problem in the United States largely because it has been politicized. For example, having an abortion has been falsely claimed to cause depression and anxiety, but there is no evidence of this.[74] In fact, those who want an abortion but are unable to get one are more likely to have adverse mental health outcomes.[75]

Of course, health misinformation is often not political and may spread without a partisan push. For example, one rumor circulating during the COVID-19 pandemic was that ingesting methanol or alcohol-based cleaning products could prevent or treat the illness. One study did find a significant increase in deaths in Iran due to methanol poisoning at the start of the pandemic. However, as its authors mention, it is not possible to know exactly how many of these deaths were due to misinformation.[76] It is possible many more people were drinking recreationally to distract themselves or even to cope with their depression during the pandemic.

We do know that many unverified cures and treatments for the virus were selling off the shelves. One of the most notorious examples of this is Miracle Mineral Solution, which had initially been touted as a cure for cancer, hepatitis, the flu, and even HIV/AIDS. Its main ingredient is sodium chlorite, which is commonly found in disinfectants. Although the FDA labeled this substance as dangerous and there was no scientific evidence that it worked,[77] the creators of Miracle Mineral Solution were making $32,000 in sales per month. When the product resurfaced in March 2020 in advertisements promoting it as a cure for COVID-19, monthly sales jumped to $123,000![78] Beyond being a waste of money, Mineral Miracle Solution was reported as the cause of death for several people who were trying to cure various ailments.[79]

Medical misinformation has long been a cause for concern.[80] Sometimes the misinformation can be relatively harmless, as is the myth that taking a small amount of vitamin C can protect you from a cold.[81] Other medical misinformation can be quite lethal. For example, one survey by the American Society of Clinical Oncology found that 39 percent of Americans believe that alternative treatments for cancer (such as dieting and using supplements or herbs) can cure cancer without any evidence-based medical treatments![82]

Additionally, ingesting random supplements not approved by one's doctor can be dangerous. One study estimated that about 23,000 emergency room visits a year were made by people taking various dietary supplements to treat their ailments.[83] Cardiovascular issues caused by taking weight loss or energy-boosting supplements were a common reason for these emergencies. The *Dr. Oz Show* was an extremely popular television program that attracted nearly 3 million views per day. The show's host, Dr. Mehmet Oz, a cardiothoracic surgeon, regularly promoted various health products; however, a study revealed that over half of his recommendations were not backed up by sufficient scientific evidence.[84]

The town of Flint, Michigan, had a tragic water crisis during which tens of thousands of residents were exposed to the harmful levels of lead in their drinking water. Efforts to fix the problem were delayed by misinformation spread by officials.[85] For example, a Michigan health official reported that his department's 2014 analysis of blood lead levels in Flint children showed these levels were "within range of years before." However, the analysis actually concluded that the blood levels were indeed "higher than usual."[86] Pediatrician Mona Hanna-Attisha was at the center of efforts to uncover the water crisis and put pressure on officials to do something, but we can't always rely on individual heroes to fight against misinformation.

Alternative medicine involves practices that are not conventionally approved by physicians and that may not have any evidence supporting their effectiveness for a particular illness. Alternative medicine is a broad category and can range from taking herbal supplements to practicing yoga to seeing a chiropractor. About half of Americans report trying alternative medicine at least once. However, about one in five Americans reports having tried alternative medicine instead of conventional, evidence-based medicine.[87] Opting for alternative medical treatments can have serious consequences. Research has found that cancer patients who opt for alternative treatment have poorer health outcomes and even die earlier than those who stick to conventional medicine.[88] While some physicians actively combat medical myths,[89] it can be hard to find reliable medical information online when the ecosystem is inundated with pseudoscience. Even published medical journals contain concerning amounts of unreliable information.[90] As we'll see in the next section, health misinformation can also have a major impact on the economy.

The Economy

Misinformation, including medical misinformation, can have dire impacts on the economy. For example, disease outbreaks that could be wiped out or mitigated by vaccines (such as measles, hepatitis, and human papillomavirus) cost the United States about $9 billion per year.[91] We will cover antivaccination misinformation and how it spreads in detail in chapter 5.

The COVID-19 pandemic severely hurt the global economy, as many businesses had to shut down. There was a great deal of misinformation regarding public health guidelines, including even something as simple as whether to wear a mask. Not only would adherence to public health guidelines have saved lives, but also it would have helped the economy. A 2020 study found that in counties with mask mandates, people felt more comfortable shopping, and their spending increased by 5 percent.[92] Billions of dollars were spent quickly developing vaccines to curb COVID-19. However, this pales in comparison to the estimated $9 trillion loss that would have occurred if we had failed to vaccinate and the virus had continued to ravage our society.[93] Many U.S. states started offering financial incentives to take the vaccine because the cost of a predominately unvaccinated population was far greater. For example, Ohio entered vaccinated individuals in a lottery in which each lucky winner would receive $1 million.[94] Unfortunately, these lotteries did not appear to significantly increase vaccination rates compared to states that did not offer such lotteries,[95] but governments were willing to invest in these efforts because of the potential positive impact.

Beyond the economic impacts of public health issues, misinformation related to finances can impact the broader economy. A 2017 survey found that 63 percent of Americans believe that fake news makes it more difficult to make important financial decisions.[96] Financial advisers have estimated that financial misinformation contributes to bad financial decisions that cost Americans $17 billion in retirement savings per year.[97]

Additionally, there is no shortage of misinformation in investment advice appearing online. For example, GameStop stock gained national attention when there was a battle between short sellers and buyers, which caused the price to skyrocket. After hovering around $20 a share at the start of January 2021, the GameStop stock price accelerated quickly all the way to $483 per share in late January before crashing to $40 just a few weeks later. During this chaos and national attention, there were plenty of accounts on social media making wild predictions about the stock in both directions, but no one actually knew how high the price could rise or when it would drop. Fake social media accounts were even created to try to manipulate the price, but it is unclear how much of an effect these had.[98] GameStop is just one salient example of how actors on the internet provide an endless stream of bad financial information, preying on those hoping to make a quick buck. It's quite likely that financial misinformation contributes to the fact that the average investor earns lower returns than the overall stock market.[99] For most people, buying a simple index fund that tracks the stock market will yield higher returns than trying to beat the market through short-term trading.[100]

Financial misinformation goes beyond individual traders making bad decisions—fake news can tank the entire market. This happened in December 2017 when ABC News falsely reported that a national security adviser would testify against President Trump.[101] After this report was released, the S&P 500 (an index of the largest 500 companies) dropped by thirty-eight points, which equated to a loss of over $341 billion.[102] The market did quickly recover after the news was corrected, but it was incredible to see how much power fake news has over our economy. Another example of fake news impacting financial markets happened on September 13, 2021. Litecoin, the cryptocurrency often called "the silver to Bitcoin's gold" (with a market capitalization of nearly $12 billion at the time), spiked 35 percent in less than an hour.[103] This price surge happened because a fake report claiming Walmart was going to accept payments in cryptocurrency went viral on social media. This hoax was even shared by the Litecoin Foundation's Twitter account![104] When a Walmart representative denied the report, the price of Litecoin came crashing down.

Finally, climate change is one of the most pressing issues in our society and economy today. Global warming due to carbon dioxide emissions could cost the United States up to $520 billion a year. Disruptions from warming temperatures include rising sea levels, more extreme weather, and a higher demand for energy.[105] Immediate action is required to combat global warming, but misinformation and disinformation slow down efforts.[106] In an attempt to provide journalistic objectivity, the media used to explicitly give equal attention to climate change deniers and those sharing scientific evidence about global warming. This has come to be known as *false balance*, where all positions are presented as equally credible despite a substantial fact-based consensus on one side of an issue. While this false balance is less explicit today, climate change deniers still receive substantial media attention, primarily on right-leaning outlets.[107] Furthermore, the internet is inundated with social media posts and articles from conservative think tanks that dismiss climate change.[108] Importantly, oil and gas companies regularly donate to politicians who vote against environmental protections, fueling their incentives to reject climate change evidence.[109] Research has shown that when people believe there is substantial disagreement among scientists regarding climate change, they are less likely to believe climate change is a problem.[110]

❦

Misinformation clearly has negatively impacted the well-being of our society. Yet it continues to be a perplexing problem to solve. As mentioned in the introduction, words related to misinformation (and the word *misinformation* itself) have been chosen as Word of the Year for several years now. The public is well aware of this problem of false information, and many researchers have been studying it. There have been several wonderful books already published on detecting false information, such as *A Survival Guide to the Misinformation Age* and *Calling Bullshit: The Art of Skepticism in a Data-Driven World*.[111] Even back in 1995,

Carl Sagan wrote a chapter in his *Demon-Haunted World* titled "The Fine Art of Baloney Detection." However, while these books focus on techniques to fact-check, debunk false claims, and correctly interpret data, they don't focus as much on how and why people come to believe false claims. Learning about the impact of social forces allows us to understand why people become biased in how they process information. Education and critical thinking are crucial tools for fighting false information, but they are limited unless we also understand the psychological and sociological processes involved with misinformation.

My central goal in writing this book is to describe why misinformation spreads and what can be done about it. The false information problem goes far beyond the overtly fake information being pushed into our information ecosystem, whether in printed form in grocery store checkout lines or in clickbait online. The crux of the issue is the psychological and sociological forces that drive people to internalize the information that confirms their beliefs and prevents them from correcting their misguided perspectives. Well-studied phenomena in social science can help us understand why people believe false information.

CONCLUSION

I began this chapter with the story of the fake country of Poyais, a salient example of how false information is not a new phenomenon. Next, I defined the various types of false information: misinformation, disinformation, hyperpartisan news, yellow journalism, and fake news. Misinformation is often a catch-all term, but it is important to be mindful of the more specific definitions when they apply (e.g., disinformation is intentionally false or misleading information). Researchers study misinformation through a variety of strategies, including surveys, polls, experiments, content analyses, and social network analyses. These research tools help us understand how much and what types of misinformation exist as well as what can be done to combat misinformation. I ended this chapter by highlighting how misinformation can negatively impact public health, our economy, and our democracy. In the next chapter, we will look at the psychological factors that can make misinformation so compelling to our minds.

CHAPTER 2

HOW OUR IDENTITIES CAN MAKE US VULNERABLE TO MISINFORMATION

Belonging to various groups can help us connect with other people, but group membership can also bias how we process information and even influence how we treat other people. Racism is one harrowing example of how humans can treat others horribly simply because they belong to a different group. After the national tragedy of the assassination of Martin Luther King Jr. on April 4, 1968, Jane Elliott, an elementary school teacher in Iowa, decided to teach her class about racism and in-group biases, using a class exercise that would garner national attention.

To illustrate racism and prejudice, Elliott told her students that eye color was associated with greater intelligence and ability. Specifically, she claimed that blue-eyed children were inferior compared to brown-eyed children because of differences in melanin levels. She also imposed several classroom rules, such as having the blue-eyed children drink out of cups instead of directly from the water fountain. These rules further established distinct boundaries between the two groups of students. Elliott reported that the brown-eyed children began to make fun of the blue-eyed children as the day went on. The blue- and brown-eyed children even segregated themselves at recess. Additionally, the blue-eyed children performed worse on tests, as their confidence was reduced. Arbitrarily claiming that one group was better (while enforcing several rules that separated the groups) certainly illustrated how swiftly prejudice could be formed. The next day Elliott told her class that the blue-eyed children were the smart ones and saw similar discrimination, though it was less extreme, as the students may have become more skeptical at this point. Finally, she admitted that she was lying and further discussed racism and prejudice with her students as they reflected on the experience.

Elliott's exercise certainly would not pass an ethics review board today and was not a formal experiment, but it did generate quite a bit of attention because of how it addressed racism and discrimination. This classroom experience was shared in Elliott's local newspaper and eventually led to her being regularly interviewed on popular outlets, from *The Tonight Show Starring Johnny Carson* to *The Oprah Winfrey Show*. Elliott used the attention she received from her classroom exercise to advocate for civil rights and against racism.[1]

In 2010, Elliott spoke at Westminster College, and I was fortunate to attend as a student. She described her classroom exercise and led a discussion about the psychology of prejudice. I was so intrigued by her talk, and I reflected on how and why group identity can bias our thoughts. As soon as the question-and-answer period started, my hand shot up. I asked her what we could do about people who are stuck in their ways and refuse to change their minds. She essentially told me that we can keep trying but that sometimes we won't be able to reach certain people.

I remember feeling dissatisfied with that answer. Of course, Elliott was correct, as there are certainly times when people will not even listen to us, let alone change their minds. However, looking back now, I realized my main interests were the specific elements that determine what we believe and the conditions that are required for us to update our beliefs. I didn't realize, as a twenty-year-old psychology major, that these interests would become a major focus of my academic career. Understanding the psychology of our beliefs is also a foundational element of this book. In this chapter, I will describe the social, emotional, and cognitive architecture that allows misinformation to thrive. More specifically, I will examine the relationships among identities, emotions, and cognitive styles, which are crucial to understand if we are to learn why our minds are susceptible to believing false information in the first place.

THE POWER OF IDENTITIES

Identities are central to our beliefs, behaviors, and associations with others. Sociologist Peter Burke has spent his career studying identities and how they influence us. He defines *identities* as "sets of meanings" tied to the roles we have, the social groups we belong to, and the traits that make us unique.[2]

The definition is broad because we can hold a wide range of identities at once. Take, for example, being a parent, employee, customer, or sports fan—or any combination of some or all of these identities—at different times of the day or throughout the course of our lives. Some of these identities may be more important than others, and many are held simultaneously (e.g., we can both be a parent and an employee at the same time). Adhering to the values, roles, and scripts of our identities helps us navigate our social worlds.[3] The values associated with our identities are compared with the input from our environments in a continual feedback loop and create a motivation to adhere to these values.[4] We can also think of our identities as structured arrays of *heuristics*, the mental shortcuts that help us make decisions in our world.[5] Once we develop these identities, we are psychologically motivated to maintain consistency between the values of the information we experience and the values of our roles or identities.

Our identities have varying degrees of influence on our day-to-day lives, depending on their salience and prominence at any given time. Identity salience refers to how likely an identity is to be active in our mind, whereas identity prominence categorizes how subjectively important a particular identity is to us.[6] Identity prominence can increase when more people close to us support a particular identity. Think back to the introduction where our friend John found a supportive community at Flat Earth meetings. Identities are also more prominent when they share meaning with other identities we hold.[7] For example, if the identities of professor, parent, and partner all share the requirement of kindness, then this overlap could make each of those identities more prominent. Additionally, people rely on us to uphold certain identities, and this social feedback can further increase identity salience and prominence.[8] Identities do not occur in a vacuum, as other people's feedback is integral to the formation and maintenance of our own identities. Not only is direct feedback important, but also our beliefs about what other people think about us—regardless of whether they are accurate—can help influence our identity processes. Sociologist Charles Horton Cooley used the term "looking-glass self" to describe how we consider the perspectives of others in our efforts to understand our sense of self. We might alter our behavior, depending on how we think others feel about us.[9]

To illustrate identity salience versus prominence, let's pretend we have a friend named Rachel who is both a computer programmer and an artist. Her programmer identity may become more salient while she is working on computer code in the office. This helps her enact roles related to her salient programmer identity, such as making sure the application she coded runs successfully. However, her artist identity is more important to her than her programmer identity. In other words, her identity as an artist has greater prominence for her. Her broad artist identity incorporates activities that she loves and values, such as playing the guitar, writing songs, and making art. Each of these more specific art-related identities (e.g., guitarist, songwriter, and painter) is also important to her and how she views herself. Thus, when presented with a day where she has a choice between working on a computer program and practicing guitar, Rachel chooses to practice guitar to maintain that prominent artist identity. Aside from her own internal motivations, she may also feel pressure to practice guitar because of the perceived expectations of her bandmates. She believes her bandmates will be displeased with her when they next play together if she hasn't practiced. On the other hand, she may also feel pressure to work an extra day, but the amount of pressure she feels may be less because she does not have as strong a bond with her boss or work colleagues as she does with her bandmates. Rachel's programmer identity is salient when she is regularly performing programmer roles at work, but because she personally cares more about her artist identity and because her identity as an artist connects to a greater number and variety of identities she adheres to, it is a more prominent identity. We are constantly juggling an abundance of identities, and the salience and prominence of each identity constantly interact to help determine how we behave.

Performing roles and enacting values that are related to our more prominent identities can make us feel better about ourselves and boost self-esteem.[10] When we behave in ways that are consistent with our identity, we feel more positive about ourselves, and when we behave in ways that clash with our identity, we feel less positive.[11] Behaving in ways that support our identities is called identity verification. The more we verify our identities, the more salient or prominent they can become. For example, cleaning our bedroom makes us feel good because we have completed a task, and it also bolsters our identity as someone who is clean. Cleanliness is often viewed as a virtue, so verifying this virtuous trait can then boost our self-esteem. However, if we strongly value our "clean person" identity and fail to clean our bedroom, then seeing our mess can make us feel bad about ourselves. This process can happen within our "clean person" identity or within a more expansive identity, such as our political identity, which we will talk about at length later. Neuroimaging research has revealed distinct brain activity patterns when participants processed identity congruent versus identity incongruent information. Specifically, brain regions involved in self-referential processing were active during identity verification, and brain regions involved in decision-making and processing error were active during nonverification.[12] Future research will continue to integrate biological markers with social science to better understand identity processing.

Identities are also important because of the connections they allow us to form with other people. We are fundamentally social creatures, so these social connections are crucial to how we navigate our environment. As illustrated with the external motivation provided by the expectations of Rachel's bandmates, people are integral to how we view ourselves and our identities. We human beings have a fundamental need to belong and connect with other people, and when this need is not met, our mental and physical health deteriorates.[13] Socially isolated people score lower on cognitive performance tasks, including reaction time and memory. Social isolation has also been linked to reduced gray matter in the brain and an increased risk for dementia.[14]

Dr. Vivek Murthy, former U.S. surgeon general, has said that loneliness is worse for our health than obesity. He views its impact on life expectancy as comparable to that of smoking fifteen cigarettes—nearly a pack—a day.[15] Rejection by others can be harmful even when it comes from complete strangers. One study had participants play a simple online game where they virtually tossed a ball to one another. In this study, the other players were actually bots that were programmed to avoid throwing the ball to the human participant. The participants reported a more negative mood after being ignored by the other players during the game. Even when the study changed the rules and allowed participants to get paid more money if they were ignored during the game, the participants still experienced a decreased mood when they were ignored by others.[16]

Just as loneliness hurts our health, an enriching social life has been shown to improve health outcomes. Indeed, a decades-long study from Harvard University concluded that healthy relationships are the best predictor of happier and healthier lives.[17] Having meaningful relationships can be a tremendous buffer against the stresses and uncertainties of life. This includes all types of relationships—from friendships to family ties to romantic partnerships. The human need for social connection is met through our relationships, and bonding over shared identities helps facilitate this social connection process.

Our need to fit in and connect with others is so strong that we will even conform to very silly behavior to enrich social connections. Psychologist Solomon Asch conducted a famous conformity study from the 1950s in which he had participants evaluate the length of lines shown on a projector. It was a very simple task, and nearly 100 percent of the participants answered correctly each time. However, Asch then asked the participants to provide their answers in a room with several people who had been coached by Asch to share the wrong answer aloud before the other participants could answer. Participants in the study then also began to give wrong answers. The desire to conform made 75 percent of the participants get at least one answer wrong throughout twelve trials.[18] The Asch study is an amazing example of how groups can influence an individual's beliefs.

A follow-up to the original Asch study evaluated how people performed when they did not have to publicly disclose their answers. One of the participants was told they were late and that they could simply write their answer on a piece of paper. That participant still witnessed the others give the wrong answer, but they almost never provided a wrong answer themselves. What seems crucial in these conformity studies is whether others will be able to explicitly judge if we followed the group. A 2024 study had participants interact on Zoom while judging moral dilemmas and found that they still conformed to the group answer even while interacting via a computer screen.[19] Our need to connect with others can directly influence our judgment and behaviors.

We yearn to connect with others. Our identities can enhance our social experience by helping us connect with like-minded individuals. Belonging to social groups can be a critical source of self-esteem that we may not be able to achieve at the same level by ourselves.[20] When we reject an identity, it can be particularly painful because we may limit or, at times, eliminate the social connections associated with that identity. For example, let's pretend we have a friend named James. He is a devout Catholic and actively involved in his church, which provides a community of like-minded individuals. He begins to read more about other religions throughout history as well as about the philosophy and psychology of belief systems. Eventually, he decides that the Buddhist religion and philosophy resonate more with him, and he no longer shares the same belief systems ingrained in Catholic ideology. This change in religious identity isolates him from his devout Catholic family and the church community. While James finds personal meaning in his newly acquired Buddhist identity, the loss of social connections from his former Catholic identity is still quite painful. To mitigate this social isolation, he may seek out friendships with those who share his new religious identity, which will only strengthen his religious conviction. We will cover in detail how personal networks influence identity in chapter 4. For now, we will focus on how the motivation to protect our identities—and the personal meaning and social benefits they provide—can bias our information processing and subsequently make us more vulnerable to misinformation.

HOW OUR IDENTITIES IMPACT OUR INFORMATION PROCESSING

The interplay between social motivation and our identities explains their influence on our behavior. The way that our identities bias our thoughts and behavior has been referred to as "identity-protective cognition."[21] As the name suggests, we are motivated to constantly protect our identities as we cognitively process—whether consciously or unconsciously—our information environments. Usually, identities and their impact on our behavior are benign. We may feel motivated to complete work-related tasks, attend social functions, or adhere to cultural norms during social situations. For example, we can think back to our friend Rachel, whose programmer identity helps her perform roles at work and whose artist identity motivates her to practice guitar and draw. This identity process is often neutral or even good, as it helps us navigate our worlds, but it can also cause problems when we feel pressure to ignore evidence that conflicts with our identities.

We can think back to the introduction to this book where we discussed our hypothetical friend John, who believes the Earth is flat. Remember how we saw John's interest in Flat Earth beliefs coincide with new social relationships with people who share that belief. This is similar to our friend James, whom we met in this chapter, whose religious identity has changed. Psychologist Per Espon Stoknes provides an insightful quote about how personal relationships may be lost if a flat-earther or religious individual leaves their community: "Say you lose faith in this thing. What then happens to my personal relationships? And what's the benefit for me doing that? Will the mainstream people welcome me back? No, they couldn't care less. But, have I now lost all of my friends in this community? Yes. So, suddenly, you're doubly isolated. . . . It becomes a question of identity. Who am I in this world? And I can define myself through this struggle."[22] Believing that the Earth is flat could be a central component of a broader, and more important, identity that involves constantly questioning dominant narratives, just as religious affiliation is often tied to more than attending church functions. Beyond losing social relationships, rejecting a Flat Earth belief may force a person to adjust their identity as a contrarian, which provides a major source of self-esteem for them. Indeed, a study found that people who attended a Flat Earth conference scored higher on general conspiracy theory belief.[23]

Flat-earthers are an evocative example of not only how identities can influence our beliefs surrounding false ideas but also how membership in a group and the identity processes associated with that group can change how we navigate our world. A broader example of how identities can bias information processing is provided when our identities intersect with those of other groups. Specifically, when we belong to a group, we can form a social identity around that group membership.[24] Social identity theory emphasizes the meanings that are associated with group membership, and identity theory emphasizes the meanings of specific role identities.[25] Both help us understand how identities influence our information processing.

Our group-based identity can motivate us to protect our group while holding negative evaluations of opposing groups. This motivation can result in a biased interpretation of how we think about different groups related to our identity.

Group identification can be formed rather quickly and immediately impact how we process our environments. Classic social psychology experiments demonstrated how research participants who were randomly assigned to an arbitrary group both evaluated that group more favorably and offered greater rewards to their own new in-group members compared to out-group members.[26] Such in-group biases were also be formed by telling participants in an experiment that they preferred one style of artwork over another.[27] Thinking about the social groups we belong to and strongly identify with can activate the same parts of the brain that we use when we think about our own individual attributes.[28] The social groups we join that we deem personally important help us form a major sense of how we see ourselves and how we form our personal identities.

While Jane Elliott's classroom exercise wasn't an experiment, the general idea was transformed into one in a 2011 study. Researchers divided the children into groups by having them wear red or blue shirts. Even though the children had not met anyone from the other group, they still quickly revealed a significant in-group bias. Specifically, the children preferred to play with children who wore the same color shirt, and they evaluated pictures of children more positively when those pictured wore the same color shirts they were wearing.[29] Not only has in-group bias been quickly observed in children, but also preference for one's own in-group has been documented in many different cultures around the globe.[30] Social science consistently reveals how easily and quickly we form in-group biases through a variety of formats and scenarios.[31]

A bias in favor of one's own team is something many sports fans are probably familiar with. I know I am always rooting for the Pittsburgh Steelers to do well. I also recognize that I have a bias toward supporting my home team and a proclivity to believe that calls made in my team's favor are more justified. I'm sure many sports fans do this as well, though it certainly is easier to see the bias in other fans compared to ourselves. A classic 1954 study asked Princeton and Dartmouth students to report how many penalties they observed during a football game between the two schools. Both Dartmouth and Princeton students thought the opposing team committed more penalties despite watching the same game.[32]

While passion for a sports team can be strong, it's also relatively harmless. Thinking a penalty shouldn't be called against your team rarely has much lasting or widespread importance. But there are some identities we hold that bias us about information that certainly has a more meaningful impact. Political identities are one such example. Political scientists Patrick Miller and Pamela Johnston Conover studied attitudes of Democrats and Republicans toward one another using a variety of survey questions. They concluded that partisan behavior is quite similar to biased support for a sports team, as members of political groups care more about preserving the status of their "team" than about giving thought-ful consideration of the political process.[33]

You might think that political identities are focused on policies, but political scientists have consistently found that people think about their politics much more in terms of identity than a collection of beliefs.[34] As I mentioned previ-ously, the more important the identity or the stronger the identity prominence, the stronger the bias to support that identity. Feeling strongly about a politi-cal party—and also about the opposing party—only increases the motivation to defend our in-group and any values associated with our in-group identity. Importantly, the strength of these identities directly correlates to the levels of polarization in our society.

HOW IDENTITY LINKS POLARIZATION AND MISINFORMATION

Political polarization continues to grow at alarming rates in the United States. An increase in partisan media and a decrease in political norms of civility and cooperation are two major factors contributing to this polarization.[35] In 1964, only about 30 percent of Democrats and Republicans reported having negative feelings toward the opposing political group. This rose over time, and in 2012, nearly 80 percent of Democrats and Republicans had negative feelings toward the opposing political group.[36] A 2022 survey found 62 percent of Republicans and 54 percent of Democrats had a "very unfavorable" view of the opposing party, whereas in a survey from 1994, less than 25 percent of survey respondents from each party said they had a "very unfavorable" view of the other party.[37] A 2018 study found that about 42 percent of Democrats and Republicans agreed that the opposing political party is "not just worse for politics—they are down-right evil." More alarmingly, around 16 percent of Republicans and 20 percent of Democrats admitted that they occasionally think the country would be better if large numbers of those in the opposing party died![38]

How does this political polarization affect information processing? When a particular identity, such as political identity, becomes especially important, it can fuse with one's overall sense of self. This makes anything that challenges that identity feel like a personal attack. Conversely, anything that affirms that identity can boost self-esteem.[39] Thus, strong partisans are motivated to interpret infor-mation in a way that supports their political group. More broadly, psychologist Jay Van Bavel and colleagues summarize how polarization is inextricably linked to the spread of misinformation: "In a polarized environment, relatively neutral information may be seen as politically relevant, information offered by one's in-group is more likely to be believed, and information offered by the outgroup is likely to be dismissed as false."[40]

This bias is easy to observe regarding explicitly political topics. For example, Democrats generally favor abortion access and gun control at significantly higher rates than Republicans.[41] Studies have shown that when Democrats and Republicans read information about abortion and gun control, they interpreted the conclusions in a way that supported their existing values.[42] Furthermore, conspiracy theories are much more likely to be believed when they support one's political group or attack the opposing group.[43] Democrats and Republicans are more likely to support their own team whenever they can, and research has shown that both groups were more likely to offer jobs and scholarships to candidates of their own party, even if this meant choosing a candidate who was less qualified.[44] Some of my own research showed that Democrats and Republicans were more likely to believe and share vague rumors, such as "(Democrat or Republican) bullies a person," if the rumor favored their own group. Information one encounters that can be linked to their identity will influence how they processes it. Because politics saturates our information ecosystem, political identity is integral for understanding the spread of misinformation. For example, there was zero evidence of significant voter fraud during the 2020 U.S. presidential election, but we saw a majority of Republicans doubt the election results.[45] It can be very frustrating to watch people reject the evidence for such a significant social event. Research on identity theory helps make sense of this process. A 2021 poll highlighted this phenomenon, as 59 percent of Republican voters said that believing Donald Trump won the election was "important" for identifying as a Republican.[46] This poll result directly links identity and misinformation.

Political identities' impact on information processing has also been observed in brain activity.[47] Neuroimaging research demonstrated that people did not pragmatically reason through information that challenged their political beliefs. Instead, the parts of the brain associated with negative emotions were highly active when reading information that challenged their political beliefs.[48] Another brain imaging study found that the participants whose emotional centers were less active while reading information that challenged their political beliefs were also more likely to change their beliefs.[49] These neuroimaging studies highlight the importance of emotion while processing information.

A 2021 study specifically showed that the parts of the brain used in thinking about ourselves were also active while making decisions. This study recorded the brain activity of UK participants who were either in favor of or opposed to Brexit. While having their brain activity recorded, the participants read a variety of headlines about Brexit—some that supported their position and others that challenged it. Then they were asked whether they thought each headline was true. The results revealed that brain areas involved with self-referential judgments were most active while participants considered whether the headlines were true. Importantly, the prefrontal cortex, the area of the brain associated with complex decision-making and reasoning, was not active during these deliberations. This provided neural evidence that people were focused more on their selves (and identities) than on objective analysis when shown opposing political information. The authors of the study concluded that "people do not necessarily systematically reason about strengths and weakness of offending information, but rather critique to reject it and protect the self."[50]

Making these identities more salient in our minds can further increase their influence. Remember, identity salience means that it is more active in our minds at the moment. For example, a laboratory study found that conservatives who were primed to think of their conservative identity were even more likely to reject the idea of anthropogenic climate change compared to conservatives who did not first reflect on their political identity.[51] A 2022 study addressed the same impact of political identity salience but from a different angle. This study found that when liberal and conservative participants were primed to think they had a more moderate political identity, they answered political questions with less bias.[52] Furthermore, both Democrats and Republicans became more polarized when they reflected on their political identity. However, when partisans were asked to reflect on their national identity, both Democrats and Republicans were less polarized.[53] This highlights the power of identity in influencing our beliefs, but it also highlights how malleable this process can be. When we have multiple important identities competing with one another, the ones most salient in our minds have the highest likelihood of influencing our behavior.[54] People who reject evidence are often not naive or ignorant; instead, their rejection of evidence is a rational, if unconscious, choice to defend the values associated with meaningful identities and the self that is derived from these identities.

When we are overly attached to our jobs, our work identities can be a source of bias as well. Scientists are at risk for this bias, although their jobs constantly require application of analytical thinking and openness to embracing new ideas—even those that conflict with prior beliefs. Identifying as a scientist may also mean that a person could become partial to a certain theoretical framework that they apply often in their research. In a classic 1977 study, a group of sixty-seven psychologists reviewed articles that described fake experiments. These fake experiments used identical methods but differed in their theoretical frameworks. Psychologists evaluated the experiments much more favorably when they supported the same framework these psychologists had used in their careers.[55] This is consistent with the studies mentioned earlier that demonstrated how both Democrats and Republicans interpreted the same information differently so as to support their respective political party and to affirm their identity within that specific group.

I discuss political identities in a significant portion of this book because political misinformation is one of the most common forms of misinformation. Anything can become politicized, and if we have political biases, then we are constantly at risk of interpreting information in a way that supports our political team. Political identities are also relatively easy to measure and are well studied in social science literature. While I often focus on American political identities here, it is important to note that the identification process can be found across political identities around the world.[56] American political identities, however, have morphed into what political scientist Liliana Mason terms "megaidentities" because of their overarching influence and connection to a person's other identities. She described these mega-identities in an interview on Ezra Klein's podcast:

> Each election becomes, not just, you know, "Well the Democrats lost, so what a bummer, you know, because my party lost. . . . Instead, it's like, "Well, my party lost, so that means my racial group and my religious group and my cultural group, and all the people that I know and all the people that I watch on TV, everything that I know, everything that makes up who I think I am, it's all gone. I lost it all. I have nothing left inside of my sense of who I am. . . . The more things that get involved in each election, the more vulnerable our self-esteem is.[57]

Again, because American political identities have these unique "mega" properties, they offer a powerful example of how identities can influence our susceptibility to misinformation. But any important identity can bias us to varying degrees. Let's dive into this process even further.

CONFIRMATION BIAS, COGNITIVE DISSONANCE, AND THE BUILDING BLOCKS OF BELIEFS

The motivation to interpret information in a way that supports our identities can be further understood through research on *confirmation bias*, a well-studied phenomenon in the psychological literature. Defined as the tendency we have to interpret new information in a way that supports our existing beliefs,[58] it can range from believing penalties against our favorite team were unfair to interpreting politically relevant information in a biased way. We are motivated to search for information that supports our identity while also selectively avoiding information that challenges our identity.[59] Confirmation bias is so strong that it can even be measured with eye-tracking data. A study found that participants' eyes avoided looking at political ads inconsistent with their partisan ideology and spent more time looking at ads consistent with their partisan ideology.[60] Confirmation bias constantly tries to reduce conflict between our identity and the information we encounter.

As if this confirmation bias process wasn't strong enough by itself, simply showing someone the same information over and over can increase the likelihood they will believe it is true.[61] This phenomenon, called the illusory truth effect, has important implications for understanding why people fall for misinformation. In a 2018 study, researchers had participants read a variety of fake news headlines and then mentioned that the veracity of these headlines was contested by independent fact-checkers. Despite being explicitly told these stories may not be true, participants in the study were more likely to believe these headlines were true when asked about them again a week later.[62] If we are always selectively attending to information that confirms our identity and we are more likely to believe information we are exposed to repeatedly, it becomes clear how easily we can be guided toward false information. This feedback loop of exposure to biased information that bolsters our identity can be thought of as a mutually reinforcing spiral.[63]

The confirmation bias process helps alleviate any discomfort we may feel when we are exposed to information that conflicts with our held beliefs. Psychologist John Petrocelli has for decades studied why people believe false and misleading things. His specific field, the science of bullshit (yes, that is the technical term), is defined as the intentional or unintentional sharing of information with little regard for truth or evidence. The science of bullshit has clear implications for understanding misinformation. In Petrocelli's latest book, *The Life Changing Science of Bullshit*, he outlines two of the strongest motivators that lead us to fall for bullshit: (1) the need to belong and feel accepted by others and (2) the need to be consistent in our thoughts and behaviors.[64]

One clever experiment showcased how the need to maintain consistency can change reported attitudes. Psychologist Thomas Strandberg and colleagues had participants report their feelings for 2016 presidential candidates Donald Trump and Hillary Clinton. When participants were not looking, the experimenter changed their answers to be much more moderate and then asked them to explain what they thought were their answers. One participant rated Clinton as much more experienced initially, but when shown their manipulated answer that Trump and Clinton were equally experienced, they reasoned, "I think they're both experienced in their field. Trump is a really successful businessman." Overall, participants changed only 12 percent of their manipulated answers.[65] Even though participants first gave quite polarizing answers, they then had to justify to themselves why their responses were actually quite moderate in order to reduce any resulting cognitive dissonance.

Social psychologist Leon Festinger first proposed the theory of *cognitive dissonance* in the 1950s.[66] It can be thought of as a psychological need that has to be satisfied lest it create significant discomfort. We experience cognitive dissonance when we become aware of any conflict between at least two thoughts in our mind, a discomfort we can resolve by changing the perspective we hold toward the uncomfortable information. For example, if we support a political candidate and read a news article about the candidate behaving poorly, we may either change our perception of the candidate or dismiss the article as inaccurate in order to avoid cognitive dissonance. Of course, cognitive dissonance can occur in many areas beyond politics and misinformation.

A classic example is the smoker who knows smoking is bad for them. Smokers do not generally want to cause harm to themselves, but the evidence is overwhelming that smoking causes a number of health issues. Thus, smokers often engage in a dissonance reduction strategy such as suppressing thoughts about the health risks or convincing themselves there are many ways they can die other than smoking.[67]

Cognitive dissonance also describes the psychological tension we can experience from conflicting thoughts about our identities and social environments. We are motivated to avoid the psychological tension caused by cognitive dissonance, and we are also motivated to constantly protect our identity. The social support intertwined with a particular identity can affect how difficult it is for us to update our beliefs. Festinger and his research team interviewed a group of cult members who believed a giant flood would destroy much of the planet on December 21, 1954. When this date passed and no flood came, Festinger and his team interviewed the members of the cult again. They found that the cult members with the strongest social support and involvement with the group doubled down on their commitment to the cult and relied on various excuses to explain why the flood did not happen. Conversely, cult members who were less socially connected with the group expressed more skepticism, and several even left the cult altogether.[68]

David McRaney, author of several books on cognitive biases, including *You Are Not So Smart* and *How Minds Change*, has a great analogy for confirmation bias: "It's useful to think of confirmation bias as the goggles we put on when we feel highly motivated by fear, anxiety, anger and so on—in these states we begin looking for confirmation that the emotions we feel are justified."[69] If our confirmation bias goggles are what we put on when we experience uncomfortable information, then we can think of our identities as the lenses in those goggles that determine the distorted view.

So far, I have focused on how identities can influence beliefs because of the motivation we have to protect the meanings and values associated with them. Next, I'll add another level of detail and discuss how we can further break down beliefs into smaller elements. The combination of these elements can predict how likely we are to believe the information we receive.

According to affect control theory, we are motivated to maintain consistency between our cultural beliefs and our social environments.[70] Whenever we experience something in our social world that conflicts with our preexisting cultural ideas, we reevaluate the situation. This is similar to identity theory, but the focus here is more on general cultural beliefs, which don't need to contain identities. For example, we may hear a story about a mother who has hurt a child. Since we have a positive evaluation of "mother" and "child," then it would cause conflict for us to think about a mother hurting a child. Thus, in order to reduce this inconsistency, we may relabel the mother in this story as a "monster." A powerful aspect of this theory is that it can break down social situations into smaller (and easily quantifiable) pieces of actor, behavior, and object. In our example here, the "mother" is the actor, "hurt" is the behavior, and the "child" is the object. When we break down situations into these more digestible concepts, we can measure and predict how people may respond during these social events. I like to think of affect control theory as a framework for understanding the building blocks of our beliefs.[71]

By breaking down beliefs into smaller chunks, we can analyze exactly what a person is changing in their mind when they experience cognitive dissonance. In the example of the mother who hurt a child, we discussed how we often recategorize the mother (or the actor) as a monster. However, research suggests that we are much more likely to reinterpret behavior instead of the actor (so we may change "hurt" to "discipline"). It may be cognitively easier for us to relabel the behavior as opposed to the identities within the social scenario.[72] Importantly, this motivation to reframe social situations influences how we share information. When the participants in a study were asked to read a short story and share what happened, they reinterpreted the events in a way that was more consistent with their preexisting beliefs. Participants read short stories involving an unemployed person, such as an unemployed person helping a salesclerk. In this same study, it was found that unemployed people were rated very negatively overall. Because people rated "unemployed person" so negatively, we might expect a reinterpretation of events that is more consistent with that commonly held negative stereotype. This is precisely what the study found, as participants were more likely to report the unemployed person was "telling something to" the salesclerk instead of "helping" them. The less positive behavior of "telling something" is more consistent with the negative evaluation of the unemployed person.[73] Thus, we unconsciously reinterpret this and share the updated story in a way that fits our prior attitudes. As we navigate our social worlds, we are constantly interpreting events in a way that fits our preexisting beliefs. The impulse to reframe information in order to be consistent with what we expect provides a fertile ground for misinformation.

Building on affect control theory, I was curious if Democrats and Republicans would interpret the same information differently in order to support their own political group. Since both political groups had more positive feelings about their own group and more negative feelings toward their opposing group, I predicted that they would interpret information in ways that would be consistent with those evaluations. In my study, I had Democratic and Republican participants rate the likelihood of various stories they read involving political group members that approximated fictitious headlines. For example, they would read that they overheard someone saying "a Republican helped a person" or "a Democrat helped a person." Even though they had no other context for this brief amount of information, both Democrats and Republicans were significantly more inclined to rate the situations as more likely when the headline supported their own group or indicated an actor from their in-group had performed a positive action. My participants also rated the likelihood of situations in which political group members hurt someone, with headlines like "a Republican bullied a person" or "a Democrat bullied a person." Again, participants supported their political group by saying the opposing political group was more likely to bully a random person.[74] Because the participants already had formed beliefs about each political group, they were motivated to interpret the story in a way that was consistent with their beliefs. My findings were consistent with an extensive literature showing how both Democrats and Republicans have consistently interpreted information in a way that is biased toward supporting their own. They reveal how our identities bias our interpretation of even simple pieces of information. This research project began as part of my dissertation, and I later analyzed how social networks play a role in political bias as well (which I'll cover in chapter 4). Believing and spreading false information is a by-product of our cognitive architecture because that same cognitive architecture helps us make sense of our social worlds.

WHY IDENTITY BIAS CAN TRUMP EDUCATION

Our identity biases are so strong that they can often override education and reasoning abilities. Research has shown that people rejected the scientific consensus on issues that conflicted with their political identity even if they also scored highly on general science tests.[75] In one study, about 90 percent of a large, representative sample of Americans believed they are "above average" in their ability to detect fake news. Of course, it is mathematically impossible for 90 percent of people to be above average, and, sure enough, nearly 75 percent of these Americans were found to overestimate their ability to determine whether headlines were real or fake. Perhaps the most troubling finding is that those who were most confident in their abilities to detect fake news were the worst performers.[76] Additionally, people who spent more time watching political media were even more inaccurate when evaluating political claims.[77] Later in the book, I will discuss how the media and our social connections can often reinforce our identity instead of providing a more balanced perspective.

Education itself is not enough to override these identity processes. A study by Dan Kahan, a psychologist at Yale University, found that religious people were far more likely to agree that human beings developed from earlier species of animals—but only if "according to the theory of evolution" is put in front of that statement. This allowed them to answer the question accurately without threatening their religious identity.[78] In other words, if the evidence in front of us is clear but it threatens our identity, we will either reject it or need a qualifying statement added to it in order to agree. This is consistent with research showing people were still biased by a statistic they read about migration even when they were told it was false.[79] Moreover, a 2024 study had Democrats and Republicans evaluate both false and factual news headlines and found that strong political bias persisted across all education levels. The study also revealed that the most biased individuals were those who believed their political group was unbiased and objective.[80] These findings are also consistent with the conclusions political scientists Christopher Achen and Larry Bartels reach in their book *Democracy for Realists: Why Elections Do Not Produce Responsive Government*: "Even among unusually well-informed and politically engaged people, the political preferences and judgments that look and feel like the bases of partisanship and voting behavior are, in reality, often consequences of party and group loyalties."[81]

While identities can sometimes override our educational background, education is still an incredible tool for protecting us against misinformation. Additionally, recognizing our psychological biases is a crucial component of a well-rounded education. We will learn more about the benefits of education and media literacy skills in chapter 6. Thus far, we have covered how identities influence our information processing. This is critical for understanding our susceptibility to misinformation. In the next section, I will describe how different elements of our sense of self can influence this identity bias process.

FACTORS RELATED TO THE SELF THAT INFLUENCE OUR IDENTITY PROCESS

Thus far, I have described the identity process as more of a static process. If we have an important identity, then that identity will influence our information processing. This influence can be so strong that it overrides our educational background because we are strongly motivated to protect our identity. It's also important to be aware of other factors that intersect with this identity process and can increase or decrease its effects. I will specifically address self-uncertainty, self-affirmation, and social identity complexity and describe how they influence our identity biases.

Uncertainty Makes Us Cling to Our Identities

All humans and animals living on our chaotic planet inevitably experience uncertainty. To cope with this unpredictable and often harsh environment, our ancestors relied heavily on heuristics and social relationships to increase their odds of survival. As I mentioned earlier, heuristics are mental shortcuts that help us make decisions based on incomplete information. If one of our ancestors heard rumbling in a bush, it was more advantageous to assume that sound was a predator about to eat them for lunch. By assuming danger through that heuristic, they could prepare themselves and increase their chances of survival. Our minds evolved to fill uncertain aspects of our environments with previous knowledge that could help us make sense of and prepare for that uncertainty. In modern times, we might not have to worry as much about being eaten by wild animals, but we still use prior knowledge to fill in gaps. We may choose to buy a product from a brand that we are familiar with instead of one we are not. Or we may not know the answer to a question we face, but we can find it in the words of someone we know and trust.

Social relationships were also crucial for the survival of early humans. A saber-toothed tiger may have had no problem eating one of our ancestors for lunch, but by working together, a group of humans vastly improved the odds of surviving the giant cat. Social psychologists have theorized that because of our reliance on social relationships, connecting with others through social bonds has been a critical strategy for the survival of our species.[82] In fact, psychologist Kip Williams argues that social exclusion can feel so profoundly painful because social connections are crucial to both our physical survival and our ability to create meaning in our lives.[83] Our brains are pattern-recognizing and meaning-making machines that have the singular task of keeping us alive. Social identities are incredibly powerful because they help us make sense of our chaotic, uncertain world and they connect us with others, and such sociality has been (and continues to be) critical for human survival.

As we will regularly see throughout this chapter and book, identities connect us to others and help us navigate our social worlds. Our social identities can also provide meaning and protect us against uncertainty itself because they provide a sense of belonging and connect us to others who share our worldviews.[84] Research has found that people identify most strongly with their identities when they are feeling uncertain.[85] Think about a time when you felt painfully uncertain during your life and reflect on how that could change your mood. When participants in one study wrote about times they felt uncertain in their lives, they identified even more strongly with their political identities.[86] Additionally, priming uncertainty was found to make participants adopt even more polarized beliefs about capital punishment and abortion.[87]

We can think about how people with a political or religious identity can be motivated to interpret the world in a way that supports their group, but this process can be observed even when study participants were randomly assigned to groups in an experimental setting. When those participants completed challenging tasks that made them feel uncertain, they also identified more strongly with their artificially generated group.[88] If being randomly assigned to a group in a room can create this in-group effect, then we can imagine how influential a fully developed identity such as religion or politics can be in our lives.

Religious identities are especially powerful in their protection against uncertainty because they can create rigid boundaries or rules for viewing the world, and this can allow people to avoid processing contradictory information.[89] For example, they may experience a traumatic event, but justification is provided because of their belief that God works in "mysterious ways." This protection against uncertainty can be observed in neural activity as well. Psychologist Michael Inzlicht led a study in which his research team recorded the brain activity of individuals with high and low religious commitment while they completed a simple cognitive task.[90] The individuals who were very religious had decreased activity in the parts of the brain associated with processing uncertainty and error detection when they made mistakes on the task. In a follow-up study, Inzlicht and his team recruited two groups of participants who strongly believed in a theistic God.[91] One group of participants wrote about what their religion explained in their lives and what it meant to them before completing the cognitive task, while the other group did not write before the cognitive task. Like in the previous study, the participants had to complete a simple cognitive task while the team measured their brain activity. However, before the participants completed that task, they spent time writing about either their religion (the experimental group) or their favorite season (the control group). The group that reflected on their religion before the experiment had reduced activity in the parts of the brain associated with processing error and uncertainty. This replicated the effects of the previous study, in which strong religious belief reduced error processing in the brain, but this error processing activity was reduced still further when they spent time reflecting on their religious identity.

Ideologies and identities do not require supernatural belief to buffer anxiety and move our attention away from uncertainty.[92] As discussed earlier, any important identity can offer protection for our sense of self by boosting our self-esteem. This identity protection process can be quite broad. For example, one psychology study found that nonreligious individuals reported a stronger conviction in science when they were stressed out before an upcoming competition compared to other nonreligious individuals who did not have to compete soon.[93] Another study found that individuals with strong political beliefs had reduced error processing when seeing abnormal playing cards, like a queen of hearts printed with black ink instead of red.[94] These neuroimaging studies revealed that a strong commitment to an ideology may change how our minds process the world. It is also possible that those who have inherent muted responses to uncertainty may find ideological commitment more attractive. One neuroimaging study found that participants who scored higher on an intolerance of uncertainty survey had even stronger neurological reactions associated with political polarization while watching political debate videos.[95] This was true for both liberals and conservatives, suggesting that those who are less tolerant of uncertainty may be especially reactive to political information challenging their identities.

When we feel uncertain, we become particularly motivated to seek out psychological appeasement through identities and ideologies so we can restore feelings of certainty and return to our baseline. Not only does uncertainty make people double down on their beliefs, but also feeling anxious can increase the likelihood they will share misinformation online.[96] This can create a vicious cycle of misinformation, in which anxious individuals assuage their uncertainty by sharing polarizing information and/or misinformation that attacks an opposing political group. When that misinformation attacks the opposing political group, it can end up upsetting those on the other side. Then those on the other side may feel threatened and share polarizing information and/or misinformation to make themselves feel better—and the cycle continues.

Self-Affirmation Can Make Us More Open to Identity-Threatening Information

Just as self-uncertainty can exacerbate our identity-protective cognition, feelings of self-assurance can make us less reliant on the identity feedback loop. We can feel more confident in our ability to handle the stresses of life through a process of self-affirmation. *Self-affirmation* involves focusing on personally important qualities, values, and accomplishments. When we affirm our own value, we buffer ourselves against identity threats and, in doing so, even make ourselves more open to identity-challenging information.[97] Experiments have shown that people derived self-affirmation from simply writing about a time when their behavior supported a value that was important to them. For example, they might have written about how their family identity is important to them and how they have acted in a way that was consistent with the values associated with that identity. Research has found that participants who wrote about and reflected on their positive qualities felt more positive about themselves afterward.[98]

This self-esteem boost from a self-affirmation exercise can also make people better able to handle identity-threatening information.[99] When people completed self-affirmation tasks during experiments, they also reported less bias toward opposing political candidates, reduced animosity toward out-group members, and increased openness to opposing political beliefs regarding abortion policy.[100] Self-affirmation also increased the acceptance of health messages and the adoption of healthier behaviors such as diet and exercise.[101] When people reflected on the times they felt accomplished or proud of themselves, they felt more capable of handling challenging information or adopting difficult (but healthy) behaviors.

Researchers have used neuroimaging to evaluate activity in the brain regions associated with self-affirmation in individuals who had previously scored high on body mass index measures and low on daily physical activity. Participants completed either a self-affirmation task, such as reflecting on important values, or a control task that did not provide any affirmation. All participants also read health messages about the harm of sedentary behavior. The neuroimaging data revealed that those in the self-affirmation group had higher activity in the ventromedial prefrontal cortex (VMPFC), an area associated with self-related processing and positive evaluation, while reading the health messages. Those in the self-affirmation group were also significantly more active one month later (as measured objectively via personal accelerometers). Thus, the researchers concluded that self-affirmation helped individuals see how information that was personally threatening could also be personally relevant and valuable to receive.[102] Another study evaluated self-affirmation's impact by scanning participants' brain activity as they solved stressful math problems. The researchers replicated the findings of increased VMPFC activity during self-affirmation primes and also found that self-affirmation was associated with lower activity in the anterior insula, a brain region associated with processing stress. Self-affirmation provided a buffer against difficult or stressful information and helped the brain attend to self-relevant information.[103] While self-affirmation does appear to show promise, it's also important to note that subsequent research has found its ability to change political beliefs is more limited than previously thought.[104] This could be due to a lack of reliability in both how self-affirmation is measured and how it is primed.

How we feel about ourselves impacts how our identities influence our behavior. Because priming self-uncertainty and self-affirmation in succession can cancel out their effects, it appears they operate on a similar compensatory mechanism.[105] I conducted a study in which the participants completed a self-affirmation or self-uncertainty priming task. After each group completed the prime, they rated how they currently felt about themselves. Those in the self-affirmation group rated themselves significantly more positively than those in the self-uncertainty group.[106] Broadly, we can think of self-affirmation and self-uncertainty as opposing forces. By using reliable methods to measure these priming forces, researchers can better determine when they are most effective. When we feel more confident about ourselves, we are more open to identity-threatening information. Conversely, when we feel more uncertain about ourselves, we are less open to identity-threatening information.

The complex interplay among emotions, self-esteem, and identities directly impacts people's motivations for sharing fake news. Studies have shown that the hatred of political opponents was a major predictor of whether someone would share fake news. People who generally scored high on the ability to evaluate the accuracy of online media would still share fake news in order to attack their political opponents.[107] This is consistent with research showing that we can feel better about ourselves when we tear down someone who belongs to our outgroup.[108] To put it bluntly, sometimes we don't care if what we share with others is true because our biggest motivator is our need to attack the bad guy and sub-sequently give ourselves an ego boost. Our own self-doubt or personal feelings of uncertainty can make us more likely to go after that quick ego boost. When we feel self-assured or self-affirmed, we tend to be less likely to be on the offensive. Think about your own social media habits. When do you post the most evocative content? When do you find yourself on social media most often? Do these times coincide with any particular emotions?

Identities provide meaning and social support. We often care more about keeping that hard-earned and valuable social support than about being correct during some abstract argument. As we've already established, human connection is a fundamental need, and social isolation is linked to poorer health outcomes. Sharing facts is unlikely to change someone's mind if accepting those facts also means they will lose status within a group they care about. How someone is currently feeling about themselves significantly influences how receptive they are to information that challenges an important identity. People may be better able to absorb negative feedback for one identity when they are receiving sufficient support for other identities via some form of self-affirmation. We must consider how multiple identities intersect in order to better understand our overall susceptibility to identity bias and the impact that identity bias has on our willingness to consider opposing viewpoints.

Social Identity Complexity Impacts Our Beliefs

The interplay of our identities can reduce their impact, as can the way each one is related to the others.[109] The perception of how our identities connect and overlap is referred to as *social identity complexity*. When people have many important identities that do not easily overlap and that are more independent of one another, they have greater social identity complexity. Remember the mega-identity concept described earlier? Mega-identities, such as political identities in some people, have overarching influence on and connection to other identities (i.e., there is low social identity complexity). If identity processes make us more vulnerable to bias when they are interconnected into more simple overarching identities, then identities that have little overlap may have less influence on us. Higher social identity complexity can be thought of as the opposite of simplistic mega-identities, in which all of our important identities (such as our political and religious identities) neatly fit within one another. It can be helpful to visualize and actually draw the spatial relations among our important identities.

We have an extreme example of low social identity complexity if all of our important identities completely overlap to form a perfect circle.

As noted previously, feelings of uncertainty can make us cling to our identities more; however, research has shown that people were less vulnerable to identity biases due to uncertainty when they had higher social identity complexity.[110] Researchers have also found that individuals identified more strongly with their political identities when their political groups had greater overlap with other important identities (e.g., religion).[111] This impact of social identity complexity on beliefs goes beyond influencing our political attitudes. For example, a 2024 study found that people were more open to information that could improve their diets when their dietary identity did not strongly overlap with other important identities.[112] Remember that identities can become even more prominent and influential when they share meanings with other identities.

We can think back to the start of this book, where we discussed our friend John, who believes the Earth is flat. Thinking the Earth is flat supports John's contrarian identity, but imagine that distrusting mainstream science is also a central value within his political and religious identities. Thus, John's contrarian, political, and religious identities all support a distrust of science. This means John has three separate identities that are all supported if he believes something that goes against the scientific consensus. Now let's say we have another friend, Sally, who also has a contrarian identity. Both John and Sally get a rise from disagreeing with others in their social group and being contrarian. However, Sally's religious and political beliefs differ from John's, and discounting mainstream science is not part of them. Sally may get a slight boost in self-esteem from supporting her contrarian identity by believing the Earth is flat. But because her other identities do not share the value of distrusting science, Sally has less motivation to keep believing the Earth is flat, since it supports only one of her identities. Conversely, John has three identities that relate to one another through their commonality of distrusting mainstream science, providing John a greater motivation to hold beliefs that provide identity verification for all of his identities. The greater the overlap of our identities, the more we are motivated to behave (or believe) in a way that supports them.

Individuals who acknowledged greater social identity complexity reported feeling closer to people who belonged to different racial and religious groups.[113] Social identity complexity was also found to positively correlate with greater support for multiculturalism. Furthermore, such social complexity was possible even for individuals who did not have significant contact with different groups, suggesting that intergroup contact is not required for higher social identity complexity.[114] While intergroup contact itself may not be required, being particularly open-minded and also being exposed to different groups could help increase one's social identity complexity.[115]

Importantly, how a person conceptualizes their own identities can be a very subjective and idiosyncratic experience. Even though Republicans and Democrats are often clearly defined groups, making many of their attitudes predictable, there is still a lot of variability within their political identities. For example, a Republican who is also a Christian may correctly believe that their political party is majority Christian. However, if they also believe that their opposing political party, the Democrats, is mostly non-Christian, despite the majority of Democrats being Christian, then this would create a simple and distinct separation: Republicans and Christians would belong in one group, and Democrats and non-Christians would belong in another. In other words, this failure to acknowledge the overlap between political and religious identities would result in low social identity complexity. Conversely, if the Republican acknowledges that both their political group and their opposing political group are majority Christian, this would reflect a higher degree of social identity complexity because it accounts for some key overlap of political and religious groups. Considering how two groups can overlap in some ways requires complexity; it is much easier (and usually wrong) to pretend that large groups have nothing in common with one another.

As I mentioned at the start of this book, at different times in my life I've identified as both a Democrat and a Republican and as both an atheist and a Christian. When I reflect on this journey, I think having once shared the beliefs of the groups I no longer identify with helps me have more empathy for them and for others who think differently than I do. I remember the different identities I had to navigate over time and the obstacles and experiences I had. Both Christians and atheists shared an identity with me at different points in my life, and it makes it easier for me to consider how both groups have common goals (e.g., both groups want to cultivate a sense of community). Sharing identities with someone may help develop social identity complexity, but it is not required. Even just reflecting on the complex facets of our own identities and those we currently share with others can be helpful. If nothing else, we all share the identity of being human. We must eat food to survive and breathe oxygen to live. We all grow and change. We experience joy and sadness. We can start at these most basic aspects of life and then move from there to more complex identities we have in common.

Psychology experiments have found that priming people to think about their multiple unique identities could reduce the impact of overall identity bias.[116] However, other research has concluded that simply listing identities did not produce a strong effect and has suggested that having participants spend more time reflecting on the uniqueness of their identities may be more effective.[117] It's important to be cautious about the results of these priming studies (and priming studies in general) because psychological priming often generates very small effects and frequently fails to replicate.[118] Future research ideally would replicate the effects of social identity complexity primes while also increasing our understanding of the underlying mechanism that explains why certain procedures work better than others. Future research on identities will also help us better understand how identities interact with one another in a dynamic process rather than remaining fixed and categorical. Understanding how multiple identities (including current and previously held identities) dynamically interact will help us better understand social identity complexity and its impact on our beliefs.

While social identity complexity itself may have a unique influence on processing information, it could alternatively be that simply reflecting on the competition of identities creates the perception that any single identity is less important. Essentially, when we have a strong identity, it narrows our processing of the world in a way that is consistent with that identity. When other self-relevant factors intercept this identity process, the impact can be reduced.

We can connect this to what we have learned about self-affirmation as well. Thinking about important values (self-affirmation) or the intersections of important identities (social identity complexity) broaden where we can derive our self-esteem from. As noted above, identities can provide us with a crucial source of self-esteem. When we have a diverse and complex set of identities, we have access to a greater reservoir of self-esteem. Indeed, research has shown that having more prominent identities that are verified predicted higher self-esteem. However, having more identities did not increase self-esteem if many of those identities were not sufficiently verified.[119] Thus, having additional sources of self-esteem through identities does provide more opportunities to bolster our self-image, but it is more important that the identities we do have are regularly supported. If our identities all overlap and we don't have other sources of self-esteem, then we are especially motivated to protect our identities, as they are our primary source of self-esteem, and this increases our vulnerability to holding polarized beliefs and believing misinformation.

REDUCING POLARIZATION THROUGH SHARED IDENTITIES

Identities formed by in-groups and out-groups and can be quite powerful in the way they bias our beliefs and behavior. We just looked at how social identity complexity can reduce this influence over us. By reflecting on the complex and even contradictory facets of our own identity maps, we can reduce the hold any single identity may have over us. Another strategy to reduce polarization through identities is to focus on a shared identity. The ability to break through group biases has been observed in those who were able to see the Earth in its entirety—literally.

Astronauts have regularly reported a sense of awe when seeing the Earth from space, and some have discussed how they felt more connected with all of humankind and the planet.[120] This emotional experience and shift in thinking has been called the "overview effect" by Frank White, who interviewed dozens of astronauts for his book published in 1987.[121] The overview effect continues to be observed in current astronauts as well. Sara Sabry, the executive director and founder of the Deep Space Initiative and the first Egyptian woman to travel in space, shared her thoughts on the overview effect and her experience seeing the Earth from space: "The Overview Effect is the experience of seeing the reality of the Earth from space, which is immediately understood to be a small, fragile ball of life, suspended in the void, protected and nourished by a thin envelope. From space, national borders disappear, and the conflicts that divide people become less important, while the need to create a planetary community with a united will to protect this 'pale blue dot' becomes both clear and necessary."[122] Hayley Arceneaux, a physician assistant and cancer survivor, described her experience after seeing the Earth while traveling on the first all-civilian space flight in 2021. "It felt unifying, but it also made me think of healthcare disparities in a different way. How can someone born on that side of the globe have a completely different prognosis from someone born over here? . . . I could see the nations all at once, and it felt more unfair than ever, the ugliness that existed within all of that beauty."[123]

While going to space seems like an incredible and transformative experience (and something I would love to do someday), the good news is that we can produce similar feelings of shared identity without any rockets. We can simply ask people to reframe how they think about themselves. If identities can bias how we process information, then we can use that same identity process to help us get back on track. Research has found that when liberals and conservatives were asked to reflect on their American identity, they reported reduced feelings of political polarization.[124] Experiments showing the power shared identity has as an intervention for reducing negative feelings toward the political out-group are found throughout the literature. Commonalities deriving from community-based identities (e.g., sports fandom, religious ties, or community arts) may be especially likely to have lasting effects in reducing polarization.[125] One clever 2021 experiment evaluated whether shared identities could help reduce polarization. In the first step, participants were matched who shared common interests, such as music preferences and hobbies, but who were political opposites.[126] Then participants saw a profile of their match and what they had in common before reading their partner's essay on wealth distribution (which countered their own political beliefs). Participants on both sides of the issue became less polarized after reading their partner's essay. This was especially true the more the participants rated feeling close to their partners after reading what they had in common. Thus, this shows the influence of connecting through the nonpolitical things we have in common in reducing polarization.

Other research has found that when Republican parents reflected on their parent identity, they were more likely to be concerned how climate change may impact their children in the future. However, this research study also found that when Republican parents were later reminded of their political identity, any increased concern for climate change was eliminated.[127] Thus, our identities are constantly competing with one another as they impact our attitudes and behaviors. Despite this complication, a major study hosted by Stanford University researchers found that highlighting a common American identity was one of the most effective interventions for reducing political animosity between Democrats and Republicans. This intervention also reduced bias when evaluating politicized facts (such as immigration and unemployment numbers under Democratic and Republican presidents). Participants in this intervention read about how important democracy has been for the success of the United States and how its loss is a major risk to the country. Participants then read about studies that show most Americans support democracy and were asked to write about their two favorite things about being an American.[128] We are motivated to hold beliefs and behave in ways that support the group identities we deem important. This can lead to polarization and a biased interpretation of information. By reflecting on broader identities we share with others, we can become more objective and less polarized in how we navigate our social worlds.

I began this chapter by describing how identity processes influence our information processing. Next, I discussed how other factors related to the self (social identity complexity, self-affirmation, self-uncertainty) can add variation to this identity process. In the final section, I will discuss how the general cognitive ability to process information can influence susceptibility to misinformation. Awareness of these cognitive components can enrich our understanding of why we are vulnerable to false information and help us brainstorm techniques to better combat it.

ARE WE BIASED BY OUR IDENTITIES OR
JUST COGNITIVELY LAZY?

Cognitive style is one of the most well-studied psychological variables in misinformation research. There is an ongoing debate among misinformation researchers about what matters more when predicting susceptibility to misinformation: identity or cognitive style. First, what do I mean by cognitive style? Daniel Kahneman, a psychologist and economist at Princeton University, discusses two main cognitive styles in his book *Thinking, Fast and Slow*. Type 1 refers to thinking quickly and intuitively, and type 2 involves more deliberative and analytical processing. Each of us uses type 1 and type 2 thinking, and both are useful in different contexts. For example, it may benefit us to use type 1 thinking when choosing what to eat so we don't agonize over hunger. When deciding what career to pursue, it might help to rely more on type 2 so we carefully consider each option. Research on cognitive styles illustrates that everyone uses both types of thinking, but some of us are more comfortable using one type over the other more often in our daily lives.[129]

The Cognitive Reflection Test (CRT) is one of the most common measures used to reveal our preference for these two thinking styles.[130] A sample question from this test is "If you're running a race and you pass the person in second place, what place are you in?" Answering "first place" might sound correct initially. But the answer is "second place," as we are now in the position of the person we just passed. In this example, answering "first place" relies more on our intuitive and quick thinking (type 1), and answering "second place" relies more on our deliberative and analytical thinking (type 2). Those who are more apt to use type 2 thinking are less likely to believe fake news headlines and are better at determining whether something is true. Psychologists Gordon Pennycook and David Rand have produced highly influential work on the relationship between cognitive styles and susceptibility to misinformation. They published a study in 2020 in which participants' accuracy in evaluating news headlines on social media was compared with their scores on the CTR. Examples of news headlines included "Pope Francis Shocks World, Endorses Donald Trump for President" (fake) and "Donald Trump Says He'd 'Absolutely' Require Muslims to Register" (real). Those who scored lower on the CRT were worse at differentiating between real and fake news headlines.[131] In a different study, Pennycook and Rand found that regardless of whether the headlines supported one's political group, higher scores on the CRT predicted a higher degree of accuracy in evaluating the veracity of headlines. They concluded from their data that "lazy thinking" predicted greater susceptibility to falling for fake news than did partisan bias.[132]

Other interesting work shows an association between belief in misinformation and self-reported enjoyment of solving difficult problems. This preference for solving difficult problems is called the need for cognition (NFC).[133] Research has shown that individuals who scored lower on NFC were more susceptible to health misinformation.[134] Another study had participants watch a video showing a burglary and then listen to a summary of what they just watched. The summary contained several pieces of misinformation, and participants who scored higher on NFC were less likely to fall for misinformation, as they remembered more of the details of the video.[135]

While some people score lower on NFC and cognitive reflection, it is important to remember that we can decide how much cognitive effort we want to give! Studies have shown that asking someone to reflect on being accurate can make them less susceptible to believing and sharing fake news. For example, Gordon Pennycook and colleagues found that when participants were asked to evaluate headlines accurately at the start of the study, they were much more accurate in rating factual and false headlines. Reading this simple accuracy prime made participants almost three times more likely to correctly discern truth from fiction. This simple nudge also decreased the intention to share fake news. The accuracy nudge was used in a real-world Twitter experiment, in which participants volunteered to share their Twitter accounts and experimenters monitored their online behavior. In one group, participants were sent a message asking them to evaluate the accuracy of headlines (the accuracy prime) and participants in the other group never received such a nudge. Those who received the accuracy nudge shared significantly less fake news on their Twitter accounts.[136]

We have already noted that we are more likely to believe information that we are exposed to via the illusory truth effect. However, a 2020 study showed how accuracy priming can also counteract this cognitive bias. The study found that people were far less likely to believe false information after repeated exposure when they were asked to behave like fact-checkers. An initial focus on accuracy seemed to prevent the repeated exposure (i.e., the illusory truth effect) from increasing their likelihood of believing the false information.[137] Sometimes we just need a nudge to be more careful in our thinking!

One of the challenges of studying the social science of misinformation is that various psychological factors may impact misinformation differently, depending on how we measure and define misinformation. Think back to all of the different subcategories of misinformation in the last chapter. There is a wide range of terms that fall under the misinformation umbrella, from hyperpartisan interpretations of the news to completely made-up stories from low-quality sources. For example, we should differentiate between the ability to detect completely fabricated news from unreliable sources (i.e., fake news) and the ability to reduce our biases when reading hyperpartisan news that aligns with our political identity. A team of psychologists led by Cédric Batailler reanalyzed CRT scores while also incorporating the overall political bias participants had while judging headlines.[138] The team came to a more nuanced conclusion and asserted that while both cognitive reflection and identity bias play important roles in understanding misinformation, they each influence different aspects of misinformation.

Social psychologist Bertram Gawronski expands on this idea that cognitive reflection predicts truth discernment and political bias predicts general belief in information. He asserts that cognitive reflection explains belief in fake news more than partisan bias does because of the conceptual overlap in studying misinformation. We are motivated to believe favorable information about our political group, which means we will increase our accuracy in judging real news that supports our identity, but we will also be more likely to believe fake news that also supports our identity. Additionally, we will be more accurate in calling out fake news that attacks our group but less accurate in judging real news that attacks our group. Because these effects cancel one another out, Gawronski cautions that "the net result is a null effect on truth discernment for both favorable and unfavorable news, but this null effect should not be confused with absence of partisan bias. It simply reflects the conceptual independence of partisan bias and truth discernment."[139]

Cognitive styles may be particularly useful for predicting if someone will put in the effort to determine if a headline is true before sharing. However, political identity bias appears to influence our broader tendency for processing information, as we want to both believe information that supports our group and ignore information that attacks our group. In my own research, I found that stronger identity prominence predicted belief in politically congruent rumors and fake news headlines. However, lower cognitive reflection predicted belief only in politically congruent fake news headlines but not in rumors! I'll delve into this research more in chapter 4, when we cover the role of personal networks in identity strength. The broader point is that both cognitive reflection and identity factors influence how we evaluate information. We can become more reflective through accuracy nudges, since we have a general motivation to be accurate instead of inaccurate. However, if the information we are evaluating poses a significant threat to an important identity, then our desire to protect our identity can become more important than our pursuit of accuracy.[140]

As this chapter comes to a close, you may wonder if learning about how our identities and cognitive styles operate can decrease our vulnerability to misinformation. There is some evidence that learning about how our minds can misguide us can help us become less likely to fall for general cognitive biases.[141] However, "debiasing" with regard to politics and misinformation has been shown to be particularly difficult, and we often fail to recognize our own biases. In chapter 6, we will look at how education and media literacy can empower individuals to protect themselves against misinformation.

CONCLUSION

In this chapter, I covered what identities are and how we are motivated to maintain consistency between the values of our identities and the ways we interact with our world. Identities provide us with meaning and self-esteem and also connect us with other people. However, our motivation to protect our identities can make us susceptible to believing false or misleading information. I also covered how various forces can increase or decrease the influence our identities have on us. Feeling uncertain can make us cling to our identities even more strongly, whereas self-affirmation can make us more open to identity-threatening information. When information impacts two important identities in different ways, we are forced to reflect on that information more carefully. Additionally, when our identity maps are more complex, we are more open-minded, since we are not narrowing our attention to a simple worldview. Finally, cognitive styles play an important role in how we process misinformation. When we are more cognitively lazy, we may be more vulnerable to misinformation. Susceptibility to misinformation is particularly dangerous during crises that affect society at large, such as a public health crisis. In the next chapter, I will apply some of the research presented here in a case study of misinformation surrounding the COVID-19 pandemic.

CHAPTER 3

POLITICAL IDENTITIES, THE RESPONSE TO COVID-19, AND HOW TO HAVE PRODUCTIVE CONVERSATIONS

I was in my seventh grade performing arts class when a voice came on the PA speaker with a brief yet ominous announcement: "A plane crashed into the World Trade Center. Students can check in with their families to make sure they are safe." I, like many other Americans at the time, felt both confused and scared. The terrorist attacks on September 11, 2001, were a generational event and created powerful memories for those who were old enough to remember the events.[1] American flags appeared throughout my neighborhood that September. At least for a few months, we were united as a nation as we faced a common enemy. Such national unity was evidenced by surveys conducted after the 9/11 tragedy that revealed nine out of ten U.S. adults were proud to be an American.[2] In 2022, just 38 percent of Americans reported being "extremely proud" to be an American, and just 65 percent of Americans reported being proud to be an American at all.[3] Additionally, a poll in October 2001 found that 60 percent of U.S. adults expressed trust in the federal government, with polls at the time showing combined average trust levels around 55 percent. This level of trust had not been observed in polling in the previous three decades, and polls since then have not been close. For comparison, a 2023 poll found that only 20 percent of Americans trust the federal government.[4]

Americans experienced an even longer period of unity after the 1941 attack on Pearl Harbor, which led the United States to enter World War II. After Pearl Harbor, an incredible 97 percent of Americans agreed with the government's decision to go to war.[5] This was a spike from June 1940, when only 35 percent of Americans agreed the United States should help with the war after Germany took over France.[6] President Franklin D. Roosevelt enjoyed an average approval rating of about 72 percent from July 1941 until December 1943. Toward the end of the war, President Harry S. Truman had an approval rating of 87 percent in June 1945 and 82 percent in October 1945.[7] These stories surrounding 9/11 and World War II might give the impression that national crises can rally a country to rise above political divisiveness. In March 2020, Americans found themselves again under attack, but this time it was from the deadly COVID-19. However, Americans did not rally together, as the pandemic only fueled further political division. In this chapter, I will examine how the pandemic became politicized and how such polarization led to misinformation. Then I will end the chapter by discussing some practical tips that can increase our chances of having productive conversations about political issues.

THE COVID-19 PANDEMIC

The coronavirus disease of 2019, better known as COVID-19, became a household word around the world in early 2020 when it sparked a global pandemic. This novel coronavirus soon became the subject of investigations across the world as scientists scrambled to study it. Early reports showed that it could be quite deadly and spread rapidly, but the specifics were still unknown. This scientific uncertainty caused some inconsistent and mixed messaging from health organizations regarding what the severity of the virus was, whether masks reduced viral transmission, and how widespread the virus would be.[8] People generally listen to experts during national crises,[9] but scientists were still learning the best ways to respond to the virus. This gave political leaders an opportunity to provide guidance to the public, as the public would also follow the cues of political elites. Unfortunately, these political elites differed substantially in their responses to the virus.[10]

Politicians' attitudes regarding the severity of COVID-19 differed despite the growing number of deaths caused by the virus. In early 2020, the United States watched as the pandemic first spread through other countries and wreaked havoc.[11] I remember seeing the horrible stories of overwhelmed Italian hospitals in March 2020, right before COVID-19 really hit the United States. Europe was reporting thousands of deaths per day in the second half of March 2020, while the United States was still reporting under one hundred per day. By April 2020, COVID-19 had spread throughout the United States, and the country was regularly reporting over two thousand COVID-19-related deaths per day.[12] Despite the global chaos of the COVID-19 pandemic, there was still stark disagreement between Democratic and Republican leaders in the United States on how to handle the virus. As a result of the political nature of the pandemic, a significant portion of the population resisted wearing masks, maintaining social distancing, and generally taking the virus seriously.[13] These dismissive attitudes remained even as COVID-19 became the third leading cause of death in the United States between 2020 and 2021.[14]

It was shocking to observe many Americans rejecting simple hygiene, such as wearing a mask or practicing social distancing, as the deadly virus spread throughout our country. These were simple, evidence-based practices that reduced the spread of the virus, and many other countries, such as South Korea and Germany, had no problems implementing them.[15] Why was there so much resistance in the United States? Once the virus became politicized, there was going to be an aversion to accepting scientific information that challenged one's political identity. The polarized political identities in the United States provided a fertile ground for the spread of COVID-19 misinformation. Sadly, as someone who studied political bias and identity, I was not surprised by this outcome. But how did COVID-19 become political in the first place?

HOW COVID-19 BECAME POLITICAL

At the start of the pandemic, Democratic and Republican leaders differed substantially in their attitudes toward COVID-19. Democratic leaders generally signaled the importance of taking the virus seriously more than Republicans.[16] For example, on January 26, 2020, Chuck Schumer, the Senate minority leader and a Democrat, wanted to declare COVID-19 an official public health emergency. This step would allow the federal government to immediately provide additional funding for COVID-19 tests and medical supplies and expand health care coverage, all of which were sorely needed during the pandemic.[17] While many Republican politicians did not endorse extra government support, COVID-19 was officially declared a public health emergency on January 31, 2020. Starting in mid-February 2020, Democratic politicians discussed the virus more frequently and emphasized threats to public health and the country's workers.[18] Conversely, Republican politicians discussed China and the impact COVID-19 had on businesses.[19] Media figures and politicians were quick to view the pandemic in partisan terms instead of focusing on unifying around a common problem. The mass media can cause further polarization because the most extreme politicians receive more attention, and the media often highlight the conflicts and disagreements between political parties.[20] Also, the pandemic allowed people to spend more time watching television and browsing social media, which provided even greater opportunities to view polarizing content.[21]

As we learned in the last chapter, the United States is extremely polarized. Because of this, it was not surprising to see the country have the greatest disagreement between liberals and conservatives on how to handle COVID-19 compared to ten other countries around the globe.[22] This makes the United States a particularly fitting case study of how partisan identities can increase susceptibility to misinformation, even during a national public health crisis.

Republican President Donald Trump held office in 2020 when the virus hit and certainly influenced the political nature of COVID-19. In the early days of the virus, he confidently downplayed the danger of the virus and regularly refused to wear a mask.[23] These cues from political leaders were internalized by everyday Americans. Polls from early February showed that concern about the virus was similar for Democrats and Republicans. However, just a few months into the pandemic, political party affiliation became the variable that best predicted whether someone would take COVID-19 seriously.[24] Cues from political leaders continued to have substantial influence. For example, people living in counties that heavily voted for Trump in 2016 were less likely to shelter in place at the start of the pandemic. However, people in these same counties began to stay home more after Trump also started taking the virus more seriously. After Trump got COVID-19 himself, polls showed that conservatives were slightly more likely to take the virus seriously.[25]

I spoke to Ben Tappin when he was a postdoctoral researcher at MIT studying the impact of cues from political leaders (he's now a professor at the London School of Economics and Political Science). I asked him about the possible link between cues from political party leaders and Americans' responses to COVID-19. Determining whether it was Trump himself or the conservative media that influenced responses to COVID-19 is a challenge, but we do see that messaging can have a substantial impact. Tappin explained: "Republican-leaning counties presumably watch more Fox News, and Trump also watches Fox News. If particular shows or personalities on Fox are changing their messaging or information provided regarding the virus (for whatever reason), this could jointly cause both the Republican-leaning counties and Trump to change their behavior—rather than Trump's position-taking being the cause of the county-level behavior."[26] While establishing the specific causal link will require additional study, the evidence illustrates how partisanship was a crucial factor leading people to believe misinformation about COVID-19. Other factors, such as communication failures by public health organizations, also had an impact (and I'll address that later in this chapter), but it's critical to emphasize the impact political leaders had on the spread of misinformation.

The United States is not unique in the degree of influence its political elites have on public opinion. An analysis of nineteen countries found that general political polarization among their citizens follows the polarization of their political leaders.[27] However, when it came to COVID-19, the polarization and partisan signaling in the United States was unique. Conservative ideology predicted COVID-19 attitudes much more strongly in the United States compared to the UK and Canada.[28] A massive sixty-seven-country study led by Jay Van Bavel found a very weak global correlation between political ideology and COVID-19 attitudes.[29] He shared his thoughts on these results: "This has led me to conclude that political ideology actually has very little to do with attitudes and behaviors related to COVID-19. It appears that Republican party leadership (Trump), partisan identification, and (social + traditional) media play a far bigger role in the US."[30]

Even beliefs about the origin of COVID-19 were politicized throughout the pandemic. The first documented cases of COVID-19 were traced to Wuhan, China.[31] However, we saw disagreement along partisan lines over what caused the virus to spread to humans. In April 2020, President Trump became one of the first prominent Republicans to suggest that COVID-19 spread from a lab leak in China, and this idea was supported by other conservative politicians and news outlets.[32] In opposition, many liberal politicians and media figures dismissed the idea of a lab and instead were quite confident the virus had zoonotic origins.[33] The initial evidence was not strong for either origin hypothesis, but that didn't stop it from becoming a partisan issue. In May 2020, 72 percent of Republicans believed COVID-19 "probably" or "definitely" originated in a laboratory in China compared to just 49 percent of Democrats.[34]

A few years after the pandemic began, there were more data to analyze about the origins of the virus. Support for the lab leak origin centers around the fact that a research lab in Wuhan, China, was studying coronaviruses. In February 2023, the U.S. Department of Energy concluded, although with "low" confidence, that COVID-19 accidentally spreading from a lab was the most likely explanation.[35] Other agencies, such as the National Institute of Allergy and Infectious Diseases, concluded that a zoonotic origin was most likely but still admitted the lab leak hypothesis was plausible.[36] The authors of a paper published in *Science* in 2022 concluded that the earliest cases of COVID-19 were geographically connected to the Huanan Seafood Wholesale Market in Wuhan, China.[37] A 2023 analysis of swabs from this same wet market found evidence of COVID-19 along with genetic material from wild animals. This suggests, but does not prove, that an animal could have been an initial host for the virus before it spread to humans.[38] In a 2024 international survey, a panel of experts within epidemiology, virology, and evolutionary genetics was asked about the origins of COVID-19. On average, they believed there was a 77 percent chance that COVID-19 had zoonotic origins and a 21 percent chance that it spread from a research-related accident.[39] While expert consensus leans toward a zoonotic origin as I write this book, we still do not have access to data that would provide clear and confident answers to the questions around COVID-19's origins. It's possible we may never know exactly how COVID-19 began.[40] The broader point is that early partisan disagreement regarding the origin of the virus speaks to how politicized COVID-19 is across many different domains.

Dr. Anthony Fauci, a physician and an infectious disease expert, became entrenched in this political divide as one of the lead members of the White House's Coronavirus Task Force under President Trump. He often found himself in the difficult position of having to correct the president's statements about the virus. For example, President Trump repeatedly claimed in August and September of 2020 that we were "rounding the final turn" in terms of the virus.[41] However, when asked about these comments, Dr. Fauci replied, "I have to disagree with that, because if you look at . . . the statistics . . . they are disturbing. We're plateauing at around 40,000 cases a day and the deaths are around a thousand."[42] This tension between the president and one of the leaders of his COVID-19 Task Force led to the politicization of Dr. Fauci himself. Democrats would applaud his disagreements with President Trump, and Republicans would criticize them. This politicization of Dr. Fauci lingered throughout the pandemic, as his approval ratings ranged from 85 percent among Democrats to 19 percent among Republicans in late 2021.[43] Sadly, Dr. Fauci and other public health officials were not just subjected to intense criticism; they were regularly harassed and at times even threatened.[44] Scientists who were not even working for any government organization were also subjected to attacks for simply sharing their expertise on the topic.[45] These types of attacks can make medical professionals and scientists less willing to publicly share important health information.[46]

As the pandemic raged on, the political divide persisted regarding the dangers of the virus. In July 2020, several months after COVID-19 hit the United States, 85 percent of Democrats believed COVID-19 was a serious health threat, but only 46 percent of Republicans agreed.[47] These attitudes directly correlated to mask-wearing preferences, as 63 percent of Democrats believed that masks should always be worn compared to just 29 percent of Republicans. Additionally, only 4 percent of Democrats said masks should never be worn compared to 23 percent of Republicans.[48] Because wearing (or not wearing) a mask is such a visible behavior, it was easier to connect it to political identities as well. In fact, those who wore masks and those who did not were more likely to cooperate in an online game when they found out the masking behavior of their partner matched their own.[49]

Even as the COVID-19 death toll rose to over 170,000 in August 2020, 57 percent of Republicans believed that the number of U.S. deaths from COVID-19 was "acceptable" compared to just 10 percent of Democrats.[50] For context, 170,000 people is the population of Jackson, Mississippi.[51] The political polarization around COVID-19 was extreme, but it's important to note that the country did have discord, though to a lesser degree, during a previous public health crisis. The United States experienced another major pandemic just over one hundred years earlier: the 1918 influenza pandemic. This pandemic took the lives of 675,000 Americans, and there were groups of people opposed to wearing masks during that time as well.[52]

Some crucial differences are evident between the media and political landscapes during the 1918 influenza pandemic and those during the COVID-19 pandemic. Some political polarization certainly existed in the early twentieth century, but when the 1918 pandemic hit, the United States was experiencing elevated national pride after World War I.[53] There also were no social media and cable television to exacerbate polarization in the early 1900s. In addition, the structure of our national health agencies was quite different in 1918. Public health issues were handled more at the local level. There was no Centers for Disease Control and Prevention in Washington, D.C., and a national public health department did not exist.[54] This lack of national media and national public health agencies was important, as Americans tend to view national organizations as more political and tend to trust their local news and their local health organizations instead.[55] The differences in mass media production and consumption, polarization levels, internet and social media engagement and access, national pride (or national landscape, since the 1918 pandemic occurred right after World War I and national pride was different), and structure and organization of governmental agencies charged with handling public health concerns resulted in greater politicization of COVID-19. This level of polarization created a barrier to rallying together as a country, as research showed that when people within a nation believed "we are all in this together," they were more likely to support public health policies, hygiene, and social distancing.[56] When reading about the 1918 pandemic, you can see several unfortunate similarities to COVID-19 in the severity of the viruses and the way some groups rejected basic precautions, but the greater polarization and media environment of the present day made it much easier for COVID-19 to become so politicized.

COVID-19 decimated the global society and economy, but there was light at the end of the tunnel.[57] FDA-approved vaccines became available in late 2020, and evidence-based treatments were being developed. However, even though vaccines and treatments would save lives and end the lockdowns, partisanship still influenced people's decisions. For example, Republicans were less likely to take the COVID-19 vaccine, although it significantly reduced the likelihood of serious illness and death.[58] Again, misinformation was a major factor in attitudes toward the COVID-19 vaccine, and the disparate sources of information were fueled by partisan media and polarized politicians.[59] For example, Tucker Carlson of Fox News emphasized the allergic reactions to the COVID-19 vaccine on his show while also claiming the promotion of the vaccine was a form of "social control."[60] Outside of Fox News, misinformation regarding the vaccine (e.g., the vaccine was not safe or properly tested) spread throughout social media.[61]

You might be wondering if this partisan difference in handling the virus led to more deaths among Republicans. Well, a study led by Jacob Wallace, a health economist at Yale University, found that Republicans were, in fact, more likely to die during the pandemic. Wallace and colleagues analyzed a massive dataset of over 500,000 recorded deaths during the pandemic among registered Democratic and Republican voters who lived in Ohio and Florida. They looked at the excess death rate, which is the increase in deaths above what is expected after controlling for age, seasonality, and geographic location. They found that after May 1, 2021, about the time the COVID-19 vaccine had become widely available, the Republican excess death rate was 43 percent higher than the Democratic rate.[62] This is a sad result of politicized health misinformation, and we will look at vaccine misinformation in depth in chapter 5. In the next section, I will show how such politicization fueled specific elements of COVID-19 misinformation.

MISINFORMATION ON COVID-19
SEVERITY AND PRECAUTIONS

We're not just fighting a pandemic; we're fighting an infodemic.
—Dr. Tedros Adhanom Ghebreyesus

Dr. Tedros, the director-general of the World Health Organization (WHO), made this remark about the COVID-19 pandemic while speaking at the 2020 Munich Security Conference.[63] As mentioned earlier, the virus was quickly politicized in the United States, creating a rich opportunity for misinformation to be spread along partisan lines. Liberal media outlets would overestimate the severity of COVID-19, while conservative media outlets would downplay its severity.[64] One salient example happened on March 6, 2020, when Dr. Marc Siegel appeared on Sean Hannity's very popular prime-time show on Fox News. Dr. Siegel claimed that the worst-case scenario for COVID-19 would be that it was just like the seasonal flu.[65] This claim was unfounded, and, unfortunately, holding inaccurate beliefs about COVID-19 impacted people's behaviors in response to the virus. A 2021 study found that having false beliefs about COVID-19 (such as equating COVID-19 to the seasonal flu) was negatively associated with wearing masks.[66]

Irene Pasquetto, chief editor of the *Harvard Kennedy School Misinformation Review*, has reviewed several papers linking conservative news and COVID-19 misinformation. Despite the challenge of determining causality, Pasquetto still concludes: "Given all the data we have seen, and all the studies we are reviewing, we can say that empirical evidence clearly shows that this social group [those who routinely watch, read, and follow far-right media and social media] tended to take the disease less seriously and delayed their own response to the virus."[67] This resulted in many tragic deaths among those who did not take the virus seriously because they believed misinformation. Mask-wearing is a simple and effective protection against respiratory illnesses, but masks were also a strong point of contention during the pandemic. That is why it's important to really examine in detail the evidence for wearing masks to prevent COVID-19.

The Evidence Showing the Effectiveness of Masks

One common piece of misinformation was that masks are not effective at all in reducing the spread of COVID-19. While masks are not perfect, there is strong evidence showing they do offer at least some protection against COVID-19.[68] First, physicians have long worn masks to protect themselves while visiting patients who are sick with respiratory viruses as well as to protect immunocompromised patients from potential exposure to any illness the physicians have.[69] High-quality masks, such as N95s, offer protection from respiratory viruses by blocking the inhalation of tiny water droplets in the air that can contain the virus.[70]

It is important to note that masks need to be made of high-quality material and worn properly in order to work effectively.[71] This is a crucial point, as several studies found that masking interventions (e.g., giving people masks or encouraging people to wear them) did not reduce COVID-19 levels in the area under consideration. Thus, a study may find that asking people to wear masks did not reduce COVID-19, but this null finding means only that the particular intervention did not work, not that masking did not protect against the virus.[72] It's not ethical to conduct an experiment where the researchers purposely expose people to a deadly virus and evaluate whether the mask helped them, which made rigorous mask studies hard to come by. Despite the lack of true randomized controlled trials, there were a few rigorous studies that demonstrated the effectiveness of masks for COVID-19. A massive study with over 340,000 participants across six hundred villages throughout Bangladesh evaluated COVID-19 rates between different groups from November 2020 to April 2021. In the voluntary mask intervention group, researchers went into popular areas of the village and provided free masks, educational materials on the importance of masks, and encouragement to share masks with others in the community. This mask intervention group was further split between those who received cloth masks and those who received surgical masks. In the control group, researchers did not give away any free masks or educational materials. They observed that mask use in the control group was only one in ten people, but this rate rose to four in ten people in the mask intervention group. Importantly, the mask intervention group saw a 9 percent reduction in COVID-19 cases, and those that used surgical masks saw an 11 percent reduction in COVID-19 cases. Researchers also observed a 35 percent reduction in COVID-19 cases among older adults in the surgical mask condition. This complex study provides pretty strong evidence that widespread mask use can help at the population level.[73]

Another study investigated COVID-19 rates among 9,850 health care workers across twelve hospitals throughout Massachusetts before and after a mandatory mask policy was implemented. The policy required both health care workers and patients to wear surgical-grade masks at the hospital. COVID-19 rates were tracked during March and April of 2020. Before the mask policy went into effect, the hospital system saw COVID-19 rates double every 3.6 days, eventually rising to an infection rate of 21.3 percent. After mask wearing was required, the infection rate stopped increasing and then rapidly decreased to 11.4 percent.[74] While this result is encouraging, it should be noted that this hospital system provided plenty of surgical-grade masks, and health care workers also generally knew how to wear them properly.

Although these studies are not perfectly randomized, controlled trials, they do show a meaningful decline in COVID-19 rates when masking significantly increases across different populations. Of course, the nuance of masking studies was lost, as COVID-19 was already quite politicized, and refusing to wear a mask became a marker of one's political identity. Despite this polarization, a 2024 survey found that 70 percent of American adults agreed is was "generally a good idea" for stores to require masking indoors during the pandemic.[75] This survey revealed that most Americans did appreciate the importance of wearing a mask to mitigate the spread of COVID-19. However, a vocal minority still opposed mask wearing throughout the pandemic, and several conservative public figures who were against mask wearing ended up being tragically impacted by COVID-19.

Consequences of COVID-19 Misinformation on the Left and Right

Former Republican presidential candidate Herman Cain, one of the most prominent Republicans to speak out against mask wearing, was later killed by COVID-19. He attended a crowded Trump rally in Tulsa, Oklahoma, in June 2020 and then tested positive for COVID-19 nine days later. He became seriously ill and spent the next few months in the hospital, where he eventually died from the virus.[76] Republican Chris Johansen, a state representative in Maine, also pushed back against mask mandates and even refused to wear a mask while working in the State House during the height of the pandemic. Both Johansen and his wife, Cindy, contracted COVID-19 in July 2021. Cindy Johansen later died from the virus, and Chris Johansen resigned from his position.[77] The deaths of prominent Republicans certainly gained a lot of attention, and as noted earlier, COVID-19 caused significantly more fatalities among Republicans compared to Democrats throughout the country. This is particularly sad, since Republicans were less likely to wear masks, avoid crowded areas, and take the virus seriously overall.[78]

In contrast to conservatives, those on the left side of the political divide were more likely to be extremely concerned about the severity of COVID-19. About 33 percent of those who described themselves as "very liberal" were "very concerned" about becoming seriously ill from COVID-19 compared to about 25 percent of people who described themselves as "liberal" or "moderate."[79] While left-leaning individuals and media took the virus more seriously, they may have contributed their own misinformation with an overemphasis on the severity of the virus. Even after controlling for age, race, gender, geographic location, and education, Democrats were still more likely to overestimate how many young people were dying from COVID-19.[80]

One could argue that such excessive caution is helpful when we are still learning about the dangers of a disease that is quickly spreading. However, some liberals did contribute to misinformation when they advocated extreme measures against the virus when the current evidence suggested such measures were unnecessary. This was clearly the case when left-leaning media outlets shamed people for going to the beach.[81] Daniel Uhlfelder, a Florida attorney and a Democrat, was a particularly vivid example of the left's hyperfocus on shutting down areas where there was a low risk of virus transmission. As a protest to the reopening of Florida beaches in May 2020, he walked around several Florida beaches fully dressed up as the Grim Reaper with a black hood and scythe.[82]

Even in the earlier days of the pandemic, scientists were confident that outdoor transmission was quite low.[83] Thus, closing parks and beaches was unnecessary and potentially counterproductive.[84] Sociologist Zeynep Tufecki often wrote articles about these excesses that included commentary from health experts. When he interviewed Julia Marcus, an epidemiologist from Harvard University, about the shaming of beachgoers, she told him: "You'd think from the moral outrage about these beach photos that fun, in itself, transmits the virus. But when people find lower-risk ways to enjoy their lives, that's actually a public-health win."[85]

Additionally, the unnecessary outdoor mask mandates and rules may have been counterproductive if they caused people to become frustrated with public health organizations. As Muge Cevik, an infectious disease expert at the University of St. Andrews, told *The Washington Post*: "Given the very low risk of transmission outdoors, I think outdoor mask use, from a public-health perspective, seems arbitrary, and I think it affects the public's trust and willingness to engage in much higher-yield interventions. We want people to be much more vigilant in indoor spaces."[86] In fact, some of these excessive policies may have exacerbated skepticism and distrust of government officials. A twenty-two-country study found that the implementation of stricter lockdown policies predicted a greater likelihood that political conservatives would believe COVID-19-related conspiracy theories.[87]

Beyond the closing of low-risk outdoor spaces, there was also a push for what Derek Thompson called "hygiene theater," a term he defined as safety protocols that "that make us feel safer, but don't actually do much to reduce risk, even as more dangerous activities are still allowed."[88] For example, virus transmission from surfaces was shown to be extremely rare; however, many businesses still made their workers constantly sanitize the surfaces within their buildings to appear extraclean during the pandemic.[89] This excessive cleanliness was not harmless, as it wasted millions of dollars and perpetuated the myth that surfaces were a major risk in the transmission of COVID-19.[90] Additionally, many restaurants switched entirely to disposable plates and utensils in hopes of reducing the spread of the virus. However, because the virus primarily spreads in the air, there is little evidence that using only disposables helped in any significant manner.[91] While such hygiene theater may have appeased left-leaning patrons, it was unnecessary and potentially harmful.

Policies that required businesses to close during COVID-19 were supported more by liberals than by conservatives, but it is also important to note that liberals are more likely to live in urban areas that are vulnerable to the spread of the virus.[92] While policies aimed at closing businesses, schools, and public places may have reduced the spread of COVID-19, they also had massive economic and social consequences. Many small businesses went bankrupt and workers lost their jobs during the COVID-19 pandemic.[93] With many people avoiding indoor places at the start of the pandemic, however, many businesses would have struggled financially even without mandated lockdowns. A Washington Post-ABC News poll from March, 2020 found that 88 percent of Americans stopped going to restaurants or bars because of COVID-19.[94] We are still calculating the costs of COVID-19 as I write this, but I think it's fair to argue that some closures were excessive based on what we knew about the virus at the time. According to Dr. Fauci, shutting down schools initially made sense, but it was a mistake to keep them closed for so long (e.g., some New York schools were closed for a year).[95] Beyond the misinformation about preventing the spread of COVID-19, there was also plenty of misinformation about treating COVID-19.

MISINFORMATION ABOUT COVID-19 TREATMENTS

Unverified cures and treatments for COVID-19 were ubiquitous throughout the pandemic. For example, a post on social media went viral because it included a picture from an old Indian textbook stating that aspirin can treat coronaviruses. This textbook was referring to the broad family of coronaviruses like common colds, not the novel coronavirus COVID-19.[96] Vitamin C was another popular treatment despite there being no evidence it helped against COVID-19. It is one of the most popular items mentioned in the FDA's warning letters about fake COVID-19 cures.[97]

One of the most notorious fake cures was Miracle Mineral Solution, which its creators had previously claimed could cure cancer, hepatitis, the flu, and even HIV/AIDS. Its main ingredient is sodium chlorite, which is commonly found in disinfectants. Although the FDA labeled this substance as dangerous and there was no scientific evidence of its efficacy, its creators were making $32,000 per month from its sale. When the product was advertised as a cure for COVID-19 in March 2020, sales jumped to $123,000![98] It was promoted by several supporters of QAnon, a right-wing conspiracy theory group.[99] Beyond being a waste of money, Miracle Mineral Solution was reportedly the cause of death for several people who were trying to cure various ailments.[100] Its major promoters and sellers were later found guilty of fraud and sentenced to twelve years in federal prison.[101]

Hydroxychloroquine, a drug used to treat malaria, was another popular false treatment for COVID-19, but it also became heavily politicized. President Trump publicly endorsed the drug and cited a study that suggested it could treat COVID-19.[102] This study, which gained the drug public support among conservatives, was later criticized because of several key problems with its methodological design. The results of the small (thirty-six total participants), nonrandomized trial should have sparked further study rather than being applied immediately in clinical settings.[103] Subsequent studies revealed the drug lacked effectiveness in treating COVID-19, and its emergency FDA approval was revoked after it was also linked to heart rhythm problems.[104] Despite these null results and potential risks, in the summer of 2021, Fox News programs were still touting the drug as a reputable treatment; in fact, hydroxychloroquine was discussed as a potential treatment significantly more often on Fox News compared to CNN and MSNBC.[105]

Ivermectin, an antiparasite medication, also became heavily politicized as a potential treatment for COVID-19 despite the lack of evidence for its effectiveness.[106] Like with hydroxychloroquine, a few small studies suggested ivermectin could help treat COVID-19, but subsequent research revealed severe methodological flaws.[107] Most people do recover from COVID-19, so if you don't run a carefully controlled study, you'll see most people improve simply with the passage of time. You might falsely assume that ivermectin made them feel better when they might have felt better regardless of whether they received it. A group of independent researchers evaluated twenty-six studies on ivermectin and COVID-19 and found that over one-third of these studies had serious problems (such as numbers being calculated incorrectly) or had even committed explicit fraud. The rest of the studies did not show any conclusive evidence that ivermectin successfully treated COVID-19.[108] Despite the lackluster scientific support, ivermectin was continually advocated by prominent Republicans such as Rand Paul and Ron Johnson.[109] In August 2021, Paul went as far as to say that "the hatred for Trump deranged these people so much, that they're unwilling to objectively study it." By this point, however, several rigorous studies had failed to show any evidence of its effectiveness.[110] Ivermectin was studied thoroughly and was frequently taken by many Americans; prescriptions for ivermectin cost Medicare and private insurers about $130 million in 2021.[111] Democrats were following the overwhelming scientific consensus that ivermectin did not treat COVID-19, and this made it a partisan issue for Paul and other Republicans.[112]

By 2022, empirical evidence showed that several medical interventions did reduce COVID-19's severity. COVID-19 vaccines were developed that significantly reduced symptom severity and the likelihood of death.[113] Drugs like Paxlovid also reduced the severity of symptoms, as did administering monoclonal antibodies.[114] So why would anyone take unproven treatments like ivermectin, despite the lack of scientific evidence of its effectiveness, when evidence-based treatments were available? David Dunning, professor of psychology at the University of Michigan, provided an important insight about this that directly connects to what we already learned about social identities in the last chapter: "People listen to people 'from their group' and whom they think they can trust. People really don't know what science is, and so do you feel you can trust the person giving you advice, rather than appraising their expertise, becomes the thing."[115] This is a critical point for understanding how beliefs form and why people become attached to believing misinformation. Most of us are not experts in virology or epidemiology. We can't explain all the mechanisms involved in recommendations to wear a mask, get a vaccine, or take a certain drug. We can't physically observe the virus in the air and watch how it responds to different treatments in real time. Most of us are putting our trust in groups we believe in. Consistent with this idea, a 2021 study found that people were more likely to share COVID-19 information when it came from a source they personally found trustworthy.[116]

These false treatments based on pseudoscience directly stemmed from a lack of trust in scientific and medical institutions.[117] Individuals who did not trust these institutions instead turned to groups of people they did trust. As discussed in the previous chapter, group membership can be a powerful force as we process information. If groups we like support wearing masks and taking a vaccine, then we are more likely to endorse those beliefs. Conversely, if groups we like are in favor of not wearing masks and of taking ivermectin, then we are more likely to endorse those beliefs and perhaps even engage in those behaviors as well.

Before the pandemic, there was already a partisan divide relating to the trust of scientists and scientific institutions. In 2018, 42 percent of Republicans had a great deal of trust in the scientific community compared to 51 percent of Democrats. This divide spread further after the pandemic began, with only 34 percent of Republicans having confidence in the scientific community compared to 64 percent of Democrats in 2021.[118] Democrats also trusted health organizations, such as the Centers for Disease Control and Prevention (CDC) and the National Institutes of Health, more than Republicans.[119] However, there was more bipartisan support for NASA, an organization that hasn't been so overtly politicized. A 2018 Pew Research Center poll found that 70 percent of Democrats and 59 percent of Republicans agreed that NASA's involvement in space exploration is essential.[120]

Those who scored higher on mistrust of scientists also were more susceptible to COVID-19 misinformation.[121] An October 2021 poll found that 84 percent of Republicans believed or were unsure about whether the government was exaggerating the number of COVID-19 deaths (compared to about one-third of Democrats).[122] The Republican Party already had a value system that endorses smaller government, so skepticism over large federally funded organizations easily followed. If the CDC and other public health organizations advocated something (e.g., that masks work or that ivermectin does not treat COVID-19), then Republicans (and anyone more skeptical of institutions) may have wanted to believe the opposite in order to support their identity and in-group. Similarly, if Democrats (and people more supportive of scientific institutions) saw such organizations advocating masks and social distancing, then they may have seen it as a badge of honor to take excessive and extreme caution.

These differences in trust can also explain why certain groups may collectively fall for different types of misinformation. If a group doesn't trust what mainstream media are saying about a particular topic, then its members may reject what mainstream media say about a different topic. Political affiliation was found to be a major predictor of believing misinformation about COVID-19, and such misinformed beliefs are strong predictors of also believing other types of misinformation. For example, one study found that believing misinformation about COVID-19 predicted believing misinformation about the Ukraine-Russia conflict even more than political affiliation![123]

In March 2020, the early days of the pandemic, Republicans trusted the CDC to handle the virus slightly more than Democrats (potentially because a Republican was president, and this was a signal of supporting their in-group): 48 percent of Republicans and 33 percent of Democrats were "very confident" that the CDC was doing a good job responding to COVID-19. By mid-November 2020 (after Republican incumbent Donald Trump lost the presidential election), 75 percent of Democrats thought the CDC was doing a good job compared to just 58 percent of Republicans.[124] In February 2021 (when Democratic President Joe Biden was officially in office), 79 percent of Democrats thought public health organizations such as the CDC were doing a good job compared to just 44 percent of Republicans.[125]

While partisanship certainly played a major role in trusting the CDC, the CDC did not help itself with its messaging mistakes throughout the pandemic.

For example, it strongly advocated *against* wearing masks early on in the pandemic. Dr. Jerome Adams, the U.S. surgeon general, tweeted a salient example of this: "STOP BUYING MASKS." He went on to say that "[masks] are NOT effective in preventing general public from catching #Coronavirus, but if healthcare providers can't get them to care for sick patients, it puts them and our communities at risk!" He later deleted the tweet. Then, just a few months later, the CDC and other public health officials switched to telling everyone to wear masks.[126] This back-and-forth messaging caused confusion and sowed distrust. Later in the pandemic, the CDC advocated returning to work just five days after a COVID-19 infection, though earlier CDC guidance suggested people should wait ten days before returning to work. The CDC also admitted that 31 percent of people infected with COVID-19 are still contagious after five days, making the change seem arbitrary at best.[127] Rob Blair, professor of political science at Brown University, said, "What we have is inconsistent messaging, sometimes from the same source. . . . What we have is utter cacophony. That's detrimental not only for the quality of the response but for trust more generally."[128]

Scientific and medical institutions have also acted in ways beyond messaging mistakes that have earned them their lack of trust. One infamous example in the United States is the Tuskegee syphilis experiment that began in 1932. This study, conducted by the U.S. Public Health Service, recruited hundreds of Black American men who had syphilis. Syphilis is a sexually transmitted infection that can cause extreme health problems and even death if left untreated. These men were told they had "bad blood"—but not syphilis—and that they would receive free medical care. The men were not told, however, that they were going to remain untreated for decades and that the goal of the study was simply to see the natural course of the disease over time. About ten years into the study, penicillin was found to be an effective treatment for syphilis, but the participants never received it despite regularly seeing doctors who were part of the study. The doctors were specifically asked not to treat the men and to keep reporting symptoms until death. The study went on for forty years before the horrible experiment was exposed in the media in 1972, after which it was finally shut down. The surviving participants and the families of those who died eventually received a settlement, and new guidelines were established for U.S. government–funded research projects. However, this was not much of a consolation to the 128 men who had died from syphilis or syphilis-related complications before the study was finally shut down. Additionally, at least forty spouses contracted the disease, and nineteen children received it at birth. Thus, the Tuskegee experiment presented Black Americans with a very legitimate reason to distrust medical science in the United States for a long time.[129] I will look at trust in more detail in chapter 5.

When you already have a deep distrust of institutions, poor messaging can only make things worse. Sociologist Zeynep Tufecki wrote that the CDC and WHO relied too heavily on fixed guidelines that "lent a false sense of precision." It would have been better for the messaging to focus on describing the known details of the mechanism by which the virus spreads. Additionally, messaging campaigns should have encouraged harm reduction instead of rigid absolutism. During a pandemic, any reduction of potential spread helps. As noted earlier, scolding people for socializing outside, which was a very low-risk activity, does not help gain any goodwill.[130] In August 2022, CDC Director Rochelle Walensky admitted that the CDC did not "reliably meet expectations," and said that her goal, moving forward, was to produce a "new, public health action-oriented culture at CDC that emphasizes accountability, collaboration, communication, and timeliness." While admitting the mistakes made is admirable, time will tell if sufficient improvements have been made before we encounter another pandemic or equivalent public health crisis.[131]

Behavioral science provides us with an understanding of why different groups of people can decide to respond to a virus very differently. Behavioral science also offers insights into what we can do to reach those who are influenced by misinformation. We learned in the last chapter that we are motivated to maintain consistency between the values associated with an important identity we hold and the information we encounter. This identity process is vital to understanding why people fall for misinformation in the first place. Furthermore, this identity process connects to strategies that allow us to have productive conversations with those who hold opposing viewpoints and that increase the likelihood we can change their minds.

COVID-19 is an unfortunate case study of how something seemingly apolitical—a virus—can become political and how our strongly held partisan identities can create bias and fuel misinformation. Unfortunately, there is no shortage of identity-driven biased information processing in our society. What can we do on an individual level to combat misinformation? I was frequently asked this question during the COVID-19 pandemic.[132] One of the best actions we can take is to have productive dialogues with those who disagree with us in an effort to reduce polarization through mutual understanding. In the next section, I will describe five steps backed by social science that can help us have productive dialogues about difficult topics. The goal of these steps should *not* be to change someone's mind. The other person may update their beliefs over time, but, ultimately, that is up to them. We cannot change someone's attitudes or behaviors, but if we follow these steps, we can increase our chances of being heard and having a meaningful conversation. If we have a meaningful conversation, we hopefully can also reduce polarization and the subsequent spread of misinformation fueled by partisanship.

HOW TO HAVE CONSTRUCTIVE CONVERSATIONS ON DIFFICULT TOPICS

Remember the 5 R's:

Respect
Relate
Reframe
Revise
Repeat

Step 1: Establish Respect for One Another

In the previous chapter, we discussed how people double down on their beliefs when they are feeling uncertain or defensive. As aggravating as it might feel to deal with those who downplay the severity of a deadly virus, the last thing you want to do when trying to share the importance of your perspective is to attack them. None of us comes fully equipped to objectively process every bit of information, and we all have different identities and networks that influence what we believe. Thus, we first need to establish a level of mutual respect and listen compassionately to others. It can understandably be very frustrating if we think that their belief can cause harm, but from a purely psychological perspective, we must account for their emotional state if we want to have a productive conversation and properly showcase our opposing viewpoint.

Clare Hooker, senior lecturer and coordinator of health and medical humanities at the University of Sydney, wrote an article in which she concludes that arguing with antimaskers and yelling at them to "mask up" is unlikely to be effective.[133] She draws on psychology that shows how important it is to make individuals feel heard.[134] High-quality listening during a political conversation decreases extreme political attitudes and increases people's willingness to reflect on their own beliefs. It consists of taking a nonjudgmental approach, giving physical cues that we are listening (eye contact, nodding, etc.), and taking time to ask if they understand what we are saying.[135] Allowing others to vent and feel validated is going to open them up to hearing our perspective. Connecting on an emotional level first is vital before we engage with the substance of their arguments. It's key that we build trust before moving forward with other communication techniques.

This process is vital, as Julia Marcus, the epidemiologist at Harvard University, found when engaging with antimaskers. She wrote a piece in *The Atlantic* about men not wearing masks during the pandemic,[136] and several men who read the piece contacted her about it. She shares how she had some productive conversations with men who were against wearing masks: "These men were universally grateful to read something about anti-maskers that didn't shame or demonize them. It made them want to hear what else I had to say about why it might be worth wearing a mask." She noted that showing compassion helped build her trustworthiness: "Compassionate public health messaging builds trust."[137] Research also has shown that credibility builds when we are able to trust the source of information.[138] Thus, when trying to reach someone who has fallen for misinformation, it is crucial they find you trustworthy.

I've found that it can be helpful to embrace curiosity as we listen to opposing viewpoints. If I keep asking myself "Why do they think this way?" it prevents my own emotions from getting the best of me. Staying curious can be a useful way to engage in "amygdala hijacking," a concept created by Daniel Goldman in his book *Emotional Intelligence*. The basic idea is that the amygdala, the part of the brain associated with experiencing emotions and preparing for threats, can be hijacked by calming ourselves and allowing other brain regions involved in reasoning to become more active.[139] If we want to have a productive conversation, we must remain calm ourselves and also keep the other person calm (so they aren't hijacked by their amygdala either). Again, mutual respect and safety are crucial for any of this to occur.

Finally, research has shown that depolarization is possible when people of different political groups get together and talk with one another. Establishing mutual respect is a crucial component for this process to be successful.[140] James Fishkin, professor of communication and political science at Stanford University, describes the key elements of his study on deliberative polling, in which he found that depolarization occurred through civil discussion: "You want to convince people that their voice matters and that they will be listened to."[141]

This work is consistent with a series of psychology studies that found people became more open-minded to different perspectives when they perceived others as responsive and supportive toward them. For example, one of these studies found that when participants felt like others were responsive to their perspective about genetically modified foods (GMOs), they became more open-minded and also less certain about their own attitudes toward GMOs.[142] The importance of feeling supported and respected aligns with research on self-affirmation. A series of experiments has shown that when people wrote about times they validated an important identity, they became more open to identity-threatening information.[143] Similarly, showing respect for others may boost how people feel about themselves, which helps them process opposing viewpoints that may otherwise threaten an important identity.[144] Finally, research has shown that when people felt psychologically safe, they were more open-minded, less defensive, and more likely to be introspective.[145]

Indeed, mutual respect and nonjudgmental communication are at the heart of deep- canvassing techniques. Deep canvassing differs from traditional canvassing because it focuses on connecting with other people before even attempting to change their minds on an issue.[146] Political scientists David Brockman and Joshua Kalla conducted influential research in which they successfully reduced negative feelings toward transgender individuals using this deep-canvassing approach. They had their team of researchers either make phone calls to or knock on the doors of random households. They instructed the canvassers to try to connect with people by listening to their opinions in a nonjudgmental way instead of just listing facts that supported a nondiscrimination law. The canvassers then shared personal stories about how transgender discrimination influenced them or someone they cared about. During these conversations, they also asked people if they had ever felt they were treated unfairly in an effort to bridge the gap between them. Ultimately, connecting with people through mutual respect and personal stories significantly reduced negative attitude toward transgender individuals.[147] Essentially, when we feel safe and supported, we are more open to being wrong and updating our beliefs.

Of course, respect goes both ways. If the person we are trying to connect with does not offer us respect, then the conversation is unlikely to be productive right from the start. Climate scientist Katherine Hayoe has spoken to many different communities about climate change and has spent years trying to reach those who disagree with her. She agrees that mutual respect is vital for productive engagement: "The key ingredient to constructive conversations is mutual respect. If it doesn't exist, and can't be fostered, step away. Your time is best spent elsewhere. There are millions of us: who knows? Someone else might be able to break through where you can't."[148]

The other person may not want to give us respect, and that severely limits our chances of having meaningful dialogue. Mutual respect is the foundation for improving our chances of having a productive conversation. Once this is established, we can try to connect to the person by applying other strategies from social science.

Step 2: Relate to a Shared Identity

As we continue to listen to the other person, we can find out if we share any
other identities with them. I have pointed out several times that identities influence our cognition. Relating to the other person through a common identity can
be a way to use this identity process to help them open their minds. Research
in social psychology has shown that highlighting a common and group identity
helped reduce division and bias.[149] Furthermore, connecting with another person
by sharing personal stories was more effective for changing their mind during a
political disagreement than simply sharing facts.[150]

As we learned in the previous chapter, group-based identities easily form and
then can generate bias. However, highlighting a shared identity can reduce that
bias. By focusing on a common identity, such as an American identity, people can
reduce feelings of political polarization.[151] As I have mentioned before, this focus
on a shared identity has been one of the strongest interventions for reducing
political bias and polarized beliefs.[152] Additionally, people are more receptive to
information that comes from in-group members and ideologically sympathetic
sources.[153] For example, Republicans were more likely to support social distancing guidelines when they saw such messaging from Republican officials.[154]
So, when engaging in these difficult conversations, it helps to give examples of
groups you both share in order to show how you might have more in common
than you think. Both liberals and conservatives were more likely to support welfare policies when they also received information that their party supported it.
Liberals would support a weaker welfare policy and conservatives would support
a stronger welfare policy if the policy was endorsed by their in-group.[155]

The ability to connect with groups of people who share a common identity
can be useful outside of politics. In fact, it was crucial to some volunteer work
I did when I lived in South Carolina. During my midtwenties, I volunteered
with my local rape crisis center, and I would regularly speak to groups around
our community. One of these groups was fraternity men at the University of
South Carolina. My close friend Patrick and I were specifically asked to speak
to these men because we were young men ourselves and could better connect
with the audience. The women volunteers had found that groups of men can
feel defensive when a woman speaks to them about sexual violence and consent.
In addition to connecting to the group through our gender, the president of the
fraternity association made sure to connect our presentation with the fraternity value of treating others with respect. Again, connecting to a shared identity is crucial in order to remove defensive biases—and this was the case here!
The audience was very respectful toward us during our presentation, and some

fraternity men even stuck around after our talk to learn more about how they could volunteer with us.

Terris King, a Baltimore pastor, connected with his fellow Christians while discussing the severity of COVID-19 by applying Christian principles to highlight public health guidelines.[156] For example, he used the story of Jesus "distancing himself from his closest disciples" to discuss the importance of social distancing. King had a background in health care, having spent thirty years working in government health care and having earned a master's and a doctorate in community health along the way. He also shared the importance of getting the vaccine with his congregants. By connecting to his community through their shared faith, he was able to effectively communicate his message and combat any misinformation regarding the virus. King's efforts likely changed a few minds, as a 2021 poll found that 61 percent of those attending religious services trusted their religious leaders as a source of information for COVID-19 vaccines. In comparison, only 41 percent of those who attended religious services trusted the news media, and 60 percent trusted public health agencies such as the CDC.[157]

Rey Maktoufi is a science communication researcher and regularly leads workshops on how to better communicate about science, including how we can connect with people who hold views opposed to our own. I attended one of her workshops in 2019, and it was a fascinating exercise. For example, I am provaccine, but during the workshop, I had to try to understand why someone might be hesitant about vaccines and also to identify what we may have had in common with one another. Specifically, I had to grapple with the understandable distrust of medical institutions and to consider what I could say that could connect with people who are skeptical of vaccines. Maktoufi asked us to find things we had in common with these people we disagree with. For example, even though the person believes something very different from what we do, we may have religion or parenthood in common. So instead of immediately jumping into a debate about the efficacy of vaccines, it is much more productive to begin by discussing things we may have in common. Then we can shift the conversation toward our feelings regarding vaccination or another, more controversial issue. Building trust over a shared identity is crucial if we want to connect with someone who holds opposing views.

When discussing the severity of COVID-19, we could share how we both are from the same community and want to protect our neighbors, are parents concerned about the future of our children, or are grandchildren concerned about protecting our older relatives. The high levels of polarization can make this process tricky, but if we can connect with the other person through a shared identity, it can increase our chances of being heard. Psychologists Dominic Packer and

Jay Van Bavel agree that highlighting a common identity can be helpful for productive political discussions. They also add an important reminder when having such difficult conversations: "It's also important to use this exercise on yourself—remember that you share common identities with people you love. Finding common ground doesn't mean you need to compromise your beliefs or values, rather it means that you understand that humans contain multitudes of identities and recognize this can allow you to humanize yourself and others."[158]

Step 3: Reframe Our Arguments to Address Others' Concerns

After relating to a shared identity, we can focus on the perspective of the other person. When we want to share facts, it is important to connect our facts to values they care about. While we can connect with others through a shared identity to build trust, we can also use identities we don't have in common as an opportunity to reach them. Whether we are talking to an auditorium full of fraternity men about sexual assault or just trying to share science with someone who finds the subject boring, it is important to consider where they are coming from and connect with them so your message lands effectively. To properly convey a message, it is crucial to understand the psychological barriers that can affect an individual's ability to receive that message. After listening to their perspective, we can also learn which concerns sparked their belief.

Emily Calandrelli is an accomplished science communicator and does a great job framing her messages for her audiences. She is a West Virginia native who graduated from MIT with both a master's in aeronautics and astronautics and a master's in technology and policy. She was a correspondent on the TV show *Bill Nye Saves the World*, is the host and producer of *Xploration Outerspace* on Fox, and also is the host and coexecutive producer of *Emily's Wonder Lab* on Netflix. I met Emily at a science communication conference in Los Angeles in 2017. She was the keynote speaker and discussed how to reach different audiences using psychological science. Specifically, she talked about how she grew up in West Virginia and how this background helped her connect with conservative coal miners in the state when discussing climate change. I had the opportunity to speak with her after her talk, and we have maintained our connection over the years. We chatted again when Emily was a guest on my podcast. Our discussion covered a variety of issues, but I wanted to follow up on concepts she touched on in the earlier conference. I asked her opinion of some of the most important elements of communicating science well, and she said: "The most straightforward strategy that I've learned is to talk about science in a way that interests your audience. And it sounds very straightforward, and very obvious. But oftentimes, we forget to care about what our audience cares about."

It does sound straightforward, but this requires us to know who our audience is and what they value, which is easier said than done. Determining what different audiences care about can be difficult if we become accustomed to speaking only to groups of people who think like us. I find it frustrating when science communicators don't make an effort to expand their audience and instead just keep engaging with people who already agree with them. It's great to share science with those who are already science enthusiasts, but I also think we should try to reach out to diverse groups of people. By failing to adapt the style of our message to our audience, we limit our ability to share information we deem important.

Emily's approach is based on some interesting sociological research on framing effects. Framing can be especially influential when focusing on differences in values due to political identity. Research has shown that liberals and conservatives differ in their moral intuitions.[159] More specifically, liberals place greater emphasis on considerations of harm/care and fairness/reciprocity than conservatives. Conservatives place emphasis more equally across in-group/loyalty, authority/respect, and purity/sanctity in addition to harm/care and fairness/reciprocity.[160] Liberals have stronger moral intuitions about protecting the environment compared to conservatives, which can explain why liberals find failing to recycle more offensive than conservatives. Conservatives are more likely to support proenvironmental legislation when it is framed in terms of purity.[161] When conservatives read messages describing how contaminated the Earth has become and how important it is for people to purify their environments, they supported proenvironmental legislation at the same rate as liberals. However, when these messages were framed in terms of harm/fairness and the ways humans are destroying the Earth, conservatives were less likely to support the proenvironmental legislation they read about.

Another framing study found that conservatives were more likely to express concern over climate change when the message was framed as a national security concern and shared by the military.[162] This is consistent with polling showing that compared to liberals, conservatives found the military particularly trustworthy compared to liberals.[163] Moral reframing can also be used to nudge liberals toward holding a more positive view of the military. Because liberals have a moral foundation that emphasizes practicing fairness and avoiding harm, they were much more likely to support military spending when they read an article highlighting the economic benefits the military offers for disadvantaged groups.[164] Subsequent research has suggested that evaluating moral attitudes toward cooperative behavior may more effectively capture differences between groups.[165] The broader point here is that it is important to frame arguments based on what each group cares about if our goal is for our message to be received and supported.

Research on framing and tailoring messages to others' identities can be applied to public health crises such as the COVID-19 pandemic. For example, someone may not want to wear a mask because they heard it doesn't work and it infringes on their freedom. According to one poll, the majority of antimaskers did not want to wear a mask because it violated their "right as an American."[166] Thus, because our antimasker is concerned about personal liberty, we can revise our own arguments to connect on that level. For example, since the virus is so contagious, not wearing a mask potentially infringes on other people's liberty by putting them at risk. If we infect someone by refusing to wear a mask, we are inhibiting their own freedom. If our antimasker is concerned about economic issues related to shutdowns due to the virus, we can talk about how more mask wearing can help keep more of the economy open. More generally, freedom of choice and the need to free ourselves from the pandemic were suggested as potentially effective messages for a more conservative audience.[167]

One study tested the effectiveness of framing attitudes toward masks with a group of conservative Evangelicals. The participants were placed into one of three conditions: one group read a religious message that connected mask wearing to "loving your neighbor," one group read a message by President Trump that connected mask wearing to being patriotic, and the final group did not read any framed message. The researchers found that those who read the religious message and the message from Trump were more likely to agree that masks are important to wear, and they were even more supportive of mask mandates.[168] These studies demonstrate that framing matters for communicating the importance of scientific issues. Framing is so effective because of the influence our identities have in our lives. If you want someone to care about a particular scientific or moral issue, you must understand their own perspective and moral concerns first so you can properly frame your position in a way that has the best chance of resonating with them.

Step 4: Revise Our Questions

The first three steps have focused on connecting with the other person and listening to their point of view. These steps all emphasize engaging and connecting with the other person on an emotional level. In this fourth step, we turn to the specific phrasing of our questions to the other person about the topic at hand. When it's our turn to speak, we must be mindful of the type of language we use before we even consider the content of our message.

Research has found that asking participants if a headline is accurate nudged them toward more careful, deliberative thinking and decreased the likelihood they would share fake news.[169] This is consistent with research showing that those who scored higher on cognitive reflection tasks were less likely to share fake news.[170] Social psychology research has shown that individuals were more likely to change their attitudes when they were engaging in more effortful and analytical cognitive processing.[171] This relates to the discussion of cognitive reflection in the last chapter: Promoting an environment of deliberation and curiosity is key for productive discussion. We will have a better chance of success if instead of activating identity salience and defensiveness, we try to center the questioning around what the individual's beliefs are and how they know these beliefs are true. But how can we prime more careful thinking in our daily lives? *How* is a keyword here.

The simple word *how* can be quite powerful in our questioning. The question "How do you know this is true?" forces people to reflect on what evidence they have instead of fluffier arguments that do not dive into substance. Psychologist Amnon Glassner and his colleagues had participants evaluate a variety of issues, such as whether using a cell phone while driving increases the likelihood of crashing one's car.[172] Participants were split into groups and were asked to explain the issue. Half of the participants were asked "How do you know?" and the other half were asked "Why do you think?" when they were questioned about the various issues. Those in the "How" group were significantly more likely to provide evidence-based answers compared to those in the "Why" group, who were more likely to provide arguments without any evidence. Related to this study, research has shown that people often relied on their intuition when they reflected on how everyday things, such as a zipper or a toilet, work. When asked to describe the mechanisms by which these things work, people ended up showing that they did not understand very much at all.[173] Focusing on "How do you know that?" can be beneficial for political conversations as well.

It can be challenging to avoid increasing polarization when talking about politics, since group discussions with like-minded individuals were found to increase polarized attitudes.[174] However, polarization was reduced when the groups were moderated to make sure the participants were both carefully deliberating as they came to their conclusions and considering that they could be mistaken.[175] More specifically, asking people to explain the details of a a policy they support made them more likely to admit what they did not understand about the policy and doing this exercise was even associated with a subsequent decrease in polarized attitudes.[176] However, a later study with a larger sample size found this reduction of polarization did not replicate, although a reduction in self-reported understanding of policy was still observed.[177] Importantly, asking participants to explain why they favored a particular policy increased political polarization in this replication study, but asking them for the details of the policy did not. This is consistent with other work that demonstrated increased polarization when people had an opportunity both to justify their beliefs and to think about their identity.[178] Thus, asking for detailed policy descriptions appears to at least add some type of buffer against increasing polarization when having a political discussion.

In essence, the "How" question forces people to reflect on the basis for their belief. Grappling with how we know something, rather than relying on simple justifications or rote talking points, immediately makes the cognitive process more complicated. Adding complexity in any capacity during these conversations is extremely helpful and is consistent with the findings of social psychologist Peter Coleman. In his book *The Way Out: How to Overcome Toxic Polarization*, he summarizes a central finding of his work that explains the cause (and potential elimination) of polarized attitudes:

> Two decades of research in our Difficult Conversation Labs in New York and Germany has found that when encounters between people with strong opposing views on complex, morally polarizing issues—abortion, free speech versus hate speech on college campuses, or Donald Trump—collapse and become cognitively, emotionally, and behaviorally more simplistic, participants get stuck, resulting in angry stalemates in which participants typically refuse to work with the other person in the future. This has been a particularly robust and central finding in our research: more simplistic, one-dimensional processing of complex problems leads to (and are fed by) more intractable conflicts.[179]

To remedy this situation, Coleman suggests that we try to increase the complexity of how we process our own world by, for example, reflecting on any internal contradictions we have and engaging in good-faith discussions with those who think differently than we do. He also offers a guide to mapping our social identity complexity (how our various identities cluster together and how we can shift our behavior when we are around certain groups).[180] We have learned that increased social identity complexity is associated with lower polarization, and it certainly seems like a fruitful avenue for reducing misperceptions. Of course, these strategies require a person to make an effort to reduce their own polarization, and many people, including those who are most polarized, may have zero interest in doing so.

Finally, when we are asking these questions and trying to correct any misperceptions, we might wonder if it is better to immediately debunk misinformation or wait until later. Research across several laboratory experiments suggests that the timing might not make much of a difference. Psychologist Briony Swire-Thompson studied the correction format while debunking misinformation and summarized her findings: "We found that the impact of a correction on beliefs and inferential reasoning was largely independent of the specific format used. This suggests that simply providing corrective information, regardless of format, is far more important than how the correction is presented."[181]

Ultimately, what matters most is asking helpful questions that inspire an individual to deliberate carefully before adopting a position. We should guide the conversation to be as reflective as possible and tackle complexity head-on. Then we can present our position in an attempt to correct any misperceptions or misinformation.

Step 5: Repeat All of the Above

Let's say we have established mutual respect, related to the other person, reframed the conversation, and been careful with our questioning. What's next? The last step is the simplest one, but it might be the most important. A single conversation is unlikely to change someone's mind. We need to develop a meaningful and long-term relationship with them based on trust, and that can take a lot of time and energy.

Dr. Karin Tamerius, a psychiatrist, helps run an organization called Smart Politics, which helps people have more productive conversations with individuals or groups they disagree with.[182] She advocates building trust as a central ingredient if these difficult conversations are to be successful: "One of the most important ways to cultivate trust is to show the other person you've listened to their point of view and understand where they're coming from."[183] Research has shown that enabling opposing groups to acknowledge one another's past suffering helped them trust and reduced conflict.[184] Additionally, empathy toward out-group members increased when they were seen more as individuals and less as part of their group.[185] Of course, these processes take time. We can't force authenticity. There are no shortcuts if we want to have productive conversations.

Even if we successfully have a productive conversation, it's unclear how long potential depolarization can last. Some research suggests that mutually respectful conversations among Democrats and Republicans did result in lower political polarization up to one week later.[186] Other research found that the positive effects of cross-partisan conversations had completely decayed at a three-month follow-up.[187] These findings are consistent with research showing the short-term effects of interventions used to teach media literacy.[188] More research is needed to figure out how long productive conversations can stay with a person and what factors can improve longevity. However, it is difficult to test these long-term effects, since running such longitudinal studies can be quite complex and expensive.

These repeated discussions have another crucial benefit: We may end up becoming less polarized and more objective ourselves. Having a discussion partner we disagree with forces us to reflect on why we hold our own beliefs. We often think of ourselves as objective and others as biased, but, of course, all humans are vulnerable to bias.[189] In fact, research suggests that our illusion of personal objectivity is one of the biggest reasons why we don't want to be exposed to opposing views.[190] All humans are biased by their very nature, and social science research reveals that we are typically not great at recognizing our own bias. Thus, one of the best antidotes to counter our own bias is having meaningful conversations with people whom we trust and who think differently than we do.[191]

Steve Sloman, professor of cognitive, linguistic, and psychological sciences at Brown University, has for several decades studied how we think and form beliefs. In a 2021 paper summarizing some of his work, he remarks: "I have come to realize that a key reason for our success as cognizers is that we rely on others for most of our information processing needs; we live in a community of knowledge. We make use of others both intuitively—by outsourcing much of our thinking without knowing we are doing it—and by deliberating with others."[192]

Personally, I feel like my own arguments that support my beliefs have been strengthened and have become more nuanced by having difficult conversations with thoughtful people who disagree with me. I generally try to follow the steps outlined here, but I admit it is very challenging to do so consistently because it takes a lot of emotional effort. I have generally found more success applying these steps during face-to-face conversations than during online interactions. Social media can introduce a wealth of other barriers to good-faith dialogue, which we will look at more in the next chapter. However, there have been times when I thought I had a productive conversation with someone only to see the perceived progress evaporate later. While I do believe these steps can help facilitate difficult conversations and promote productive disagreement, they are not a panacea and lead me to address some important limitations.

IMPORTANT CAVEATS

I must mention a few important caveats regarding the techniques I laid out in this chapter. The first is that "empathy is not endorsement." This is a quote from activist Dylan Marron, who speaks with many different people who hold hostile viewpoints toward him on his podcast *Conversations with People Who Hate Me*.[193] Some of the people he talks with hold homophobic and bigoted viewpoints. Whenever he has these conversations with people, he always notes that trying to listen to their perspective does not mean he endorses anything that they are saying. He's just trying to connect and empathize with them as another human being. This distinction is important: While I am sharing ways to use social science techniques to improve our conversations in hopes of reducing polarization, this does not mean you are endorsing someone's views by listening to them without challenging what they are saying. The five steps I have discussed only illustrate the psychological mechanisms involved that can increase our chances of having a productive conversation. No one should feel expected to engage in emotionally exhausting conversations. Finally, while some people like Dylan Marron aim to build bridges with people who share harmful beliefs, it is also okay if we do not want to. Robert Jones Jr., an activist and the author of *The Prophets*, succinctly summarizes an important boundary for civil discourse: "We can disagree and still love each other unless your disagreement is rooted in my oppression and denial of my humanity and right to exist."[194]

There are ways to combat polarization and the spread of false information other than by speaking to individuals one by one. We can volunteer our time or money to help a group that supports causes we believe in. We all have different strengths. Some of us might want to focus our attention on trying to connect with those who think differently than we do. Others of us may want to advocate in ways that do not require as much potential confrontation. Each strategy is valid and crucial if our goal is to reduce polarization and susceptibility to misinformation.

While the strategies in this chapter will help generate more productive conversations and reduce polarization between different groups, there are some key limitations. First, I want to highlight again that these strategies are not meant to change someone's mind. They only increase the likelihood that our perspective will at least be heard. Ultimately, it's up to the other person to update their beliefs when or if they want. We already talked about the importance of repeating these strategies to maximize their effects. It's also important to understand how fickle these strategies can be in increasing our chances of being listened to and having a meaningful impact. One study had participants of different political ideologies discuss their perfect day through video chat. After this friendly and nonpolitical discussion, the participants reduced their negative views of out-group members. However, when these same participants discussed their disagreements regarding politics, this reduction of polarization disappeared. This suggests the limitations of one-off conversations and the importance of establishing meaningful relationships that lead to continued discussions if we want to reduce partisanship. Having a friendly conversation with a stranger who thinks differently than we do may increase our feelings toward them. However, having a contentious conversation with a stranger probably won't.[195] Furthermore, even when polarization is reduced, there still may be some topics for which it is especially difficult to find common ground. A 2023 study found that reduced polarization did not reduce support for antidemocratic norms (as when the in-group party ignored judges who were appointed by the out-group).[196] Future research will further determine what elements are most connected to political polarization, but as it stands now, we should express some caution regarding the larger impact of reduced polarization.

Finally, sometimes the person we are trying to connect with may not be reachable. While the steps I laid out are meant to increase the likelihood of having productive dialogue, they are certainly not going to work every time. Ultimately, we need to decide what the most beneficial use of our time is. If someone is not open to listening to our perspective, then we can move on to another person who may be more open. Our goal may be to have a thoughtful discussion of ideas, but the other person may view such a discussion as an opportunity to affirm their worldview and identity, not to listen to or learn from alternative viewpoints. Dominique Brossard, professor of life sciences communication at the University of Wisconsin, has studied communication strategies for years. She agrees that there are limitations to changing minds: "If I know somebody is really set in their beliefs, at the end of the day, you're not going to change the way people think."[197]

Even scientists like Katharine Hayoe who dedicate their lives to trying to connect with different communities agree that we need to pick our battles: "Contrary to what many think, I don't spend my time talking to dismissives and my only helpful tip is this: don't bother, unless (a) you enjoy arguing and never getting anywhere or (b) there are other people listening who need to know there are solid answers to their objections."[198]

CONCLUSION

In this chapter, I focused on how political identities can influence our information processing by using the COVID-19 pandemic in the United States as a case study. After reviewing how COVID-19 became politicized, I linked this polarization to various examples of misinformation throughout the pandemic. The COVID-19 pandemic offers an important and serious example of how group biases can influence how we process information. Finally, I applied social science research to develop a sequence of five steps that increase the likelihood we can reduce polarization through productive conversations: *respect, relate, reframe, revise,* and *repeat.* Next, we will see how individuals' personal networks can influence their beliefs, how false information spreads, and what we can do about it.

CHAPTER 4

HOW SOCIAL NETWORKS AND SOCIAL MEDIA SPREAD MISINFORMATION

It's exhausting loving someone and watching them get sucked into this cycle you can't break.

This is how Emily (a pseudonym) described her experience watching her husband, Peter (also a pseudonym), become a conspiracy theorist after spending time on social media. When the pandemic forced them to work from home, they began spending more time online. Emily recalls that Peter spent more time on social media sites and online message boards and then started to say things that shocked her. His comments ranged from xenophobic remarks to claims that Tom Hanks was a pedophile (this claim was one of the popular conspiracies from QAnon supporters in 2018).[1]

QAnon is a far-right conspiracy group that follows tips on an online message board from an anonymous person who uses the name Q. Q claimed to have insider knowledge about the U.S. government and even made specific predictions: for example, that Hillary Clinton would be arrested on October 30, 2017 (she was not). It didn't matter that these claims were false because supporting Q became integral to people's identities. Unfortunately, the QAnon community also propagated false claims about rampant fraud during the 2020 presidential election and was linked to the insurrection attempt on January 6, 2021.[2]

Emily was among the many people who shared their stories on Reddit's online group QAnon Casualties. The group's goal is to provide support for those who have had loved ones fall deep into the conspiracy theories connected with QAnon. Many of the posts discuss a similar situation in which a loved one has been spending more and more time on social media and developing more and more extreme beliefs that are devoid of reality.[3]

QAnon Casualties had only a few hundred members for the first few months of 2020. However, interest in the group rose quickly as people searched for a supportive community of those who shared their experiences. In January 2021, the online group's membership had grown to over 50,000 members. This growth accelerated even further after the January 6 insurrection attempt. In just one month, the group grew from 59,000 members on January 6 to over 127,000 members on February 6. By February 2022, the group had over 230,000 members.

Though Q did not make any new posts in 2021, there were still plenty of public figures who jumped at the opportunity to become a new leader for QAnon supporters.[4] Q (or someone claiming to be Q) started posting online again in mid-2022 but was met with far less enthusiasm, as many more leaders had already emerged in the broader conspiracy movement by that time.[5] With or without Q, conspiracy groups fueled by misinformation take advantage of people's need for social belonging. In this chapter, I dive into how social connections—both online and offline—influence our beliefs and susceptibility to misinformation.

Identities help us navigate the social world. They help us not only process information, as we discussed in the previous two chapters, but also navigate our social interactions. In addition to seeking out information that is consistent with their identities, people seek to be around people who share identities with them. These two forces can interact and strengthen one another. As we have been discussing throughout the book, social forces such as identities matter, and they can influence our beliefs significantly. In this chapter, we'll see how our personal networks and social media experience create yet another feedback loop that is critical for misinformation.

BIRDS OF A FEATHER FLOCK TOGETHER

Decades of social science research have shown that people prefer to associate with individuals with whom they share characteristics. In sociology, this is called *homophily* and provides empirical support to the expression that "birds of a feather flock together."[6] While it is seen in demographic characteristics such as age, race, gender, and education, homophily is also seen in people associating with those of similar emotional states, media preferences, and cultural beliefs.[7]

Homophily is so strong that it can even be observed in the brain activity of those within a network of people. Psychologists from Dartmouth College conducted a study in which they measured the brain activity from forty-two graduate students as they watched scenes from various movies.[8] Some of these students were friends, and some were strangers. The researchers found that those who were friends had significantly more similar brain activity compared to those who were strangers! These findings suggest that we are more likely to associate with people who not only like the same things we like but also perceive and react to the world in ways similar to our own.

This preference to be around people who share our beliefs can be observed geographically as well. A 2014 poll found that 35 percent of Democrats and 50 percent of Republicans preferred to live in places where others shares their political beliefs,[9] and a 2016 poll found that over 40 percent of Democrats and Republicans agreed it would be easier to get along with a new neighbor who shared their political affiliation.[10] A 2021 study evaluated the geolocation data of registered Democratic and Republican voters in the United States to see how many of their neighbors shared their beliefs. Their results showed that the median Democrat and Republican interacted with someone in their neighborhood from their opposing party just three in ten times. Furthermore, 10 percent of Democrats live in areas that provide them with virtually zero exposure to Republicans.[11]

Sociologists Byungkyu Lee and Peter Bearman analyzed social network data across almost thirty years and determined that the political diversity in personal networks in the United States had decreased significantly over time.[12] This decrease—thanks to rising political polarization—can even impact how Americans spend Thanksgiving. Those who celebrated Thanksgiving with politically mixed company spent significantly less time together than people at politically homogeneous dinners. Researchers found these decreases in the duration of Thanksgiving dinner were three times greater when the participants had traveled from cities that had heavy political advertising. These effects remained even when controlling for distance traveled and demographics, including age and race.[13]

In early 2022, NPR documented the stories of people moving to new locations in the hope of finding places more amenable to their politics. One Republican who had just moved from California to the Dallas–Fort Worth metropolitan area in Texas remarked: "People weren't wearing masks—nobody cared. It's kind of like heaven on earth."[14] Even within Texas, there were liberal havens such as the city of Austin. A Democrat who had just moved from Indiana to Austin remarked: "We as Democrats felt very out of place [in Indiana]. If people in public were talking about politics, it was always a Trump view. We heard 'Those damn liberals' a lot."

These relocations will create a cycle in which different areas will become more and more politically homogeneous. Geographic sorting according to political beliefs will also have a clear impact on future elections. Political scientist Larry Sabato has been studying these trends and has revealed some startling conclusions. For example, in 2004 just 6 percent of counties across the country had a landslide victory, categorized as one presidential candidate receiving 80 percent or more of the vote. This trend has increased over time, and 22 percent of counties had one presidential candidate get 80 percent or more of the vote in the 2020 election.[15] The geographic sorting between the political left and right seems poised to impact our society even more in the coming years.

WHO WE KNOW IMPACTS WHAT WE BELIEVE

How does our innate push to surround ourselves with people like ourselves impact our susceptibility to misinformation? First, because we tend to associate with people who are more like us, we may not generate an accurate representation of groups we do not belong to. Additionally, this self-selection process can further perpetuate polarization and misinformation by removing ideologically diverse people who might challenge us on our beliefs (or at least expose us to different viewpoints). We can also more easily connect with those who think similarly to us, which builds our relationships as we continue to interact with them more.

Such homogeneity and lack of exposure to people with different politics could explain why both Democrats and Republicans have inaccurate perceptions about their political in-group and out-group. Political scientists Matthew Levendusky and Neil Malhotra found that both Democrats and Republicans overestimated how extreme the average member of the opposing political party is, while they also underestimated how extreme members of their own party are.[16] This is consistent with findings from More in Common, an international nonprofit organization that researches the perception gaps between different groups of people. For example, one study conducted by the organization found that Republicans thought only about half of Democrats (54 percent) agree that they are proud to be American but acknowledge the flaws of the country. In reality, 82 percent of Democrats agreed with this statement (a 28 percent gap).[17] Overall, the participants in the study estimated that 55 percent of Democrats and Republicans held extremist views, but these estimates were 25 percent higher than the actual proportions of participants who held extreme views. Economist Stefanie Stantcheva and colleagues also found that Democrats and Republicans have inaccurate views on policy.[18] For example, a 2020 study asked Democrats and Republicans how likely it is that someone born in the bottom U.S. income bracket will remain there. Democrats thought 37.4 percent of people are stuck there compared to 29.5 percent of Republicans, while the correct answer was 33.1 percent.

A large study that collected data from over ten thousand participants in twenty-six countries found in-group members regularly overestimated how much their out-group dislikes them.[19] A 2022 study found that perceived polarization in the United States reduced the overall trust participants had in their fellow Americans as well as the likelihood they would cooperate for charity.[20] If we can't trust or cooperate with those who think differently than we do and we falsely assume they dislike us more than they actually do, then it is easy to imagine how this cycle will continue to spiral.

Not only does a lack of ideological diversity bias our perceptions of the world, but also we may adjust our beliefs to mirror those of people close to us. We can think back to the classic Asch conformity study, in which people were more likely to give an incorrect answer to a simple question after seeing other people give the incorrect answer first. We are already finely attuned to considering other people as social connections that are very important to us in general.

Cass Sunstein, director of the Program on Behavioral Economics and Public Policy at Harvard Law School, published some formative work in the area of group polarization in the early 2000s. He argues that individuals are more likely to become polarized in ideologically homogeneous groups for three reasons. First, when we are in these like-minded enclaves, we are mostly exposed to arguments that fit within a particular framework. The sheer amount of information is biased in favor of the views of the group we find ourselves in. Second, because we want to be liked by others, we adapt our positions to fit the role of the group. If the group is aligned on an issue, then we are motivated to strongly commit to it. The third and final reason is that we become more confident in our beliefs when those around us support our ideology. When we lack meaningful opposition to our beliefs, we feel more confident that we are correct, and we are less likely to introspect, leaving us open to the sort of polarization that can fuel extremism.[21]

We also may be less likely to express our true opinions if we are surrounded by people who think differently than we do. Indeed, research has shown that people were more likely to share secrets with another person if they believed their secret would be met with approval.[22] We have probably all been in a situation where a contentious topic came up, and we simply avoided adding our dissenting opinions to prevent conflict. Psychologist Stephan Lewandosky argues that people may feel more comfortable expressing their opinions in a group of like-minded people, but these homogeneous discussion groups could also promote polarized attitudes.[23]

Psychologist Jessica Keating and her team found that having political discussions with people who share our beliefs led to increased polarization afterward. Both Democratic and Republican participants reported higher levels of polarization after they discussed a political question with those who shared their beliefs compared to the control group. The control group wrote their answers to the question but did not talk with anyone.[24] This shows the power of social influence. We already saw that simply thinking about one's political identity can increase polarization and that this effect can be exacerbated by having conversations with like-minded people who affirm our beliefs. Keating and colleagues also found that participants were not aware of how this group discussion influenced their beliefs. The participants recorded their attitudes before and after the group discussion and then were asked what their attitudes were before the group discussion. Interestingly, participants misremembered their prediscussion attitudes, thinking they were more extreme than what they actually recorded, so as to match their increased polarization after the discussion.

Psychologists John Blanchar and Catherine Norris conducted a longitudinal study of Joe Biden voters and Donald Trump voters and measured their attitudes toward voter fraud as well as their social network composition over time. They measured network composition in four different waves from October 2020 until December 2020, choosing time points right before, during, and after the election, which was being disputed by Donald Trump. They also used the level of support for either Donald Trump or Joe Biden among their friends and family to determine network homogeneity. The researchers found that Trump voters were more convinced of misinformation regarding widespread voter fraud as their networks became more homogeneous. Conversely, Biden voters became slightly more convinced that the election was legitimate as their networks became more homogeneous.[25]

How close we are to people in our networks is also a significant factor in determining the extent to which social influence can impact our beliefs. For example, people may share information with their friends and family because they think it can help them, but the information may be inaccurate. John Cook, a psychologist at Monash Climate Change Communication Research Hub, agrees that close ties can be powerful for combating misinformation: "Friends and family are the most trusted sources of information about climate change. When people see their friends and family pushing back against misinformation that can be effective in stopping the misinformation from spreading."[26]

In the digital age, we can experience even greater exposure to the beliefs of our family members. Online messaging apps such as WhatsApp make it easy to keep in touch with our loved ones, but they also enable us to share false information easily.[27] Correcting our friends and family when they share misinformation can be difficult. Sometimes it is easier to just let their misguided comments slide than to get into an argument. A 2019 UK-based survey found that only 21 percent of people said they corrected someone who shared false information.[28] A 2019 Pew Research Center survey found that about half of U.S. adults reported they had avoided a conversation with someone in their network because they were worried this person would interject fake news or misinformation while they were talking.[29] The same survey found that about half of U.S. adults stopped following someone on social media because they shared misinformation.

The networks we maintain and curate can influence what we believe. A longitudinal study that measured ideological beliefs over time found that people changed their political and religious attitudes to match those of the people in their networks.[30] Social scientists have found a clear association between a shared political agreement in one's personal networks and increased belief strength—as well as more negative feelings toward those in their out-group.[31] Having fewer conversations with those who do not share our political views is associated with a reduced likelihood of endorsing political compromise.[32]

Those around us who agree with our ideology can help maintain and validate our beliefs. Sociologist Peter Berger popularized this idea in the 1960s when he described how religious ideology can act as a "sacred canopy" because it provides meaning and explanation to our chaotic and uncertain world.[33] This is consistent with the way important identities, in general, can provide us with meaning and help us navigate uncertainty. Berger argues that this sacred canopy was maintained by a "plausibility structure," which is the social reinforcement that supports our ideological beliefs. Having people around us who can validate our beliefs increases the plausibility that we are correct and helps remove uncertainty and doubt.

Berger asserted that such social reinforcement (i.e., the plausibility structure) could be quite broad, although he did not specify how close the social ties needed to be to maintain the sacred canopy. Sociologist Christian Smith expanded on Berger's sacred canopy idea and found evidence that those who were especially close to someone had an even stronger impact in terms of supporting their beliefs.[34] Smith called the close associates who support our beliefs our "sacred umbrella." The term *sacred umbrella* provides a nice visualization of how such close social support protects us against the raindrops of uncertainty that might poke holes in our ideology or misinformed beliefs. As long as we have some people close to us who provide the necessary social reinforcement of our beliefs, we can confidently maintain our ideology (even the parts that are fueled by misinformation). The importance of having people close to you who share your ideology is illustrated by a study that found having higher percentage of close associates who share one's religious beliefs predicted a lower level of unhappiness.[35]

Consistent with the idea of this sacred umbrella, I argue that the type of relationship matters and has an impact on whether or not it influences our beliefs. We are motivated to protect our identity, and this motivation is further enhanced when those close to us reinforce our natural inclination toward identity-protective cognition. Conversely, greater heterogeneity in one's networks prevents identities from being as important to us, since we are exposed to a wide range of people. In my own research, I wanted to study how meaningful social relationships can impact beliefs. Joseph Langston, Heather Powers, and I analyzed social media posts from Christians who described how they were formerly nonreligious.[36] In almost half of these self-reported stories, the Christians mentioned that some meaningful social connection had influenced them. For example, one respondent shared how a family member persuaded the respondent to go to church with them. After spending time integrating within the church community, they started believing in God and identifying as religious. Of course, major limitations of this study are the self-report data and its focus on only one type of belief.

In a subsequent study also published in 2019, sociologist Matt Brashears and I evaluated whether personal network composition predicted belief strength.[37] We asked people to list the names of those with whom they discuss "important matters," as researchers can capture participants' meaningful social connections by finding out with whom they discuss "important matters."[38] Measuring meaningful social connection is important because other research has shown that greater political homogeneity within more distant network connections did not seem to increase polarization.[39]

We also asked about the political and religious identities of those with whom the respondents discussed "important matters." so we could calculate a measure of how much ideological homogeneity existed in their close networks. For example, if a Democrat listed five people in their close network and also identified all five as Democrats, then we calculated that the network was 100 percent politically homogeneous. We found that network homogeneity predicted belief strength across religious and political identities. For example, we found that a Christian with a completely Christian network was more likely to hold more extreme attitudes on abortion and homosexuality. We also found that Republicans were more likely to have more extreme views on gun control, and Democrats were more likely to hold more extreme views on government spending when they had personal networks that were 100 percent politically homogeneous. The major takeaway was that having just one person in your close networks who had a different religious or political identity predicted significantly less polarized beliefs.

My study with Matt Brashears focused only on belief strength, so then I wanted to test whether network homogeneity predicts susceptibility to different types of misinformation. Psychologist Cecilie Steenbuch Traberg and I conducted a study that investigated the relationship between someone's personal network structure and their belief in both rumors and fake news headlines.[40] Half of the fake news headlines and rumors we used were biased against Democrats, and half were biased against Republicans. For example, "Donald Trump Jr. Sets the Record for Most Tinder Left-Swipes in One Day" was a fake news headline that was biased against Republicans. A fake news headline biased against Democrats was "92 Percent of Left-Wing Activists Live with Their Parents, Study: 1 in 3 Also Unemployed." The rumors were broad and vague (e.g., you overheard a story about a "Democrat (or Republican) bullying someone."

We collected personal network data from a sample of Democrats and Republicans and had them evaluate how likely they were to believe and share these fake news headlines and rumors. Both Democrats and Republicans who had completely homogeneous networks were more likely to believe and share fake news headlines and rumors when they supported their political in-group. We also evaluated whether the strength of one's identity was influenced by their network composition. Indeed, we found that those with more homogeneous networks had stronger political identities. Our research shows not only a clear relationship between network homogeneity and misinformation beliefs but also the relationship between network homogeneity and identity importance.

In addition to network homogeneity, we evaluated whether cognitive reflection scores predicted belief in politically congruent fake news and rumors. In chapter 2, we learned that cognitive reflection measures how analytical one tends to be in their thinking. In this study, we found that cognitive reflection predicted only belief in fake news headlines and only in Republicans. Cognitive reflection scores did not have any impact on political rumors for either group. We concluded that identity-based networks may have a broader impact on beliefs associated with protecting an identity (both rumors and fake news) but that analytical thinking may matter more when distinguishing a true headline from a false one. We need to consider both individual characteristics, such as cognitive reflection, and social characteristics, such as identity-based networks, as we try to understand why people hold certain beliefs. We have covered how political beliefs can be depolarized when there is diversity in our networks. However, the studies here have focused mostly on the general correlation between network structure and beliefs. While research studies cannot ethically force a person to cut ties with people in their personal networks, experiments can try to foster friendly discussions and test whether these positive social interactions with ideologically diverse people can reduce polarization.

EXPERIMENTAL TESTS OF NETWORK EFFECTS

A 2023 study used an experimental design to test whether more misinformation was shared in a politically homogeneous or heterogeneous network. The subjects were divided into two groups—one in which their politics were mostly in agreement and one in which their politics were mixed—and then allowed to share headlines in an artificial online environment. False information that aligned with the group's politics was much more likely to be shared and then reshared throughout the politically homogeneous network. When false information was shared in the politically mixed group, it did not spread nearly as far.[41] These results show that (1) we are more comfortable sharing false information in ideologically homogeneous networks and (2) as our networks become more homogeneous, we are also more likely to believe misinformation.

Political scientists Matthew Levendusky and Dominik Stecula evaluated whether discussions within a politically heterogeneous group decreased affective polarization compared to a homogeneous discussion group or an apolitical control discussion group.[42] They recruited people who identified as Democrat or Republican and who were similar in terms of age, gender, race, and education. The participants recorded their political beliefs before being randomly assigned to the politically heterogeneous group, the politically homogeneous group, or the apolitical group. To spark the discussion, the politically homogeneous group read an article about how the United States is becoming more polarized. The politically heterogeneous group had an equal mix of Democrats and Republicans and read an article about the areas in which Democrats and Republicans have common ground. Finally, the apolitical group read an article describing the Jersey shore as a nice location for vacation. Each person was instructed to share their thoughts about the article, followed by a fifteen-minute open discussion. After the discussion, participants again reported their political attitudes.

The researchers found that participants in the heterogeneous discussion group significantly reduced their negative feelings and increased their feelings of trust toward the political out-group. Additionally, 49 percent in this group said they would not be upset at all if their children married someone from the other party, and just 33 percent in the control group agreed with this statement. However, these effects were smaller in those who had stronger political identities.

When these attitudes were measured one week later, the decreased polarization remained in the heterogeneous discussion group. The researchers also tested potential reasons why discussion reduced polarization. They found that heterogeneous discussion increased perceptions of common ground, understanding of the other side's perspective, and feelings that the other side respects their beliefs.

This robust study offers clear evidence that simply talking to each other can reduce polarized beliefs. However, the researchers were careful to mention several important caveats. The first is that the emphasis on civility is crucial for depolarization to occur. A 2017 study evaluated whether polarization from like-minded discussions could be reduced when norms of deliberation and civility were employed.[43] The researchers recruited Finnish participants who had strong feelings about the Swedish language being the official national language in Finland (this issue was particularly polarizing at the time of the study). The researchers surveyed their participants' attitudes on the issue ahead of time and then had them discuss the issue with a like-minded group of six people. In one condition, there was a passive observer who watched as the group discussed the language issue. In the other condition, there was an active facilitator who enforced deliberation, reasoned justifications, and respect for other opinions. Although the members of each group were in complete agreement, the attitudes of the group with the active facilitator were significantly less polarized compared to the free discussion group.

In February 2020, I attended a workshop put on by the organization Braver Angels, which has the goal of getting Democrats and Republicans together to reduce polarization. The workshop was very similar to the studies I described earlier in that it had people from the left and right come together and chat in a friendly and moderated forum. Even though we were discussing difficult topics, the moderation and norms of civility made me feel comfortable expressing dissenting viewpoints. I certainly remember feeling a sense of hope after the workshop, and I'm sure I would self-report being less polarized as well. Even though I had strong political disagreements with other people at the workshop, I felt they genuinely listened to my perspective and respected me as a person.

LIMITATIONS AND CAVEATS OF DEPOLARIZATION WORKSHOPS

The success of these depolarization experiments highlights the importance of mutual respect for productive dialogue. While under certain conditions we can see reduced polarization and susceptibility to believing misinformation, it is important to note the serious limitations on our ability to import these laboratory effects into the real world. If there isn't an environment that creates the norm of civility—e.g., by having an active facilitator or maintaining civility during an experiment—then there may not be any incentive to treat each other with respect and engage in a productive discussion.

There is also a major self-selection bias in these real-world workshops, as people like me who are interested in depolarization are the most likely to attend. Those who are comfortable in their ideologically homogeneous networks may have zero interest in engaging with different people. Additionally, marginalized groups may understandably cut ties with those who support political issues that negatively affect them. If a political party is trying to take away your rights, then you may not want to interact with strong supporters of that party.

When I worked at Vanderbilt University, I was affiliated with the LGBT Policy Lab, where I was able to study how political networks may have impacted LGBTQ+ individuals differently. Sociologist Tara McKay and I evaluated how LGBTQ+ networks differed before and after the 2016 presidential election, since political beliefs may be particularly salient in their lives and their impact on these networks might differ from their impact on networks outside the LGBTQ+ community.[44] On the one hand, some LGBTQ+ people may have been more likely to cut ties with those they disagree with politically, since such disagreements can involve whether LGBTQ+ individuals are granted the same rights as their heterosexual counterparts. On the other hand, some LGBTQ+ people may have already cut ties with those who were against LGBTQ+ rights or developed coping strategies to maintain the relationship, since closeness often prevents us from dropping people from our networks.

We analyzed longitudinal network data that were collected a few months before the 2016 presidential election in November 2016 and again a few months after Donald Trump took office in January 2017. We compared a sample of cisgender heterosexual participants and LGBTQ+ participants. We found that LGBTQ+ participants were more likely to drop kin in their networks who had politically different views than were their heterosexual counterparts. While kin are usually close to us and remain in our networks over time, we still found that kin were not as protected from being dropped during this particularly polarizing time for our LGBTQ+ participants. Thus, when discussing extreme beliefs and misinformation, it is important to consider how these may impact different groups. Generally, diverse viewpoints among one's close ties can reduce polarization and susceptibility to misinformation, but if someone is harmed by the beliefs that a person in their network holds, they may simply cut ties with that person.

THE INTERSECTION OF IDENTITIES AND NETWORKS

While a variety of factors can determine the composition of our networks, the key takeaway here is that our social networks do impact our beliefs. This can be problematic when beliefs are based on false information. Exposure to fake news can also generate a dangerous cycle within one's social networks. When someone is exposed to fake news, it can create more polarized beliefs, which can lead to sharing more fake news, which can generate more polarization in one's network. Moreover, the mere exposure of fake news in one's network can increase the sharing of fake news even if the person doesn't believe much of the fake news they are sharing.[45] Shared agreement within interpersonal networks can bolster our beliefs, and this process can continually perpetuate itself as our more extreme attitudes connect us with more like-minded people and repel us from those with opposing views.[46]

This process of sharing information (or misinformation) connects to identity processes. We learned in chapter 2 that we are motivated to maintain consistency between the values associated with our identities and our environments. This is a feedback loop that can bias us to behave or to interpret information in a way that supports our identities. The more important this identity is to us, the more we are motivated to support it (i.e., identity verification). Now let's consider other people in our networks who share the same identity. This provides another layer to the identity verification process because not only are we acting in a way that confirms our identity but also the people around us (who share the same identity) are verifying their identity! This creates a process of mutual identity verification and can solidify our identities and those of people in our networks.[47] Remember our identities help us navigate our social worlds and provide us with a source of self-esteem. Mutual identity verification can be positive when two people are supporting one another's identities and strengthening their relationship. For example, research has shown that couples have more satisfying relationships when each partner feels their identities are being supported.[48] If you view yourself as a creative and caring person, you will feel better when your partner regularly validates and supports those characteristics. Unfortunately, mutual identity verification can also strengthen identities in a way that facilitates belief in misinformation. For example, if two friends share a political identity, then they may share biased information that supports their political group, which also represents a shared identity. This process becomes more of a problem when many people who share an identity all verify one another's identity by sharing misinformation throughout their networks.

As I read about mutual identity verification, it reminded me of something I learned while studying neuroscience: "Neurons that fire together, wire together." This famous expression was inspired by the work of neuroscientist Donald Hebb and is often attributed to Hebb himself. However, Carla Shatz is thought to be the first to publish this phrase in an article in *Scientific American*.[49] Hebb described how neurons that activate at the same time will start to connect with one another in *The Organization of Behavior*, first published in 1949.[50] When neurons connect, they facilitate learning new things. Think about all the neurons involved in riding a bicycle. You need to connect how stable the bike feels to your steering of the bike. Eventually, you learn that you steer a certain amount to remain upright. The neurons responsible for observing the stability of the bike could connect to the neurons that guide your muscles so you can adjust your steering. There may be groups of intermediary neurons connecting the neurons responsible for vision and the neurons responsible for balance, but as all these groups of neurons fire together, they strengthen their connection and their ability to complete a particular task. This process becomes easier and easier as you practice because the different neurons are strengthening their connection as they are working toward a common goal.

I think this idea can be applied to sociology when we consider how identities and networks influence our behavior. As we've discussed earlier, people can become more entrenched in their beliefs through a process of mutual identity verification. As they behave in ways that verify and strengthen their identities, they also are supporting the shared identities of those around them. This provides greater social support for the particular identity of each person involved. As this process happens over and over again, people's identities become more important to them, which further motivates them to interpret the world in a way that supports this increasingly important identity. When someone in the group shares information that validates the group identity, it is like a cluster of neurons firing together. The identities are solidified more with each verification.

This reinforcing model is particularly relevant to social media, where our exposure to various sources of information is instantaneous and nearly infinite. Identities and identity-based networks play a crucial role in how we process information. Social media allow us not only to find others who share our identity very easily but also to express our identity with great efficiency. We described previously how powerful our identities are in processing information. They are crucial for us to navigate our world, but they are also incredibly flexible and easy to activate. A social psychology study showed how easy it was for participants to quickly switch between identities that were active in their minds when primed by the researchers. The researchers connected this to social media by stating how our social identities are now only a click away from being activated and made salient in our minds.[51] In the next section, I will cover the unique role of social media in creating and spreading misinformation across our networks.

SOCIAL MEDIA'S ROLE IN MISINFORMATION

Social media already have a massive influence on our society, and this influence seems poised to grow even further. When the Pew Research Center first asked about social media use in 2005, only 5 percent of U.S. adults reported using at least one social media website. Adoption rose sharply over the years, and the center's 2021 survey found that over 80 percent of Americans ages eighteen to forty-nine reported using at least one social media site.[52] If people were using social media only to watch cat videos and post pictures of their food, the concern about social media influence might be much less. However, despite these platforms being rife with misinformation, over half of Americans get their news from social media, and this proportion is sure to increase over time.[53] American adults ages eighteen to twenty-nine are about twice as likely to get their news from social media as older generations.[54] A 2024 study found that nearly two-thirds of young women who use TikTok intentionally use it to find health information.[55] The trend of younger generations getting their news from social media will continue to widen the gap between social media and print newspapers as news sources.[56]

As of 2023, Facebook was still the most prominent social media company, with over 3 billion active monthly users.[57] Instagram and TikTok had 2 billion and 1.5 billion active monthly users, respectively.[58] These visually focused platforms have an active user base that shares news and political opinions.[59] Twitter has a smaller user base of "only" 330 million active users. As an important side note, I wrote the bulk of this book before Twitter became X. Since most of the research I cite in this book was also conducted when it was Twitter, I'll usually refer to the platform by that name. Writing about social media is challenging because they change so rapidly, but it's clear that their impact will only become more significant with time. While more and more people use social media for news, 81 percent of Americans agree that inaccurate news on social media is a very big or moderately big problem.[60]

Social media provide an opportunity for more people to make like-minded connections, but they also facilitate ideological extremism and misinformation. Misinformation and polarization would exist without social media, of course.

False information was spread by digital means long before social media existed. Some of you may remember those chain emails that promised good luck if you shared the original email with your contacts and bad luck if you didn't. That early example illustrates just how quickly misinformation can spread thanks to the internet, and social media make this process even faster with a much wider audience. Social media's ability to connect us with others and find information quickly is precisely what makes these platforms so vulnerable to spreading misinformation. Low-quality information is fast and cheap on social media, but in-depth knowledge is expensive and slow.[61] Such low-quality information can be spread online from more distant connections, or weak ties. Social media allow us to maintain contact with our strong ties, but they also allow us to frequently interact with our casual and weaker ties. Weak ties are important because they are more likely to be a source of new information compared to strong ties.[62] Does this opportunity to connect to new ties increase knowledge, or does it just provide opportunities to confirm previous beliefs?

Social media continue to be a major resource for discussions about politics, and this trend only seems to be growing. Twitter accounts have a biography where users can briefly share a few things about themselves, such as their favorite sports teams or the kind of work they do. As another marker of polarization, the proportion of Twitter users who list a political identity in their bio has increased consistently since 2015, and in 2018, political identity became even more common than religious identity.[63] Social media make it easy to find people who think just like we do and easy to ignore those who do not by allowing us to easily connect with those who share our beliefs and avoid those who challenge them.

A common phrase shared online is that people should "do their own research" while searching the internet. This sounds great, but do people know how to evaluate information properly? It is important to consider what it means to do "good" research online. Philosopher Neil Levy describes an important distinction between the types of research done online and how they differ in terms of the information we seek and the openness with which we approach the process.[64] Exploratory research entails diving into complex information with an openness to updating our beliefs based on new evidence. The focus is on the process rather than the result. This contrasts with truth-directed inquiry, which is focused on finding a particular result and makes it easy to simply confirm prior beliefs. If we lack meaningful expertise in a certain area, which all of us do for most things, we can be quite easily misled in our search for truth. Furthermore, the very process of finding an answer ourselves can bias us into thinking it is correct. Have you ever built a piece of furniture yourself? If so, you may feel a particular attachment to it, since you put in the work to make it whole. Indeed, there is a psychological bias that leads us to value items more if we helped create them.

Sociologist Francesca Tripodi describes "the IKEA effect of misinformation" in her book *The Propagandists' Playbook*.[65] According to the author, "Searching for more info makes audiences feel like they are engaging in an act of self-discovery when they are actually participating in a scavenger hunt engineered by those spreading the lies."[66] When we add all of the cognitive and identity biases in processing information, the odds are stacked against our being able to reliably seek out true information online.

So, given how easy it is to find bad information online, are people still pretty good at spotting misinformation? Well, a study led by Ben Lyons, professor of communications at the University of Utah, suggests there is still a great deal of room for improvement. He and his team evaluated just how good people are at differentiating between true and false headlines. They presented participants with headlines that appeared like they would in one's Facebook newsfeed and asked them if they were true. They also asked how confident people were in their ability to discern truth from fiction. About 75 percent of participants thought they had a higher ability to spot fake news than their results showed; 90 percent of participants believed they were "above average" in identifying false information.[67] Lyons noted a worrying implication of these results: "If people incorrectly see themselves as highly skilled at identifying false news, they may unwittingly be more likely to consume, believe and share it, especially if it conforms to their worldview."[68] Later in the book we will look at people's ability to distinguish between true and false and some education programs that can potentially help improve this skill. For now, we will focus on how and why social media can facilitate the spread of misinformation.

In a groundbreaking study, communication scholars Eran Amsalem and Alon Zoizner evaluated whether social media use increases political knowledge. They recruited nearly half a million participants in seventy-six countries and used both observational studies and experiments. In short, the observational studies revealed no evidence of increased political knowledge, and the experimental studies revealed very small increases. The authors concluded that social media's contribution to making society more politically informed is "minimal."[69] Again, when we consider how easy it is to find low-quality information online, it makes sense that this may cancel out those who do use social media to become more politically informed. Research has also found that those who got their news from social media were more likely to report holding conspiracy beliefs and believing misinformation, but this relationship was conditional on the participants already being attracted to conspiratorial explanations for events. In other words, social media use was correlated with believing misinformation—but only when accounting for people's proclivity for believing misinformation.[70] As we'll see throughout this chapter, social media exacerbate various vulnerabilities to believing and spreading false information.

Finally, it's important to define the types of misinformation online when we consider the scope of the problem. One major study analyzed how much misinformation the average Twitter user was exposed to during the 2016 presidential election. The research team categorized misinformation as information shared by sources that frequently share false information, as determined by a consensus of journalists, fact-checkers, and academics. For example, Infowars and Zero Hedge were two of the most popular low-quality websites identified in this study. The results showed that 80 percent of the false information the team identified was seen by less than 1 percent of Twitter users. Furthermore, 0.1 percent of Twitter users shared over 80 percent of this misinformation.[71] This suggests that misinformation, defined as links to low-quality websites, is seen by only a small percentage of social media users. Another team of researchers analyzed how often a sample of 3,500 Facebook users shared links from websites that frequently post false information. They found that over 90 percent of these Facebook users did not share a single link from fake news websites. They also found that the Facebook users who were over the age of sixty-five were seven times more likely to share fake news articles compared to the youngest age group in their study.[72]

Another characteristic of fake news is that it is largely found in niche spaces but is less common more broadly. This adds some perspective when we consider the larger information ecosystem: Most people are not sharing overtly false information online. However, sharing broadly misleading and hyperpartisan content may be much more common. Research has shown that exposure to hyperpartisan information predicted a greater likelihood of believing falsehoods and also significantly reduced trust in media.[73] Additionally, 40 percent of Americans have said the inaccuracy of social media news is what they dislike the most when using social media to follow current events.[74] Thus, social media might not cause people to believe or even be exposed to a large amount of explicitly false information. The larger concern with social media is more that they can expose many people to misleading information while also fueling polarization and mistrust.

Can we identify specific factors within social media that explain how and why this polarization and mistrust occur? Fil Menczer is a computer scientist at Indiana University and one of the leading experts on social media and misinformation. In a keynote speech at the 2020 International Conference on Computational Social Science, he summarized four main factors that make us vulnerable to manipulation by social media: *information overload, engagement bias, echo chambers*, and *manipulation*.[75] I agree that these four categories explain why we are vulnerable to manipulation, and I also think they do a great job laying out the main factors that explain why social media has a unique influence on the spread of misinformation more broadly. I'll describe each factor, summarizing some of Menczer's points on manipulation. However, I'll also cover how each factor relates to misinformation and extremism.

Information Overload

A wealth of information creates a poverty of attention.
—Herbert Simon

Herbert Simon won a Nobel Prize in 1978 for his work on the decision-making process of economic organizations and was an influential social scientist. As noted in this quote, much of his work focused on how information can influence our attention resources.[76] He coined the term "attention economy," which he used to evaluate the influence of overstimulation on our cognitive processing.[77] Specifically, he theorized that an environment saturated with information can exhaust our attention. Simon passed away in 2001, shortly before the internet's popularity exploded, taking over our lives with its immense ability to overload us with information.

Social media provide us with an abundance of information about whatever our unique interests may be. There were an estimated 147 zetabytes of data on the internet in 2024.[78] A zetabyte is a trillion gigabytes, and a gigabyte is equal to about 250 songs or 50,000 emails.[79] Growth is expected to continue rapidly, especially with artificial intelligence becoming more accessible and ingrained in our social technologies. The sheer volume of information can be quite useful for connecting with like-minded people and finding out niche information, but it also means we are completely inundated with information.

There is so much information on the internet that no one can possibly read even a significant fraction of it. In fact, studies have shown that about 59 percent of links shared on Twitter were not clicked on at all.[80] Even when we do click on links, we tend not to spend much time digesting the information. I kept track of the traffic on my old blog and was pleasantly surprised that I had over 2 million clicks in the few years it was active. However, I also saw that the vast majority of people stayed on the page for only a few seconds. I tried to be mindful of how long I spent on articles while browsing the web and realized I, too, quickly read through many of the pages I opened without really digesting the content. This pattern of superficial browsing seems to be pretty common, as a study found that 55 percent of people spent fifteen seconds or less reading online articles.[81] We may see someone post something interesting on social media and click on it. But then we skim it for only a few seconds if nothing further captures our attention beyond what was in the headline and the first couple of sentences. It's easy to see how this tendency allows inaccurate and incomplete information to spread. Simon's theory of information and attention only continues to become more relevant as the amount of information available grows exponentially.

A team of computer scientists evaluated whether Simon's ideas about the economy of attention applied to the sharing of memes on Twitter. Indeed, their simulations found that the information-rich environment of Twitter means only a few memes can go viral. The structure of Twitter combined with our limited attention was enough to explain why there is typically a winner-take-all result for viral information on Twitter. Only a few things dominated the information ecosystem even when the team's model accounted for each meme having an equal likelihood of going viral at the start.[82] Xiaoyan Qiu, Fil Menczer, and their team later investigated whether the quality of the meme made a difference in this process.[83] Even when they adjusted their models to prefer higher-quality and more accurate memes, there was not much improvement in the typical quality of what was shared most in the high-information/low-attention simulation. They concluded that "even when we want to see and share high-quality information, our inability to view everything in our news feeds inevitably leads us to share things that are partly or completely untrue."[84]

Computer scientist and social media researcher Kate Starbird notes that quickly sharing important information about crises and disasters was more challenging before social media and the internet. However, now the ease of information access creates new problems: "In the connected era, the problem isn't a lack of information but an overabundance of information and the challenge of figuring out which information we should trust and which information we shouldn't trust."[85]

When we think about something like the COVID-19 pandemic, the amount of new information can be truly overwhelming. In the first four months after COVID-19 was defined in the literature, nearly eight thousand academic articles were published about it.[86] And these were just the academic articles found by scientific search engines. On May 8, 2020, there were over 600 million tweets about COVID-19 or coronavirus![87] Even before we consider all the social biases that impact how we process information, we have to acknowledge just how much of an impact information overload has on our thinking and on online behavior itself.

When we consider the massive amount of information online, we should also think about who is most likely to pollute this already heavily congested information ecosystem with bad information. As we'll see in the following sections, people do not always share information online because they believe it to be true. In fact, research suggests that people report intentions to share false information online at much higher rates than intentions to share information they consider to be true.[88] Ideological extremism significantly predicts how likely low-quality news information is on Facebook, and low levels of social trust predict sharing low-quality information on Twitter.[89] Lower levels of cognitive reflection also predict higher likelihoods of an intention to share misinformation online.[90] As noted above, research has found that just 0.1 percent of Twitter accounts were responsible for 80 percent of the fake news shared on Twitter.[91] Social media make it easy to spread bad information, but most people online are not substantially contributing to this problem.

The overwhelming nature of social media can inspire people to more readily share low-quality and false information. In addition to the sheer amount of information, the engineering of social media websites aims to make us spend more time online. These websites have the incentive to keep us online as much as possible because doing so generates ad revenue. What effect does this focus on perpetual engagement have on our information processing?

Engagement Bias

Frances Haugen, a data scientist and former employee of Facebook,[92] made international news in 2021 when she became a whistleblower, sharing thousands of Facebook's internal documents. In testimony before the UK Parliament and the U.S. Congress, she claimed that Facebook's algorithm "amplifies divisive, polarizing, extreme content" in order to keep people on the platform longer.[93] Additionally, she argued that the social media company focused more on realizing profits than on properly dealing with the extremism and misinformation on its platforms: "The thing I saw at Facebook over and over again was there were conflicts of interest between what was good for the public and what was good for Facebook. And Facebook, over and over again, chose to optimize for its own interests, like making more money."[94]

Such claims were also consistent with previous criticism by Facebook whistleblower Sophie Zhang, who worked for the company as a data analyst.[95] Mark Zuckerberg, CEO of Facebook, responded to Haugen's testimony: "Many of the claims don't make any sense. I think most of us just don't recognize the false picture of the company that is being painted."[96]However, he also admitted that "no matter where we draw the lines for what is allowed, as a piece of content gets close to that line, people will engage with it more on average."[97]

What do the data reveal about engagement bias on social media? Are sites like Facebook amplifying extremism and misinformation? Or are such concerns overblown? We'll review some of the recent studies to obtain a more complete picture. Engagement bias is the result when programmers develop social media algorithms that pay more attention to and amplify information that is already more popular.[98] These algorithms detect what is being interacted with or "engaged" with and then show such content to us. Of course, the exact details of how these algorithms work are often unknown, since their codes are often proprietary and unknown to the public.

Computational social scientist Sinan Aral describes how social media constantly engage us (and hype us up) in a reinforcing cycle in his book *The Hype Machine*. He describes the "like" button as the engine for the attention economy. Not only do we get a rewarding boost when we get "likes" on our own social media, but also the "likes" we give to others curate the content we see, which will result in us "liking" more and more similar content to keep us online longer. Aral recounts an experience he had that led him to directly quantify the power of a single "like" or thumbs-up on our online behavior.[99]

Aral was going to post an online review of a restaurant and was planning to give it three stars for adequate service and food. But as he was about to start writing his review, he saw another review that gave it five stars for quality food for the price. He paused for a second and agreed it was a good value, and, ultimately, in his review he gave the restaurant four stars. He reflected on how this positive rating influenced his own rating and decided to chat with a few of his colleagues about it. They decided to run a study to see just how often a simple upvote can influence subsequent upvotes on social media.

Aral notes that this process of experiencing something in real life and then deciding to run a study related to it is often how social science ideas are generated. I agree with this experience, and it also highlights how important it is to have a diversity of people in science so they can each generate different ideas that help us understand our world. Of course, this is not limited to social science, as other scientists choose their topics based on their interests and experiences as well.[100]

What did Aral and his team find with their upvotes study? Using data from over 100,000 online comments, they randomly upvoted, downvoted, or did not interact with these random comments; the last group acted as the control. The comments that had a single thumbs-up ended up having 25 percent total upvotes compared to the control comments.[101] This finding was replicated by the work of Fil Menczer and colleagues when they evaluated social influence on articles from low-credibility sources. They found that people were more likely to share articles from low-credibility sources after they saw that many other people had "liked" and engaged with these articles. Interestingly, they noted that people were also less likely to flag social media posts for misinformation after many people had interacted with them.[102]

Many of the properties that make social media so addictive also allow misinformation and disinformation to spread like wildfire online. Computer scientist Soroush Vosoughi and his team analyzed over 100,000 tweets from 2006 to 2017. After fact-checking the tweets using six independent fact-checking websites, they found that tweets with false information spread significantly faster and further compared to tweets with true information. Political misinformation was by far the most popular category of misinformation in their dataset (followed by urban legends, business, and terrorism/war). Misinformation also was found to be classified as more novel than true information was. Furthermore, fake news stories inspired tweets that contained fear, disgust, and surprise. True information rarely spread to more than 1,000 people, but false information regularly reached between 1,000 and 100,000 people. More specifically, true information took about six times longer to reach 1,500 people on Twitter compared to false information. Vosoughi and his colleagues also looked at the influence of bots on the spread of false information. They concluded that bots influenced the spread of true and false information roughly equally and that the tendency of real human beings to share false information was still the main driver of misinformation online.[103]

Vosoughi's study is consistent with other research that found social media posts expressing moral outrage were most likely to spread virally online and we receive more engagement online by attacking political outgroups.[104] Tristan Harris, cofounder of the Center for Humane Technology, summarizes this outrage machine with a concise and descriptive quote: "We are being rewarded for being division entrepreneurs."[105] Politicians have learned to capitalize on the outrage machine as well. Research has shown that politicians who used more polarizing rhetoric online received significantly higher engagement and that viral social media posts from their campaigns predicted large boosts in fundraising.[106] This creates a clear incentive for politicians to post extreme or misleading content online in hopes of increasing attention and fundraising.

৶

Amplification is key to the spread of misinformation, according to communication scholar and professor Claire Wardle.[107] This is why her education nonprofit, First Draft News, works with people in the media to try to prevent the accidental spread of misinformation.[108] Specifically, the group notes that it may not always be a good idea to reshare something incorrect in order to correct it. We saw in chapter 2 that simply increasing how often we see information makes us more likely to believe it is true. Therefore, we should be extra careful about sharing things on social media in an attempt to debunk them. Let's imagine a scenario in which we are on Twitter and we see a tweet sharing false information. Perhaps we feel like we should "quote tweet" that fake news, thereby sharing the original tweet along with the correction. The problem with this strategy is that we are amplifying that original message to all of our followers as well. Yes, some people may read through our argument carefully and update their beliefs. Others, however, who may never have seen that piece of information may be exposed to it, and they may be motivated to believe it if it comports with the values associated with their identity. This is especially problematic if the account sending the misinformed tweet we shared has only a few followers and we have many more. As we learned earlier, the algorithm promotes social media posts that are being engaged with, so we give misinformation an algorithmic boost even if we intend to debunk that misinformation and prevent its spread. If we give the false content an algorithmic boost, more people will see it, and the more people see information, the more likely they are to believe it to be true.[109] Thus, sometimes when we see false information online, the best course of action is to do nothing at all.

When thinking about algorithmic bias, it is important to remember that these algorithms are often in flux, which can significantly impact the amount of misinformation on a social media platform at any time. Major events like the U.S. presidential election can also influence the amount of fake news online. For example, a 2019 study found that people clicked on more fake news links leading up to the 2016 election. However, after the election was over, fake news declined on Facebook but continued to rise on Twitter.[110]

We already know that people are more likely to "like" and share something when they see others are doing the same. So once something is engaged with online, it is far more likely to be shared. We also know that sensationalist and fake news is most likely to be shared online. Furthermore, the process of getting "likes" motivates us to stay and post similar things, since getting "likes" activates the reward centers of our brains. A neuroimaging study had teenage participants play around with a pretend Instagram website while researchers recorded their brain activity. When those teenagers received a lot of "likes" on their own photos, the nucleus accumbens, a part of the brain associated with processing rewards, was significantly more active.[111] Sean Parker, who was Facebook's president when it was just five months old, shared some concerning comments about the development of the website: "The thought process that went into building these applications, Facebook being the first of them, . . . was all about: 'How do we consume as much of your time and conscious attention as possible?' . . . And that means that we need to sort of give you a little dopamine hit every once in a while, because someone liked or commented on a photo or a post or whatever. And that's going to get you to contribute more content, and that's going to get you . . . more likes and comments."[112]

People are motivated to find information that is consistent with their beliefs, and the algorithms on social media make this process much more efficient. Facebook and Twitter aim to show us additional content that they believe we will engage with.[113] However, this also means that "liking" just one source of misinformation can result in a cascade of additional misinformation. Elon Musk bought Twitter in October 2022 and was quite transparent about the outrage feedback loop built into his social media platform when he tweeted: "Trashing accounts that you hate will cause our algorithm to show you more of those accounts, as it is keying off of your interactions. Basically saying if you love trashing *that* account, then you will probably also love trashing *this* account. Not actually wrong lol."[114]

Platforms like Facebook and Twitter are not merely passive tools but active participants in shaping our online experiences, prioritizing content that keeps us engaged—even if it deepens divisions or amplifies falsehoods. This engagement bias highlights why it's inaccurate to view social media as an equitable and free marketplace of ideas, as low-quality and emotionally charged information is far more likely to be shared and amplified. Once anything is engaged with, it is more likely to be amplified and "liked." If social media form a town square, then those who spread extremism and misinformation have megaphones, while everyone else has to talk normally. Those who spread sensationalism online will receive an algorithmic boost more often than those who do not.

This algorithmic boost is quite separate from freedom of expression on the internet. Internet scholars have pointed out how this freedom of expression does not also include freedom of reach via an algorithm boost.[115] Again, algorithmic amplification tends to favor those who spread low-quality information, so social media companies should be mindful of who receives their built-in boost. This is also why I have an inclination to be skeptical of public figures with massive online followings because spreading low-quality content will gain more attention and followers than sharing thoughtful, nuanced perspectives. Personally, I find that smaller accounts tend to share the most interesting content, whereas larger accounts can be quite formulaic (of course, there are plenty of exceptions to this as well). Furthermore, we are attracted to information that confirms our prior beliefs, so it's easy to see how social media can provide a skewed version of reality. While engagement bias broadly explains how social media platforms amplify divisive and sensational content, its impact of on political polarization and the spread of extremism warrants a closer examination.

ENGAGEMENT, POLITICAL POLARIZATION, AND EXTREMISM

The algorithms of social media websites strongly reward polarizing content, and this can create a clear bias regarding the kinds of politics we see online. Facebook data analyzed by the Pew Research Center showed that posts with more disagreement were more likely to be "liked," commented on, and shared compared to posts without disagreement.[116] Tweets about political issues that contained words expressing anger or disgust were more likely to be shared.[117] A 2021 study evaluated both Facebook posts and Twitter tweets from news media accounts and U.S. congressional representatives. The researchers found that referencing the political out-group significantly increased the odds of an item being shared and that the largest predictor of Facebook shares and Twitter retweets was negative references to the political out-group.[118] Researchers from Google and George Washington University analyzed how extremism, misinformation, and hateful content spread across different social media platforms. They found that extremist groups used mainstream social media (such as Facebook) to find new members and that once members gained trust, they were recruited to visit less-moderated platforms to openly discuss hateful and extreme content and false information.[119]

The engagement bias built into social media can also make us feel like people are much angrier than they are in reality, since outrage is constantly boosted by the algorithm. A 2019 study from the Pew Research Center found that the most prolific political Twitter accounts made up just 6 percent of all Twitter accounts. However, this same small group of users created 73 percent of tweets about politics and 20 percent of all tweets on the entire platform.[120] Most Twitter users (60 percent) did not even follow any political leaders, but those who did follow any political elites mostly followed politicians who agreed with them.[121]

Additionally, even though only about 6 percent of Americans identify as very liberal or very conservative, over half of these same highly active Twitter accounts identify as such.[122] So when we see political content online, we often see a very skewed perception of political opinions. It's easy to imagine someone from our political group sharing one of these extreme opinions from the opposing group.

If that extreme opinion is indicative of the types of interactions we have with those with differing politics, it is no wonder that polarization can thrive.

Social media also can reward expressions of outrage. A 2021 study found that Twitter users who expressed moral outrage were rewarded via positive social feedback from their networks. Having higher positive social feedback also increased the likelihood of expressing outrage in the future.[123] The social feedback we receive can be quite powerful because we, as a species, generally place such a high value on feedback from other people. Here we see a dire feedback loop from social media: Our outrage is rewarded by showing our posts in more newsfeeds (via an algorithmic boost), and this also produces support from our in-group. Research has shown that expressing oneself online reinforced our political identity, which in turn increased our ideological conviction.[124] Again, social media provide us with ample opportunity to verify our identity and solidify the accompanying beliefs. Thus, it is easy to see how social media users are motivated to keep spreading outrage, hyperbolic claims, and often misleading information to continue the reinforcement feedback loop.

The internet makes it easy to share misinformation that confirms our beliefs and the beliefs of our online social networks. Psychologists Pierce Ekstrom and Calvin Lai evaluated whether people would be more likely to share information that supported their ideology. Participants in this study read summaries of two policies that have deep partisan slants: banning assault weapons and increasing the minimum wage. Participants were separated into different groups, and the summaries concluded that both the weapons ban and the minimum wage increase had either positive or negative outcomes. It was not that surprising that both liberals and conservatives said they would share information that was consistent with their beliefs. However, the interesting finding of this study was that participants said they would share such information even if they thought the information may not be accurate.[125]

This is consistent with work on the motivation of partisans to share fake news to attack members of their out-group. A 2021 study analyzed Twitter data and found that people who had the most hate for their political opponents were also the ones most likely to share fake political news because they perceived it as useful for attacking their opponents.[126] This aligns with other work showing that some people shared misinformation even if they knew it is false because it connected them with their social group.[127] The rewards of verifying an important identity are incredibly motivating for sharing false content online. A series of experiments found that people were more likely to share false information when they received positive social feedback for doing so. This happened even when the individuals sharing the fake news knew the information was false.[128]

Partisan social media accounts amplify their messages through their homogeneous audiences because people will share content they agree with. A 2022 study investigated how much network amplification occurred with tweets about COVID-19 in prominent liberal and conservative Twitter accounts. The researchers found more sharing within the networks of prominent conservative social media accounts compared to liberal accounts. The authors concluded that this asymmetry in sharing could explain in part why conservatives were more susceptible to COVID-19 misinformation.[129] An analysis of the Twitter activity of Republican and Democratic politicians found that they had similar rates of sharing untrustworthy links online in 2016 (though Republicans were still more likely to share low-quality news sources). However, over the next few years, Republicans doubled their sharing of untrustworthy news sources, while Democrats' likelihood of sharing untrustworthy news remained stable.[130] Again, this is why identity and politics are so crucial for understanding misinformation.

We continue to see that people are strongly motivated to support their identities and that this motivation can trump motivations to share accurate news online. Additionally, certain groups may be especially likely to see low-quality information in their networks and subsequently share these mistruths. Elon Musk's decision to change the verification process on Twitter severely limited users' ability to find reputable information on the website. Now anyone with $8 can get a blue check and verification badge added to their name, regardless of whether they are using their real name or have any affiliation with a reputable media organization. Paying for a blue check also gives users a massive algorithm boost. NewsGuard analyzed 250 of the most engaged posts on Twitter that were spreading misinformation about the Hamas terrorist attack that took place in October 2023. Of these 250 viral posts, 74 percent were from "verified" accounts.[131] The social media landscape changes rapidly, and such changes can have massive impacts on our information ecosystem. In addition to considering the impact that platform design has on the spread of misinformation, we should pay attention to individual characteristics that make us particularly susceptible to sharing false information.

We have already seen in chapter 2 that lower scores on cognitive reflection predict higher likelihoods of sharing fake news online. Other work has found that those who scored higher on the need for chaos were more likely to share hostile political rumors.[132] A higher need for chaos is defined broadly as a desire for a fresh start by tearing down the existing order and structures. It's usually measured by how much someone agrees with statements like "I think society should be burned to the ground." Some people simply do not care whether what they share is true as long as it causes disorder in the world. Importantly, this need for chaos is related to a motivation to gain social status, as a study found that those high in this need for chaos were more likely to support a new society that benefits them. Among this group of people, their goal of spreading disorder is far more important than any goals related to accuracy or civility. Unfortunately, because of the algorithmic design of social media platforms, such toxicity can be promoted by increased engagement through "likes," comments, or the sharing of scandalous content. A 2021 study found that initial toxicity in Facebook comments on political articles increased the likelihood of subsequent toxic comments.[133] People who used Reddit and were usually civil were more likely to act in a toxic manner when they engaged in a Reddit community in which users behaved more hostilely toward one another.[134]

Even if we don't interact with those who just want to see the world burn, engaging in online debates can be extremely emotionally draining. A 2017 study found that participants perceived more political disagreement on social media compared to face-to-face political conversations.[135] It is much easier to fail to recognize the human being behind the text when you are behind a computer screen too. People also overestimated how outraged others online were.[136] Furthermore, when people were exposed to online criticism of their political group, they reported more negative feelings toward the opposing political group.[137]

To address the causal issue further, a 2020 study randomly assigned participants to deactivate Facebook for a month and measured their polarization levels before and after. Participants who were in the group without Facebook reported reduced knowledge of current events, but they also reported reduced polarization on policy issues. However, they did not have reduced polarization in their attitudes toward their political out-group. The researchers also found that those without Facebook spent more time watching television and doing things with friends and family, so this could still have exposed them to political echo chambers if their social networks were largely politically homogeneous. Importantly, the group without Facebook reported increased mental well-being as a result of taking a month's break from Facebook.[138] When looking at Twitter use over time, another study found that increased Twitter activity was associated with decreased mental well-being, increased feelings of outrage, and increased polarization.[139] A 2024 study found that Twitter/X users who unfollowed politically extreme accounts were significantly less polarized, had more satisfaction on the platform, and were more likely to share higher quality news. Being bombarded with political content that makes us feel angry, anxious, or depressed content can certainly be a drain on our mental health.[140]

It's easy to keep scrolling online, since the interface of modern social media is designed to never run out of material. Early Facebook users may remember a time when they had to click on "load more posts" after they got to the end of their newsfeed. It was an implicit reminder that maybe we should take a break. But now, thanks to the infinite scroll technology developed by engineer Aza Raskin, we can keep seeing new content seamlessly and endlessly.[141] Interestingly, he now regrets his role in developing such technology and tweeted in 2019: "One of my lessons from infinite scroll: that optimizing something for ease-of-use does not mean best for the user or humanity."[142]

In a 2021 interview, Raskin shared that he rarely uses social media now, and when he does, he tries to reflect on how he is feeling and whether he is the best version of himself while online. I think this is a good technique for anyone, and I've tried to be mindful of my own social media use.[143] I also have frequently taken breaks from social media and have always enjoyed my time away from it. I still log in to Facebook every few weeks to check what my friends are up to. I also log in to Threads or Bluesky almost daily to keep up with news and recently published academic articles and to share my own research with a broad audience. I've found LinkedIn a useful way to keep up with academic work. However, I really try to be mindful of the time I spend on social media and find it helpful to set a timer on my phone so I don't get sucked into the infinite scroll. Mastodon, a decentralized social media site that looks like Twitter but has no ads or attention-seeking algorithm, since no one owns it, is an encouraging alternative to the outrage or hype machines that dominate our information ecosystem. However, these decentralized sites lack the infrastructure and support that major companies like Meta have. While I enjoyed using Mastodon, I admit that I began to prefer Meta's Threads because it offered a better user experience and a much broader user base. Bluesky might be an ideal balance of decentralization and support. It is run on open source and decentralized software which provides users with a great deal of personal control, but still has a paid staff that helps develop the platform. Threads and Bluesky are connected to the fediverse, which allows many social media platforms to communicate with one another (i.e., you can post on your Threads or Bluesky account and people on Mastodon can still view your content). The fediverse is just one way the social media landscape may drastically change in just the next few years.

Social media platforms that are funded by companies try to keep us online as much as possible because attention and time spent online can be sold to advertisers (and our posts can be used in AI training models). Beyond showing us the most outrageous things, social media companies specialize in personalizing what we see on their sites to keep us online even longer. Such personalization of online information, like Google search results and social media newsfeeds, has been referred to as filter bubbles.[144] The algorithmic personalization from filter bubbles further helps us form homogeneous networks when we primarily follow people online who share our beliefs Social and digital environments that constantly affirm our existing beliefs are often called echo chambers and have regularly been discussed in mainstream media as a major source of polarization and misinformation. In the next section, we will look at how we measure online echo chambers and how much of an impact they have on us.

Echo Chambers

It's common for mainstream media to describe echo chambers as contributing to polarization and misinformation. But how are these echo chambers defined, and what role do they have in spreading extremism and misinformation? Kathleen Hall Jamieson and Joseph Cappella wrote an early and influential book on echo chambers, in which they define an echo chamber as "a bounded, enclosed media space that has the potential to both magnify the messages delivered within it and insulate them from rebuttal."[145]

As discussed the start of this chapter, exposure to similar viewpoints can bolster our beliefs. Social media make this process extremely efficient. We have a lot of control over our social media news feeds, so we can unfriend or unfollow people who disagree with us very easily. Michael Slater, professor of communication, has proposed a reinforcing spirals model, in which we are motivated to seek out media that affirm our identity.[146] This process is especially likely to occur during times of threat and uncertainty and can reinforce existing beliefs and identities. These reinforcing spirals offer a clear bridge between media consumption— whether social media or traditional media—and our existing identity process. In this way, social media facilitate our natural preference to associate with those who think like us. Social media provide an accessible opportunity to develop a shared social identity with other people, and then we become more invested in protecting that identity.[147] Let's consider some of the work outlining the specifics of how online echo chambers work and what effects they can have on us.

The nonprofit human rights organization Global Witness decided to test how two new Facebook users might see very different results about climate change, depending on what pages they initially "liked" on the platform.[148] The first user they created, whom they named Jane, "liked" a page belonging to Net Zero Watch, which strongly campaigns against net zero initiatives (i.e., those intended to cut greenhouse gas emissions) and regularly posts disinformation on its page. After "liking" this initial page, Jane was given Facebook recommendations to "like" other pages. Of the eighteen recommended pages the researchers tracked, seventeen contained disinformation regarding climate change (e.g., global warming is natural and humans do not have any influence).

Facebook does have a system that attempts to find climate misinformation on its platform, and as a way to combat the misinformation, it provides a link to a site with reliable climate information beneath misleading posts.[149] When analyzing the posts from these climate misinformation pages, however, just 22 percent of their posts with explicit misinformation were flagged by Facebook. Over 99 percent of scientists agree that humans have a direct impact on climate change, but climate disinformation is still common on social media.[150] As we'll see in the next chapter, fact-checking on social media yields mixed results, but this example shows that most climate misinformation might not even be flagged at all. A new Facebook user may start curating an echo chamber full of misinformation and rarely see any warnings.

Global Witness also simulated a new Facebook user named John, who "liked" the page of a reliable source of climate change information, the Intergovernmental Panel on Climate Change. The researchers found that every page suggested to John contained much more reliable information regarding climate. This study by Global Witness is more of an illustration of how online echo chambers form than a rigorous experiment, since it has only a few simulations of one topic. This certainly suggests that we can quickly become trapped in echo chambers, and this process is fueled by the way social media websites are programmed. This illustration is also consistent with my own anecdotal experience: When I "like" something on social media, I am immediately fed similar (and often more sensationalist) content.

But what does the peer-reviewed evidence on echo chambers and social media reveal? A robust literature review of echo chamber research across seven different countries found that around 5 percent of people consumed news from sources that were ideologically biased in one direction. Among these countries, the United States stood out, with more than 10 percent of Americans living in echo chambers.[151] There is a great deal of research focusing on English-speaking samples, and while I admit my book is largely focused on U.S. studies, it is the case that the United States continues to have significant polarization compared to many other countries, which constantly makes its citizens a target for misinformation and disinformation.

Facebook researchers conducted a study in 2015 to assess how many cross-belief posts were seen by Democrats and Republicans. They found that only about one-fourth of the content shared by Democrats and Republicans was ever seen by people who belonged to the opposing political group. When focusing on

Twitter, a 2015 study found that 85 percent of retweets occur between Twitter users who share similar ideological views.[152]

There are plenty of examples of groups of people with shared beliefs congregating on social media. A team of computer scientists found that by analyzing only the network structure of a Twitter account, they could predict that individual's political beliefs with 95 percent accuracy.[153] A large 2021 study evaluated over 100 million pieces of content across Facebook, Reddit, Twitter, and Gab to see if social media users tended to congregate due to like-minded views toward political topics. They found significant homophily regarding several political issues (such as gun control, vaccination, and abortion). However, Facebook had significantly higher ideological segregation compared to Reddit.[154] Reddit does not provide the same type of personalized feeds that Facebook does. This highlights the necessity of evaluating social media sites separately, as each may yield different results based on its various structures.

Homogeneous networks on social media do significantly increase the speed at which information (or misinformation) can spread.[155] When many people interested in the same topics are interacting with one another, each reinforces the others' social media activity. When looking at fact-checking of low-credibility articles on Twitter in 2018, researchers found that fact-checking mainly occurred in the periphery of networks. In the densest homogeneous spaces, fact-checking was rare.[156]

A team of computational social scientists analyzed Facebook accounts that followed either science news or conspiracy theory public pages. Accounts that followed both types of pages tended to follow other, similar pages, creating homogeneous Facebook content (or an echo chamber). The authors concluded that people often gather in groups with shared interests, which strengthens their existing beliefs and creates echo chambers. This leads to division and extreme viewpoints. As a result, the quality of information suffers, and biased stories spread, fueled by unverified rumors, mistrust, and paranoia.[157] This process can be self-fulfilling as well because those who believe conspiracy theories can be socially excluded by those who do not, which motivates the believers of conspiracy theories to spend more time with the community that supports them.[158] Furthermore, the connection between two social media users will fade when one account regularly shares fake news and the other does not.[159]

Online political networks that are tightly clustered can determine how viral something on social media can become.[160] If we share something relevant to only a small portion of our networks, then fewer people may "like" or share it. However, a network of like-minded people will be more likely to "like" and share something relevant, which can have cascading effects. When people are politically active, they can form smaller groups that are closed off from politically different people. These smaller groups may not be exposed to information that challenges the group identity, which can increase the spread of misinformation.[161] Cass Sunstein, the legal scholar mentioned earlier, predicted some of these problems back in 2007: "There is a general risk that those who flock together, on the Internet or elsewhere, will end up both confident and wrong, simply because they have not been sufficiently exposed to counterarguments. They may even think of their fellow citizens as opponents or adversaries in some kind of 'war.'"[162]

Computational social scientist Mohsen Mosleh and colleagues wanted to test whether partisanship influenced the accounts or people we followed on social media. Their research team created fake Twitter accounts that leaned either Democratic or Republican and also varied in how strongly partisan they appeared to be. For example, the weaker partisan accounts simply listed Democrat or Republican in their Twitter bio, but the stronger partisan accounts had a big picture of either Biden 2020 or Trump 2020 in their profiles. These fake accounts followed their participant sample, and the researchers could test whether they were more likely to follow back a bot that shared their politics. The results showed that both Democrats and Republicans were about three times more likely to follow back an account that shared their politics. There was not a significant difference between Democratic and Republican participants. The bot's strength of partisanship did not appear to matter much; basic signifiers of belonging to one's group were enough to seriously increase the likelihood of a follow-back.[163] This study showed how likely we are to add people who share our beliefs to our online networks and also how easily we can add like-minded accounts to our own information bubble, whether they are real people or not. We have frequently discussed how political identities make us vulnerable to misinformation, but let's consider the question of whether political groups differ in their misinformation-sharing behavior.

Low-quality and biased media can be found all over the political spectrum.[164] While both the left and the right can be vulnerable to fake news, some studies have shown a tilt toward conservatives sharing more low-quality information online.[165] Engagement of low-quality conservative news sources does seem especially high on Facebook, but it is hard to know whether higher levels of activity indicate support of or opposition to the content. Also, higher levels of conservative engagement may reflect a small subset of highly engaged people who spend a lot of time online rather than any kind of broader group.

Brendan Nyhan, a political scientist and one of the leading scholars on political misperceptions, comments, "There is a thriving pages ecosystem on Facebook, where highly emotive content performs well . . . and many conservative publishers appear at the top of the highest-performing pages and URLs for generating engagement." However, he also is clear that such activity does not always indicate one's voting behaviors.[166]

Additionally, other research has found that it was not accurate to lump all conservatives together as being more likely to share more misinformation than liberals. A small subset of conservatives who had a lower level of conscientiousness and a higher need for chaos were responsible for sharing a great deal of false information.[167] Scoring low on conscientiousness means you have less of a tendency to be organized and goal-directed and to adhere to rules and norms. So far, we do see a pattern in which people associate online with those who share their beliefs.

What happens when you ask people to leave their social media bubbles and directly expose themselves to opposing viewpoints? In a landmark study, sociologist Christopher Bail and colleagues evaluated whether exposure to opposing views on social media really reduced polarization. There is so much discussion about echo chambers on social media, but would increasing interactions among individuals with differing viewpoints be beneficial in reducing extremism and subsequent misinformation? Bail and his team used a sample of Democrats and Republicans who were regular Twitter users. In one condition, these partisans were paid to follow a Twitter bot that shared messages from their opposing group (these tweets were from public figures and media organizations). In the other condition, they had no such exposure to the other side. After one month, the study evaluated whether attitudes changed as a result of having more exposure to the other side. Republicans in their study did significantly change their attitudes: They became even more polarized against Democrats than they were at the start of the study! Democrats also became more polarized, but their change did not reach a level of statistical significance. The mere exposure to the opposing ideology did not necessarily reduce polarization and make others see the good or logic in opposing viewpoints, and, in fact, it even increased polarization![168]

As mentioned earlier, a close and meaningful relationship can increase the likelihood of reduced polarization. Brief exposure to Twitter accounts promoting an opposing ideology is certainly not a close or meaningful relationship. Previous research has shown that increasing the salience of an identity increased its effects on us. So merely seeing these political tweets could have activated the political identities of those in the study, thereby increasing their polarized attitudes. In short, there simply was not any productive dialogue taking place in a way that would motivate any chance of depolarization.

Bail and colleagues did not specifically include political accounts that attacked the opposing group. However, political attacks are commonplace, and we've learned that making political identities more salient increased their impact. Perhaps it is unsurprising that two experiments found exposure to explicit political attacks via social media increased political polarization in both Democrats and Republicans.[169] As discussed in the section on engagement bias, some people shared misinformation because it attacked their out-group and connected them to their in-group.

Bail expanded on this major study in a book on social media and polarization titled *Breaking the Social Media Prism*. He argues that social media don't necessarily create polarization, but because of the nature of the platforms, they can exacerbate it. Social media ultimately allow us to express identities in a way that makes extremism much more efficient. He concluded:

> Having carefully reviewed the literature—and studied thousands of people as they use social media over multiple years—I think that our rapidly shortening attention spans are only part of the story. The deeper source of our addiction to social media, I've concluded, is that it makes it so much easier for us to do what is all too human: perform different identities, observe how other people react, and update our presentation of self to make us feel like we belong. . . . The great tragedy of social media, which has critical implications for political polarization, is that it makes our tendency to misread our social environment even worse.[170]

I agree with Bail that the human element is crucial but often underappreciated when we discuss social media's role in polarization and misinformation. Social media and algorithmically programmed echo chambers make problems worse, but the uncomfortable truth appears to be that the vulnerability to polarization and misinformation lies within our own psychology and motivations for social connection.

Bail notes two significant aspects of social media that increase ideological extremism: normalizing extremism and muting moderates. As noted in the section on engagement bias, the most emotional and sensationalist content is rewarded by the algorithmic processes that try to keep us actively online as much as possible. So we constantly see the most extreme content that gets the most engagement, whereas the more nuanced and moderate takes are hidden. Bail draws on the work on the false consensus effect, which incorrectly convinces individuals that their viewpoints are more common than they actually are. Social media make this process even easier because those taking extreme positions start to believe that most people share their views when they are constantly exposed to other extreme views.[171] Additionally, because the most politically active are also the most likely to share political viewpoints on social media, this can further create the illusion that one's social media feed is more partisan than it really is.[172]

People with more moderate views are not going to be rewarded by an algorithm that promotes extreme attitudes. Beyond this, moderates are more likely to feel worn out by social media compared to other partisans,[173] and they are significantly less likely to post about politics on social media.[174] This, of course, is quite understandable, as moderates can easily be attacked from both sides when presenting their viewpoints. It can be much easier to curate your social media following so it contains only those who agree with you (except for the occasional follower who posts attacks on your opponents, which you can share with your in-group for positive feedback).

As already mentioned, it is quite common for people to avoid talking to others who they believe spread misinformation. Experiments have shown that Democrats and Republicans refused payments in order to avoid having both political and nonpolitical conversations with the other side.[175] Those who are less ideologically attached to a particular issue might be able to identify politically biased misinformation more readily, but they simply may determine it is not worth the emotional effort to get into a debate, and instead they remain silent.

We have described evidence of how social media bolster our tendencies to become polarized and share misinformation. It is also important to note research that has shown the limits of social media in their ability to create and sustain echo chambers purely because of their design. For example, a 2021 study found that Facebook's algorithm did indeed reduce the likelihood of seeing posts from outlets whose ideology differs from our own. However, this study also found that when people were exposed to news that contrasted with their ideological bias, they reported fewer negative attitudes toward their opposing political group.[176]

As we noted earlier, even in the highly polarized United States, only about 10 percent of Americans are in news echo chambers.[177] While this proportion can still have quite a strong impact on society, it is important to be mindful that most people consume diverse news and maintain diverse social networks. A 2019 study looking at over 640,000 Twitter accounts found that 40 percent of the accounts in their sample followed no political figures at all.[178] Of the 60 percent that did follow at least one political account, the majority followed both conservative and liberal accounts.

This finding that most people consume diverse political content is consistent with a study that evaluated the web-browsing behavior of over a million people. The researchers found that those who didn't use social media at all ended up being exposed to less ideological diversity than those who did use social media. Using search engines to find articles that support your beliefs can create a more effective echo chamber than the exposure that comes with being on social media, despite the tendency for homophily.[179] Indeed, people were more likely to consume media and follow current events when they were supportive of their political group.[180]

YouTube is another online platform that has been targeted as a potential hotspot for the spread of misinformation. *The New York Times* featured a story about a young man named Caleb Cain, a moderately liberal person who went down the YouTube rabbit hole and started spending more time watching conspiracy theory videos. He eventually became an extremist who endorsed conspiracy theories before finding another rabbit hole of more reasonable content.[181] It's a powerful story in which we see how someone can become radicalized by algorithms showing more and more extreme content.

What do the data reveal about this phenomenon of radicalization? A 2019 study analyzed over 300,000 YouTube videos from far-right YouTube channels and found that individuals who watched moderate and more fact-based content on YouTube did sometimes move on to watching more extreme content. However, this happened to only about 1 out of every 100,000 users.[182] This unlikely radicalization pattern was consistent with the 2019 study of political scientists Kevin Munger and Joseph Phillips. While their study revealed no shortage of extremist content on YouTube, the authors found that the vast majority of extreme political videos were viewed by new users seeking such content, not by those trapped in the algorithm.[183] This conclusion is similar to that of a study that analyzed the activity of over 10 million Facebook users. These researchers found that the vast majority of evidence for ideological echo chambers resulted because Facebook users decided to click on certain content they liked, not because such content was pushed to them by the algorithm. This is also consistent with a 2023 study that found exposure to partisan and low-quality news on Google was driven primarily by the choices of the person and not the platform's algorithm.[184]

Additionally, a 2022 research study led by data scientist Annie Chen found that evidence of YouTube users going down the rabbit hole and watching more and more extreme content was quite rare. Viewers of the most extreme and potentially harmful content were strongly concentrated among those who already had high levels of extreme views.[185] Importantly, the data for this study were collected in 2020, so they cannot be used to address the algorithmic issues that might have occurred in the earlier days of YouTube. There was less research done in the early days of YouTube (the company started in 2005), and the company has since made an effort to change its recommendations for suggested videos, suggesting there was a problem that needed to be addressed.[186]

It's possible that if the current recommendation algorithms were eliminated from all social media, then misinformation and extremism wouldn't significantly decrease, since enough people would seek it out on their own.[187] Computational social scientist David Lazer is a coauthor of several major papers that investigated the impact of Facebook's algorithm on political polarization during the 2020 U.S. elections. He sums up the findings by stating that the Facebook algorithm helps people find politically congruent content by "making it easier for people to do what they're inclined to do."[188]

Measuring the impact of echo chambers on our beliefs can be quite challenging, and the research I have described here does paint a complex picture. On the one hand, yes, there does appear to be a small group of people who consume only ideologically similar content, and this same group appears to be responsible for the majority of false information being spread online. On the other hand, most social media users are not polarized and not interested in sharing fake news. I do not think it is accurate to claim that social media and their algorithms drive polarization and misinformation by themselves. Instead, it is the availability of and access to emotionally charged content on social media that exacerbate existing tendencies to share bad information online.

MANIPULATION

So far, I have discussed the properties of social media that enhance the likelihood of believing and sharing misinformation. The last factor is manipulation: the process of using social media to run disinformation campaigns. One of the most well-known tools to spread disinformation online is the fake social media account programmed to share certain content—what is commonly referred to simply as a *bot*.

How common are bots, and do they have any influence on our beliefs? We can identify bots by tracking how often they post online, what accounts they follow, and whether they change their locations frequently. If we have enough data from accounts we know are bots, we can analyze the common patterns of these bot accounts and try to detect other bots online.[189] Researchers from Carnegie Mellon University analyzed over 200 million tweets that mentioned the coronavirus and found that 45 percent appeared to behave like bots.[190] In another study, researchers from the University of Southern California analyzed a month's worth of data leading up to the 2016 presidential election and estimated that about one-fifth of the tweets related to the presidential campaign were generated by bots.[191] Further, a 2017 study found that bots on Twitter accounted for up to 15 percent of its active users and that these bots regularly amplified false information related to the candidates.[192] Social media's global influence also allows for global efforts to spread disinformation. For example, social media bots were identified that spread misinformation about the Brazilian election of 2018,[193] and a massive 2019 study found that either political groups or government agencies used social media for manipulation attempts in seventy different countries! Various social media platforms were used for these disinformation campaigns, but Facebook was used most often, with fifty-six of the seventy countries using Facebook for their propaganda and disinformation.[194]

A 2018 study analyzed over 13 million tweets and found that people on Twitter shared almost as many low-credibility tweets from bots as they did tweets from real humans.[195] This certainly suggests bots can have a major impact on our social media experience. Political activists have been open about how automating programming helps them amplify their message on social media by tweeting hundreds or thousands of times a day.[196] Bots have become sophisticated enough that it can be pretty challenging to determine if they are real or not.[197] You can see how well you can identify bots using Clemson University's *Spot the Troll* quiz. I had my students at Vanderbilt University take this quiz, and they could identify the fake account only about 50 percent of the time. Our ability to detect bot accounts is also influenced by our ideological bias, as we are more likely to think social media accounts are bots when they share information that opposes our beliefs.[198]

Disinformation is spread not only by automated bot accounts but also by real humans who are employed to run fake accounts that share false information and/or propaganda. The Internet Research Agency (IRA) is a notorious Russian organization that spreads misinformation on social media.[199] It gained notoriety after Donald Trump won the 2016 presidential election because the IRA had spread misinformation to help Donald Trump and attack his opponent, Hillary Clinton.[200] Over 30 million people shared posts from IRA accounts on Facebook and Instagram between 2015 and 2017.[201] The main goal of these IRA posts was to further polarize Americans by spreading sensationalist and false information, as researchers have identified both left-wing and right-wing bot accounts created by the IRA.[202]

One of the IRA's attempts to generate civil unrest in the United States resulted in two opposing protests in Houston, Texas.[203] The Facebook page Heart of Texas had almost 250,000 members who regularly espoused conservative ideology and goals. Russian trolls used this page to advertise a protest against Islam that would take place in Texas on May 21, 2016. At the same time, other

Russian trolls infiltrated the Facebook group United Muslims for America and called on its over 325,000 followers to support Muslims in Texas with a protest at the same location as the anti-Muslim protest. Neither group discussed the other protest, and the Russians successfully created a potentially chaotic event completely through social media manipulation. When the two groups met at the same place, there was certainly plenty of conflict and verbal attacks. But thankfully the altercation did not result in physical violence. This story highlights just how powerful online manipulation can be. As I have documented earlier, social media already provide a fertile ground for the spread of misinformation and extremism. It is unsettling how easy it can be to take advantage of this environment to promote one's political goals.

While the number of bots and their sophistication certainly seem concerning, other research has suggested that social media bots may not have had a significant impact, since they predominantly interacted with individuals who were already extremely polarized.[204] Additionally, we still are improving our bot detection methods, and researchers would benefit if they had access to more data from social media companies.[205]

As mentioned earlier regarding the Twitter study conducted by Saroush Vosoughi and colleagues, real human beings are still the main drivers of misinformation online. A study looking at bot accounts on Twitter that tweeted fake news regarding Brexit found that the bots by themselves did not produce widespread diffusion. However, they could bolster and expand the hyperpartisan news that was already shared by human accounts.[206] Again, bots are not responsible for spreading large amounts of misinformation on social media, but they can effectively pollute the ecosystem by bolstering bad information. As already noted, disinformation campaigns can use fake social media accounts to generate division between opposing sides. A 2020 study found that fake Twitter accounts on opposing political sides were coordinated by the same group or organization.[207] Just as we are more likely to believe something when we are exposed to it repeatedly, we are more likely to share something online when we are exposed to it repeatedly.[208] These fake accounts capitalized on the existing biases that make us vulnerable to misinformation and extremism.

While using some computer programming to automate tweets can flood the information ecosystem, the depth of manipulation significantly increases when we consider the impact when organizations with access to our social media data use this information to target us with more detailed manipulation, an action referred to as microtargeting. Cambridge Analytica is a political consulting firm that became infamous after the 2016 presidential election because of its microtargeting using Facebook data. The saga started off innocently enough with a

Facebook application called This is Your Digital Life. When 270,000 users signed up to take this personality quiz on Facebook, they agreed to the terms and conditions: The company agreed to pay them a few dollars to take this quiz, but by doing so, they also enabled the company to access their own Facebook data as well as the data of their Facebook friends. This provided data for about 87 million Facebook accounts (those who had not consented to the use of their data).[209] So now the app developer of This is Your Digital Life had a massive dataset of personality characteristics tied to what people "like" on Facebook. With so much data, researchers could use a few statistical techniques to predict what personality traits and political affiliation a person would likely have based on their Facebook activity.[210] For example, "liking" Hello Kitty on Facebook meant that a person was significantly more likely to be a Democrat, an African American, or a Christian.[211] The app developer then shared the Facebook data with Cambridge Analytica. Facebook contended this action violated its policies and later made it more difficult to share data collected on its platform.[212] However, the main point was that Cambridge Analytica now had access to a ton of detailed data to use however it liked.

Political campaigns, including Donald Trump's campaign, used Cambridge Analytica's services to better tailor their social media advertising.[213] People on Facebook could now be targeted by political campaigns with different political ads, with content that depended on what they "liked," in an attempt to change their vote. This sounds scary, but the impact of such elaborate tactics to create custom advertising might have been quite limited in the real world. There is evidence showing that people are more likely to click on Facebook ads if they are tailored to their personalities.[214] For example, ads for beauty products were more effective for introverts when they said "beauty doesn't have to shout" compared to ads for extroverts that showed a woman dancing in a crowded party. The advertisements that matched specific personalities did get significantly more clicks and purchases. Additionally, there is some research showing that personalized microtargeting could impact people's attitudes toward bipartisan public policy more than advertising aimed at a general audience.[215]

However, getting people to buy products and change their self-reported attitudes toward nonpolarizing policies is much different than getting them to change their voting behaviors. Senator Ted Cruz's 2016 campaign for the presidential nomination also used Cambridge Analytica for political advertising but stopped after it was found that "more than half the Oklahoma voters whom Cambridge had identified as Cruz supporters actually favored other candidates."[216] Furthermore, as political scientist Brenden Nyhan notes: "Most forms of political persuasion seem to have little effect at all." In short, most people are already very committed to their political party, and it is very hard to get significant change from advertising.[217] This is consistent with a massive Twitter study that investigated the overall impact of Russian foreign influence accounts on the 2016 elections. This study, led by political scientist Gregory Eady, surveyed respondents and tracked their Twitter feeds over time. The researchers found that exposure to Russian disinformation was extremely concentrated: Only 1 percent of users accounted for 70 percent of exposures, and these exposures were heavily concentrated among Republicans. They concluded that exposure to these disinformation accounts did not have a significant influence on attitudes, polarization, or voting behavior.[218] All the research here was conducted around the time of the 2016 elections, and the social media and data landscape is always changing. While some disinformation campaigns may have had a limited impact on actual behavior, it is important to be aware of these strategies, as they could grow to have more of an impact in the future.

We often see political motivations behind disinformation campaigns, but there are also financial motivations. As I discussed in chapter 1, there is evidence that fake social media accounts posted about Gamestop stock in early 2021. A 2020 analysis of Twitter activity found many examples of bots promoting smaller stocks (which could be because smaller market capitalization stocks could be easier to manipulate compared to those of major companies). The researchers concluded that it is difficult to know just how much influence these bots have on people, as they would have to see the tweet and change their attitude toward the stock. However, they noted that one area that needs greater exploration is how such bots impact algorithmic trading processes. For example, a financial analyst may use an algorithm that tracks positive versus negative sentiment about a particular stock on social media. It is unclear how many of these programs can accurately filter out bots.[219] As already discussed, bots can play a role in amplifying a certain narrative that is already accepted by humans. Thus, if there is already some negative sentiment about a stock, bots can make it appear that such sentiment is much more widespread, which can create a spiral of further negative sentiment (and further decreased share price).

This spiral process (facilitated by engagement bias) is another reason why real humans often do much of the heavy lifting in the spread of misinformation. We are already primed by our identities to seek certain types of information. Furthermore, if a public figure we follow presents a certain narrative that aligns with our beliefs, we may then work to find "evidence" that can support that assertion. Kate Starbird calls this collaborative process "participatory disinformation."[220] Unfortunately, this process was common after the 2020 presidential election when former President Trump and his allies made bald assertions of election fraud without evidence. Matthew Masterson was a senior cybersecurity adviser at the U.S. Department of Homeland Security's Cybersecurity and Infrastructure Security Agency and failed to detect any widespread election fraud. He even called the 2020 election "as smooth a presidential election as I've ever seen." When asked what to do about the disinformation spread by public figures regarding the election, he spoke about the hope of increasing media literacy: "That goes to the real heart of this question around how do we respond to disinfo and how do we build resilience as a society to it, to have the ability to say, 'I don't have to rely on a post on Facebook or a tweet. I can turn to other sources of information to get the facts that I need.' "[221] I'll cover media literacy efforts in detail later on. For now, it is important to realize that online manipulation can come from a variety of sources: bots, microtargeting, and even real public figures using their massive platforms to spread disinformation.

As I have noted, it is hard to calculate with precision how bots and disinformation campaigns change minds. However, this might not matter for those who are using manipulation and disinformation, as filling the internet with so much misinformation has consequences by itself. By simply polluting the information ecosystem with so much bad information, they reduce trust in our democratic institutions.[222] When we can't trust each other, society suffers. Exposure to online misinformation has also been linked to increased political cynicism.[223] Sarah Kreps, director of the Tech Policy Institute at Cornell University, wrote a book titled *Social Media and International Relations*, in which she looks at how social media influence global politics. She agrees that chaos is the biggest consequence of bad information polluting the internet: "Misinformation does not necessarily succeed by changing minds but by sowing confusion, undermining trust in information and institutions, and eroding shared reference points that are the basis of coherent foreign policy."[224] For those who wish to cause harm, sowing general distrust and hopelessness is a much more significant outcome than whether a particular group believes something that is false.

Data from a massive twenty-six-country study revealed that those who used social media as more of a news source were more likely to distrust news from the traditional media.[225] Increased social media use has also been linked to decreased trust in the government and political institutions.[226] A 2021 experiment revealed a causal link between social media and trust. The researchers found that participants who saw tweets from their political out-group attacking their in-group reported lower levels of trust compared to a control groups that was not exposed to any tweets.[227] Another study revealed that this relationship may have depended on preexisting partisan biases. Specifically, they found that after controlling for social media use, Democrats and Republicans had more trust in the government when their party was in power. Conversely, when their opposing party was in power, they reported less trust in the government.[228] As discussed throughout this chapter, social media make it very easy to find content that supports our beliefs, so our social media feed may show us more examples of the government operating effectively when our party is in power and more examples of the government having problems when the opposing party is in control.

Finally, it is important to consider a crucial reason that people are vulnerable to manipulation in the first place. Christopher Bail's book *Breaking the Social Media Prism* includes a very interesting section about his interviews with people who posted very strong partisan content online. He summarizes a common theme in his qualitative data:

> One of the most common things I observed after studying extremists on social media is that they often lack status in their off-line lives. Many people with strong partisan views do not participate in such destructive behavior. But the people who do often act this way because they feel marginalized, lonely, or disempowered in their off-line lives. Social media offer such social outcasts another path. Even if the fame extremists generate has little significance beyond small groups of other outcasts, the research my colleagues and I conducted suggests that social media give extremists a sense of purpose, community, and—most importantly—self-worth.[229]

Shruti Phadke and her team of computer scientists found, like Bail did in his interviews, that social status is a key factor in predicting whether someone joins an online conspiracy group. Specifically, they found that those who joined conspiracy theory groups on Reddit were more likely to have experienced ostracism from nonconspiracy-theory Reddit groups before joining conspiracy groups. Connections with members of these conspiracy groups were also a major predictor of whether someone would join them.[230] This provides some quantitative data to support the interviews from Bail's work. It is also consistent with my work on atheists who became religious. Finally, an experiment by psychologists Damaris Graeupner and Alin Coman found a causal link between social exclusion and belief in conspiracies. In their study, they had participants write about themselves and then told them people would read these self-summaries and decide if they would want to work with them in the future. The feedback they received was fake and designed, so some of the participants were told that other people would want to work with them in the future (inclusion condition) and other participants were told that other people would not want to work with them (exclusion condition). After receiving this feedback, those in the exclusion condition were significantly more likely to endorse conspiracy theories compared to those in the inclusion condition. When people felt their sense of self was threatened, they looked for alternative sources of meaning to feel better about themselves.[231] Since we are social animals, the power of social influence is often inescapable.

Taking an even broader view, as long as there are unhappy and lonely people in the world, there will be vulnerable targets for misinformation and disinformation. Alice Marwick and Rebecca Lewis published an excellent report describing many key factors involved with media manipulation and disinformation online.[232] One of these factors is adjacent to loneliness but focuses more on the broader perception that the fabric of society has broken down. This mismatch between how we think society should function and how we feel it is functioning was defined as *anomie* by Emile Durkheim in 1893. Feelings that society has broken down have been associated with embracing political extremism and believing misinformation.[233]

When people feel society has no place for them, they can easily find groups online that will provide them with a sense of meaning. Unfortunately, many of the groups outside of the mainstream tend to be extreme, which creates a significant vulnerability to believing misinformation to support one's identity. In short, social influence is a major factor in the genesis and maintenance of our beliefs. We may like to think that our beliefs are the product of careful reasoning and analysis, but there is robust evidence that who we know can predict what we believe. If we feel marginalized and isolated from one group but another group welcomes us with open arms, we are far more likely to identify with the more welcoming group. Furthermore, as we have learned about identity, the self-protective factors we use to maintain our identity are quite powerful. Thus, if we must decide between rejecting our social group and identity and updating our beliefs that were based on misinformation, it can be a very easy decision to hold onto our misinformed beliefs.

CONCLUSION

Social science can teach us how misinformation spreads through our networks and why we believe it. We prefer to spend time with like-minded people, and doing so can strengthen our identities, which can make us more likely to believe information that supports those identities. Political polarization in the United States continues to grow as Democrats and Republicans are spending more time among those who share their political affiliation. Interventions such as depolarization workshops can help, but they have major limitations in scalability and the types of people who would seek out such workshops. Overall, we should be mindful of how our identities can be supported by our networks (i.e., mutual identity verification) and how this process can create a feedback loop that biases how we process information.

Social media accelerate this mutual identity verification process by making it easy to find others who support our beliefs. They also can overload us with information, which creates an environment that makes it easy for false and low-quality information to spread. Further, they promote posts that are popular, which can create a self-reinforcing spiral in which certain viewpoints are heavily promoted in our personalized feeds. The interaction of echo chambers and social media is complex; however, these ideological silos do make it easier for misinformation and extreme beliefs to spread. Finally, social media provide flexible tools to manipulate people through disinformation campaigns. It is hard to quantify just how much impact disinformation campaigns can have on our beliefs, but they certainly pollute the information ecosystem and can prey on socially isolated members of our society.

While determining the exact impact social influence has on our beliefs is an ongoing research question, the evidence is clear that who we associate with does have some impact on our beliefs (and the likelihood we will believe misinformation). In this chapter, I provided a broad overview of how interpersonal networks and social media influence our beliefs. In the next chapter, I will analyze how social networks influence our beliefs, using attitudes toward vaccination as a specific case study. Additionally, I will cover how the erosion of trust in scientific institutions is a significant factor when studying the misinformation landscape.

CHAPTER 5

THE SPREAD OF VACCINE MISINFORMATION AND HOW WE CAN REBUILD TRUST IN INSTITUTIONS

It's almost like thinking like you have this cheat code to keep your kids healthy from disease and allergies. . . . You feel like you've got a way to game the system to avoid all that. It does kind of become a large part of who you are.

These are the words of Lydia, a mother of three who used to be opposed to vaccination.[1] She recalled how concerned she was after her eight-week-old daughter screamed in a way she hadn't heard before after getting three vaccinations. She asked a nurse if her daughter was hurt, and the nurse simply said that such a reaction was normal. Lydia wasn't reassured. She felt brushed off and started to question her decision to vaccinate her child: "So then you start thinking, did I just hurt my child? . . . They give you an answer that the other people couldn't give you or didn't give you. And so now you don't have any trust."

While searching for people who wouldn't brush aside her concerns, Lydia found an online community of people who were skeptical about vaccination. They listened to her concerns and validated her emotions. Lydia spent more time in these online forums and became more convinced about the dangers of vaccination. As we learned in the last chapter, it is easy to find answers online that align with what you are looking for, and once you find support, you can easily remain in a digital bubble that affirms your beliefs. Lydia eventually decided not to give her daughter any more vaccinations.

Years later, after Lydia had two additional children, she didn't plan on vaccinating them at all. The online communities she visited made her feel confident about her vaccination stance and supported her antivaccination identity. However, once the COVID-19 pandemic hit, she started wondering about other pandemics and whether other diseases that had previously been eradicated by vaccines—vaccines that her children had not received—could spread. Out of concern for her children's well-being, she again searched online, but this time she ventured outside her online echo chamber. She wanted to verify some of the claims that had made her and others within her online communities so hesitant about vaccination. More specifically, she recalled that many people in her vaccine-skeptical groups claimed the dangerous ingredients in vaccines could pass through the blood-brain barrier. However, Lydia found in her research that this claim was not true at all.[2] If her online communities were wrong about that, she wondered what other beliefs they espoused may be incorrect. "That really was the catalyst to just keep going with the research and consider for a moment that maybe I'm wrong, even if it is embarrassing, even if it is uncomfortable, that I could be wrong about more things."

After more investigation, Lydia updated her beliefs about vaccines and decided to vaccinate her children. She also decided to go to nursing school and hopes to help parents who may have similar concerns about vaccines: "I want to be able to tell new parents how to handle anti-vaccine rhetoric [and] how to dismantle it and see it for what it is . . . without making them feel like they're talked down to or dumb."

This story reflects the influence online communities and social networks can have on our beliefs. However, it also demonstrates how being dismissed by a health care provider can reduce our trust in them. In this chapter, we will dive into how social connections—both online and offline—influence our beliefs, using vaccination as a poignant and timely example. I will end the chapter with a discussion of the need for trust in scientific and medical institutions and the ways they can rebuild some of the trust that has been lost.

VACCINE MISINFORMATION IS A MAJOR PUBLIC HEALTH ISSUE

Writing this book during the COVID-19 pandemic certainly influenced my decision to focus on COVID-19 and vaccine misinformation. Vaccination was a contentious topic long before COVID-19, and it will probably continue to be so for the foreseeable future. We will likely see another pandemic at some point, and we also will likely see the same vaccine hesitancy fueled by misinformation. Before diving into vaccine misinformation, it's important to understand the background of vaccines and how they have evolved to where we see them today.

Before vaccination, people were protected from smallpox using a related technique called *variolation*, in which they were deliberately infected with a small amount of the virus (often from open lesions) so their bodies would develop a natural immunity to it. Variolation occurred in Africa, India, and China for potentially hundreds of years before the concept came to Europe and North America in the eighteenth century. A man named Onesimus helped popularize the idea of variolation among early American settlers. He was kidnapped from West Africa and enslaved by the settlers of the Massachusetts Bay Colony. When Boston minister Cotton Mather purchased Onesimus in 1706, he asked the West African if he had been exposed to smallpox (since slaves were more valuable if they had a natural immunity to the virus). Onesimus answered "Yes and no" and explained the process of variolation to Mather. The minister was intrigued by this inoculation method and tried to advocate its use to local medical authorities. While many dismissed the medical advice of Onesimus due to racism, a Boston physician named Zabdiel Boylston decided to inoculate a group of 280 people using Onesimus's method. These inoculated people were found to be six times more likely to survive smallpox compared to those who got the disease without any variolation.[3]

By the late 1700s, variolation had become more popular in Europe and the American colonies. In 1798, British physician Edward Jenner published a scientific paper describing how exposure to cowpox could protect against the much deadlier smallpox. Variolation involved direct exposure to a virus to receive protection from that same virus, but Jenner used viral matter from a different virus, the weaker cowpox virus, to provide protection against smallpox. This process is considered the world's first official vaccine, and it was much safer than the variolation technique.[4]

Vaccination has become a widespread and common medical procedure for various diseases over the years and continues to save lives and reduce the economic impact of widespread illness by billions of dollars.[5] Scientists estimate that vaccines will save between 80 and 120 million lives globally between 2000 and 2030.[6] Techniques of vaccination have advanced throughout the decades, and now scientists can use messenger RNA (mRNA) vaccines to provide instructions to your cells so they learn how to make antibodies against a particular virus. This mRNA technique was developed throughout the 1990s and early 2000s, and in 2020, it was approved for use in the mass production of the COVID-19 vaccine, which became widely available in 2021.[7] Scientists have estimated that from December 2020 to September 2021, the COVID-19 vaccines saved between 170,000 and 305,000 lives in the United States.[8] Another recent vaccine breakthrough is the malaria vaccine. In 2020, malaria killed 627,000 people, most of whom were living in Africa. The new vaccine does not offer perfect protection, but it is estimated to have reduced hospital admissions from severe malaria by 30 percent.[9]

While vaccines provide many lifesaving benefits, they are not completely without risk. For example, the COVID-19 vaccines did result in rare cases of myocarditis (inflammation of heart tissue) in children and teenagers.[10] Serious cases of myocarditis can require hospitalization, but the condition will often resolve on its own.[11] No deaths from COVID-19 vaccine–related myocarditis had been reported as I wrote this book.[12] Among the female children who received the COVID-19 vaccine, there were 32 cases of vaccine-related myocarditis per 1 million vaccinations. For male children, there were 179 cases of vaccine-related myocarditis per 1 million vaccinations. Any risk of myocarditis is still scary for parents, but when we look at the data on the effects of COVID-19 on children, we see that the typical myocarditis symptoms from the vaccine are much less severe than the myocarditis symptoms from the virus itself. Furthermore, it's been estimated that for every 1 million vaccinations, 203 hospitalizations (and 1 death) were prevented in male children and 172 hospitalizations (and 1 death) were prevented in female children. Getting the COVID-19 vaccine still prevents the rare death in children. Both the severity of nonfatal illness and the rate of death were less in those reporting symptoms from vaccination compared to unvaccinated groups. These results are focused on children, since that was where most myocarditis cases were observed, but as mentioned earlier, the vaccine saved hundreds of thousands of adult lives too. From a pure risk-versus-reward calculation, getting the COVID-19 vaccine provides a higher likelihood of a better outcome.

COGNITIVE AND POLITICAL BIASES RELATED TO ANTIVACCINE BELIEFS

We humans are not always great at calculating risk and conducting statistical analysis, even before we introduce any potential biases from our social groups. One example of this is the base rate fallacy, which means we ignore the general prevalence (or base rate) of something and instead focus on specific cases.[13] For example, let's pretend we have a friend named Bret who thinks a vaccine does not work because he knows several people who got the vaccine and still got the virus. If he ignores the fact that vaccinated people are still less likely to get seriously ill overall compared to vaccinated people, he is committing a base rate fallacy.

When thinking about vaccines, we should try to think about it as a risk-versus-reward calculation. Of course, this is easier said than done. Not only are we not always great at calculating statistics, but also we are vulnerable to plenty of other types of biases. Another common psychological bias, loss aversion, can explain why some people avoid getting vaccinated. Loss aversion is the tendency to focus more on avoiding a loss than on considering potential gains. If you hear that a vaccine causes a negative side effect, then you are exhibiting loss aversion when you focus more on that side effect than on the potential benefits of the vaccine.[14] Furthermore, if you are loss averse, you may not want to expose yourself directly to any risk by getting the vaccine; instead, you decide to take your chances with the virus, since that could be perceived as more random. The very idea of using a needle to inject something foreign into our bodies can create a strong emotional response, and this fear of needles is linked to some vaccine hesitancy.[15] Personally, I am not a huge fan of needles either, so I certainly would prefer not to get shots if I weren't already so convinced of their importance! Thus, before we even get into the social factors that spread vaccine misinformation, a few psychological factors keep antivaccine sentiments prevalent in our minds.

The World Health Organization listed vaccine hesitancy as one of the top ten threats to global health in 2019. Even before the COVID-19 pandemic hit us, there were serious concerns about antivaccination beliefs.[16] Measles infections have grown 30 percent globally, and vaccine hesitancy has been linked to measles outbreaks in the United States.[17] In 2019, Washington state declared a public emergency due to a measles outbreak that occurred because over 40 percent of kindergarten children in Clark County had not received their measles vaccinations before going to school. Alan Melnick, Clark County's public health director, summed up the problem well: "When you have large numbers of unimmunized people, and you introduce measles into that population, it's like putting a lighted match into a can of gasoline. . . . It will just spread pretty quickly."[18] Unfortunately, the resources spent on creating these vaccines tend to far outweigh the resources spent on communicating their importance.

The U.S. government poured billions of dollars into developing a COVID-19 vaccine, but it provided almost zero funding for efforts to understand the social science of vaccine hesitancy.[19] In fact, Dr. Francis Collins, who in 2021 left his position as director of the National Institutes of Health (NIH), shared quite a concerning comment regarding vaccine hesitancy in the United States: "You know, maybe we underinvested in research on human behavior. I never imagined a year ago, when those vaccines were just proving to be fantastically safe and effective, that we would still have 60 million people [in the United States] who had not taken advantage of them because of misinformation and disinformation that somehow dominated all of the ways in which people were getting their answers. And a lot of those answers were, in fact, false. And we have lost so much as a result of that."[20]

I, along with many other social scientists who study misinformation, predicted the COVID-19 vaccine hesitancy. Once the response to the virus was politicized, it was likely the vaccine would be as well. I wrote about this in a 2020 article, in which I described how mask wearing had been politicized and how vaccines would be too if COVID-19 remained a polarized topic.[21] Sadly, my worry came true. It would have been nice if public health officials had listened to social scientists' ideas for improving their messaging regarding the vaccine before it became available. In the next section, I'll discuss which groups of people are more likely to spread misleading or false information about vaccines and whether antivaccine attitudes are a new phenomenon. I'll also outline a few of the mistakes made by public health officials and describe how they were used to fuel vaccine misinformation.

WHO SPREADS VACCINE MISINFORMATION?

Vaccine hesitancy fueled by misinformation was surprising to the former NIH director, but antivaccination sentiment has a long history. After the success of the smallpox vaccine, the UK and several states in the United States started mandating vaccination for the general public in the 1800s. Some antivaccination groups formed in both countries as a response because they did not trust the government and thought such measures were an affront to their personal liberty.[22] Because the vaccine was so effective at preventing smallpox outbreaks, these antivaccination groups remained in the minority. However, vaccination rates did drop, as did the frequency of outbreaks. People seemed to become complacent when the risk of serious disease was less imminent.

In 1948, a vaccine for diphtheria, tetanus, and pertussis (DPT) was licensed in the United States. It was widely used and reduced the rates of these diseases considerably. However, in the 1980s, several antivaccination programs were aired on television, including one titled *DPT: Vaccine Roulette*, which featured stories of children who took the vaccine and later developed serious neurological issues. These programs helped inspire support for powerful antivaccine lobbying groups.[23] Later research published in 1990 asserted that there is no causal relationship between the vaccine and brain damage.[24] But, unfortunately, once a powerful and emotional story is heard, it can be tough to unhear.

One of the most infamous and influential cases of vaccine misinformation is a medical paper by Andrew Wakefield and colleagues, in which they claimed to have found a link between the measles, mumps, and rubella vaccine and autism in children. The paper was immediately met with scrutiny, as the research study on which it was based had a small sample size of just twelve children; in addition, all the children had been referred to a group that was already looking at vaccine side effects, so they did not constitute a random or population-based sample.[25] Years later Wakefield's paper was retracted after it was found that the data were also fraudulent. In one example, Wakefield's paper claimed all twelve children were "previously normal" before the vaccine, but five of the children had documented preexisting developmental concerns. Furthermore, three of the families were excluded from the analysis when their children took the vaccine and did not report any immediate symptoms. This exclusion of important data made it appear that each child had symptoms that developed much closer to the vaccine date.[26] Although his paper was ultimately retracted after these problems came to light, Wakefield himself became a popular figure in the antivaccine movement and an activist against the medical establishment more broadly.[27] There is a natural overlap between activists against vaccines and activists against the medical establishment, as both groups can position themselves as the "outsider" confronting a major institution. In fact, the anti-GMO website March Against Monsanto had an advertisement for an antivaccine documentary,[28] clearly reflecting knowledge of its audience, as there is sufficient overlap in the messaging of antivaccination influencers and anti-GMO influencers. Both groups ignore the scientific consensus and promote the faulty reasoning that something is better simply because it is "natural."[29] We can reflect on the many, many dangerous animals, plants, and conditions that demonstrate how natural is not always better or more healthy for us.[30]

Both antivaccine and anti-GMO groups are also critical of the greed and corruption of large companies. Antivaccine groups condemn the money Big Pharma makes, but many antivaccine influencers profit greatly from their work as well. Research has shown that in 2022 a group of prominent antivaccine influencers were making at least $2.5 million annually from the Substack blogging platform and might have been making as much as $12.5 million per year.[31] This creates quite a financial incentive to continue creating antivaccine content. Joe Mercola is one of the most successful social media influencers and also profits from his own health supplement store. A 2017 affidavit revealed that he had a net worth of over $100 million.[32] The COVID-19 pandemic became another profitable opportunity for Mercola, as he wrote a best-selling book full of COVID-19 misinformation. For example, in the book he insists that public health measures taken to combat COVID-19 will be permanent, although restrictions started falling quickly in mid-2021, after the vaccine became widely available. Additionally, Mercola suggests COVID-19 was planned but cites weak evidence for this outrageous claim. One such piece of purported evidence that he cites is the pandemic preparedness event hosted by the Johns Hopkins Center for Health Security in 2019, which could have meant COVID-19 was planned.[33] Importantly, Mercola does not recommend that you take the COVID-19 vaccine, but he does suggest that you buy his supplement on his website to protect yourself from the virus.[34]

The authors of many of these antivaccine posts talk about how "they" (referring to some vague medical institution or government group) don't want you to know something. There seems to be something particularly enticing about this "forbidden knowledge" that is being kept from you (but that the influencer somehow has special access to). Remember Lydia, whom we met at the start of this chapter; she shared that she felt like she had a "cheat code" that allowed her to protect her children by not adhering to conventional medicine. Victoria Parker, Jeffery Lees, Anne Wilson, and I completed a study that found that this type of forbidden knowledge framing is particularly attractive and valuable.[35] Also, many of these influencers start from a kernel of truth. As I discussed in chapter 3, some of the COVID-19 restrictions (such as closing down beaches) did go overboard. The COVID-19 pandemic was an understandably stressful event for many people. If a social media influencer is blaming the effects of the pandemic on a group you already don't like (such as the government), it becomes clear why these antiestablishment takes were so wildly successful. As we will see later in this chapter, who you listen to about a scientific topic largely comes down to trust. If you are already skeptical of the government and someone is blaming it, as well as scientific institutions, for a pandemic, you may trust that source more. However, if you already trust scientific institutions, you may view a pandemic as a random event that government and medical institutions try to combat (despite not always doing so perfectly). Additionally, if you are already opposed to vaccines, you may be more likely to ignore bad behavior by your own group if it bolsters negative views of vaccines.

For example, antivaccine influencers regularly used someone's tragic death or health crisis to promote their agenda by speculating that the COVID-19 vaccine was to blame. Damar Hamlin, a twenty-four-year-old professional football player, collapsed after making a tackle in a game on January 2, 2023. He had suffered cardiac arrest and required immediate CPR on the field. Just minutes after the news broke, antivaccine influencers were already speculating on social media that his condition was caused by the COVID-19 vaccine.[36] Hamlin was struck in the chest at precisely the worst time and location, which can trigger cardiac arrest. It's extremely rare, but it can happen—and did happen to Chris Pronger, a professional hockey player who was struck in the chest with a puck during a game in 1998. There was never any empirical evidence to link Hamlin's cardiac event to the vaccine, but publicly speculating that there could be a link is an effective way to gain attention on social media (and paid subscribers to one's newsletter).

Antivaccine influencers also exploited the death of journalist Grant Wahl. He died suddenly at the age of forty-nine when an aneurysm in his heart ruptured while he was covering the World Cup in Qatar. There was zero evidence that a COVID-19 vaccination caused this event, but that didn't stop wild speculation that Wahl died from the vaccine.[37] Alex Berenson, a novelist without any medical background, would regularly make inaccurate statements about COVID-19, including a claim that the United States would never surpass 500,000 deaths due to COVID-19 (the country surpassed over a million deaths in December 2022), and he was also highly critical of the COVID-19 vaccine.[38] Immediately following Wahl's death, Berenson suggested on his Substack that the vaccine could be to blame. He has over half a million followers on Twitter, and tens of thousands of people pay $6 a month to subscribe to his newsletter.[39] Antivaccine influencers have a clear financial incentive to spread outrage and misinformation, as it garners a great deal of attention and numerous subscribers.

Mainstream media and medical institutions are certainly not immune from spreading vaccine misinformation themselves. Tucker Carlson had one of the most popular cable news shows on television before he was fired in 2023.[40] During the pandemic, he regularly cast doubt on the COVID-19 vaccine, suggesting that it doesn't work and that maybe "they" are keeping that information from you. "If the vaccine is effective, there is no reason for people who've received a vaccine to wear masks or avoid physical contact. . . . So maybe it doesn't work, and they're simply not telling you that. Well, you'd hate to think that, especially if you've gotten two shots. But what's the other potential explanation? We can't think of one."[41]

The potential explanation here is simply that the vaccine helps but is not a panacea. Again, it is about managing risks. Getting the vaccine in conjunction with mask wearing had the biggest impact on slowing the spread of the virus. Casting doubts without providing the full picture is a common tactic used by those spreading vaccine and other medical misinformation. Of course, some mainstream media hosts directly shared false information. Fox News host Will Cain claimed that the COVID-19 vaccines are more dangerous for children than actually getting the COVID-19 virus. This is not true at all.[42]

Those at the other end of the political spectrum are not immune to spreading vaccine misinformation either. For example, President Joe Biden said that people who received the COVID-19 vaccine could not spread the virus. Again, while the vaccine reduces the spread, claiming its effectiveness in such binary terms is not accurate.[43] President Biden "should not be so firm" in his phrasing, said epidemiologist and infectious disease researcher Tara C. Smith. "Vaccination does significantly reduce transmission from vaccinated breakthrough cases but does not completely eliminate it."[44]

Rachel Maddow, host of one of the most popular liberal news and opinion shows on television, echoed these sentiments about the COVID-19 vaccine preventing transmission.[45] She presented the vaccine as a silver bullet, while medical scientists were much more cautious about whether the vaccine alone would be enough to end the pandemic and were focused on getting more people vaccinated in hopes of preventing further mutation of the virus. Maddow also claimed that "Trump never encouraged Americans to get the vaccine" while he was president, but there were several examples of President Trump praising the vaccine and even telling people to "get their shots." Later on, Maddow did correct herself, but we have already learned how hard it is to correct misinformation once it has spread.[46] Similar to what we saw with mask wearing, the lack of humility in the messaging from public officials and certain liberal media figures probably decreased their perceived trustworthiness even further. It's also possible that the levels of distrust in institutions, disinformation spread by political leaders, and overall polarization were so high that better messaging wouldn't have mattered much, but it would have been nice to have seen significantly better communication by public officials throughout the pandemic.

On the conservative side, we saw the denial of the effectiveness of the COVID-19 vaccine, but on the liberal side, we saw an exaggeration of its effectiveness. It's easy to see how misleading statements on each side can be used by the other to continue the spread of polarization and misinformation. Such misinformation did appear to have grave consequences, as there were several tragic examples of prominent conservatives who refused to get the vaccine and were later killed by the virus.[47] Are vaccine attitudes mostly a political issue now, or do other factors also play a major role?

WHAT FACTORS CONTRIBUTE TO VACCINE HESITANCY?

Now that we've had a brief overview of the history of vaccination and vaccine misinformation, we can look into the social and psychological factors that predict vaccine attitudes. Because the COVID-19 pandemic was so politicized, it is important to look first at vaccine attitudes before the pandemic. A 2017 Pew Research Center survey found that 88 percent of Americans agreed that the benefits of vaccines outweigh the risks and that 91 percent agreed that their preventative health benefits are "high" or "medium" (73 percent for high and 18 percent for medium).[48] A massive 140-country study from Gallup and Wellcome Global Monitor found that in 2018, 84 percent of participants agreed that vaccines are effective.[49] General vaccine hesitancy has historically been spread rather equally across political ideologies, but conservatives have been slightly more inclined to agree that vaccination should be a parent's choice.[50] Before COVID-19, antivaccination views were generally connected to holding extreme beliefs at either end of the political spectrum.[51]

However, there is evidence that vaccination is becoming more and more of a partisan issue that isn't relegated to the extremes. After the politicization of COVID-19 and the COVID-19 vaccine, we did see a partisan split in vaccination attitudes. A Pew Research Center survey in August 2021 found that 86 percent of Democrats and 60 percent of Republicans had had at least one shot of the COVID-19 vaccine.[52] By August 2021, there were several months of data showing the effectiveness of the vaccine in preventing significant illness and death from the virus.[53] This partisan split remained consistent throughout the pandemic as the vaccine became more readily available and even as booster shots became available. A January 2022 survey found that 62 percent of Democrats had gotten their COVID-19 booster shots compared to just 32 percent of Republicans, even though the booster shots were providing impressive protection against serious illness.[54] This partisan bias persisted in the fall of 2023, when 79 percent of Democrats said they would "probably or definitely" get the new COVID-19 booster shot compared to just 39 percent of Republicans.[55] A 2022 study found that about half of Americans who had used dating sites said that COVID-19 vaccine status is at least somewhat important to see on dating profiles.[56] Skepticism toward the vaccine is also associated with higher distrust in institutions, which correlates with conservative politics, but there is certainly distrust in institutions from both the right and the left.[57] I will discuss trust further at the end of this chapter. In short, the politicization of the COVID-19 vaccine accelerated trends of distrust in institutions, and those beliefs tend to correlate more with those on the political right.

Unfortunately, this political polarization regarding the COVID-19 vaccine has spread to other vaccines as well. Before the pandemic severely hit the United States, a February 2020 poll showed that 58 percent of Democrats and 54 percent of Republicans said they had received a flu shot in the last year.[58] A 2021 study found that Democrats were significantly more likely to get a flu shot compared to

Republicans: 68 percent of Democrats said they had received a flu shot or would get one compared to just 44 percent of Republicans.[59] This is consistent with my own research, which shows that Republicans are more skeptical of both the COVID-19 vaccine and vaccines in general. Additionally, my research has revealed that those who are skeptical of the COVID-19 vaccine are also more likely to be skeptical of vaccines overall (I'll present this research in more detail in the next section on social networks). Furthermore, a 2022 study found that for 23.8 percent of participants their confidence in vaccines decreased between 2019 and 2022. However, it's important to note that overall confidence was still fairly high. The vaccine confidence scale used in this study had scores ranging from 5 to 25; in 2019, the median score was 22, but in 2022, it was still 20, even with the drop in confidence.[60]

Recent research looking at the voting patterns of Democrats and Republicans revealed that political polarization on childhood vaccination bills has significantly increased in the past twenty years and is almost as partisan as the voting patterns found on abortion bills.[61] Vaccination has become so politically charged that a 2023 study found Republicans reported a lower intention to take a hypothetical vaccine that would prevent Alzheimer's disease. The study also found that Republicans were especially unwilling to get a vaccine if it was funded by Democrats.[62] Finally, a 2023 study found that 53 percent of Americans reported some hesitancy to vaccinate their dogs and 37 percent believed vaccines can cause cognitive issues in their dogs, such as canine autism![63] Unfortunately, antivaccine attitudes have spread across many domains and are not expected to decrease anytime soon.

While partisanship is a growing concern, research has found that the strongest predictor of antivaccine beliefs is holding misinformed vaccine beliefs. Such misperceptions about their harm explain antivaccine beliefs more than politics, education, religion, or other demographic variables.[64] Conversely, certain personality factors, such as having a higher level of prosocial attitudes, may predict a higher likelihood to vaccinate.[65]

Finally, there is a spectrum of vaccination attitudes. It is imprecise to categorize anyone who has concerns about vaccination or who has not gotten the COVID-19 vaccine as an antivaxxer. As I mentioned earlier, there is an overlap between those opposed to the COVID-19 vaccine and those opposed to vaccines in general. However, some people generally favored vaccines but wanted to wait and see if there were side effects from the COVID-19 vaccine before getting it. Furthermore, there is a clear distinction between those who are unilaterally and passionately opposed to vaccination (the antivaxxers) and those who are open to vaccines but more skeptical.[66]

Overall, the people who legitimately identify as antivaccine make up a very small group. A 2022 study measured how many people held extreme negative sentiments toward vaccines and found that just 7 percent of Americans, 7 percent of

Canadians, and 3 percent of the British fall into the category of "anti-vaccine."[67] Again, it's important to remember that the vast majority of the global population believes vaccines are safe and effective medical procedures. However, a significant number of people express some hesitancy toward vaccines without being full-on antivaccine. This same 2022 study found that 23 percent of Americans, 11 percent of the British, and 15 percent of Canadians would fall into the "vaccine-hesitant" category. People in this group recognized some of the benefits of vaccines but were concerned about potential side effects and were less supportive of vaccine mandates. It is inaccurate to broadly label them antivaccine simply because they express concerns.

It was unfortunate to see people condemning all unvaccinated people for simply ignoring the scientific consensus instead of appreciating the variety of reasons why someone may not be vaccinated. Pediatrician Rhea Boyd noted the same unfortunate phenomenon on Twitter during her interview with science journalist Ed Yong, in which they discussed the nuances of the unvaccinated. Boyd makes several important points, such as the simple inability of many lower-income people to get the vaccine. Further, they might not be able to take time off from work or might not have transportation to get the COVID-19 vaccine even though it is available to them. Boyd notes that the most extreme antivaccine advocates are a small minority, but because they are so loud, they leave the impression that anyone who hasn't gotten the vaccine because of access or any other reason is part of their group.[68] A 2021 study supported Boyd's argument when it found that those who did not get the COVID-19 vaccine were twice as likely to be uninsured.[69] While dispelling myths and communicating the effectiveness of vaccines can be helpful, improving access to vaccines is an important factor in increasing vaccination rates.[70] Thus, increasing access to health care is crucial to increasing vaccine uptake and doesn't require as much effort as trying to change minds. When we consider those who hold antivaccine views, it's important to note that such beliefs do not exist in a vacuum. Furthermore, just like the viruses themselves, antivaccination beliefs can spread quickly through personal networks.

HOW DO VACCINE BELIEFS SPREAD THROUGH PERSONAL NETWORKS?

In the last chapter, I discussed how personal networks can influence our beliefs and susceptibility to misinformation. Because of the importance of vaccination for public health, many studies have investigated the specifics of relationships that influence these vaccine beliefs. The spread of vaccine beliefs through our personal networks provides a great case study of this process.

As we saw in the last chapter, our personal networks can influence our beliefs, and vaccine attitudes are no exception. A 2021 meta-analysis combed through the results of eleven different studies that analyzed the vaccine beliefs of the participants as well as the vaccine beliefs of those in their personal networks. The conclusion drawn from the analysis of these studies was that social networks do matter: Positive vaccine attitudes in one's network predict positive vaccine attitudes, and negative vaccine attitudes in one's network predict negative attitudes.[71] Studies focused solely on the COVID-19 vaccine also found that the vaccine attitudes in a participant's network significantly predicted their own vaccine attitudes.[72] A 2021 study found that people generally trusted their personal networks: 61 percent of the respondents who were not vaccine-hesitant agreed that friends and family were trustworthy when it came to the COVID-19 vaccine, and a slightly smaller proportion of the vaccine-hesitant respondents (51 percent) trusted their friends and family about the vaccine, but this was still significantly higher than the 39.6 percent who trusted government officials about the vaccine.[73] Generally, those who had lower trust in medical professionals also expressed higher vaccine hesitancy and lower trust in any information source.[74]

COVID-19's political nature explains the partisan divide over the COVID-19 vaccine. A longitudinal study found that Democrats expressed positive sentiments toward the vaccine over six months, while Republicans expressed negative sentiments that became even more negative with time.[75] These conflicting beliefs are consistent with findings that homogeneous network structures can reinforce our identity strength and the consequent impact on our identity-related attitudes. Vaccination itself can create in-group and out-group dynamics, as experiments have shown that participants who are vaccinated are less generous toward those who are not vaccinated while playing a game.[76]

When vaccination becomes politicized, vaccine attitudes can correlate strongly with political identification. The COVID-19 vaccine has become so partisan that state vaccination rates correlate with their 2020 election results more than they correlate with their 2000 election results.[77] One study did find that provaccine messages from Republican politicians did slightly increase Republican intentions to get vaccinated.[78] However, one of the study's coauthors admits that even if Republican politicians had started openly supporting the vaccine, it may not have made a large difference. This is because attitudes on how to respond to the virus may have already been formed rather strongly in the months before the vaccine was available.[79] This could explain why President Trump's public endorsements of the vaccine may have been too little, too late in terms of changing this partisan split on vaccine attitudes.

In my own research, I found that social network composition predicted attitudes toward both COVID-19 and general vaccines.[80] Specifically, when at least half of one's close network got the COVID-19 vaccine, people were twice as likely to get the vaccine themselves compared to those who had no one in their network who got the vaccine. I also found that people with more Democrats in their personal network were more likely to get the COVID-19 vaccine and had more positive attitudes toward the COVID-19 vaccine. Conversely, having more Republicans in one's network was associated with more negative COVID-19 attitudes. In addition, I found that having a greater proportion of Democrats in one's network predicted having significantly more positive attitudes toward vaccines in general (i.e., having more confidence in the safety and effectiveness of vaccines). Conversely, having more Republicans in one's network predicted having more negative attitudes toward vaccines in general.

Friends and family have regularly been shown to predict various attitudes, including those on vaccination, but I wanted to analyze whether certain types of relationships had more influence than others. For example, would a friend who is vaccinated have a more significant impact than a family member? After all, we choose our friends but can't choose our family. But perhaps our family and close friends already have vaccine attitudes similar to ours. I found that the more distant ties, or "non-kin others" (people who are not close friends or family, such as neighbors or coworkers), were especially likely to predict vaccine attitudes. Specifically, among unvaccinated, non-Democratic connections, the non-kin others had the strongest negative impact on vaccine attitudes. A person whose network had more non-kin others (neighbors, coworkers, and other people they knew) who were also unvaccinated and not Democrats was particularly likely to have more negative attitudes toward vaccines. It could be that the attitudes of these weaker ties were most predictive of the participant's own vaccine attitudes because many of the other social connections were already well established.

Not only do we see positive relationships between networks and vaccine beliefs, but also we can trace the impact of these networks' effects over time. For example, a 2017 study tracked the social networks and vaccination behavior of Harvard students during the H1N1 influenza pandemic. The researchers found that these students were more likely to vaccinate when their friends did and that this effect spread throughout the broader network. They used the term "dueling contagion" to describe the spreading of vaccination behaviors as opposed to the spreading of the flu virus.[81] A 2021 study found that higher levels of conspiratorial thinking were associated with lower levels of vaccination intention.

However, when participants believed those close to them approved of vaccines, the conspiracy theory mentality no longer predicted vaccination intent. This shows the power of the attitudes of those close to us, even when we account for relevant personality factors.[82] Even if we have a proclivity for believing conspiracy theories, we are still heavily influenced by the actions (and perceived actions) of those around us.

Using a clever study design, researchers experimentally evaluated whether homogeneous networks would produce a more efficient spread of health behaviors.[83] All participants joined an online health community, in which all members posted their own exercise activity and diet. One group of participants was randomly assigned to a health community that mostly shared their characteristics, including age, weight, and gender. The other group's online community was random and diverse. Thus, the researchers could experimentally see whether a more homogeneous network would have a greater influence on health behaviors. The study revealed that those in the more homogeneous network adopted health behaviors significantly more quickly than those in the heterogeneous network. People are influenced by others, but this effect is amplified when they can relate more to others. Social influence can be so strong that simply sharing the information that most people intend to get vaccinated can increase vaccination intention.[84] However, this positive effect was mostly found in those who were uncertain. If someone's mind was firmly made up one way or another, the intentions of their social group had minimal impact on their own intentions. Furthermore, another study found that vaccination intentions did not change much in a sample of people aged eighteen to thirty when they were told most young people were vaccinated.[85] Social influence from our networks can have a powerful effect on our behaviors and beliefs, but this process is strongest when the social influence comes from those most like us and those closest to us.

A longitudinal survey tracked the COVID-19 vaccination intentions of a group of Polish citizens over time. About 27 percent of those who said they did not want to get a vaccine did end up getting vaccinated a year later, and 57 percent of those who were not sure about getting vaccinated in the first survey also got vaccinated a year later; the remaining 16 percent were planning on getting the vaccine. What changed their minds? Well, among those who said they did not want a vaccine, 50 percent said their own health and safety was the reason for changing their mind, while desire for travel was the reason for 26.6 percent. Among those who were unsure about getting the vaccine, 69 percent said their own health and safety was the reason for ultimately getting the vaccine, while 12.6 percent said it was their desire to achieve social safety and herd immunity.[86] Another longitudinal survey found similar results, as 21 percent of those who said they would not get the COVID-19 vaccine or who were hesitant about getting it ended up changing their mind. Interestingly, 17 percent of those who changed their mind said they did so because a family member convinced them it was the right decision. This was compared to 10 percent who said a health care provider influenced them and 5 percent who said a close friend influenced them. Of those who changed their mind, 52 percent admitted that they learned or heard something that persuaded them to change their mind.[87] As we've seen throughout the book, we often don't like to admit how much impact social influence has on our attitudes, so this number may be even higher. However, I am a little surprised that this many people even admitted an external force impacted them.

If those opposed to vaccination are having their beliefs reinforced in their echo chambers, what is the best way to present provaccine information? As we learned in the previous chapter, forming an emotional connection is critical to depolarization. This is why depolarization can occur among one's close connections, but exposure to the opposing viewpoints of online strangers rarely changes someone's mind. Consistent with these findings, research has shown that simply presenting antivaccine parents with facts about vaccines will not change their minds.[88] Instead, if you connect to antivaccine people in your network with empathy and respect, they are much more likely to actually listen to your message. This is quite similar to the way vaccine advocacy organizations communicate. One study that analyzed the communication styles of these provaccine groups found that their strategies included "communicating with openness in an evidence-informed way; creating safe spaces to encourage audience dialogue; fostering community partnerships; and countering misinformation with care."[89]

Eve Dubé has confirmed the effectiveness of such strategies in her research. She works at the Quebec National Institute of Public Health, where she studies different social factors that contribute to disease prevention. In an interview with *The Atlantic*, she notes that "being harsh or not listening" is "not helpful." As I mentioned earlier, there are few people who are stridently opposed to vaccines, with many more adopting a less harsh stance. Dubé states, "Most of the people who have doubts and concerns are feeling unsure, perhaps because they've heard some troubling stories about injuries and illnesses believed to be linked to vaccines, or because they might believe that vaccines are being pushed on patients by the pharmaceutical industry." She suggests that we ask open-ended questions to "get at where this is coming from." Finally, she recommends applying personal narratives to connect with the other person and to try to relate to them.[90] Her advice seems to follow my own steps, set out in chapter 3, for listening and relating to the person you wish to have a productive conversation with. For example, if someone is concerned about child vaccination and you are a parent as well, you can approach the conversation as an attempt to figure out what is best for your children. Relating to a shared identity can help reduce those initial barriers.

Of course, these types of communication strategies are going to be far more effective if the person we are trying to reach actually cares what we think of them. Sociologist Brooke Harrington argues that simply being nice won't matter if our goal is to reduce misperceptions about vaccines. People care about validation from the groups that matter to them, which sociologist Robert Merton called "reference groups."[91] If you are not within someone's reference group (e.g., their church, family, or political group), then your opinion of them may not matter all that much. In fact, if you belong to an opposing group, your disapproval of them may only serve as a banner of pride. Thus, Harrington doesn't think it is useful to go around being nice to everyone just to increase vaccination rates among those who are extremely vaccine-hesitant. Instead, it may make more sense to identify influential members of groups that overlap with those who are skeptical of vaccines. If these influential members can be convinced to reach out to the vaccine-hesitant, this will be far more powerful than a random encounter with a stranger who is being nice and opposing their views.[92]

Indeed, contemporary social science finds that these reference groups matter for vaccination attitudes as well. A 2021 study had Democratic and Republican participants view messages from either Democratic or Republican leaders who were advocating getting the COVID-19 vaccine. Unvaccinated Republicans who saw Republican leaders endorse the vaccine expressed a 5.7 percent higher intention of getting the vaccine compared to a control group that did not see any vaccine endorsement. Conversely, Republicans who saw a Democratic leader endorse the vaccine reported more negative attitudes toward the vaccine as well as a lower likelihood to recommend the vaccine to others.[93] Another 2021 study investigated whether unvaccinated Christians would express more positive vaccine attitudes if shown a religious leader endorsing the vaccine. Consistent with the previously mentioned study, Christians who saw someone from their own in-group endorse the vaccine expressed a higher intention to get the vaccine and a higher intention to tell their personal networks to vaccinate.[94]

While connecting over shared identities can certainly help, it is important to remember that people are certainly more complex than their identities. Dr. Kimberly Manning, an Emory University professor of medicine, spent time reaching out to those in her community who had doubts about the COVID-19 vaccine. She contends that being kind and avoiding shaming can go a long way toward connecting with people, and the evidence we have covered thus far certainly supports her personal experience. She shares her thoughts on connecting to people in her community:

I'm a Black American who is connected through a lot of people in my community. It started to make me tired to keep reading these reasons why people thought that Black people wouldn't get vaccinated. Specifically, I've even had some personal frustration just with the use of the word "hesitancy." Because, to me, to be hesitant suggests that you're afraid. Minority people are not a monolith. And every individual person has different reasons why they feel the way that they do. I started recognizing, after I would have conversations in the community with people, that the simple question of asking people, "Why? What is your biggest concern?" And listening to their answer and then having a respectful conversation—it goes really far.[95]

Black Americans have legitimate concerns regarding the health care system, stemming from their history of being unethically experimented on by science and health professionals (as during the infamous Tuskegee syphilis study[96]). So listening to their concerns and not shaming them can go a long way if your goal is to help them see why a particular medical intervention will help them. Beyond mutual respect and shared identities, there are specific messages that have been shown to be more effective when having conversations with those who have doubts about vaccines.

We previously learned that discussions can be more effective if they are reframed to match the other person's concerns and values. I summarized this as "Reframe," the third step in having more productive dialogue, in chapter 3, and reframing our conversations can help us address vaccine hesitancy. Boston University political scientist Matt Motta is one of the world's leading experts on identifying the social and political determinants of antivaccine attitudes. When I asked him about some of his recent research on vaccine messaging, he said: "We're doing what some call 'message matching': the idea that people are psychologically more open to changing their minds when counter-attitudinal information is presented in a way that *affirms* their previously-held beliefs in some area related to (but conceptually distinct from) the misinformation itself."[97] So if someone is concerned that "foreign substances" in the vaccine can make them sick, then you can describe how the virus itself can wreak all sorts of havoc on their body. Then you can describe how this vaccine will help protect them from that harm. If we want to have a positive conversation about a hot-button issue with someone, we need to meet them where they are and address their concerns. While personal networks clearly play a large role in the formation of our beliefs, they are not the only type of network we encounter. The social networks from social media also have a unique role in the spread of misinformation. Again, we will use vaccine beliefs as a case study illustrating how social media can help spread misinformation.

HOW DO VACCINE BELIEFS SPREAD
THROUGH SOCIAL MEDIA?

I remember looking through Twitter when President Biden tested positive for COVID-19 in July 2022. Many replies to conservative public figures sharing the news mocked the president for getting vaccinated after he got sick. This simplistic view of the vaccine in binary terms, as I have already discussed in relation to the mainstream media, was apparent on social media as well. It was further spread by the mechanism of social media, which reward sensationalism over nuance. Tweeting "vaccines don't work" is quick and punchy and fits into a hashtag (#vaccinesdontwork was a real and quite popular antivaccine hashtag on Twitter). In contrast, a tweet that says "The vaccine reduces severe symptoms and the likelihood of death from COVID-19, but it doesn't completely prevent getting the virus or spreading it," just doesn't sound as catchy. Researchers have also found that bots have amplified both pro- and antivaccine messages on social media, fueling false equivalency and conflict.[98]

When President Biden tested positive for COVID-19 again just a few days after recovering, this further added to the online chaos. This second positive test was likely due to the treatment Paxlovid, which can create a mild rebound in the virus and lead to another positive test.[99] Of course, this nuanced point about the properties of a new treatment is not as exciting compared to dunking on the president for "getting COVID-19 twice" after his vaccine and booster shots. It was misleading to say that he got COVID-19 twice, and it would have been more accurate to focus on the weird quirk of the Paxlovid treatment (this rebound also occurred with Dr. Anthony Fauci).[100] But as we learned in the previous chapter, social media incentivize sensationalism, and echo chambers can easily form. Once people become trapped in an algorithm that continues to show them a skewed version of reality, polarization and misinformation can easily continue to spread. As we have seen, certain strategies can improve our likelihood of productive in-person conversations and meaningful connections, but interacting with other people on social media can be a very different experience.

How do social media treat vaccine misinformation? We already learned that social media make it easy for misinformation in general to spread, so we would not expect any exception for vaccine misinformation. A 2022 survey of health care workers found that 73 percent believed that social media spread misinformation and negatively influence patient care. Specifically, 72 percent believed that social media misinformation negatively impacts patients' decisions to get vaccinated.[101] It is difficult to measure exactly how much true or false information is on the internet for any one subject. However, researchers have tried different techniques to measure how much pro- and antivaccination information is online. A group of researchers led by data scientist Neil Johnson studied the spread of pro- and antivaccination views while analyzing the activity of 3 billion Facebook users. They found that about 6.9 million users shared provaccination views compared to 4.2 million users who shared antivaccination views. However, the antivaccination users were spread across significantly more clusters (a cluster is a Facebook page where many members interact with one another). The researchers also found that antivaccination views were increasing much more quickly than provaccination views and that antivaccination users interacted much more frequently with undecided individuals. They suggested that antivaccination views could "dominate" in a decade if this trend continues.[102] A 2021 study that investigated over 4 million vaccine-related tweets on Twitter yielded similar results: 34 percent of these tweets referenced the COVID-19 vaccine positively compared to 25 percent that referenced it negatively.[103] A longitudinal Twitter study compared vaccine-related tweets before and during the COVID-19 pandemic and found that vaccine opposition increased by 80 percent on Twitter once the COVID-19 pandemic reached full force.[104] As we learned in the previous chapter, bots can help amplify certain messages, and research has found plenty of vaccine misinformation amplified by bots.[105] It doesn't matter if minds are changed; simply adding to the chaos and reducing trust in institutions can be a sadly achievable goal.

Of course, humans are still the main drivers of misinformation, and antivaccine influencers can reach quite large audiences with their social media posts. Researchers found that just twelve of the top antivaccine activists accounted for 65 percent of the antivaccine misinformation on social media![106] I will discuss the big business side of this later in the chapter, but you can probably imagine that such a massive reach is good for their bank accounts.

It is hard to know exactly how many people were influenced by all of the vaccine misinformation on social media. One study did find that skepticism toward the COVID-19 vaccine was predicted by exposure to online misinformation and reliance on social media for news.[107] However, this relationship was significantly reduced for participants who reported higher levels of news literacy.[108] Furthermore, a 2020 study investigated whether President Trump's antivaccine tweets (before the COVID-19 pandemic) increased vaccine hesitancy among his supporters who saw them. The researchers found that his antivaccine tweets did increase antivaccination attitudes among his supporters; however, liberal voters who saw these tweets did not become more provaccine.[109] As we learned in chapter 3, listening to cues from political leaders and people in our own group can be particularly effective.

An intensive longitudinal study tracked 2.6 million Facebook users' posts about vaccination for over seven years. The results, published in 2018, showed that vaccine echo chambers did emerge and that vaccine attitudes did become more extreme over time. Most of the Facebook users in the study consumed information that was either in favor of or opposed to vaccines.[110] However, a 2020 experiment found that pro- and antivaccine participants did not become more polarized when they were exposed to a series of online messages that supported their current position.[111] Echo chambers can go beyond simply exposing people to the same message over and over again. Furthermore, many people already seek out information that confirms their views, limiting the impact of social influence. For example, the Vaccine Adverse Event Reporting System (VAERS) has been exploited by antivaccine activists to spread vaccine misinformation. VAERS is comanaged by the FDA and the Centers for Disease Control and Prevention and is designed to help collect self-report data about the potential side effects of vaccines. Because this website relies on unverified stories, it is easy to share false or misleading information about a vaccine if one has an agenda.[112] For example, in one account in VAERS, a person reported that their grandmother died just hours after getting the COVID-19 booster shot. This story was picked up by Tucker Carlson of Fox News and used to question the safety of the vaccine. However, it was irresponsible to assume the vaccine caused the death of this elderly person without additional evidence.[113]

One study does offer a silver lining for the spread of vaccine misinformation: It may have little impact on actual behavior regarding vaccines. When participants repeatedly saw false information about vaccines, they did start to believe this information was more accurate (consistent with the illusory truth effect we learned about earlier). However, even when they thought these stories were more accurate, they did not report being significantly less likely to get the COVID-19 vaccine.[114] Psychologists Ciara Greene and Gillian Murphy did find, however, that showing false information about the problems of a COVID-19 contact tracing app did reduce people's willingness to download the app. We can see links between fake news and behavior, but it is very complicated to link exposure directly to behavior.[115] Though it's hard to quantify exactly how much antivaccine misinformation can change attitudes and behaviors, one study did estimate that exposure to false information about the COVID-19 vaccine reduced one's likelihood of getting a vaccine by 6 percent.[116] Even if antivaccine misinformation has a small impact on people's beliefs, this could still be a significant problem, given the sheer volume of false information online. Additionally, antivaccine rhetoric is predicted to grow rapidly in the next decade, since it is so emotionally evocative and sensationalist—both properties that social media strongly reward.[117]

Mere exposure to misinformation can increase skepticism about certain topics, even if the change in behavior is more subtle or not immediate. A series of experiments exposed participants to both the scientific consensus view on climate change (global warming is at least partially influenced by human activity) and the contrarian view (humans have no impact on global warming).[118] When participants saw both sides presented to them in the experiment, they reported a lower estimate of the weight of the scientific consensus, regardless of the lack of adequate expertise in the contrarian sources. Simply exposing people to the idea of a potential "balance" of ideas can make them more skeptical of the scientific consensus; however, the researchers found that when they explicitly focused on the scientific consensus and the weight of evidence in detail, the reduction of belief in the scientific consensus disappeared.

Psychologist David Rapp, one of the authors of the climate change study, makes an important point about the issues with false balance online and in the media: "Climate change is a great case study of the false balance problem because the scientific consensus is nearly unanimous. If 99 doctors said you needed surgery to save your life, but one disagreed, chances are you'd listen to the 99. . . . But we often see one climate scientist pitted against one climate denier or down player, as if it's a 50-50 split."[119] This finding is related to what we learned previously about the illusory truth effect. Just hearing something over and over can increase the likelihood that we think it can be true. Rapp also agrees that this can make the issues of false balance even worse: "People think anything they can easily recall is likely to be true. If that's false or misleading information that the media parroted or gave a platform to, the person will still give weight to it if it crops up again later because they've heard it once before."

Rapp recommends continuing to remind people about the consensus view, as his study shows that can help. Of course, a presentation that showcases the weight of evidence leading to a clear consensus may not be as exciting to watch as a heated debate between two sides. Unfortunately, since the news media and social media are incentivized by clicks and attention, framing an issue with a clear scientific consensus as more of a battle can be more compelling. In fact, research shows that knowing less about a scientific issue is associated with higher confidence.[120] It's easy to see how the incentive structure of social media accelerates this process. A study found that simply sharing such articles on social media made people feel like they know more about the topic, regardless of whether they have even read the article they shared![121] Furthermore, polemical claims such as "vaccines are bad and don't work" are catchy, while addressing the nuances of the risks and rewards of vaccines just doesn't garner as many clicks. Given how quickly social media can spread misinformation, are there any evidence-based techniques that can help?

WHAT CAN HELP COMBAT MISINFORMATION ONLINE?

Social media do appear to accelerate misinformation in certain areas, but as we learned previously, it's very difficult to quantify the impact they have overall. The difficulty in measuring social media's impact also makes finding evidence-based solutions quite challenging. However, I will present three broad ideas that can help with social media's misinformation problem: improving fact-checking, adding friction, and increasing data transparency. These apply to online vaccine misinformation and also to online misinformation in general.

Fact-Checking

With all these social biases we have to protect our beliefs and identities, you may wonder if fact-checking even works. The short answer is, Yes, they work! The longer answer is, yes, they work, but many factors contribute to their effectiveness. Overall, meta-analyses have shown that exposing people to reliable and independent sources (fact-checks) does help improve their accuracy when rating various sources of information. However, fact-checks are less effective when people are more partisan and when they do not trust the source of the fact-check. Democrats are more likely to support fact-checking than Republicans, and this relates to the partisan differences in institutional trust, which we will get to shortly.[122] A Pew Research Center survey found that, overall, 71 percent of American participants (57 percent of Republicans and 87 percent of Democrats) agreed with social media companies fact-checking politicians. Furthermore, both political groups agreed that it is better for social media companies to be more accurate in labeling false information, even if it takes more time.[123] Partisanship is an important factor to consider, but there is overall support for fact-checking practices online.

When evaluating fact-checking effectiveness, there are other factors to consider beyond political beliefs. For example, fact-checks are often less effective when the study design resembles more of a real-world scenario of fact-checking compared to what you would see on social media.[124] Furthermore, the effectiveness of fact-checks fades over time. A large study investigated how effective fact-checks for COVID-19 misinformation were over time for thousands of participants from the United States, Great Britain, and Canada. While reading factual articles about COVID-19 (the fact-checks) did reduce misperceptions initially, the researchers found that such effects lasted only for a few weeks.[125] As we found in chapter 3, repeating information is often necessary for someone to update their beliefs. Of course, a change in attitudes does not always predict a change in behavior. Another COVID-19 fact-checking study found that having people read fact-checks reduced their false beliefs regarding the virus. However, being exposed to these fact-checks did not increase vaccination rates.[126] Research suggests that fact-checks are more effective when they are provided early on, offer an alternative explanation for the myth, include a credible source, and repeat the correction at least twice.[127]

Fact-checking can take a variety of formats beyond just sharing a fact-based article or summary. You don't have to necessarily be an expert if you cultivate a group that celebrates evidence-based inquiry. For example, several moms created a group on Facebook that encourages good-faith debate about vaccines to correct any misperceptions.[128] Kate Bilowitz, one of the group's moderators, claims that the group's combination of empathy and fact-checking has changed the minds of hundreds of vaccine-skeptical people.[129] If you are an expert in an area, sharing your knowledge and correcting any popular myths in your area will help! One study has shown that people do trust scientists they know and that scientists can use their social media effectively to conduct science outreach and correct myths. The authors referred to trusted scientists among their local networks as "nerds of trust."[130]

Scientists publicly sharing their expertise is characterized as science communication, and I am certainly a big fan of this practice. There are now many diverse scientists who can reach millions of people through social media. Additionally, there are more and more opportunities for scientists to practice their communication skills. I've enjoyed attending various science communication conferences over the years. One simple, fun, and effective activity at these events is the Jargon or Awesome game. Participants summarize their main research topic in one minute, and the crowd holds up signs that say "Jargon" or "Awesome" as they do so. It is a great way to get real-time feedback on whether your words are actually getting through to a diverse group of people. I led a science communication workshop at a conference of the American Society for Virology and had the scientists in the room play Jargon or Awesome; they seemed to enjoy it as well! When scientists can succinctly explain their research to a broad audience, it greatly expands the number of people who can learn about their work.

Widespread fact-checking on social media would, of course, present several logistical challenges. Social media companies would need to constantly monitor the false information on their platforms and deploy resources to generate a fact-checking force. As mentioned earlier, fact-checking is less effective when people do not trust the fact-checkers. Who fact-checks the fact-checkers and which fact-checkers companies decide to use are crucial factors as well. There are a few ideas that can help with these challenges though. For example, research has shown that layperson crowds strongly agreed with professional fact-checkers when analyzing how trustworthy different news publishers are.[131] A study comparing Wikipedia articles found that those with ideologically diverse editors are of higher quality than those with ideologically homogeneous editors.[132]

Instead of relying only on a few fact-checking organizations, platforms could have crowds evaluate the trustworthiness of individual articles. Research has found that crowdsourced fact-checking influenced participants' decisions as a helpful guiding cue, but they still relied on their own evaluation and additional research.[133] A 2021 study found that politically balanced groups of laypeople provided accuracy ratings for news headlines that closely aligned with those of professional fact-checkers. The researchers concluded that crowdsourced fact-checking could effectively complement traditional fact-checking methods.[134] While crowdsourcing offers some promise, it would have to account for potential political bias within the crowd, since an article's perceived veracity could depend on how many people from each political group rate it. Also, crowdsourced fact-checking is significantly more accurate when crowds have high political knowledge.[135] Other scholars have suggested we reward social media users who regularly share reliable information with a badge, but this would work best if users trust those who are bestowing such rewards.[136]

Another issue with fact-checking is that pressure to spend resources on fact-checks is rarely proactive. A deadly pandemic forced policymakers and the public to put more pressure on social media companies to consider their role in the spread of misinformation. However, misinformation about a danger that occurs more gradually, like climate change, has been less of a focus. The threat of global warming is more psychologically distant than a pandemic that is killing people right now, so that may explain why there has been less pressure to fact-check climate misinformation. Similar pressure may be needed to get social media companies to place more emphasis on fact-checking disinformation regarding climate change.[137]

More research needs to be done to understand how to properly and ethically administer fact-checks, but the silver lining here is that presenting people with accurate information does increase the likelihood they will update their beliefs in a way that is consistent with the best evidence available. The major advantage of fact-checking is that it doesn't attempt to remove what can be discussed online; instead, it's all about sharing more expert knowledge with users on the platform. Major social media companies like Facebook, Twitter, and TikTok have tested various fact-checking programs, and as I write this book Facebook and X (formerly Twitter) have moved toward crowdsourced fact-checking instead of fact-checking labels from organizations.[138] As mentioned in the poll above, most people generally agree that fact-checking is worthwhile, but the major challenge making sure it is done well. Beyond fact-checking, most Americans even agree that social media companies should engage in some content moderation, even if it also reduces some freedom of expression.[139] Whether social media companies implement content moderation or fact-checking, it is crucial that they be as transparent as possible about their methods in order to increase trust and their perceived legitimacy.

Friction

As we learned in the previous chapter, misinformation is more likely to spread quickly on social media. If sensationalist and fake content spreads so quickly online, one option is to adjust the algorithm and somewhat slow down the transmission of viral content. This idea of slowing down information online has been referred to as adding *friction* and is supported by several prominent experts who study misinformation on social media.

In chapter 4, we saw how the internet creates an information overload, and this structure allows low-quality information to flourish. Computer scientist Fil Menczer argues that adding friction to slow down virality may be one of the best ideas we have for combating misinformation online. Adding friction doesn't have to deal with the challenges of moderating content or even adding fact-checks. It simply changes the speed at which information travels online. He proposes that social media companies force users to solve short puzzles before posting something online, slowing down the process. This may curb the influence of bots as well. Other ideas include requiring small payments to post (which would also reduce the need for ads or profit from user data), but, of course, this would be a major disadvantage to those with less money. Social media can also implement "proof of humanity" in the form of some type of documentation that also curbs bots. In addition, this requirement may force people to think twice about harassing others online, as they are not 100 percent anonymous and can be identified if they behave in a criminal fashion.[140] Adding friction may reduce the total amount of information online as well as slowing it down. Menczer argues that "with less information people would be able to pay more attention to what they see. It would leave less room for engagement bias to affect people's decisions."[141]

Social media companies can add friction without changing their algorithm by asking people to reflect on what they post before they post it. Psychologist Lisa Fazio has found that simply asking people to explain why they think a headline is true makes them less likely to share false information.[142]

Renée DiResta, former research manager for the Stanford Internet Observatory, and Tobias Rose-Stockwell, a technology writer, also advocate friction. They agree that high-velocity information facilitates the spread of misinformation online. They acknowledge that lower velocity can still allow posts to go viral online, but the process will involve more checks and balances. DiResta and Rose-Stockwell suggest that identifying content that is spreading quickly and then fact-checking it could help mitigate the spread of misinformation, although this proposal involves all of the challenges of implementing fact-checking programs within social media. They also point out that accuracy nudges could help, as research has shown that simply asking people if they were sure they wanted to share their post before sharing it reduced the sharing of false and sensationalist content.[143] However, subsequent research suggests that these accuracy nudges may have little effect on decreasing misinformation overall.[144] As usual, more data are needed to figure out the conditions in which such nudges are most effective.

I was fortunate enough to see Facebook whistleblower Frances Haugen speak at a conference at the University of Cambridge in 2023. She gave a great talk on social media's role in spreading misinformation. When she was asked if there were any easy-to-implement technical solutions that could help reduce the negative impact of social media on our society, she suggested that social media companies simply decrease the scrolling speed between posts if people are spending too much time on the platform. A simple nudge that may eventually make a user choose to do something else with their time does seem like a good idea, even though it may reduce overall engagement by a small amount.[145]

Of course, the details would need to be worked out for any strategy that could potentially reduce virality and subsequent misinformation. Importantly, companies may not want to implement any strategy that reduces overall engagement and overall profits. However, there is bipartisan support for algorithmic regulation, so it is possible we could see social media companies being forced to slow down their sensationalist content.[146] A major benefit of focusing on the properties of the algorithm is that it is content-agnostic and can be adjusted to treat interactions more like real life, where certain conversations do not have a megaphone compared to others. It's certainly worth investigating further, especially if we could learn more about how these algorithms work. This leads me to my last suggestion that can help: data transparency.

Data Transparency

It would be great to know exactly how social media companies' algorithms work in order to evaluate their exact role in spreading false information. However, these companies are pretty cautious about what data they share with independent researchers. They argue that they don't want their data to fall into the wrong hands, as happened with the Cambridge Analytica scandal discussed in the previous chapter. This is a legitimate concern, but perhaps a proper vetting process would allow for greater access to data.

Historically, Twitter has been one of the most easily accessible data sources for academics, so that is why it has been such a popular subject for research studies.[147] Twitter's application programming interface (API) made it easy to retrieve and analyze public tweets and was a great tool for independent researchers. However, things change quickly in the social media world, and as of 2023, Twitter no longer offered free access to its API.[148] YouTube, Facebook, and TikTok have been quite restrictive regarding independent research and access to their data.[149] TikTok's lack of transparency is particularly concerning after a 2022 study revealed that nearly 20 percent of the search results for major news events on its website contained misinformation.[150] Companies are open to collaborating with select academics, but this, of course, limits the types of questions that can be asked.

Even when Twitter had a more open approach to data sharing, there were still important limitations. For example, researchers did not have data on which users saw specific tweets or even how many people saw a certain tweet. Political scientist Joshua Tucker worked with Facebook on a study using data related to the 2020 U.S. presidential election. He admits his partnering with Facebook allowed him to have access to data that weren't available for the 2016 election, but he concedes that "regardless of the importance and value of this particular effort, the model—while replicable for other focused studies—is by design not something that is scalable to meet the ongoing needs of the larger research community."[151]

Tucker and law professor Nathaniel Persily say that YouTube and TikTok are still far behind in their data-sharing programs compared to Twitter and Facebook. They note how independent academic researchers have to rely on the goodwill of these companies to access their data. Again, social media companies do not have much of an incentive to share data that could potentially make them look bad or expose them to accusations of violating user privacy. Tucker and Persily argue that a federal agency should be empowered to help accelerate the process of getting giant tech companies to share some of their data with academics. A team of leading misinformation researchers cowrote an article in 2022, in which they conclude that we need lawmakers to push for more access to social media data in order to keep users safe.[152]

As I write this book, some progress is being made, as social media companies are becoming more open to sharing their data with academics for independent analysis. A few dozen of the world's leading misinformation researchers shared what specifically they could do with additional access to data in an article published in the *Harvard Kennedy School Misinformation Review*.[153] These scholars argue that greater data transparency and access are crucial because they allow researchers to measure the true reach of misinformation, understand user behavior differences, and conduct unbiased randomized controlled trials. Furthermore, access to detailed user data helps researchers study how different demographics experience social media, detect manipulation by bots and trolls, and assess the effectiveness of platforms' actions against misinformation. Finally, social media data should belong to the people who created the data, as they represent a type of digital library of humanity. It's important to keep in mind that even if fact-checking, friction, and data transparency are all implemented fully, they still do not address why people are motivated to seek out alternative sources of information online to begin with. Why do people distrust scientific institutions? That is the focus of my last section.

TRUST IN SCIENTIFIC AND MEDICAL INSTITUTIONS

So far, we have covered how personal network structure and social media can influence attitudes toward vaccination. We have also addressed how various strategies can improve productive dialogue and reduce the spread of misinformation through social media. Vaccine misinformation can spread rapidly online, but why do people feel motivated to seek out these alternative sources of information to begin with? To determine how vaccine misinformation spreads, we must focus on an issue central to the spread of scientific misinformation: the lack of trust in scientific and medical institutions.

When we look at survey data on trust in scientific and medical institutions, the results are mixed. A 2022 study by 3M and Ipsos found that trust in science and scientists remained quite high, even after the COVID-19 pandemic.[154] Across seventeen countries (and 17,000 participants), researchers found that 90 percent of participants agreed that they trust science at least somewhat. This is even higher than the 84 percent that responded positively to the same question in 2020. When asked if they trust scientists, the overall positive average was still 86 percent.

Another large 2020 survey asked similar questions across twenty different countries. However, the participants were asked if they had "a lot" of trust in scientists to do what was right for the public. Across all countries, the median proportion of those trusting scientists "a lot" was just 36 percent.[155] A December 2021 survey found that just 29 percent of Americans expressed "a great deal of trust" in both scientists and medical scientists.[156] Additionally, 22 percent of Americans said they did not have too much confidence or they had no confidence at all in them (with 77 percent agreeing they had a fair amount of confidence). Scientists fared much better than elected officials, as only 2 percent expressed a great deal of confidence in the latter (a whopping 76 percent said they did not have much or any confidence in public officials). A 2024 poll found that just 60 percent of Americans agreed that science benefited them.[157]

When looking at the polarized United States, we do see a stark difference in scientific trust by ideology: In 2021, 43 percent of Democrats said they trusted scientists to do what is right for the public compared to just 13 percent of Republicans.[158] Again in 2021, 64 percent of Democrats said they had a great deal of confidence in the scientific community, whereas only 34 percent of Republicans said the same.[159] Democrats also trusted health organizations [such as the Centers for Disease Control and Prevention (CDC) and NIH] more than Republicans. However, there was more bipartisan support for NASA, which is less overtly politicized.[160] As I noted previously, people are more likely to listen to scientific information provided by groups they trust. If Republicans are already more skeptical of institutions as part of their ideology, then they may be more likely to associate more with those outside of the mainstream. Conversely, Democrats who embrace institutions more may have an easier time accepting science as an institution within their ideological stance.

It would be too easy (and incorrect) to simply say this lack of trust in science is unilaterally caused by partisanship. After all, many Republicans still trust their personal doctors and engage with modern technology just like everyone else. Their lack of trust is more about their feelings toward particular institutions than their feelings toward the scientific process. It would also be too easy to simply write off those who don't trust science as uneducated. Many highly educated people will reject scientific evidence when it conflicts with their personal identity. That is why I argue the overly cynical framings that claim large groups of people simply embrace ignorance are incomplete. For example, *The Death of Expertise* is a popular book by Tom Nichols that was published in 2017. He makes some important comments about how experts are not valued in our society and how confirmation bias contributes to this problem. However, he places far too much blame on people's ignorance when he states, "Americans have reached a point where ignorance, especially of anything related to public policy, is an actual virtue." He goes on to say that "to reject the advice of experts is to assert autonomy, a way for Americans to insulate their increasingly fragile egos from ever being told they're wrong about anything."[161] Again, this is an incomplete analysis, and I argue it's also unhelpful. Moving specifically to vaccines, there were mainstream articles that tapped into a similar vein, such as Charlotte Alter's "Nothing, Not Even Hard Facts, Can Make Anti-Vaxxers Change Their Minds," published in *Time* magazine.[162] Several of these articles cited a 2014 study that found provaccine messaging may not change minds.[163] However, the authors of that study even concluded that other types of messaging strategies may be more effective and that future study is needed. Ironically, forcing a single study to fit one's agenda is quite an anti-intellectual act by itself.

Such extreme rhetoric aimed at those skeptical of vaccination became more prevalent during the COVID-19 pandemic only after vaccination itself became a hotly contested political issue. One extreme example of this is how large media outlets published opinion pieces that questioned whether those refusing the COVID-19 vaccine should be denied medical care.[164] This, of course, is an extremely unethical position and would also have ended up hurting many people who were unable to receive the vaccine for various reasons.

Establishing an us-versus-them dynamic is not helpful when trying to explain the merit of one's position. As we have seen throughout the book, polarizing two opposing groups can make people double down and support the side they are on. Writing an article that attacks people who are against vaccines might yield plenty of "likes" and be shared in one's social media bubble. However, such polemical rhetoric is unlikely to change hearts and minds. When attacked, people are more like to become defensive and to retreat to their side, making such antagonistic articles counterproductive. We can have productive discussions with those who think differently than we do by trying to understand where they are coming from. To do this on a deeper level, we can examine the cause of their lack of trust.

Philosopher Maya Goldenberg wrote an excellent book titled *Vaccine Hesitancy: Public Trust, Expertise, and the War on Science*,[165] in which she makes the case that addressing the crisis of trust is vital for understanding vaccine hesitancy. She argues that when members of the general public form their attitudes toward vaccines, they rely on "epistemic trust," which involves trusting someone in their "capacity as a provider of information."[166] In other words, they base this trust on how competent they believe the person to be. Additionally, trustworthiness relies on the perception of the expert's character.[167] If someone doesn't think an expert is honest or has good intentions, then they lose trust. For example, if someone does not disclose a clear conflict of interest, then that would reduce the perception of honesty. Expressing empathy is a final component of trust. When health care providers expressed empathy, their patients had more trust in them and in their advice on vaccination.[168] Several studies have found that people tend to develop trust in persons who belong to their group, including their political in-group.[169] Again, whether we believe information ultimately depends on inter-personal relationships based on mutual trust and on shared identities—and even the interaction of both of these processes.

If someone is already critical of medical institutions, then an antivaccine influencer can focus on this shared interest (and identity) to build trust. Goldenberg makes this point using Andrew Wakefield as an example, as Wakefield took time to listen to the concerns of parents and regularly questioned the medical establishment. He set an example for many of the modern-day vaccine skeptics who scold the health care system for its financial interests while conveniently ignoring their own financial interests in growing their Patreon or Substack presence. As mentioned earlier, Joe Mercola was critical of those taking the COVID-19 vaccine but had no issue selling his own COVID-19 supplements. What I find interesting is how trust can determine how much leeway you might give someone's claims. Mercola's COVID-19 supplement was backed up by only a single non-peer-reviewed article that discussed which molecules might bind to COVID-19's spike protein. In contrast, he found that the mounds of peer-reviewed scientific evidence supporting the effectiveness of the vaccine were not sufficient.[170] This seems inconsistent, but if you trust Mercola and don't trust medical establishments, then his weak scientific evidence gains much more credibility, and the robust evidence from the medical establishments is far less compelling.

The purposeful twisting of facts to fit one's narrative has been recently deemed *malinformation*. Malinformation doesn't need to rely on anything false; instead, it consists of real information that has been cherry-picked to deliberately mislead.[171] This is a common tactic used by antivaccine influencers, as they can claim they are using "scientific evidence" but omit how they are purposely distorting such evidence. When someone is deliberately trying to mislead us, possibly because of their financial incentive to do so, it can be almost impossible to have a productive conversation with them. While there are some effective techniques to improve dialogue, the fact is that some people are heavily disincentivized to change their minds on a topic. If someone is spreading malinformation and/or lobbing personal attacks at those they disagree with, we may not want to waste time trying to have a good-faith discussion with them. With that important caveat out of the way, let's focus on people who might be open to new information but who distrust scientific institutions.

What is driving this distrust of scientific and medical institutions? Heidi Larson, professor of anthropology, risk, and decision science at the London School of Hygiene and Tropical Medicine, published a book titled *Stuck*, in which she delves into the details of vaccine hesitancy around the globe. She describes where vaccine rumors start and how they persist and agrees that trust is fundamental: "Vaccine acceptance is about a relationship, about putting trust in scientists who design and develop vaccines, industries that produce them, health professionals who deliver them, and the institutions that govern them. That trust chain is a far more important level of acceptance than any piece of information. Without these layers of confidence, even the more scientifically proven and well-communicated information may not be trusted."[172] She goes on to write that vaccines will always have some degree of risk and that trust is crucial in the calculations people make to determine whether the risk is worth the reward.

The health care system has legitimate problems that can make it understandable why people have lost trust. I already discussed the legitimate problems with the way health organizations like the CDC handled COVID-19, which further lessened people's trust. Scientific institutions can do a better job at building trust as well. As mentioned earlier, the United States spent tons of money developing a vaccine but *far* less money on applying the social science of messaging and learning why some people may be skeptical of the vaccine. Again, blaming this solely on misinformation misses opportunities to correct the problem.

Research has shown that industry-funded studies can be biased in favor of the companies' products and do not always include research that would be most useful for public health.[173] Ultimately, such studies are vulnerable to bias when health care corporations are incentivized by profit without enough checks and balances. One harrowing example of this was culminated Johnson & Johnson and three drug distribution companies—AmerisourceBergen, Cardinal Health, and McKesson—had to pay $26 billion to settle claims regarding their role in the American opioid epidemic. The distributors were accused of having careless procedures that allowed their addictive painkillers to be obtained too easily and through illegal channels. Johnson & Johnson was accused of downplaying the addiction risk of its drugs in its marketing campaigns.[174]

When we focus on vaccines, we see that corporate greed directly affects who obtains access to lifesaving vaccines. Science reporter Amy Maxmen wrote an excellent article addressing vaccine inequalities that were observed during the COVID-19 pandemic.[175] She first notes how the richest countries had over 75 percent of their populations fully vaccinated against COVID-19 by the middle of 2022. Conversely, the poorest countries had just 15 percent of their populations fully vaccinated by then. She goes on to note how this trend was observed with other treatable diseases, with the wealthy countries being protected much more rapidly. These treatments and vaccines could be deployed to many more people in poorer nations, but intellectual property barriers slow down global adoption. For example, Moderna's mRNA vaccine patent made it harder for countries in Africa to develop and deploy their own mRNA vaccines. While supply chain issues might have had more of an impact than intellectual property issues, the fact is that health care companies like Moderna became fantastically wealthy during the pandemic.[176] When companies are massively profiting from selling treatments or vaccines and poor countries are not receiving anything near equal treatment, it is fair to ask if these companies could have helped more people by taking a slight pay cut. We should not be overly critical of those who do not trust institutions when the institutions themselves have earned some of that distrust.

The failures of institutional leadership and the greed of health care companies seem to bleed into the negative perception of everyday working scientists, who are the ones on the ground collecting data. It's not the scientists' fault that their boss is trying to alter the scientific message for greed or political convenience. I've seen several scientists advocate vaccines on social media, only to be called "paid shills" by various commenters. I've been called that myself! The joke that usually follows is that we are still waiting for these alleged checks from Big Pharma.

Even if many scientists advocate something like vaccines, some people may still prefer alternative medicine because of the money made by the big pharmaceutical companies. However, it is important to note that alternative medicine is a booming business also motivated by profit. For example, the global homeopathy market was estimated to be worth about $18 billion in 2021.[177]

Okay, so there are some legitimate reasons to distrust the medical establishment, and antivaccine influencers have exploited this skepticism (along with parents' concern for their children's safety) to promote their own brands. As long as there is a crisis of trust, there is space for charlatans to sell their own false information. So how can we rebuild trust?

REBUILDING TRUST

Maya Goldenberg concludes her book by describing how scientific institutions can rebuild trust by making an effort to fix their relationships with the general public. I agree with this point because, as I've described throughout this book, social factors are key to the formation and maintenance of our beliefs. Trust is about social relationships. However, there is no easy answer to the question of how we can regain trust. The onus is on the institutions that lost trust to regain it themselves. Still, we can identify a few specific areas that institutions can work on to improve their relationships with the community. I think it helps to categorize how we can rebuild trust on three levels: the micro (interpersonal), the meso (community), and the macro (societal).

Starting with the microlevel, we can think back to the beginning of this chapter when Lydia felt she was brushed off by the nurse after expressing concern about how her baby was reacting to the vaccine. Would Lydia have avoided the antivaccine rabbit hole if the nurse she talked with had met her concerns with genuine empathy and patience? Or perhaps Lydia was already going down that path, and this interaction sent her over the edge. Either way, the point is that health care professionals do have a strong influence on the public. In 2020, most people (85 percent) still trusted their personal doctors about health topics in general—and even about something as political as the COVID-19 vaccine.[178] Lydia isn't the only person who felt disrespected by a health care professional, as studies have shown that vaccine-hesitant parents regularly reported feeling dismissed by primary care providers.[179] Health care providers are often overworked already, but perhaps they could be given additional training on how to better communicate vaccine concerns (or they could be provided with easily accessible resources if they encounter vaccine-hesitant patients).

This process requires the expert in question to learn what the individual is concerned about and then address their needs instead of assuming to know what is best for them. Again, the focus here shouldn't be just on correcting misinformation. There first needs to be a genuine effort to connect with and listen to people in the community. The first step toward a productive conversation is establishing mutual respect. Improving interpersonal relationships takes time and energy, but if scientists and health professionals make a concerted effort, I believe we will see a major improvement in trust. Researchers at Yale University helped create a video to train health care providers to better communicate with their patients about vaccination concerns. One of the key actions recommended by this video is to "express empathy by validating the patients' concerns."[180]

In addition to expressing empathy, another useful tactic, and one that has been well received, is admitting (and embracing) any existing uncertainty or ambiguity. In fact, a 2024 study found that scientists who expressed uncertainty while discussing new scientific findings were protected against a loss of trust if the evidence changed later.[181] When an expert admits that they do not know something during an interview, it may not yield as many clicks as if they had confidently made declarative (or even polarizing) statements that could be cropped into attention-grabbing sound bites. However, embracing uncertainty when warranted is likely to increase their credibility and trustworthiness overall. In an interview in 2024, Dr. Francis Collins admitted that public health officials did not pay enough attention to the social science of communication during the COVID-19 pandemic. Specifically, he said he wished he had more explicitly addressed the uncertainty around public health recommendations during the pandemic because scientists were still learning about the virus.[182] I agree with Dr. Collins that embracing the evolving nature of scientific knowledge about the virus (and subsequent recommendations) would have increased the credibility of and trust in public health officials.

Beyond improving communications from major health and science organizations, scientists can have a significant impact on the microlevel by sharing their expertise with people they know. We can think back to the "nerds of trust" study that described how scientists are trusted in their local networks on social media. Republican or politically moderate scientists could have a particularly strong impact in rebuilding trust among conservatives through public outreach, since they can relate to a shared identity.[183] Not only can scientists make an impact on an individual level, but also they can share their knowledge with larger groups of people in both online and offline settings and thus reach a wider audience.

Moving on to the mesolevel, we can identify how individuals, communities, and institutions can work together in order to rebuild trust. Political scientist Robert Putnam argues that individuals can learn to trust other people by joining social clubs and engaging in their local communities. This social connectedness and learned trust in one another are vital for producing trustworthy institutions.[184] When focusing on how to rebuild trust in scientific institutions, I found that biologist Jonathan Berman makes an important point in his book *Anti-Vaxxers: How to Challenge a Misinformed Movement*.[185] He sees community-based strategies as beneficial for connecting with people who hold different attitudes about vaccination. Some of these strategies involve health officials working with popular figures in their local community (such as pastors) to share health information, but in addition to sharing information, these community leaders also listen to the concerns of different groups. When people hear messages from prominent individuals in their group, it can be more effective, but success also relies on putting in the work to establish various community relationships. A research study led by sociologist Rashawn Ray that investigated the COVID-19 racial disparities in Detroit suggested that public health officials should set up testing and vaccination sites in local community pillars such as churches, barbershops, and hair salons.[186] Not only does this make access to medical care more convenient, but also working with the community directly could reduce medical distrust. If more people learn about what these health or science organizations do from those who are actually doing the work on the ground, then they might be less likely to demonize and distrust them. One study found that conservatives were more supportive of the CDC after they received information about the specific role the CDC has in responding to infectious diseases.[187] Ethan Zuckerman, author of *Mistrust: Why Losing Faith in Institutions Provides the Tools to Transform Them*, agrees that involving citizens is key to rebuilding trust. Additionally, he argues that "the answer is not a PR campaign on trusting American institutions again; it's actually looking at badly broken institutions like policing and prisons, abolishing the broken ones and creating new community safety institutions in their place."[188]

Maya Goldenberg writes that if a public health organization wants to implement something like vaccine mandates, these mandates must meet the needs of the community. Specifically, she states that this process should involve a thoughtful and transparent description of why certain vaccines should be required. Again, having public health organizations foster meaningful relationships with community leaders is key. Finally, as noted earlier, scientists are trusted within their local networks and can make an impact by sharing their knowledge with their local communities. In addition to knowledge, they can share information about the process of doing science. Academia and science can have some pretty weird rules (such as those concerning our volunteer peer-review system for scientific articles). I think if we make an effort to describe this process, be honest about its flaws, and share why we still trust it, it will help regain trust.

Finally, when we think about the macrolevel, there are a few things scientists can improve on as well. Scientists and science educators should spend more time on the scientific method used to produce knowledge. Many people still don't understand how scientists come to a conclusion and why that conclusion can change. Even if this method is taught in school, it is not reaffirmed in many media outlets. Public health officials and scientists on television are asked more about the result rather than the process itself, which is understandable because the media may prefer a simple answer that fits into a sound bite. However, I wish science communication and media would focus more on the complexity of science. Highlighting this complexity should help build trust in scientists and scientific institutions. Astrophysicist and science communicator Katie Mack has succinctly summarized how the chaos of science actually helps us trust the result when there is consensus:

> Personally, I think part of the problem around people's trust or distrust of science is tied up in the notion that the scientific community is some kind of organized cabal—the main reason the consensus of the sci community can ever be trusted is that it's a big chaotic mess![189]

Anyway there are a lot of ways in which the scientific community messes up (systemic bias, historical and ongoing harm to certain communities in certain areas, all sorts of bad things to do with academic/research culture), but scientific consensus by fiat is not one of them.[190]

Additionally, psychologist Carol Tavris has made a great statement summarizing the self-correcting nature of the scientific process: "I see science in general as a form of arrogance control, in the sense that it's one of the most organized methods we have of forcing us to put our beliefs to the test, and forcing us to face dissonance if the test does not confirm what we believe."[191]

The scientific method constantly tests what does and does not work. On top of that basic self-correcting principle, many scientists test one another's theories and share their results in a global network of knowledge. At the end of all of this testing and shared information, we can see if one answer is confirmed overall, which is what we call scientific consensus. Science as an institution is too decentralized across the globe for any malicious cabal to dictate the answers. Of course, scientific institutions are nowhere close to perfect, leaving plenty of room for improvement, and these improvements may also build trust.

The first area in need of improvement is the scientific institution itself. Two strategies involve advocating for open science practices and improving the peer-review system. Open science, as the name suggests, is a focus on data transparency, scientific rigor, and the open sharing of materials. For example, scientists could share their predictions and propose experimental designs *before* they collect their data. This holds them accountable for their predictions based on their theories and eliminates the practice of ignoring results that do not confirm hypotheses.[192] Unfortunately, the publishing system still rewards positive results over null results. However, a null result can still offer important insight into something that did not work.

While open science is advocated within academic circles to improve their own disciplines, political scientist Tamarinde Haven and colleagues argued that open science would also help improve trust among nonscientists as well.[193] I agree with them that removing the opacity of science and explicitly supporting research integrity would help the image of science. This is just one technique, but it is powerful because not only does it have the potential to enhance the public image of science, but also it could improve scientific rigor.

The peer-review system, in which academics review scientific papers to determine if they should be published or not, also has a great deal of room for improvement. One major issue is the entire funding model. Academics who publish articles and academics who review papers do not get paid any money for their time. The giant publishing companies make all the money. Even articles resulting from federally funded studies, whose funds come at least in part from taxpayer dollars, are kept behind paywalls so readers must be pay additional money to access them. Thankfully, the Biden administration is removing this policy as to federally funded research, so it will be available immediately when it is published and without a paywall.[194]

The act of reviewing is seen as a service to the community, but when scientists are already overworked, there is less motivation to provide free labor. This results in serious delays in publishing. I once waited six months after I submitted an academic paper before I heard anything from the journal. The journal told me that they could not find enough suitable reviewers for my paper, and, sadly, since they could not review my paper, there was no hope of publishing my study in that journal. Six months of waiting wasted. Some journals are now paying academics for their time reviewing articles. Editors can still make sure the reviews are of sufficient quality and invite only people who have given quality reviews. Furthermore, the peer-review system for publishing articles and awarding grants must continue to reduce as much bias as possible. We know that learning the name of the person submitting the article or grant can influence how well the reviewers evaluate the work. For example, one study found that academics were much more likely to reject an article for publication when they were told it was written by a scientist early in their career than when they read the same article but were told it was written by a Nobel Prize winner.[195] The process will take time and experimentation to improve, but a new generation of academics is working toward solving these problems.[196] As academics look to keep improving their system, I hope they can be open about the progress they make as well as their missteps. Transparency and authenticity might be scary at first, but they vastly improve trustworthiness.

Diversity and inclusion are other elements crucial to regaining trust in medical and scientific professionals. Earlier I mentioned how Dr. Kimberly Manning helped her community by directly engaging with its members because people trusted doctors who could share their experiences. People of color have legitimate reasons to be skeptical of the medical community. Supporting more people of color at all levels of health care can help address the needs of different communities that historically have been neglected. Marginalized groups also reported better treatment by physicians who shared their demographics.[197]

Research has shown that, in addition to the benefits realized when health care providers are able to connect with diverse groups of people, science itself can benefit from diversity. Studies have found that teams of employees were more productive when they were more diverse. Being able to bring in many different perspectives and think proactively about cultural problems helped with solving complex problems.[198] Similar benefits were observed in areas of science with more diverse research laboratories.[199] In addition to facilitating problem-solving, diversity in science can inspire a wider range of topics that people are interested in and balance out the biases found in more homogeneous groups.[200] If all the researchers on a team think the same way, they are vulnerable to group bias, just like any other group. However, when people come from different backgrounds, they may want to study elements that have been previously ignored. For example, women's reproductive health has greatly benefited from the presence of more women in medical science, since they ask research questions that men might not even think to investigate.[201] In fact, scientific teams composed of men and women create research that is more novel and cited more often than all-male or all-female teams.[202]

Finally, the health care industry itself needs reform to regain trust. As I mentioned previously, there are legitimate criticisms of the health care industry and health care officials. Economic partnerships between industry and academia boost medical innovation but can erode public trust due to conflicts of interest involving institutional decision-makers. To maintain trust, institutions should disclose financial interests, separate facilities, restrict information sharing between investment and research staffs, and ensure oversight by independent panels, thereby promoting progress while maintaining public confidence in research.[203] Fixing these problems requires courageous and thoughtful legislation developed collaboratively by our lawmakers and industry leaders. Of course, this is easier said than done, but we can collectively put pressure on our politicians to hold companies accountable and advocate greater transparency in their practices.

CONCLUSION

I began this chapter with a brief overview of the history of vaccines as well as vaccine misinformation. I then described how vaccines have been politicized, which has accelerated the spread of vaccine misinformation. The politicization of COVID-19 helped speed up this process as well, but there was already a growing distrust in institutions. I also explored the impact of social networks on susceptibility to and dissemination of misinformation. Who we associate with can impact what we believe about vaccines, as the number of vaccine-hesitant (or even explicitly antivaccine) people in our networks consistently predicts how antivaccine we will be ourselves. Social media further expedite this process, as it is easy to find people who support our views in the vast reaches of the internet, and antivaccine influencers will happily spread misinformation for profit. A few techniques that could help combat vaccine misinformation (as well as general health misinformation) are implementing fact-checking strategies, adding friction to social media, and increasing the data transparency of social media companies. Finally, rebuilding trust in institutions would decrease people's susceptibility to believing false information about science and health-related topics. When we consider how to rebuild trust, we can focus on the micro-, meso-, and macrolevels. Scientists and public health officials can improve their messaging strategies and use more empathy to connect with people. Additionally, institutions can make a greater effort to engage with their local communities. Finally, the institution of science can improve by adding transparency, fixing the problems within the publishing system, and advocating diversity. So far in the book I have focused on various social and psychological factors that contribute to the spread of misinformation. As we near the end of this book, I will focus on how we need to promote information and media literacy to prevent ourselves from becoming susceptible to believing false information in the first place.

CHAPTER 6

HOW EDUCATION, MEDIA LITERACY, AND PREBUNKING CAN COMBAT MISINFORMATION

On September 13, 2021, pop singer and rapper Nicki Minaj tweeted to her 22 million followers that the COVID-19 vaccine had caused impotence in her cousin's friend. Additionally, this unidentified man's fiancée called off the wedding—not only because of the impotence but also because of the swollen testicles, which she attributed to the vaccine's side effects. It's an absurd story, as there is still zero evidence that the COVID-19 vaccine causes such side effects. Yet, due to Minaj's fame, her tweet made international headlines. We looked at the skepticism about vaccines in the last chapter, so here I'd like to highlight another tweet Minaj made shortly afterward. In it, she said she would get the vaccine only once she felt she had "done enough research." One has to wonder what Minaj's research process entails if it previously led her to believe the COVID-19 vaccine caused impotence and swollen testicles.[1]

"Do your own research." This phrase has been used so frequently online that it's become a meme on social media as well as a topic of interest for academics and journalists. If you haven't seen this phrase, it is frequently associated with conspiracy theories, misinformation, and speculative investments. It is common within antivaccination echo chambers, and it is also popular in discussions of health misinformation or pseudoscience more broadly. For example, someone may share a miracle cure for some ailment on social media but then end the post with the caveat or instruction that people should do their own research as well. "Do your own research" is often abbreviated as DYOR and can also be found within online spaces that discuss investing. Sometimes you will find DYOR at the end of risky investment advice posted online. For example, someone will declare that a penny stock or cryptocurrency is the next big thing but then caution people to DYOR before buying.[2]

Of course, the issue with doing one's own research is that many people do not know how to research properly. And even if they can effectively research one area, they may be biased in how they research another area that overlaps with an important identity. For example, a person might have no problem critically analyzing some miracle cure for acne, but they might stop researching the moment they find a YouTube video that supports their critical view of vaccines. In fact, a 2023 study found that individuals with positive perceptions of doing their own research were more likely to believe misinformation about COVID-19 and to have less trust in science. The authors of this study concluded that support for doing your own research may be an expression of antiexpert attitudes rather than a willingness to examine information carefully.[3]

Throughout the book, we have learned about how our identities and networks motivate us to believe information that is consistent with our worldview. This can certainly reduce the accuracy of our own research if the process is about confirming what we thought was true instead of exploring the evidence in an honest and balanced way. The concept of doing one's own research can also be thought of as a rejection of expert and mainstream sources (which could also affirm one's contrarian identity, for example).

What does exploratory, honest, and unbiased research look like? While we can never fully escape our biases, we can certainly improve how we investigate information online. As we'll see in this chapter, research has shown how greater media and information literacy skills can reduce susceptibility to false information. Identity processes are crucial for understanding why we believe misinformation, but a conversation about combating misinformation is incomplete unless we also address the tools that help us properly discern truth from fiction. The tools I will cover in this chapter include higher education, critical thinking, media and information literacy, and prebunking techniques. Then I will end the chapter with a discussion of the broader systemic challenges of implementing educational techniques that help prevent the spread of misinformation.

HOW OFTEN DO WE SPOT MISINFORMATION?

As I've described throughout the book, misinformation is widespread in our information ecosystem. While explicitly false or fake information is less common than biased or misleading information, there is still a lot of bad information polluting our ecosystem overall. Furthermore, we are generally not great at recognizing our own limitations. I previously mentioned a study that evaluated how accurate people were at discerning whether headlines were real or fake. The study also compared participants' accuracy to their confidence in their ability to detect fake news. About three-fourths of the people believed they were better at detecting fake news than their accuracy scores indicated.[4] People are generally bad at objectively evaluating their misinformation detection abilities. Another study found that individuals who were overconfident about their reasoning abilities were more likely to score lower on an objective reasoning test that required deductive reasoning to rate the validity of syllogisms. Furthermore, overconfidence in their reasoning ability predicted a greater endorsement of conspiracy theories. This cautions us not to do our own research because we may severely overestimate our ability to detect fake news and our general reasoning ability.[5]

Misinformation spreads from all demographics. While older adults are less familiar with social media and may have difficulties spotting sponsored content or images that have been manipulated, people of all ages struggle to identify fake news.[6] A 2021 study led by psychologist Didem Pehlivanoglu evaluated the ability of young adult college students and older adults (aged sixty-one to eighty-seven) to detect fake news. Both age cohorts evaluated the accuracy of a series of real and fake headlines. The study found that both groups had similar abilities to distinguish factual from fake news headlines. However, when only comparing the oldest participants (seventy years of age or older) to the college students, the most elderly sample was less able to distinguish true from fake news, and this decreased ability was correlated with reduced scores on measures of overall analytic reasoning.[7] Thus, people of all ages struggle to detect fake news, but those who are among the oldest people and who also have decreased cognitive faculties may be especially vulnerable.

A large study run by the Stanford History Education Group assessed how well high school students could detect several forms of false information.[8] The 3,446 students who participated in the study came from both urban and suburban school districts in fourteen states. Overall, the authors of this study concluded the results were "troubling." When the students were shown a grainy video of alleged ballot stuffing during a 2016 Democratic primary, 52 percent believed it to be true. Only *three* students made an effort to find the original source, and a quick Google search could have revealed that the video was shot in Russia and had nothing to do with any Democratic primary. Furthermore, two-thirds of the students were unable to spot the difference between news stories and ads they saw (e.g., when sponsored content was visible on the ads). Finally, the students read about climate change on a website that had direct ties to the fossil fuel industry, but 96 percent did not consider why this may have been a conflict of interest. Instead of focusing on this connection, which could have generated clear bias, many students commented on the design and aesthetics of the website. Additionally, only 38 percent of the students said they had learned how to analyze media messages in high school. Despite spending up to eight hours a day online, many teenagers still have not learned basic information and media literacy skills.[9] While media literacy can be improved globally, some countries are doing a better job than others. A major analysis of international education, press freedoms, and institutional trust found that the United States ranked fifteenth out of forty-four countries (Finland, Denmark, and Estonia topped the list). The United States has a serious problem with media literacy.[10]

A 2021 national survey found that seven in ten voters supported teaching media literacy in schools.[11] This survey also found that media literacy programs had majority support on both sides of the political aisle, with 82 percent of Democrats and 64 percent of Republicans supporting them. A 2022 national survey found that more than eight in ten people agreed that schools should require media literacy education for all students.[12] While this sounds like a good thing to do on the surface, it is important to remember that teachers are already overworked and underpaid. While additional media literacy education would be great for young people, there are certainly major challenges if we are to make this a reality. I will discuss broader issues with media literacy at the end of the chapter. First, let's break down how we measure different types of media literacy.

MEASURING MEDIA AND INFORMATION LITERACY

The ability to analyze information from our media landscapes is clearly important, but how exactly do we define media literacy? Researchers have frequently cited communication scholar Patricia Aufderheide's definition of *media literacy*: "the ability of a citizen to access, analyze, and produce information for specific outcomes."[13] Media literacy is often used as a catch-all phrase to refer to greater competence in objectively analyzing the media environment people are exposed to. Of course, this is still quite broad, so researchers have attempted to narrow the concept to more concrete domains and measurable skills.

Many researchers study media literacy with self-reported measures of how well a person claims to critically evaluate the media they consume.[14] Communication scholar Erica Weintraub Austin developed a series of media literacy questions that predicted how likely someone was to adhere to scientific guidelines during the COVID-19 pandemic.[15] These included questions asking whether the participants agreed with statements such as "I think about who created the news I am seeing" and "I compare news information from different media sources."

News literacy is closely related to media literacy, but as the name suggests, it has more of a focus on critically evaluating news and news organizations.[16] Like media literacy studies, news literacy studies have relied on self-report questionnaires. For example, researchers have asked how the profit motives of news companies impact the way they cover stories, how sensationalist their coverage might be as a result of these profit motives, and how this can impact the beliefs of people who watch these news stories (or just the news).[17]

Finally, information literacy is similar to media literacy but is focused more on how we broadly process and evaluate the information we encounter. It goes beyond just understanding how media or news may be biased and describes our ability to critically examine the information that we are exposed to. This can include both digital and traditional media environments.[18] A paper written by communications scholar Sonia Livingstone and colleagues succinctly described the main difference: "Media literacy sees media as a lens or window through which to view the world and express oneself while information literacy sees information as a tool with which to act upon the world."[19]

Information literacy can be measured using more objective measures that don't rely on self-report data. One particular measure, developed in a study led by Bojana Boh Podgornik, professor of scientific and technical informatics, has people complete a brief multiple-choice test based on a broad understanding of information literacy.[20] For example, participants read pieces of information from a bibliographic database and answered specific questions about the author and format of information that was shown. Additionally, participants read short summaries of articles and reported on which elements were opinion- and fact-based. Participants were also asked about their knowledge of the basic Boolean operators that can be used in search engines (e.g., AND, OR) and the sequence of components of research articles (e.g., abstract introduction, material and methods, results, discussion, conclusions, and bibliography). A major strength of this test is that it covers many different types of information literacy and measures them in an objective manner.

But do objective measures work better than subjective, self-report evaluations? A study led by communication scholar Mo Jones-Jang examined whether scores on self-report media literacy questionnaires were as accurate as this objective information literacy test in predicting participants' ability to distinguish true from false headlines. The self-reported media and news literacy questionnaires did not effectively predict participants' accuracy in evaluating fake news. However, good scores on the objective information literacy test described here did successfully predict a stronger ability to spot fake news.[21] This study highlights the importance of not relying too heavily on self-reported competency questions. It is simply much easier for someone to claim they critically evaluate information than for them to actually do so.

Another common way to measure information and media literacy is to ask participants to evaluate the accuracy of headlines or news articles. Asking participants to identify which headlines are false and which are factual provides a more objective way to measure literacy. Of course, if these headlines are political at all, they can create bias, so researchers need to account for other factors that can influence how someone will evaluate information.

When talking about media literacy, broadly defined, it is important to make sure people are learning tangible and testable skills. While there are slightly different types of literacy that emphasize critical thinking in slightly different domains, for efficiency I will continue to refer to the ability to critically evaluate information as *media and information literacy* for the rest of this chapter. Possessing objectively measured media and information literacy skills is important if we are to detect misinformation. Is it possible to effectively teach skills that bolster media and information literacy? And if so, what techniques work best?

EDUCATION'S ROLE IN DETECTING MISINFORMATION

Former Vice-President Al Gore gave a talk regarding political polarization at Vanderbilt University while I was working there in 2021.[22] As I watched this event, I wondered what solutions he might propose for the hostility in our current political climate that fuels misinformation. When the interviewer asked Gore what we can do to combat misinformation, he highlighted the importance of education and said people need to read more books. While this is an admirable goal, I found his answer a bit superficial.

Of course, education and reading are all well and good, but what specifically can help address the problem of our polluted information ecosystem? Does formal education even protect against misinformation, given all the other social and psychological biases we have? Back in chapter 2, I covered a few studies that showed how identity biases can override education levels. Specifically, even if someone scores high on general science literacy, they will still reject the scientific consensus on a topic if it conflicts with their political ideology.[23] It would be incorrect to conclude that education does not protect us against misinformation; however, the role it plays depends on multiple factors.

Higher levels of formal education do predict a lower likelihood of falling for general fake news, COVID-19 misinformation, and conspiracy theories.[24] Scoring higher on a science knowledge test is also associated with a lower likelihood of believing vaccine misinformation.[25] Overall, we can conclude that higher education is an important buffer against susceptibility to misinformation. Again, personal identities and other idiosyncrasies play large roles in how people process information. Additionally, there are plenty of examples of incredibly smart and thoughtful people who have not attended college and can identify fake news quite well. In social science, we look for the largest predictor variables with the understanding that there are often many other variables involved and exceptions to the general rule.

But why exactly does attending college predict at least some decreased likelihood of believing false information? Psychologist Jan-Willem Van Prooijen conducted two studies to analyze which factors could explain the link between education and conspiratorial thinking. He found that those with higher education were less likely to believe in simplistic solutions for complex problems and were also less likely to feel powerless (which made them also feel more in control). Conspiratorial thinking is predicted by a tendency to look for simple patterns to explain complex situations as well as by a decreased feeling of personal control. Education often teaches that there are no easy solutions for complex social problems and can help a person realize their value in society. Thus, education helps develop comfort with complexity and feelings of self-worth, which provide a buffer against believing in conspiracies that prey on the vulnerable. Additionally, education is linked to higher analytic thinking, which helps bolster aversion to simple solutions.[26]

In chapter 2, I discussed how cognitive reflection, the tendency to process information more analytically and deliberately, is associated with a decreased likelihood of believing fake news. While the debate continues as to what types of misinformed beliefs are most predicted by a lack of cognitive reflection, the results are pretty clear that higher cognitive reflection does help protect against believing misinformation.

One of the largest predictors of cognitive reflection is a higher level of education.[27] Psychologists Stefan Stieger and Ulf-Dietrich Reips found that higher levels of formal education predicted the highest scores on the Cognitive Reflection Test (CRT). Education even predicted slightly higher levels of cognitive reflection than previous experience completing the CRT in other studies! Economist Shane Frederick, the creator of the CRT, also found that education level strongly correlated with cognitive reflection scores. Furthermore, students at the most academically rigorous universities, such as the Massachusetts Institute of

Technology, score especially high on the CRT. Scores on scholastic exams such as the SAT and ACT also positively correlated with scores on the CRT.[28]

Importantly, cognitive reflection is a malleable skill. Simply asking people to be more reflective can boost their ability to distinguish true and fake news.[29] While some people who score higher on cognitive reflection may be more likely to seek out higher education, higher education also incentivizes students to think more carefully as they go through their classes. Frederick defined the cognitive reflection measured by his CRT as "the ability or disposition to resist reporting the response that first comes to mind."[30] Ideally, college will help people develop the practice of critically questioning information they see in the world. As mentioned earlier, education helps develop an appreciation for and comfort with complexity.

THE BENEFITS OF THINKING MORE CRITICALLY AND WITH MORE HUMILITY

While general education and analytic thinking are helpful, educators and scholars have been working on developing specific skills that are more focused on protecting ourselves against misinformation. One of these skills is critical thinking. I hear critical thinking often brought up as almost a panacea to many of the world's problems . . . including misinformation.

But what exactly is critical thinking, and how can it help? *Critical thinking* has been defined as "the ability to analyze and evaluate information."[31] This is rather broad, and scholars have not agreed on a more meaningful definition of the term. This lack of academic consensus on a clear definition makes it hard to build on previous research and test different interventions. One way to measure critical thinking is to have people evaluate arguments and explain their strengths and weaknesses. One such task has the person read a fictional letter to the editor that argues a specific area should prohibit overnight parking. The participant then must evaluate whether the arguments presented do or do not support the author's contention in a response essay. This essay test is evaluated by three independent judges who are trained to evaluate strong argumentation. Participants who can sufficiently explain the strengths and weaknesses of the overnight parking article are deemed to have stronger critical thinking abilities. One 2021 study led by psychologist Anthony Lantian found that those who scored high on this critical thinking essay test were less likely to believe in conspiracy theories.[32] In the same study, researchers asked people if they believe they have good critical thinking ability as a subjective measure of critical thinking. There was no significant relationship between self-reported critical thinking ability and belief in conspiracies. Thus, an objective essay test could capture critical thinking abilities more accurately. Of course, having participants write an essay and having judges rate the essay require far more time and resources than simply asking participants self-report questions.

There are examples of successfully teaching critical thinking skills to people in a scalable fashion. A study led by psychologist Ben Motz examined whether an online critical thinking lesson could improve people's ability to detect logical fallacies while reading arguments. The group of participants in the critical thinking condition completed the thirty-minute lesson, which described how to identify logical fallacies and poor reasoning within arguments. The other group completed a lesson about general psychology concepts. Those who were in the critical thinking condition scored significantly higher on a subsequent test that evaluated logical fallacies within arguments.[33] Of course, these skills apply to evaluating general arguments, not necessarily misinformation.

Psychologist Lauren Lutzke led a study that directly evaluated whether critical thinking could reduce the belief in and spread of fake news by having participants read false and real information about climate change.[34] Specifically, the participants were told they would read and evaluate Facebook posts about the subject, and then they were divided into research and control groups. Lutzke and her collaborators had their research participants read a set of guidelines that prepared them to think critically about evaluating information on social media (e.g., "Is the post politically motivated?"). However, those in the control condition were not given information on evaluating social media content. The participants who were primed to think critically about social media posts was less likely to trust, "like," or share the false headlines they saw. This study showed that simply nudging people to think more carefully about social media content and specifically giving them guidelines about what to look out for could reduce their vulnerability to misinformation.

Despite these encouraging examples, research on critical thinking lacks studies that replicate such findings on a larger scale. Furthermore, if educational interventions make people cynical about all media, they may not bother to differentiate between good and bad information. Indeed, a study led by psychologist Alan Bensley found that higher levels of cynicism (a general lack of faith in people) predicted greater susceptibility to conspiracy theories and unsubstantiated beliefs. However, a generally skeptical attitude (a comfort with questioning one's assumptions) did not.[35] Furthermore, a 2023 study teased apart underlying factors that comprise different types of skepticism and made an important discovery.

Skepticism motivated by a concern that the media landscape is biased against a group one belongs to (identity-motivated skepticism) increased susceptibility to believing misinformation. Conversely, skepticism motivated by a general concern with accuracy, even if that means updating one's beliefs (accuracy-motivated skepticism), predicted a reduced susceptibility to false information.[36]

Critical thinking is challenging to operationalize and measure in a consistent manner because it is such a broad concept and has been defined differently by many people. Scholars also disagree on whether critical thinking is a domain-specific skill or a general skill that can transfer across different areas and on whether it is a trait.[37] Finally, people can be selective when they apply their critical thinking, making it difficult to create a measure that can predict how someone will process all types of information. For example, a study found that individuals who do well while critically examining general health and nutrition information can still fall for misinformation regarding vaccines.[38] While critical thinking certainly has some promise as a concept that can protect against misinformation, further work is required to understand exactly how and why it helps.

Another type of thinking that can help protect against extremism and misperceptions is *intellectual humility*, which has been defined as "recognizing the limits of one's knowledge and awareness of one's fallibility."[39] It has not been shown to positively correlate with additional formal education. Instead, it has been linked to curiosity and an intrinsic desire to learn new things.[40] Those who scored high on questions that measure intellectual humility were more tolerant of those with different religious beliefs, less politically polarized, and more open to interacting on social media with those who think differently than they do.[41] Additionally, higher scores on intellectual humility questions were positively associated with being critical of health misinformation and being more likely to get the COVID-19 vaccine.[42] In fact, one study found that scoring high on intellectual humility predicted positive attitudes toward vaccination even more than political orientation or education.[43]

Like with critical thinking, there is some evidence that intellectual humility can be taught and developed in a classroom setting.[44] Students who took a class on developing intellectual humility rated the other students who had been in class with them as more intellectually humble after taking the class, but they didn't rate themselves as more humble afterward.[45] This suggests that the class did increase its humility overall but that participants were hesitant to claim they had achieved humility themselves, since doing so would suggest a lack of humility. Asking participants to think about themselves from the perspective of a distant observer was associated with higher self-reported intellectual humility, as was simply having participants read about the benefits of being intellectually humble.[46]

Despite these early results showing intellectual humility can be learned, there are a few key limitations as well. First, many of these studies rely on self-report data, which could be biased. Having participants rate other people's humility in addition to their own does mitigate this issue somewhat, but this still relies on self-report data. Additionally, there is little research on how long the effects of interventions to boost intellectual humility last. Finally, and perhaps most importantly, individuals may be quite selective in their intellectual humility. It may be quite easy to admit to being wrong about a topic not central to our identity. However, we may find it much more challenging to be open to being wrong about new information that directly conflicts with our ideological stance.[47] Critical thinking and intellectual humility do still offer promise though. Processing information critically and remain aware of our limitations are important skills to develop. With the overabundance of information on the internet, it may be especially useful to help people filter out bad information in the first place.

CRITICAL IGNORING AND LEARNING
HOW TO DO OUR OWN RESEARCH

There is growing support for a technique that prevents people from being overwhelmed with bad information in the first place. It is called *critical ignoring* and is defined as "choosing what to ignore, learning how to resist low-quality and misleading but cognitively attractive information, and deciding where to invest one's limited attentional capacities."[48] Anastasia Kozyreva, a misinformation researcher at the Max Planck Institute, was the lead author of a 2023 paper that separated critical ignoring into three main parts: using self-nudges, not feeding the trolls, and practicing lateral reading.

We have already learned that social media are in the business of trying to capture as much of our attention as possible. This subsequent information overload can allow bad information to spread quickly, since we cannot properly evaluate everything that comes into our newsfeed. The first component of critical ignoring consists of nudging ourselves to regain control of our information environments. These self-nudges can involve setting time limits on social media apps and even periodically deleting them from our phones. If social media are going to continue to be in the business of capturing attention, then we can limit the impact of this information overload by simply trying to create boundaries around our use of these platforms. We can make an effort to avoid websites and (social) media that can stir up strong emotions in us. We can set timers and use other technologies to prevent ourselves from spending too much time on websites designed to capture our attention for as long as possible. Of course, there is a trade-off, as studies have shown that removing social media entirely is associated with less awareness of political events but also results in increased happiness and decreased polarization.[49]

Related to this self-nudging practice is the next component of critical ignoring: avoiding feeding the trolls. This is a common suggestion for anyone who spends time on the internet, but it really is important to remind ourselves just how consequential it is. Trolls are individuals who interact with other people online in bad faith and often have a goal of annoying them. They can spread lies or faulty arguments about sensitive topics that can make us want to engage with them. However, as we have learned, there are people who simply have a high need for chaos (or benefit from having more people click on their articles), and their incentives are not aligned with thoughtful dialogue. Of course, it is important to fact-check and share good information, but we need to be selective with our time. It is simply not productive to try to engage with someone who is not open to learning from us or from anyone else. Social media make it easy to block these bad-faith individuals and move on with our lives. I feel quite confident that you will not miss debating trolls online and will enjoy spending that time on anything else. Finally, if we engage with trolls, that engagement is rewarded by the social media algorithms, and this can promote their misinformation to more people. If we don't engage with the emotionally evocative content they share, then it may just stay within their network and not spread across multiple networks.

The last component of critical ignoring, and perhaps the most complex, involves learning a specific technique called *lateral reading*, which involves evaluating online information by checking its credibility across multiple other sources. Instead of just trying to dive into the details of an article you see online, you need to assess what other reputable sources are saying about the author of the article and the website that hosts it. For example, let's pretend we are browsing social media and a friend of ours shares a news story that captures our attention. As we read the article, we can try to determine if the arguments make sense and appropriately cite primary sources. This is a vertical approach to reading, since we are staying within the confines of the article (or website). On the other hand, we can use lateral reading, which involves our checking how other legitimate sources evaluate this website and author. If the consensus from other sources is that the original author or website has a reputation for spreading false information, then we can proceed with caution and skepticism. If other sources claim they are legitimate, then we can feel more comfortable with the article's claims.

Professional fact-checkers tend to do lateral reading when evaluating claims, but untrained students tend to evaluate claims through vertical reading.[50] Studies have shown that teaching this lateral reading technique increased the ability of both students and adults to judge the credibility of online information.[51] Mike Caufield, a research scientist at the University of Washington's Center for an Informed Public, developed a simple acronym, SIFT, that summarizes this broader lateral reading process. First, we *stop* reading and reacting and instead focus on what we need to know to determine if the source is legitimate. Then we *investigate* the source through lateral reading techniques. After that, we *find* better coverage that gives us additional detail. And, finally, we *trace* claims, quotes, and media back to the original context.[52] This SIFT process is similar to that used to evaluate scientific claims. Science articles published in journals with strong standards for blind peer review and corrections of errors should earn more trust. However, when evaluating the validity of scientific claims, we must keep in mind that science can be a dynamic process with constantly changing information.[53]

Lateral reading and fact-checking generally require us to compare the claim in question with the consensus of reputable sources. But how do we know if a source is reputable? Fact-checking websites like PolitiFact and Factcheck.org are helpful resources, but they are not perfect and certainly cannot cover every news story. The News Literacy Project, a nonprofit and nonpartisan education organization, suggests we check whether the website in question has transparent reporting practices and takes accountability when it makes a mistake.[54] Value checking is another tool we can use to evaluate the credibility of a source. It identifies the ideologies and subjective background of that source, which give us additional context when determining how its values may bias its perspective.[55] Transparency of methodology, sources, and funding are also some of the main factors considered by the International Fact-Checking Network.[56] In the endnotes of this chapter, I've included links to additional free online resources that help increase media, news, and information literacy.[57]

In addition to checking the journalistic standards of each website, we can use resources from other nonprofit educational organizations like AllSides and Ad Fontes Media.[58] AllSides uses academic research, independent review, and politically balanced community feedback to assign political bias ratings to popular news sources.[59] Ad Fontes Media has a politically balanced group of analysts who evaluate the bias and reliability of news and news-like sources.[60] It's encouraging to see a great deal of overlap in the results of these two organizations. For example, both organizations agree that CNN has a left-wing bias and Fox News has a right-wing bias. Both organizations also agree that Reuters and *The Wall Street Journal* score high on reliability and low on partisan bias. Importantly, these organizations use general heuristics and rough tools to determine bias. Individual stories and authors may be more or less biased than the overall news platform. Thus, figuring out if a news story is true takes work from multiple angles, and even that may not result in clear-cut answers. However, there are definitely steps we can take to make sure we are consuming and sharing as much reputable information as possible.

When we see someone claim to do their own research, it's key that we reflect on what proper research entails. It requires carefully vetting each source and author and using multiple websites to determine what is likely true. Doing our own research can be a good thing; it is important that we have a healthy amount of skepticism instead of assuming that whatever is shared in our social media newsfeed is correct. However, taking a more scientific and objective approach to finding the truth is crucial. This requires an exploration of facts without looking for justification for a particular answer that we have already decided is true.[61] Remember the study I mentioned at the beginning of this chapter, which found that those who had positive attitudes toward doing their own research were also more likely to believe in misinformation. It's simply too easy to Google something, immediately find any information that supports a prior belief, and feel satisfied that we did our own research.

Finally, no media literacy program is complete without a discussion of how our personal identity biases can influence us. I hope that learning about our identity processes can help us be more mindful of the forces that could impact how we consume and interpret information. Specifically, we can think back to the technique of writing our own identity map. Psychologist Peter Coleman suggests this exercise as a tool for depolarization, and I also think it can help us reduce our susceptibility to misinformation. By learning how our identities intersect and conflict, we can reflect on how those forces may motivate us to judge material with less objectivity. Ideally, we should reflect on what set of identities and meanings provides us with a critical amount of self-esteem and how we may be biased in our information processing when it comes to this particular social identity cluster. This, of course, requires honesty and intellectual humility, so we could have a friend who thinks differently than we do help us by evaluating our maps to keep us honest and humble.

In addition to being vigilant about our own identity-fueled biases, we can regularly ask ourselves a few questions when we form our beliefs. Biology professor and science communicator Melanie Trecek-King came up with a great list of six simple questions that can help us check if our beliefs are true.[62] The first one is "How sure are you that the belief is true (0–100 percent)?" You may remember that we discussed the importance of asking "How" when we evaluate other people's beliefs, and that same approach can and should be applied to ourselves (though it would also be useful to ask another person these questions while trying to have productive dialogue). The second question is "What is the source of the belief?" This aligns with what we have been learning in this chapter about checking our sources and doing lateral reading. The third question is "What are your reasons for believing it's true?" Again, the question makes us reflect on what analysis we used to check the veracity of the claim (or whether we did any analysis at all). The fourth question is "How could you figure out if it is true?" This allows us to be proactive and use our media literacy techniques to fact-check ourselves. It is also helpful to use a trusted friend who thinks differently than we do as another source of information. The fifth question is "How would you feel if you were wrong?" This is an important question because it forces us to reflect on why this belief matters so much to us. We can think back to making our identity maps and reflecting on how our identities may bias our judgments and information processing. Finally, we can ask ourselves "What evidence would change your mind?" This also forces us to reflect on how our identities may be biasing us and to grapple with what evidence is necessary to be persuasive. Again, having a conversation with an honest and trusted friend is helpful here. Trecek-King teaches these types of techniques in her classroom and on her website.[63]

While these media literacy programs have been shown to be effective, one common limitation is the resources required to implement them. Getting people to take even a short course can be difficult, since they need to both find the course and be motivated to finish it. And you also need someone to teach it. Another promising approach for teaching media and information literacy is gamified versions of prebunking techniques, which are both fun and scalable and do not require the presence of a real-time instructor.

THE BENEFITS OF PREBUNKING

Instead of trying to debunk misinformation, which can be a never-ending game of Whac-A-Mole, researchers have been looking at the promising technique of prebunking. This technique is called *prebunking* because it involves giving individuals tools to identify and protect themselves against misinformation before they are even exposed to it. This idea of preparing people to fight against misinformation builds on the inoculation theory developed by social psychologist William McGuire in the 1960s. He theorized that if people were be first "vaccinated" with a smaller dose of typical persuasive techniques, then they could become less vulnerable to persuasion.[64] Psychological inoculation presents the individual with a "weakened" version of a typical argument that spreads misinformation. This practice focuses on helping people more critically evaluate *future* information rather than on trying to correct misperceptions after the fact.[65]

On the one hand, I find it interesting that psychological inoculation and the work surrounding it have gained popularity while we also are faced with a wave of vaccine misinformation regarding COVID-19. However, it's also somewhat unfortunate that the name of the original theory may sound nefarious to those already skeptical of vaccination. The origins of this theory did not focus on vaccine beliefs at all and instead analyzed cultural truisms, such as the effects of brushing one's teeth.[66] For example, participants first read or wrote about why it is helpful to brush their teeth regularly and then read a counterargument as to why regular brushing is not important. Being exposed to the evidence behind these truisms reduced the likelihood of they would change their attitudes when they read the counterarguments. For the sake of simplicity and consistency, I'll refer to this process of teaching people about misinformation (e.g., psychological inoculation) as prebunking for the rest of this chapter.

What exactly does prebunking look like? Prebunking involves warning someone of possible manipulation, such as a faulty argument that is used to promote a false claim. Then the person is shown a refutation of this particular manipulation attempt. For example, a 2017 study led by Sander van der Linden, a psychologist at Cambridge University, evaluated whether exposing people to a misleading argument about climate change could inoculate them against subsequent climate change misinformation.[67] Participants in the inoculation condition read about a petition signed by over thirty thousand "scientists" who did not believe humans were causing climate change. They then read about how this petition was flawed because many of the signatures were fake and because subsequent analysis revealed that less than 1 percent of the signatories had any background at all in climate science. Participants also read that 97 percent of actual climate scientists agree that human-caused climate change is happening. Participants in the control condition just read about the petition without any description of why it was flawed. Participants in both conditions had initially estimated that about 72 percent of scientists agree that global warming is caused by human activity. Those who just read about the fake petition without refutation decreased their estimate of scientific consensus by nearly 9 percent. Those who read that the petition was flawed and that 97 percent of actual scientists agree about our role in climate change increased their estimate by over 12 percent. This study showed that exposing people to misleading climate change information shifted their beliefs away from the scientific consensus. However, when they were equipped with the tools to refute the misinformation, their beliefs aligned more closely with the scientific consensus.

These passive inoculation techniques have shown effectiveness in reducing misinformed beliefs about the HPV vaccine, biotechnology, and animal research.[68] For example, providing people with scientific evidence about the safety of vaccines reduced their likelihood of believing vaccine misinformation that was later presented to them.[69] Additionally, a 2023 study found that participants who were taught the logical and factual errors of antivaccine talking points were less likely to be persuaded by antivaccine misinformation.[70] However, other research has shown that providing this kind of passive inoculation made some people double down on their antivaccine beliefs.[71] We will see later that the source of the inoculation can be very important for its effectiveness.

Other research has shown that directly explaining how human activities influence climate change can be quite effective in reducing misconceptions. This prebunking technique, aptly named misconception-based learning, examines the details of scientific misunderstandings and teaches how scientific processes work.[72] One study had two groups of college students learn about climate change during their introductory atmospheric class. One group learned about climate change using more traditional materials that covered the basic science of how greenhouse gases impact the atmosphere. The misconception-based learning group explicitly focused on common misunderstandings in addition to learning about the basic science. For example, they learned how some people can confuse a hole in the ozone layer with the greenhouse effect and why this misconception is incorrect. In another example, they learned how some people think all radiation is the same and why this is not the case. Both groups of students had increased their understanding of climate change at the end of the course. The students in the misconception-based learning group were more accurate when asked about climate change two weeks after the course ended. This study showed that directly tackling the misconceptions may have had more lasting effects than merely covering the material in a more traditional format.[73] Other forms of misconception-based learning include learning from the faulty reasoning of psychic ability, UFO sightings, and paranormal activity and then applying these lessons to other forms of pseudoscience and misinformation.[74] These more active types of prebunking often require a classroom setting, which limits scalability. Misconception-based learning also can focus on a select number of misinformation topics instead of the wider range of techniques used in the spread of misinformation.

Technique-based prebunking equips people with tools to fight against misinformation more broadly and offers some promising scalable solutions. Even better, many of these techniques avoid exposing people to identity-threatening information, since they focus on the general mechanism of misinformation rather than a specific topic. They also can be proactive, as learning about the techniques will prepare people for future misinformation they encounter. Furthermore, these technique-based interventions have been developed to be more engaging than simply reading text.

Several popular and effective prebunking tools have been developed into scalable and engaging activities. University of Cambridge psychologists Jon Roozenbeek and Sander van der Linden, in collaboration with a Dutch media platform, developed an online game that also serves as a prebunking tool. The game, called *Bad News*, takes only about fifteen minutes and teaches the player common misinformation techniques.[75] Players take the role of a fake news creator, and their goal is to gain as many followers as possible while also maintaining sufficient credibility. They have the option to choose what they post in this fake social media world, and each choice impacts follower count and credibility. The game offers commentary on how each social media post aligns (or doesn't align) with the six common misinformation techniques: impersonation, emotional content, polarization, conspiracy, discrediting opponents, and trolling. Players earn a badge after completing each segment in the game based on one of these techniques.

Impersonation deals with how misinformation can spread when people pretend to be legitimate and credible accounts. In the game, players have the option to pretend they are NASA's social media account, and to get engagement, they share a post about how a meteor is hurtling toward Earth. If you have spent any time on Twitter, you probably have seen fake accounts of celebrities tweeting similar absurd things or sharing some suspicious links to click on. The next technique looks at how emotions like fear and anger are used to also drive engagement. If players share "The Lives of Ants: The Joy of Routine" in the game, then they see that they don't gain many followers. However, if they post "They Test Anti-Aircraft Guns on Innocent Puppies," then they receive much more attention from their followers. Throughout the game, players lose credibility if they keep sharing outrageous things. In the polarization segment, players learn how to post about an event in a way that riles up a political group. For example, they can post about how the government is corrupt due to right-wing or left-wing politicians. Players gain additional followers if they use bots to share their polarized message even further. In the conspiracy segment, players learn that posts that are too devoid of reality (like aliens building pyramids) may turn off some of their followers. However, if they talk about a real event, like the COVID-19 pandemic, but add in a conspiracy, like the virus being a plot by Big Pharma to make more profits (I've seen this one in real life), then they gain more followers. Next, players learn how they can continue growing their following by discrediting opponents. For example, if a fact-checking organization debunks one of their posts about COVID-19, they can simply make up a story that the organization didn't pay its taxes and is corrupt. This shifts the attention to the organization, which has to defend itself instead of the veracity of its claims. Finally, players have the option to share some trolling posts with their audience. For example, they can photoshop a picture of a plane crash to make it seem more suspicious. Trolling here refers to inciting a reaction from one's target audience by using bait. By making the plane crash look more suspicious, players get the Aviation Disaster Committee to respond, which also allows their followers to attack the committee's comments. Players' scores depend on how many followers they earn by sharing online misinformation in the game.

Participants who played *Bad News* in a research study were significantly more likely to identify misinformation techniques compared to a group of participants who did not play the game.[76] Participants who played the game also increased their confidence in spotting misinformation. A follow-up study with fifteen thousand participants showed that those who completed the game were better at spotting fake headlines and misleading tweets.[77] For example, to test whether they learned from the impersonation segment, participants read a tweet from a Twitter account that pretended to be the television company HBO by slightly changing its Twitter username. The *Bad News* game is a pretty straightforward intervention that helps people identify broad misinformation techniques. A subsequent prebunking game developed by Roozenbeek and van der Linden called *Harmony Square* focuses on political polarization and misinformation. Research on *Harmony Square* revealed results consistent with those of *Bad News*: Participants who played the game were better able to identify real examples of misinformation, including politically polarized messages.[78]

Bad News received various accolades, including the Brouwer Trust Prize, awarded by the Royal Holland Society of Sciences for social impact initiatives that bridge science and society.[79] The European Commission deemed it "one of the most sustainable paths to combating fake news."[80] *Harmony Square* and *Bad News* have been translated into dozens of languages, and they are regularly used in classrooms and by other media literacy organizations. The elements of these prebunking games can also be distilled into short thirty-second videos, which have been shown to increase people's ability to differentiate between trustworthy and untrustworthy content. These videos capture the viewers' attention with opening lines like "Great. Looks like the trick worked." That particular video goes on to describe how emotional language is used within online misinformation. A massive study paid for these short prebunking videos to be shown as advertisements on YouTube, where nearly 1 million people saw them. The study also kept track of YouTubers who did not see the ads. Within a day of each ad, You-Tubers were shown another ad that asked them a simple question about the type of manipulation tactic shown in the previous ad. The individuals who watched one of the prebunking ads were significantly more likely to correctly identify the manipulation technique presented.[81] This study highlights that even a simple, engaging video can effectively teach people to recognize misinformation tactics that they may be exposed to online.

Despite all of these positives, it is important to consider the limitations of these prebunking games, as they certainly are not a panacea. One major limitation is that they require people to have enough motivation to engage with them in a meaningful way.[82] Additionally, the effectiveness of these prebunking strategies does wane over time. The positive effects of playing *Bad News* have been found to significantly fade after a couple of months. However, this decrease in effectiveness can be curbed if people receive regular reminders of what they have learned, or boosters, to maintain the effects.[83] Finally, prebunking games may increase skepticism about both true and false information.[84] So far, we have focused on more of the microlevel processes of information and media literacy. Next, we'll cover the larger structural problems that create barriers to the adoption of widespread efforts to increase media and information literacy.

INFORMATION AND MEDIA LITERACY'S LIMITATIONS AND UNANSWERED QUESTIONS

While these media literacy interventions show promise on the individual level, it is crucial to consider their limitations and unanswered questions. I will focus on four major limitations that media and information literacy programs need to address: source effects, English-speaker and Western bias, economic incentives, and the need to consider macrolevel solutions. Finally, I will end the chapter with a discussion of how incorporating identities and social networks can bolster the effectiveness of media literacy.

Source Effects

As we've learned throughout the book, people are more comfortable around and have more trust in those who share their identities. Who is behind a media literacy intervention or program is extremely important for how it will be received by a particular audience. Some prebunking games, like *Bad News*, seek to be politically neutral. Playing the game provides the option to attack both liberal and conservative ideologies. *Bad News* also uses politically neutral pop culture references (like *Star Wars* and *Family Guy*) to make its educational points. However, this is not the case for all media literacy interventions and programs.

Research has shown that inoculation techniques were more effective when the source of the inoculation was viewed positively.[85] It is vital to consider how identity interacts with information processing when trying to develop media literacy tools and education techniques. Studies that accounted for cognitive factors as well as identity factors more accurately explained an individual's susceptibility to misinformation.[86] Furthermore, a study led by psychologist Cecilie Steenbuch Traberg found that both liberals and conservatives were more likely to believe misinformation if it came from a politically congruent source. The more credibility a person believes a source has, the higher the likelihood they will believe the misinformation shared by it.[87]

We have already seen the challenges the Centers for Disease Control and other health organizations faced when they tried to share information about COVID-19 and the subsequent vaccines. If you already don't trust a source (or if a group you do trust doesn't like or trust that source), then its education attempts are going to be far less effective. Researchers can develop incredible media literacy and prebunking tools, but if people do not trust the groups that created or funded these tools, then their impact will be quite limited. Trust is foundational for media and information literacy programs.

This is not to say that people listen only to those who share their group identity. Despite waning institutional trust, there is evidence that people value the input of independent fact-checkers overall. A 2023 study found that those in a balanced sample of Democratic and Republican participants were significantly less likely to share false headlines when the news source was rated not trustworthy by professional fact-checker organizations.[88] Another study found that both Democrats and Republicans felt less favorable toward politicians after reading they had lied, according to fact-checkers. In this study, participants saw PolitiFact assessed the veracity of the claims, so at least that source was viewed as reputable by this sample.[89] In a large field experiment, political scientists Brendan Nyhan and Jason Reifler evaluated whether sending letters to state legislators warning about the reputational risks of being caught lying by fact-checkers would actually decrease the likelihood these politicians would spread false claims. They found that the politicians who received these warnings were significantly less likely to receive a negative fact-check afterward.[90] This study was conducted in 2014, so it would be interesting to see if the results could be replicated in these increasingly polarized times. However, these studies, along with the success of general prebunking interventions, do suggest that people can adjust their behavior when considering the credibility of different claims. Finally, because people are motivated to share fake news that attacks an out-group, they may still share news they know is fake even if they have strong media literacy overall.[91] Identity is going to be a central issue when we consider why people share fake news and whether they pay attention to fact-checkers.

English-Speaking and Western Bias

It is important to note that many of these studies on media literacy and various education interventions have significant selection bias. It's estimated that over 80 percent of research on misinformation focuses on populations within the Global North (i.e., the richest and most industrialized countries, which also tend to be in the Northern Hemisphere).[92] Social science research overall tends to have an overrepresentation of English-speaking, educated participants from Western countries. This is because many researchers are from the same areas, and it is easier to collect data in your own language and region.[93] College students are an accessible population to study because they are already on campus and professors can offer extra credit instead of monetary payment. This has improved in the past few years, as online participant recruitment platforms have emerged where a much larger group of people beyond college students can find research studies to complete. While this does increase diversity somewhat, there are still major limitations. Older individuals who have lower media and digital literacy are some of the most vulnerable to misinformation. Yet they are also the least likely to seek out these online platforms where they can participate in research studies.[94] If someone is going to run a study on media literacy, then they should carefully think about how their participant group may bias or limit the results.

The diversity of populations is crucial because each population may interact with its media landscape differently. As of 2024, it was estimated that 62 percent of the global population uses at least one social media website. However, there are vast differences among countries.[95] For example, Facebook is extremely common in the United States, with about seven in ten Americans using the social media platform.[96] However, its use is far less common in African countries: It's been estimated that 20 percent or fewer of the people in Nigeria, Kenya, and Cameroon are on the platform.[97] While only 40 percent of Americans use the messaging and social media app WhatsApp, it's been estimated that over 90 percent of people in Italy, India, and Brazil use the platform.[98] The dynamics of Facebook and WhatsApp are quite different. For example, WhatsApp can expose people to misinformation more directly, since they are directly messaged. Interviews with teachers in France, Romania, Spain, and Sweden also demonstrated that each country has a slightly different conception of what constitutes fake news.[99] For example, the Romanian teachers were the only group to mention hyperpartisan news as a part of fake news. All groups of teachers agreed that greater media literacy would be helpful and also shared unique barriers they faced in each country that would need to be addressed in order to improve the media literacy education they would be able to offer. The French teachers, for example, described the challenges they had trying to bring digital tools into the classroom in order to discuss visual and audiovisual components of media literacy.

Accessibility is another important variable to consider when looking at different countries, as there are data showing that countries with less access to the internet may have reduced digital literacy.[100] Different regions may use different types of social media or messaging apps, which vastly change their information diet. The resources spent on fact-checking tend to disproportionately focus on content in English. For example, 87 percent of Facebook's spending on combating extremism and misinformation is focused on text that is written in English. However, Facebook claims that about two-thirds of people who use Facebook primarily use a language other than English![101] Additionally, even when social media companies do establish fact-checking procedures for different languages, the lack of resources poured into them can limit their effectiveness. For example, Facebook failed to identify 29 percent of COVID-19 misinformation in English. When searching for COVID-19 misinformation in Spanish, Facebook failed to identify 70 percent of that misinformation.[102]

We previously met Frances Haugen, the whistleblower who exposed Facebook's unethical practices. She noted that the profit motives of the company made it simply less interested in protecting the safety of non-English-speaking users. She told *60 Minutes*: "Every time Facebook expands to a new one of these linguistic areas, it costs just as much, if not more, to make the safety systems for that language as it did to make English or French. . . . Because each new language costs more money but there's fewer and fewer customers. And so, the economics just doesn't make sense for Facebook to be safe in a lot of these parts of the world."[103]

The lack of support due to cost certainly links to the economic incentives we have discussed previously. However, there are many people working on media literacy around the globe. As already mentioned, Finland ranked first in an international survey of media literacy. What exactly is that country doing that helps its citizens discern truth from fiction? Well, its government launched a major media literacy program back in 2014, a couple of years before fake news became a major trope of the 2016 presidential election in the United States. Media literacy classes have continued to be widespread in Finland's public schools and colleges, and in 2016, it introduced a critical thinking part of its public school curriculum that focuses on developing skills needed to spot disinformation online.[104] Educators, government officials, and academics all work together to educate the public. Finland also maintains a strong trust in its mainstream media, with nearly seven in ten of its citizens saying they trust the news media. In comparison, just 32 percent of Americans say they trust the news, according to Reuters's "Digital News Report 2023."[105] Thus, Finnish citizens may feel less of a need to search for alternative sources of news that may overwhelm them with misinformation. In the last few years, countries around the globe have been trying to learn from Finland and follow its successful blueprint.[106]

In other parts of the world, media literacy and fact-checking organizations try to focus their efforts on specific regions. The United States has many great media literacy programs based at universities that freely share media literacy tools on their websites, including Civic Online Reasoning from Stanford University and Project Look Sharp from Ithaca College.[107] Portal Check is a great resource for Spanish-speaking people and those from Latin American countries. It provides resources that are tailored for diverse groups of people, including materials for the general public, journalists, content creators, teachers, and government officials. Africa Check is a fact-checking organization that focuses on debunking false claims from public figures all over Africa. You can easily search for the country whose fact-checks you are interested in viewing, and you can also read about media literacy programs in the area.[108] Political scientist Jeremy Bowles led a study in which he and other researchers collaborated with Africa Check to determine if their debunking materials helped improve media literacy among a sample of South African participants. Participants in the treatment condition regularly received debunking information in the form of podcasts or text-based fact-checks via WhatsApp. Participants who received these fact-checks were significantly better at identifying misinformation on social media than those in the control condition, who did not receive the fack-checks.[109]

As mentioned earlier, *Harmony Square* and *Bad News* are now available in dozens of languages, as they are being implemented by more and more researchers around the world. *Bad News* has been found to enable people across Sweden, Germany, Poland, and Greece to detect misinformation.[110] Other scalable interventions have been developed as well. For example, in a large study led by

Andrew Guess, a media literacy intervention was administered to thousands of participants from the United States and India. It took the form of a passive inoculation, in which participants read tips for checking the credibility of online information (e.g., look at other sources to confirm). The researchers found that participants in both countries were better able to distinguish true and false headlines when they were in the media literacy intervention condition.[111]

After focusing only on U.S.-based misinformation myself, I wanted to participate in research with a more international perspective. I was thrilled to be able to do just that when I accepted a postdoctoral research position at the University of Notre Dame. I worked with Tim Weninger, a professor in the Computer Science and Engineering Department, who was leading a major research project on media literacy in Southeast Asia and needed a sociologist to join his team.

During my time at Notre Dame, we developed different media literacy tools that were designed for a Southeast Asian audience. We used a media literacy website, Literata.id, which was developed by IREX, a global education organization that creates both online and offline media literacy content. The website covers various topics from algorithmic biases to confirmation biases. Its focus is to help people learn how easily misinformation can spread online, and it provides practical tips to identify and avoid falling prey to common manipulation tactics. We also developed an inoculation game, *Gali Fakta*, while working with Moonshot, a global company that fights extremism and violence around the world.

Gali Fakta is a WhatsApp-inspired game, in which you interact with characters who share misinformation with you. We designed the game in this style because WhatsApp is an extremely popular social media platform in Indonesia, with over 86 percent of the population using it.[112] WhatsApp allows users to easily control what information they see and who they connect to. Many WhatsApp groups are private as well, making it difficult to quantify just how much misinformation is on the platform. WhatsApp does not have the same type of algorithmic boost for emotional content that other, broader social networking sites have so users can focus more on interacting with their close friends and family. Information from stronger ties may be shared more frequently on WhatsApp, and we have learned that information from more meaningful ties can be more influential.

Gali Fakta begins by describing the goal of the game: to help your friends and family identify misinformation. The characters in the game pretend to be your family and friends, and they message you when the game starts. You can choose among different text responses to them as you try to get them to reflect on whether what they shared is legitimate. *Gali Fakta* allows you to choose among several topics you are interested in: health and beauty, news and politics, and online investment schemes. For example, suppose your little sister messages you about some acne cream that a celebrity endorsed. She shows you the link, and you can point out to her that there is no scientific research supporting the cream and that the "celebrity" endorsing the product is actually a fake account.

We worked with Indonesian TikTok influencers who helped spread our media literacy tips and website on their platforms. We also paid for advertisements to show our media literacy website and game on Twitter, YouTube, and Facebook. About 500,000 people clicked on the links to play *Gali Fakta* or learn about media literacy on our website.[113] *Gali Fakta* won a Webby Award (like an Emmy but for work done on the internet) for its sleek design and positive feedback from those who played the game.[114]

But did this intervention actually help people detect misinformation? In a pilot study, we surveyed a group of Indonesian participants who had the option to spend time on our media literacy website and play *Gali Fakta*. Those who spent time on our website and played the game were significantly less likely to report that they would share misinformation online than those who did not.[115] While this suggests that engaging with our media literacy materials may make people more cautious about their online behavior, this was just an initial test and not a randomized design.

We then conducted a study with one thousand Indonesian participants who were randomly assigned to play either *Gali Fakta* or *Tetris* (the control condition). Participants then evaluated the accuracy of seven true and seven false news headlines currently circulating in the Indonesian media. Some of the headlines focused on COVID-19. For example, one false headline claimed that the COVID-19 vaccine created more COVID-19 variants. However, we also had more apolitical headlines, such as the false claim that buying gasoline with a total that ends in an odd number could save you money. The true headlines were matched to cover similar topics. We also assessed self-reported media literacy and various demographics.

What did we find? Well, participants in the *Gali Fakta* condition were significantly more likely to rate the false headlines as more inaccurate. Additionally, they reported a decreased likelihood of sharing false headlines. Crucially, playing *Gali Fakta* did not increase skepticism of all headlines, as those who played *Gali Fakta* were not significantly more likely to rate true headlines as more inaccurate or reduce their intent to share factual headlines. These results all held while controlling for age, gender, education, urban versus rural status, income, religion, and political ideology. Finally, we found that playing *Gali Fakta* did not increase self-reported media literacy behaviors (e.g., agreeing with statements like "I think about how someone creates news that I see").

Thus, our prebunking game designed for Indonesian participants successfully increased their ability to detect misinformation and decreased their intent to share false headlines.[116] Essentially, by playing a game that forced them to reflect on the veracity of what their friends and family members shared online, they became more skeptical of misinformation. These results are certainly encouraging, but our prebunking game is just one tool that can help combat misinformation.

I asked Tim Weninger, the leader of our international media literacy project, what things he thought are important when establishing an international media literacy research program. He shared that sincerity is crucial. He then went on to give an example of how he flew from South Bend, Indiana, to Washington, D.C., just for a fifteen-minute meeting with the U.S. Agency for International Development. Doing so demonstrated his sincerity and commitment to the project, and it was eventually funded by the organization. When we work with different cultures, it is imperative that we make a strong effort to understand their perspectives and work with them in a meaningful way. When I joined Notre Dame, part of my job was seeking out and building connections with scholars actually living in the countries we were studying so that we could make sure their perspectives were represented in the articles that we published together. We don't want to use a different country for a new publication and leave our collaborators in the dust. We also share our research with Indonesian academics, journalists, and other industry leaders so they can use our findings to help their own communities directly.

In January 2024, I was able to travel all the way to Jakarta, Indonesia, to present my work on *Gali Fakta* at a conference. It was the first time I had ever visited Indonesia (or Asia), and it was by far the longest trip I've ever taken. I literally flew around the world, as I crossed over the Atlantic Ocean on the way to Indonesia and over the Pacific Ocean on the way home to the United States. I had an amazing time meeting my collaborators in person rather than over Zoom; talking to various other academics, industry leaders, and journalists from the area; and learning firsthand about Indonesia's culture, politics, and media ecosystem. I'm thankful that my time at Notre Dame allowed me to connect with many different scholars and educators around the world, and I plan on maintaining these connections throughout my career.

Our team at Notre Dame is just one group working toward establishing a more global approach to media literacy and misinformation. I hope to see greater collaboration among researchers in different countries, as misinformation is certainly a global problem. Creating parity among those studying misinformation around the globe is one macroissue that can help combat that misinformation. There are other structural considerations we must look at as well.

Economic Incentives

Conversations about media literacy and misinformation are often incomplete without acknowledging the economic incentives and challenges involved. As I mentioned earlier, it would be great if teachers could build media and information literacy into their lesson plans. However, they are already overworked and underpaid. Supporting teachers and media literacy nonprofits is certainly a critical step that anyone can take. There are also groups such as Media Literacy Now that advocate support for teachers so they can teach media literacy in schools.

While developing prebunking games is much cheaper and more scalable than paying teachers to teach a class, this also comes with its own financial challenges. As already discussed, video advertisements describing online manipulation tactics show promise, but someone still must pay for these ads. There are great organizations—such as Civic Online Reasoning at Stanford University and Project Look Sharp at Ithaca College, mentioned earlier—that provide free media literacy resources and lesson plans to anyone who visits their websites. Of course, people must still find them or incorporate them into their lesson plans. Furthermore, media literacy programs can vary in quality and may be focused more on corporate interests than on global education.[117] Media studies scholar Nolan Higdon highlights some of these challenges in his book *The Anatomy of Fake News: A Critical News Literacy Education.* He notes that some media literacy programs are focused more on promoting corporate software products and can even harvest user data for profit.[118]

While media and information literacy efforts can require considerable amounts of money, spreading misinformation can be extremely profitable. We've already seen various examples of social media influencers who are making millions of dollars by sharing fake news and/or pseudoscience on their platforms. Emotionally evocative and sensationalist information is promoted by social media algorithms, which give misinformation an unfair advantage over nuanced factual information. Social media boost sensationalism because they are designed primarily to attract as much human attention as possible in order to sell advertisements.

Carl Bergstrom, biologist and misinformation researcher, has studied how individuals, by sharing information, help form the groups and the communication patterns of complex social networks.[119] These are his thoughts about online advertising and social media: "Our message is not 'ads are evil.' It's that social media etc. been designed largely to sell ads, which means it is not designed with care to facilitate the spread of reliable information, let alone improve human well-being."[120]

Social media accounts that regularly share extreme, misleading, and sensationalist content generate a lot of attention, which means advertisements on the platform get seen much more frequently. The Center for Countering Digital Hate estimated that just ten of the most extreme accounts on Twitter could generate $19 million annually in advertising revenue for the platform.[121] Social media companies may lose advertisers if they have too much polarizing, sensationalist, and/or extreme content on their platform, but that same type of content also generates a substantial amount of activity and engagement. I've been focusing on social media since their influence on our information ecosystem only continues to grow. However, it is important to note that news programming on traditional media has also suffered from an overreliance on advertising revenue, which incentivizes sensational and partisan news coverage instead of nuance and substance.[122] When any medium's main revenue source is directly linked to the number of people watching, captivating content becomes significantly more profitable than purely informational content.

Psychologist Jon Roozenbeek, creator of several major media literacy games, wrote an article in which he argues the importance of addressing structural problems in addition to employing individual strategies like the prebunking games.[123] After reading his paper, I asked him what specific issues should be addressed from an economic perspective, and he responded:

> I think it's useful to take a critical look at how tech companies and content creators make money with news and information, and identify perverse incentives. 'Ad-tech' is shorthand for the technologies around how advertisements reach people. The basic incentive is to ensure maximum engagement with ads, which often means displaying them near engaging content. Unfortunately, some of the drivers of online engagement are negative (for example negative emotions, outgroup animosity etc), and so the algorithms that optimise for ad viewing time might learn that showing people harmful content increases ad revenue. To which extent this is the case is difficult to know because these algorithms are confidential, but I do think this conversation is probably best had out in the open.[124]

Ultimately, we must grapple with the fact that many of these social media companies profit from capturing our attention in order to sell advertising. At the very least, it would be helpful to know more about how these companies operate. As noted in the last chapter, many researchers agree that greater data transparency on the part of these major social media companies would significantly increase our understanding of how their algorithms boost misinformation and how we can better develop tools to educate people about this. The financial perspective is one major structural issue, but media literacy can benefit from considering macrolevel solutions.

The Need to Consider Macrolevel Solutions

In this chapter, I have described the utility of media and information literacy as well as various techniques and interventions that can increase such literacy. However, much of the work summarized in this chapter, and in the overall academic literature on misinformation, views the problem and solution in terms of an individualistic framework. There are many experiments that describe how groups of individuals process information and how media literacy interventions, like prebunking, can help. As we start to conclude this chapter and this book, I'd like to call attention to the extreme importance of addressing the broader macroforces as well.

Nick Chater, professor of behavioral science, and George Loewenstein, professor of behavioral economics, have written an excellent article describing the importance of macrolevel structural components for solving social problems.[125] They describe how individual-focused interventions can be highly effective and yet still have quite a small impact. For example, many advertising campaigns have focused on the need for individuals to reduce plastic waste in our environment. A classic and particularly salient example of this is the "Crying Indian" ad released in 1971, in which an actor who is wearing Native American attire (but isn't Native American himself) paddles his canoe across a lake that becomes more polluted over time. When he reaches the shore, he walks toward a highway, and a passenger in a car throws a paper bag at him. The actor portraying the Native American cries a single tear. The ad ends with the message "People start pollution. People can stop it."[126]

Chater and Loewenstein state that individualistic interventions (e.g., prebunking) are common in the misinformation literature but that there is not enough emphasis on system-level solutions. However, they also do not provide any explicit solutions beyond briefly suggesting further regulation of social media. It is very difficult to conceive system-level solutions and support them with strong evidence. An experiment that attempts to change how someone answers a question on a survey can be designed and carried out relatively cheaply and easily. However, when we think about something as complex as government regulations, it becomes much more challenging because we cannot control for the nearly infinite number of variables involved.

Of course, focusing on how "people" start and stop pollution is misleading and incomplete. Corporations are vastly more responsible for plastic waste and pollution than any individual.[127] The antipollution ad campaign was run by Keep America Beautiful, an organization founded by major corporations such as Coca-Cola and Dixie Cup Co., and its ads conveniently place the responsibility of protecting the planet on individuals instead of large companies.[128] Meanwhile, major corporations have been actively lobbying for laws that allow them to produce more single-use products, which can increase their profit margins.[129] Individual recycling won't prevent the massive plastic pollution problem, as most plastics can't be recycled and there is a lack of proper infrastructure and economic incentives to make it work.[130] It's estimated that just 5 percent of plastic is actually turned into new things, and this proportion is likely to decrease rather than increase over time. Even when we see promising results from individualistic interventions, like painting footsteps toward a trash can in order to reduce littering, they are ultimately difficult to implement at scale.[131]

What does work to reduce plastic waste is forcing companies to create less waste in the first place. This requires regulation, which can be difficult, given the large lobbying firms of these corporations. However, there is evidence that when a regulation is passed that is focused on reducing plastic waste, it can make a significant difference. For example, when a plastic bag ban was implemented in San Jose, California, there were 89 percent fewer plastic bags caught in storm drains and 60 percent fewer in rivers.[132] If corporations bear the responsibility of producing polluting materials through taxation or other disincentives, then they are much more motivated to change their products so they cause less pollution.[133] Research has also shown that when people connected proenvironmental behaviors to their identity, they were more likely to support a carbon tax.[134] Thus, people became more likely to support system-level change when such solutions became part of their identity.[135]

As mentioned in chapter 5, greater data transparency on the part of social media companies is one clear system-level solution supported by the consensus of academics. However, things like the degree of content moderation come with considerable challenges and disagreement. While building friction into these platforms to slow down the speed at which extremist rhetoric and misinformation can spread, far more study is required to learn the pros and cons of such an action. Other system-level solutions involve building relationships between scientific institutions and their communities in order to bolster trust, as we also saw in chapter 5. Building more relationships among industry, academia, and government would also help. Again, putting these ideas into practice becomes tricky once we think about the details, but it is not an insurmountable problem.

We need more people to think deeply about how to develop evidence-based system-level solutions for social problems, including misinformation. As I wrap up this chapter, I'll address the importance of identities and networks in media literacy and consider how we can link the micro-, meso-, and macrolevel components of media literacy.

IDENTITIES, SOCIAL NETWORKS, AND MEDIA LITERACY

Our own individual media and information literacy is important, but an even greater power lies in our ability to help foster such literacy in others. We have already learned how powerful networks can be for bolstering our identities and spreading our beliefs. I asked Tim Weninger how we can combat misinformation more broadly, and, he highlighted the importance of the social component: "To me, the question of fake news is a question of values. What kind of information do we, as a society, value? If we don't value truth and fact, then we won't get any. No amount of take-downs and supply side controls is going to curb its spread. It doesn't work for drugs, it didn't work for alcohol or smoking, and it won't work for fake news either. Our best bet is to try to change the culture and the values of the consumer and that's what I'm trying to do."[136]

So how do we instill these values of caring about truth over fiction? Well, there is growing research on the concept of postinoculation talk, which could provide some answers. We've already seen how beliefs can spread rapidly online and how our personal connections can influence what we believe. So far, we have focused mostly on misinformed and inaccurate beliefs. However, social influence can also promote more positive behaviors. Just like widespread vaccine adoption can protect the public against a virus, widespread information literacy can protect that public against misinformation.[137] Because prebunking techniques have reduced people's willingness to share false information, they can have cascading effects in cleaning up the information ecosystem.[138]

As I've argued throughout the book, we are social creatures, and these social processes are integral to our belief formation. When one person bonds with another over a shared identity, this can increase their confidence in beliefs associated with that identity. This appears to be true when talking about the prebunking process as well. Simply participating in a prebunking intervention has been found to make people more likely to discuss what they have learned.[139] For example, one study had participants read rebuttals to arguments regarding salient political issues like the legalization of marijuana. Those in the inoculation

condition read a message that was consistent with their own attitudes on the subject before reading a counterargument later on in the study. Those in the control
condition simply read a historical account of the process of regulation before
reading the counterargument. In a follow-up study, those in the inoculation condition reported discussing the issue more with friends and family than did those
in the control condition. Not only did the inoculation lead to additional talk
afterward, but also this discussion was linked to greater resistance to the counter-
message.[140] When participants discussed what they learned in the inoculation
message with their personal networks, they bolstered their attitudes toward the
topic at hand (e.g., legalization of marijuana) and became even more resistant to
propaganda or misinformation disputing their position.

A study led by business professor Lindsay Dillingham applied inoculation
theory to financial markets.[141] During a market downturn, fear may motivate
people to sell their investments and leave the market. However, this is a bad idea
because the market will likely bounce back, and by staying out of the market,
an individual misses the opportunity to buy investments while they available at
discounted prices.[142] Dillingham had participants in her study learn about the
importance of staying in the market as the inoculation intervention. However,
some participants were asked to share this knowledge with other people (postinoculation talk), and other participants were not. The participants who were
asked to share their knowledge of the importance of staying in the market had
significantly stronger convictions about staying in the market compared to those
who were not asked to talk about what they learned. This showed that inoculation effects can be enhanced if people share that knowledge with others, as they
also internalized what they'd learned.

Leading researchers on prebunking techniques Josh Compton, Sander van der
Linden, John Cook, and Melisa Basol wrote a great review paper on inoculation
research in which they advocated more research on the social network effects of
media and information literacy interventions: "Because inoculation theory leads
to more talk, and more talk can lead to more acceptance of science, inoculation
theory-based messages about science could be particularly powerful, and future
research should focus on the best way to design such inoculation messages."[143]

Designing games for certain audiences may increase the likelihood people will
want to talk about them afterward. I conducted a study that evaluated potential
levels of postinoculation talk for our *Gali Fakta* game. The Indonesian participants played *Gali Fakta* (a media literacy game designed for an Indonesian audience), *Harmony Square* (a media literacy game not designed for an Indonesian
audience), or *Tetris* (our control condition), and then I asked them how much
they enjoyed the game they played and how likely they would be to share what

they learned with their friends and family. They rated *Gali Fakta* as more enjoyable to play than *Harmony Square*. Additionally, they reported a significantly higher greater desire to share what they learned from *Gali Fakta* compared to those who played *Harmony Square*.[144] This showed that a game designed with a certain group of people in mind could increase the likelihood they would want to share what they learned with their networks.

There are still many unanswered questions regarding postinoculation talk. How often do people share fact-checking and media literacy techniques with their networks? If they do, what conditions impact the likelihood they will share? A study led by Irene Pasquetto investigated some of these questions using WhatsApp data.[145] As mentioned earlier, WhatsApp is a unique social media website, as it focuses more on closed, intimate connections and less on open networks of information. Pasquetto and her team collected WhatsApp data from over a thousand Indian and Pakistani participants, including the strength of the ties they communicated with and whether those ties agreed with them politically. Participants then were shown various headlines that spread misinformation. After viewing the misinformation, participants read a fact-check that debunked the false information they had just seen. However, in the experiment, the participants were split into four groups, and the suggested source of the fact-check was different for each. One group was asked to imagine this fact-check was from a strong tie, a second group was asked to imagine it was from a weak tie, a third group was asked to imagine it was from a politically congruent tie, and a fourth group was asked to imagine it was from a politically incongruent tie. The participants were then asked how likely they were to share this fact-check with others on WhatsApp. Those who imagined that a strong tie sent them the debunk were significantly more likely to reshare it. Additionally, those who imagined a tie who shared their politics sent them the debunk were significantly more likely to reshare it. Thus, this experiment revealed that the strength and shared identity of ties significantly impacted how likely someone was to reshare debunking information. Again, trust was crucial here. I know I am certainly more likely to reshare something online if it's shared with me by someone I trust.

Learning these media literacy skills clearly helps protect us against misinformation. However, simply believing we have an identity that values evidence is not enough because it is too easy to fool ourselves into believing we are behaving in an objective manner. Let's call this identity that values evidence our truth-seeking identity. Even if we really try to be objective and fact-check ourselves, we are constantly battling the automatic tendency for our brains to want to affirm (and verify) our identities. For example, we may start fact-checking something online but stop once we find a source that aligns with our political identity. In this scenario, our truth-seeking identity and political identity are competing against one another. We may resolve this conflict by doing a brief fact-check that sufficiently satisfies our truth-seeking identity, but that is not actually sufficient, and we still believe something false because it supports our political identity. If we do this process alone, then there is no one else to call us out on our faulty fact-check, and we may believe we did a good job avoiding misinformation. Furthermore, we may develop overconfidence regarding our ability to distinguish truth from fiction, since it feels like we did the fact-check even though, again, it wasn't a good one. Remember, research earlier in this chapter revealed that most people think they are better at spotting fake news than they really are and that those who are the most overconfident in their reasoning ability also are more likely to believe conspiracy theories. Our brains are always looking for ways to confirm our identities and give us an ego boost, but that process doesn't always help us accurately process information.

We will have much greater success in truly developing any truth-seeking identity if we regularly bring other people (especially those who think differently than we do) into our fact-checking process. Other people we trust can hold us accountable as we try to navigate the truth. Of course, it's not practical (or possible) to always ask people to help us fact-check. One way we can still use social influence to our advantage when we are alone is to imagine a friend of ours is watching us as we go through the process of checking if a claim is true. So the next time you fact-check something, try to imagine your truth-seeking friend is evaluating your technique. Try to reflect on whether they would consider your fact-check sufficient. This imaginative technique will be more effective if you regularly discuss your fact-checking process with this trusted friend so you can constantly be reminded of what they are looking for. This can be a mutually beneficial process, as you can help call out any potential biases for them as well. Psychologist Adam Grant had a great quote summarizing this idea: "Great minds don't think alike. They challenge each other to think differently. The people who teach you the most are the ones who share your principles but not your thought processes. Converging values draw you to similar questions. Diverging views introduce you to new answers."[146]

When we think about our personal networks, it is crucial that we expose ourselves to diverging viewpoints so we can avoid falling into simplistic explanations of the world that are often fueled by misinformation. Having people around us who share principles of evidence-based inquiry and intellectual humility also aids this process. Ideally, you and a close friend of yours can mutually verify your truth-seeking identities, but these identities are focused on the principle of evidence-based inquiry. Your other values can differ, and as noted throughout this book, having ideologically diverse associates is a great way to protect ourselves from believing falsehoods. As we saw in chapter 4, having an ideologically diverse personal network decreases the likelihood of our being polarized and believing misinformation.

By regularly practicing these healthy media behaviors with our networks, we can create and maintain a new identity as someone who carefully and thoughtfully processes the information we see online. As we practice habits that align with our values, they can strengthen our important identities.[147] By including other people in this process, we can further protect ourselves from believing misinformation. We've learned throughout the book how powerful identities can be. Thus, when we use that power to work on developing an identity that carefully processes information with others who can hold us accountable, we are better equipped against misinformation. Learning how our identities and networks influence us, which could be summarized as *social identity literacy*, should be integrated into a broader media literacy curriculum. Developing our media literacy skills, reflecting on our identity biases, and cultivating diverse interpersonal relationships that can hold us accountable are all vital to protecting ourselves from becoming misguided.

CONCLUSION

Giving people the tools to inoculate themselves against misinformation and disinformation can be individually empowering. We learned about the benefits of teaching people various information and media literacy skills. Prebunking games are scalable and fun techniques that can share many of the lessons about media literacy. Individually focused solutions, such as improving media literacy and prebunking, also require a discussion of how microfocused techniques fit within a broader sociological context. Source effects, economic incentives, and an English-speaking and Western bias currently limit the promotion of global media literacy. We should also consider how broader, structural impacts can facilitate better media literacy in our society. We can apply what we have learned about identities and personal networks to better improve our own fact-checking process as well. Including other people in our efforts to spread media literacy techniques and hold ourselves accountable is crucial if we want to try to protect ourselves from misinformation as much as possible. Misinformation is a complex, multifaceted, and global problem that requires information and media literacy that considers micro-, meso-, and macrolevel approaches.

CHAPTER 7

THE FUTURE OF MISINFORMATION

I began this book by describing the various types of misinformation and their potential harm. In chapter 1, I explained how *misinformation* is a broad term that includes any type of false or misleading information. However, there is also purposeful, agenda-driven false information, which is called *disinformation*. Both misinformation and disinformation are not new, but they have received more attention thanks to both the polarizing 2016 U.S. presidential election and the growing impact of social media on society. Misinformation has consistently had significant impacts on our economy, health, and politics, and there is now a global consensus that it is a major problem.

Acknowledging the issue of misinformation is a good start, but what factors contribute to believing false information, and what can be done about it? We began to answer those questions in chapter 2 by diving into the literature around identities and how they influence our susceptibility to misinformation. Identities provide us with meaning and self-esteem and connect us with other people. We are constantly motivated to maintain consistency between the values of our identities and the way we engage with the world. This also means that we may be biased in processing information so that it affirms an important identity, and this bias could guide us to believe something misleading or untrue. To tackle the misinformation problem, we must grapple with how our identities impact both how we behave and what we believe. We can start to do this by mapping out which identities are important to us and reflecting on how our identities intersect with one another. When we feel sufficiently affirmed by one identity, we may be more open to objectively processing information that challenges another identity. If we want to connect with another person who thinks differently than we do, we can relate to a shared identity in hopes of depolarizing the conversation.

I wrote this book throughout the COVID-19 pandemic, and it was an unfortunately clear example of how identities can make us vulnerable to misinformation. In chapter 3, I described how COVID-19 was politicized early in the pandemic and how responses to the virus divided along partisan lines. The greatest predictor for many attitudes and behaviors surrounding the virus was one's political affiliation. While identities can close our minds to new information, the same identity processes can be used to open them up again. I concluded the chapter with five strategies that can help us have more productive conversations with those who think differently than we do: establishing mutual respect, relating to a shared identity, reframing the conversation to address the others' concerns, revising our language, and repeating all of the above as necessary. Using these techniques can lead to reduced polarization through mutual understanding. Having a productive dialogue with someone we trust can also help us identify when we too may have fallen for misinformation.

Exploring how to have productive conversations with other people allowed us to transition into investigating how our personal networks influence our beliefs in chapter 4. We learned that we tend to associate with those who share our beliefs and that this can exacerbate polarization and susceptibility to misinformation. Not only are we exposed to more homogeneous information in ideologically homogeneous networks, but also repeated exposure to similar ideas can strengthen our identities through mutual identity verification. Social media make this mutual identity verification process even easier, as we can readily find many different people who can verify and support our beliefs. Social media also involve information overload, engagement bias, echo chambers, and manipulation, which further promote the spread of misinformation.

I continued the discussion of personal networks and social media in chapter 5, using the antivaccine movement as a salient example. We reviewed the history of the antivaccine movement and saw that it existed long before COVID-19. Echo chambers and social media certainly made it easier for vaccine misinformation to spread, and we considered a few possible solutions that would counter online misinformation. If social media companies were more transparent with their data and algorithms, researchers would be able to understand the problem of misinformation much better. Collaborations among the tech industry, independent scholars, and policymakers would also help. Additionally, since misinformation is much more likely to go viral compared to factual information, slowing down viral content with friction could be beneficial. Rebuilding trust in institutions is also central to combating health misinformation, and I ended the chapter with several ideas for rebuilding trust at the micro-, meso-, and macrolevels.

In the first five chapters, I focused on the social and psychological components of misinformation. In chapter 6, I added the vital component of education to the conversation. We covered the complex role of formal education in susceptibility to falsehoods and learned how cognitive reflection, critical thinking, and intellectual humility all protect against misinformation. More specifically, learning information and media literacy skills (such as lateral reading) can help people better evaluate information online. Learning these skills can take time, but prebunking games are highly effective and scalable options for promoting media literacy. Prebunking, while showing promise, has several major limitations and challenges that hamper our attempts to promote media and information literacy: source effects, economic incentives, an English-speaking and Western bias, and the need to consider systemic, macrolevel solutions. Finally, I ended the chapter by integrating research on identities, social networks, and media literacy. Combating misinformation requires many different types of solutions working at different levels and from different angles.[1]

We can share what we've learned about the psychology and sociology of misinformation with our personal networks to increase the reach of this research. If you share just one technique you learned in this book with another person, you are helping influence members of our society to be more thoughtful about how they consume information. You don't have to wait until a friend or family member shares misinformation online. And as we learned previously, challenging people online can be counterproductive. Resources that explain how to critically evaluate sources or how to become mindful of our own identity biases may stick with a person if they are shared by someone they trust. As I wrap up this book, I'd like to look forward to what I believe will be important challenges for combating misinformation in the future.

✧

The real problem of humanity is the following: We have Paleolithic emotions, medieval institutions and godlike technology. And it is terrifically dangerous, and it is now approaching a point of crisis overall.
—Edward O. Wilson

This quote by Edward O. Wilson has always resonated with me because it succinctly summarizes why we are dealing with so many problems in our society today. We can also apply this quote to the problem of misinformation. Our minds have evolved to identify patterns in our environment, which helps us cope with feelings of uncertainty. We also want to connect to others and feel like we belong. These are powerful forces built into our psychology that can bias how we process information. If the information helps us reduce uncertainty and helps us feel connected to others, then we are much more likely to believe it, regardless of its veracity. I applied the research on identity theory and personal networks in describing this emotional susceptibility to misinformation. Our identities and personal networks help us connect to others and make sense of the world; however, they also can make us vulnerable to believing false information if that information supports our identity frameworks and helps us connect with others. In this book, I have mostly focused on the social psychological mechanisms that explain our susceptibility to misinformation, but we also must consider the structural, macrolevel forces that contribute to this problem.

In chapters 3 and 5, I discussed some of the problems with our institutions, such as lack of transparency, poor messaging, and even corruption. These problems reduce the amount of trust we have in our medical and scientific institutions and in our government. In the United States, both major political parties raise funds by appealing to the emotions of their supporters, which creates some vicious in-group and out-group dynamics. It's classic politics to galvanize as much support from your side as possible, even if this means attacking the other group rather than focusing on policy (just look at the fundraising emails from either Democrats or Republicans).

The American electoral system also rewards extremism fueled by misinformation. As journalist and political analyst Jamelle Bouie argues: "The problem is that the American political system, in its current configuration, gives much of its power to the party with the most supporters in all the right places."[2] For example, manipulating the boundaries of electoral districts so one politician has an advantage (i.e., gerrymandering) and so certain voting districts have unfair levels of power (i.e., malapportionment) is a major issue within the American political system.[3] We should also be mindful of which group has the most power when we evaluate the overall impact of political polarization.[4] There are many areas of government design being addressed by political scientists and other groups, but my point here is that there are larger issues we need to address when considering the spread of misinformation. If a political party has the incentive to appeal to its most extreme factions, then this is going to be a major barrier to preventing bad information from polluting our society.[5] We should remember that the American system is still a work in progress, and democracies are still fairly new inventions when we consider the timeline of human beings and civilization.

Finally, the other macrolevel force that influences misinformation is our "god-like" technology. As I end this book, I'd like to discuss a few technologies that have major roles in our information ecosystem. We have already seen how social media and their efficient capture of our attention can exacerbate polar- ization and widely spread false information. As social media become more and more optimized to keep us online, I worry that these negative outcomes will become even larger problems. In addition to experiencing more sophisticated online disinformation campaigns, we may become further isolated from people who think differently than we do, as our social media feeds may continue to cater to our "likes" and show us what the algorithm believes we will click on more. For example, the increasingly separate media diets and social networks for liberals and conservatives may continue to create very different realities, depending on which end of the political spectrum one falls.[6] This divided real- ity can still be quite profitable for social media companies, as they can show us just enough of the other side to keep us outraged—and online. Social media and tech companies and their leadership will have a huge amount of influence over society in general but especially over the way we deal with the issue of misinfor- mation. The onus is not entirely on social media companies, as we can also be more mindful of how we spend time online and take control over what data we share with these platforms.

Text-based misinformation is extremely common on social media and is also the easiest type to study. Researchers can quickly create surveys that ask people to evaluate factual versus false headlines, and they can download and analyze large amounts of text data from websites (e.g., tweets). However, videos and pic- tures also spread a lot of misinformation. Given the incredible popularity of the newer video-based social media platforms like TikTok, it will be very important for researchers to place more emphasis on video-based misinformation.

Beyond just studying video and picture formats, we must address how new technologies will make it easy to pollute our ecosystem even further with doc- tored and even wholly fake pictures and videos. Faked images are commonly called *deepfakes*, since they are created via deep learning technology. You can already find deepfakes of public figures and politicians saying outrageous things. Right now, deepfakes can be fairly easy to spot, and the most convincing ones still require some technical skill to create. However, fake pictures and videos will become more advanced to the point where most people will not be able to tell what is real and what is fake. I remember that in early 2022, my students at Vanderbilt could easily spot examples of deepfakes in MIT's quiz on detecting deepfakes. But in 2023, the website was updated, and the deepfakes were already much harder to detect.[7] Research has shown that simply exposing people to deepfakes creates uncertainty and erodes trust in what is real.[8] Not only will fake videos erode trust, but also bad-faith actors can use fake videos to advance their own agendas.[9] People were already starting to use these technologies to spread misinformation widely as I wrapped up this book in 2024, and I fear it will be much worse by the end of the decade.

Artificial intelligence (AI) poses another major concern related to the spread of misinformation and is already having a significant impact. First, it's important to define what we even mean by AI. I asked Tim Weninger, my Notre Dame colleague and professor of computer science, to give a simple but accurate definition of AI, and he told me, "Artificial intelligence is the set of technologies and computer systems that appear to have humanlike intelligence." For years, AI-powered chatbots have been able to generate text that sounds like it was created by a real human, but they still have serious limitations when we compare their conversation to that of a real human. However, in recent years AI chatbots have become much more sophisticated and much more useful for completing tasks. The deepfakes described here are another subset of AI that focuses on creating the most realistic video possible from the audiovisual data provided. AI is already being used to create fake pictures online, as computers can take the aggregate of many pictures and create a new one.

In 2022, Facebook claimed that two-thirds of online influence campaigns used profile pictures that were generated by artificial intelligence.[10] If someone wanted to create a fake social media profile using someone else's picture, you could reverse-search the image to see where that picture came from. However, with AI image creation, you can now make virtually infinite numbers and types of images that appear to be new. If you ever get spam messages on social media and don't get any hits when you reverse-search the profile picture, it is likely an AI-generated image. An encouraging 2023 study found that people could be effectively taught to identify fake social media accounts after taking an educational quiz (the *Spot the Troll* quiz). The results even revealed that people were more cautious with their social media sharing behavior after taking this quiz.[11] Of course, a significant challenge when learning how to identify fake accounts is that such accounts will only become more sophisticated and more challenging to spot as this technology evolves.

Beyond fake profiles, we are now seeing examples of fake videos in the form of news reporters sharing false information online.[12] For example, there was a deepfake video of the president of Ukraine claiming he was going to surrender to Russia.[13] There have also been deepfakes of news reporters from a made-up news outlet (Wolf News) sharing Chinese propaganda.[14] These deepfake videos are still fairly easy to spot, but, again, I worry that they will become much more sophisticated in the not-too-distant future. Teaching people how to spot fake images will be very helpful as part of an evolving AI literacy curriculum. Computer scientist Adnan Hoc and I are working on a research project that teaches

people how to recognize fake images through common manipulation techniques.[15] So far, we found that simply providing people with a brief explanation of
how to recognize common manipulation techniques significantly improves their
ability to detect fake images.

As AI becomes more advanced and more pervasive, its ability to deceive will
have a greater impact on society. ChatGPT is one salient example of how AI may
drastically change our information landscape. It was developed by OpenAI and
uses AI to answer questions in the form of a chatbot. Many other AI chatbots
are being developed, but ChatGPT gained international attention as I wrote
this book, so that is why I focus on it. A user can simply type a question into
ChatGPT, and it will provide an answer, using AI to search for and summarize
information on the internet. It operates like autocomplete while texting, but it
generates entire paragraphs instead of a single word. The writing is clear, and it
is sometimes quite accurate. As of 2023, it had already passed a Wharton MBA
exam and had been added as a coauthor on several scientific papers in recogni-
tion of its help writing or editing the paper.[16] ChatGPT can be great for getting
concise answers to specific questions, especially technical questions such as how
to code something properly. It's also useful for analyzing or editing text or code
that you share with it.

However, there are plenty of examples in which ChatGPT was quite inaccurate. For example, it provided a different (and incorrect) answer each time someone asked, "What was the first TV cartoon?"[17] There are also instances where
ChatGPT confidently provided an incorrect answer and even cited sources that
do not exist.[18] Despite these major limitations, the potential is massive. It's possible that instead of using Google or Bing to search through a variety of links,
people will rely on AI-generated answers. As I write this, search engines are just
now starting to incorporate AI chatbots with their search results. So, instead of
getting a diverse set of links when you search for something, you will get a brief
essay based on AI's decision-making. Google already provides short summaries
of search results, but these slick-looking AI chatbots give a strong impression that
a competent robot has properly searched the answer for us. Not only can AI be
used to create propaganda and disinformation, but also it may completely change
how we search for information online, even when we are trying to be objective.
Research has found that conversing with AI chatbots that are programmed to
affirm our beliefs exacerbated our biases.[19] Another study found a negative correlation between heavier use of AI tools and critical thinking abilities, especially
in younger people.[20] This further highlights the importance of developing up-to-
date media literacy skills that evolve with technology, avoiding an overreliance
on AI tools, and constantly reflecting on our own social and cognitive biases.

ChatGPT will also make it easier to pollute the internet with false but credible-sounding news outlets without needing to pay people to write daily articles.[21] Research has already shown that people were successfully persuaded on political issues by ChatGPT-generated messages.[22] The AI-generated messages were as persuasive as messages written by humans (however, both types of messages had only a small effect; as we've learned throughout this book, it's hard to change people's minds). AI can even learn how to make dating profiles that are rated as more human than human-generated profiles![23]

We are just starting to learn how artificial intelligence will impact our society. Even if ChatGPT's messages have a small effect, they are also quite scalable and could have a significant impact when deployed at a larger scale. Bots and automated accounts are rather clunky, and research has shown their impact on beliefs was probably pretty small. However, AI can grow and evolve, and it may learn how to best persuade humans in a way that is more effective than human efforts. If AI social media accounts can successfully trick people into thinking they are real people, their impact may become even stronger. While AI can help create disinformation content, however, there is still the major challenge of actually distributing the content to large groups of people. As I write this book, I agree with computer scientists Sayash Kapoor and Arvind Narayanan that the immediate risk of AI chatbots is that people will assume their output is accurate instead of doing their own fact-checking.[24] They point out that ChatGPT can provide helpful medical information sometimes, but it can also be wildly inaccurate.

I asked ChatGPT 3.0 if it contributes to misinformation. ChatGPT replied:

> As an AI language model, my responses are based on the text data that I have been trained on, which includes a wide range of sources of information. My responses are generated based on statistical patterns and linguistic rules and do not represent personal beliefs or opinions.
>
> However, it is possible that my responses may sometimes provide incomplete or inaccurate information, especially if the source of the text data used to train me was biased or contained errors. It is important to note that I do not have the ability to verify the accuracy or reliability of the information I provide and I cannot be held responsible for any actions taken based on my responses.
>
> Therefore, it is important to use critical thinking and cross-check information from multiple sources before making decisions or drawing conclusions, especially when it comes to important topics such as health, finance, or politics.

So, yes, ChatGPT "understands" that it can be inaccurate and stresses the importance of using critical thinking and media literacy skills. I asked ChatGPT this question in 2023, so its answer was from one of the earlier versions, but I was glad to see it was explicit about its limits and encouraged critical evaluation of its responses from the start. Hopefully, this book and other resources can help us understand our own biases and better equip us to counter the negative outcomes of rapidly advancing technology such as AI. Future work should also strive to understand how people interact with AI chatbots and what elements influence how much we rely on them as well as their perceived trustworthiness. My colleagues at Notre Dame and I have found that chatbot answers with at least one citation are rated as more trustworthy, but this citation impact quickly reached its upper limit, as trustworthiness did not increase if the output had five citations.[25] Understanding our own psychology and how it interacts with advanced technologies will continue to be a challenge for the well-being of our society.

Although AI is definitely a concern worth monitoring, we should remember that major media outlets still produce an overwhelming amount of biased and misleading information in our society.[26] Additionally, as we saw with the Israel-Hamas conflict on Twitter/X, out-of-context images and video game footage claiming to be videos of war still spread rapidly.[27] Thus, using very simple technology and preying on people's biases, we can still spread massive amounts of misinformation and disinformation.

It's important to note that all of the AI discussed here still falls under the umbrella of "narrow AI," which focuses on singular tasks and can achieve only limited goals (such as pulling text from the internet to answer a question or organizing audiovisual data to make a fake video). These narrow AI tools can perform certain tasks better than humans, but, again, that type of intelligence cannot reach across other domains. Artificial general intelligence (AGI) is theorized to be able to complete all the tasks a human can do across many domains. It is quite a broad term, and it may be more useful to define AI competency by describing how well it can perform certain types of tasks.[28]

The nefarious potential of AI can feel quite dystopian, but we must remember that it can also be used productively. Researchers have developed an AI chatbot that can help improve the quality of contentious political discussion. They found that when one person used it during a dialogue, their discussion partner rated the conversation more positively.[29] Another intriguing study found that participants reduced their belief in various conspiracy theories after having a three-round dialogue with an AI chatbot that debunked and fact-checked their faulty arguments.[30] I have already described the benefits of prebunking games, and AI chatbots could be a vital educational tool. Additionally, ChatGPT has been found to be a tool for helping with (not replacing) human fact-checking. A 2023 study gave ChatGPT twelve thousand different claims from PolitiFact to evaluate, and it accurately categorized 72 percent of them.[31] Other AI applications have been used to improve the reliability of search results so less false information appears in the search results and to detect fake news shared on social media.[32] My colleagues in the Computer Science Department at Notre Dame are developing a system that can detect whether an image has been doctored. Called the Misinformation Early Warning System (MEWS), this system could help identify a false image that is spreading on social media, which could be crucial for countering disinformation during elections or health crises.[33]

As image- and video-altering technology continues to advance, the ability to identify the original source will be crucial. Blockchain technology is a tool that could help prove the authenticity of a source. This idea of having a digital signature that proves the authenticity of a document has been around since the early 1990s—long before the advent of Bitcoin and other cryptocurrencies.[34] Blockchain creates a distributed list of *blocks* of data that are publicly viewable along a *chain*. These blocks of data are secured and authenticated by various cryptographic designs. Blockchain technology can record digital transactions and information in a fully transparent and secure network.[35]

Blockchain technology would allow a source of information to be publicly verifiable as original and legitimate. For example, a reporter could share the digital signature behind a video they took so you would know which one was the original and which one was faked. Benjamin Gievis is the cofounder of BlockExpert, which aims to authenticate news sources using blockchain technology. He commented on BlockExpert's impact on online information: "We're not here to say if this is good news or fake news. We're here to say this is authentic news. It's been recorded, it's being tracked, here's where it came from, here's its value. It's a new way of seeing the story."[36]

Authenticating the source of news will not eradicate the problem of misinformation, but transparency could help rebuild trust. This public and transparent nature of the blockchain is how various criminals have been caught trying to steal and launder money through cryptocurrencies.[37] Once you know someone's blockchain address, you can track their activity. Of course, for this to happen, people must agree to share their addresses so the public can verify their information, the public must trust that these addresses do indeed belong to the people named, and overall people must trust the blockchain method. None of these conditions is a given by any means. Whether it is accomplished using blockchain or some other technology, I do think it will be helpful to be able to directly check the original source of information in a transparent fashion.

It can be overwhelming to think about how technology will impact the future of misinformation, but that doesn't mean we can't learn to live with these tools in a healthy way. Cristina Tardaguila, founder of the Brazilian fact-checking agency Agencia Lupa, provides a hopeful and important perspective by comparing misinformation to email spam:

> Recall email spam. It was a problem in the late 1990s, a real hassle, but not anymore. What happened? Journalists wrote about it, educating people not to simply click, not to engage, and did it repeatedly. We need to spread the same gospel with disinformation, teaching people what fake news is, what it looks like, and that it can be real but in a misleading context. There is the technology part, just as there was with excluding spam from our inboxes. The same effort needs to go into filtering out what should not get through. Finally, we can create liability, responsibility, as we did when we made companies add "unsubscribe" lines to their emails.[38]

Not only does this quote by Tardaguila provide us with a concrete example of where we have made great progress on a digital problem by minimizing its harm, but also it addresses the importance of updating technology, adjusting policies, and adapting our own behavior. Technology experts learned how to better identify spam and likely malicious emails to keep users' emails safer. We should try to constantly improve our technologies and be mindful of how they can spread bad information.

However, we also should constantly be mindful of how our own human tendencies can contribute to the problem. Technologies can exacerbate any issues that stem from our own psychology as well as the institutions we build. Nathanael Fast, business professor and founder of the Psychology of Technology Institute, stresses the importance of understanding how our psychology interacts with technology: "To have the intelligent societal conversation about AI that we so desperately need, we require a much greater understanding of how human psychology drives our adoption of technology and how technology, in turn, shapes our psychology."[39] This sentiment was widely shared when I attended a conference on science communication and AI at the University of Notre Dame in 2024. Understanding the human element will only become more important as AI and other technologies evolve. As technology rapidly advances, social scientists, computer scientists, industry leaders, and policymakers must all work together to give us the best chance of mitigating the spread of misinformation.[40]

The problem of misinformation can feel overwhelming, especially when we consider how technology is poised to create new challenges. This can make us feel cynical sometimes, but I do think there are objective reasons for feeling hopeful about the future. Writing this book exposed me to many amazing people from different backgrounds who were using various approaches to solving the problem of misinformation. Academics, policymakers, and industry leaders are making a concerted effort to clean up our information ecosystem as best they can. This is certainly a daunting task, but it is not insurmountable, and a dystopian ending is not inevitable.

We should remember how young both humanity and our society are. We are slowly learning more about our own psychology and the ways we can mitigate our biases (formal social science is only a few centuries old). We are slowly trying to improve our institutional systems as well. As we learn more about our psychology and improve our institutions, we will be better equipped to handle our rapid technological advancement. World-changing technological developments will continue to occur at a rapid pace, and our understanding of psychology and sociology must match these advancements. The human element will be fundamental as we navigate the future of misinformation. This leads me to address why I chose the title of this book.

The title of this book is *Misguided*, which refers to the way various psychological and sociological forces influence (and guide) how we process information. Specifically, I have focused on how we are motivated to maintain consistency between the values of our identities and how we process information. Connecting with other people through shared identities provides us with a sense of belonging and meaning. The problem arises when these otherwise helpful tendencies can lead us to reject evidence in order to protect our identities and the social connections that come with them. That commonality is central to all of us, even though we might join various groups that motivate us to adhere to different values. Two people who have opposing political identities still share the same motivation to protect their identities and connect with their groups. Thus, when it comes to processing information objectively, we are all occasionally misguided by the social and psychological forces that help us stay connected with other people.

Susceptibility to misinformation is the conflict between our social motivations and reality. Life can be chaotic, tragic, and uncertain. Our social connections can keep us grounded and provide us with meaning as we navigate the absurdities of reality. To completely eradicate our susceptibility to misinformation, we would also have to eliminate fundamental aspects of our ability to connect with other people. Meaningful social connections are central to living a happy and fulfilled life. We cannot and should not try to eliminate our tendencies to connect with others through shared identities. Instead, we can work within this framework, better understand our psychology, and help guide one another away from misinformation.

NOTES

INTRODUCTION

1. *Behind the Curve*, dir. Daniel Clark (Delta-v Productions, 2018).
2. Dana Schwartz, "Director of *Behind the Curve* Shares How to Argue with People Who Believe the Earth Is Flat," *Entertainment Weekly*, March 1, 2019, https://ew.com/movies/2019/03/01/behind -the-curve-netflix-interview/.
3. "Social Media Seen as Mostly Good for Democracy Across Many Nations, but U.S. Is a Major Out- lier," Pew Research Center, December 6, 2022. https://www.pewresearch.org/global/2022/12/06 /social-media-seen-as-mostly-good-for-democracy-across-many-nations-but-u-s-is-a-major-outlier/.
4. "Journalists Highly Concerned About Misinformation, Future of Press Freedoms," Pew Research Center, June 14, 2022, https://www.pewresearch.org/journalism/2022/06/14/journalists-highly -concerned-about-misinformation-future-of-press-freedoms/.
5. Associated Press, "Dictionary.com Chooses 'Misinformation' as Word of the Year," VOA News, December 30, 2018, https://www.voanews.com/a/dictionary-com-chooses-misinformation-as-word -of-the-year/4674053.html; Shannon Bond, "'Disinformation' Is the Word of the Year, and a Sign of What's to Come," NPR, December 30, 2019, https://www.npr.org/2019/12/30/790144099 /disinformation-is-the-word-of-the-year-and-a-sign-of-what-s-to-come; "Oxford Dictionaries' Word of the Year Is 'Post-truth,'" BBC News, November 15, 2016, https://www.bbc.com/news/uk-37995600.
6. Nancy L. Rosenblum and Russell Muirhead, *A Lot of People Are Saying: The New Conspiracism and the Assault on Democracy* (Princeton University Press, 2019).
7. Marten Scheffer et al., "The Rise and Fall of Rationality in Language," *Proceedings of the National Academy of Sciences* 118, no. 51 (2021): e2107848118, https://doi.org/10.1073/pnas.2107848118.
8. David Rozado et al., "Longitudinal Analysis of Sentiment and Emotion in News Media Headlines Using Automated Labelling with Transformer Language Models," *PLOS One* 17, no. 10 (2022): e0276367, https://doi.org/10.1371/journal.pone.0276367.
9. Michael Barlev and Steven L. Neuberg, "Rational Reasons for Irrational Beliefs," *American Psychol- ogist*, April 15, 2024, https://doi.org/10.1037/amp0001321.
10. Jan E. Stets and Peter J. Burke, "Self-Esteem and Identities," *Sociological Perspectives* 57, no. 4 (2014): 409–33, https://doi.org/10.1177/0731121414536141.
11. Mihaly Csikszentmihalyi, *The Evolving Self: A Psychology for the Third Millennium* (HarperCollins, 2009).

12. Csikszentmihalyi, *The Evolving Self*, 66.

13. Svetlana V. Shinkareva et al., "Representations of Modality-Specific Affective Processing for Visual and Auditory Stimuli Derived from Functional Magnetic Resonance Imaging Data," *Human Brain Mapping* 35, no. 7 (2014): 3558–68, https://doi.org/10.1002/hbm.22421.

    My next project in cognitive neuroscience would have been studying how the brain processes different types of emotional mental imagery. I did collect some pilot fMRI data where participants imagined different scenarios while we recorded their brain activity. I never did anything with these data, but I did publish a paper describing the statistical process I used to create that mental imagery stimuli. Matthew J. Facciani, "Developing Affective Mental Imagery Stimuli with Multidimensional Scaling," *Quantitative Methods for Psychology* 11, no. 2 (2015): 113–25, https://doi.org/10.20982/tqmp .11.2.p113.

14. Susan J. Lee et al., "Fetal Pain: A Systematic Multidisciplinary Review of the Evidence," *JAMA* 294, no. 8 (2005): 947–54, https://doi.org/10.1001/jama.294.8.947; Dave Levitan, "Does a Fetus Feel Pain at 20 Weeks?," FactCheck.org, May 18, 2015, https://www.factcheck.org/2015/05/does-a -fetus-feel-pain-at-20-weeks/.

15. Diana Greene Foster and Katrina Kimport, "Who Seeks Abortions at or After 20 Weeks?," *Perspectives on Sexual and Reproductive Health* 45, no. 4 (2013): 210–18, https://doi.org/10.1363/4521013.

16. There continues to be a lack of evidence demonstrably proving fetal pain at twenty weeks. However, this is a complex topic. Research published after my scientific testimony debates the importance of a developed cortex and thalamocortical tracts to experience a form of fetal pain without the capacity for self-reflection. See Stuart Derbyshire and John C. Bockmann, "Reconsidering Fetal Pain," *Journal of Medical Ethics* 46, no. 1 (2020): 3–6. Even with this uncertainty, it is problematic to create laws that are founded on a position that lacks scientific evidence, especially when they directly impact vulnerable people. As I have mentioned in this introduction, the most perplexing and aggravating aspect of my experience was how quickly research from peer-reviewed scientific articles was dismissed while an extremely biased documentary from the 1980s was viewed as legitimate evidence.

17. Christopher Z. Heaney, "Manipulative Silent Scream," *Harvard Crimson*, March 11, 1985, https:// www.thecrimson.com/article/1985/3/11/manipulative-silent-scream-pbto-the-editors/.

1. THE SCOPE AND CONSEQUENCES OF MISINFORMATION

1. Maria Konnikova, "The Conman Who Pulled Off History's Most Audacious Scam," BBC, January 27, 2016, https://www.bbc.com/future/article/20160127-the-conman-who-pulled-off-historys -most-audacious-scam, and *The Confidence Game: Why We Fall for It . . . Every Time* (Penguin, 2017).

2. David Sinclair, *The Land That Never Was: Sir Gregor MacGregor and the Most Audacious Fraud in History* (Da Capo, 2004).

3. Daniel Balliet et al., "Ingroup Favoritism in Cooperation: A Meta-Analysis," *Psychological Bulletin* 140, no. 6 (2014): 1556–81, http://dx.doi.org/10.1037/a0037737.

4. Robert B. Cialdini, *Influence, New and Expanded: The Psychology of Persuasion* (HarperCollins, 2021).

5. Gordon Pennycook et al., "A Practical Guide to Doing Behavioral Research on Fake News and Misinformation," *Collabra: Psychology* 7, no. 1 (2021), https://doi.org/10.1525/collabra.25293.

6. Brendan Nyhan and Jason Reifler, "When Corrections Fail: The Persistence of Political Misperceptions," *Political Behavior* 32, no. 2 (2010): 303–30, https://doi.org/10.1007/s11109-010-9112-2.

7. Jianing Li and Michael W. Wagner, "The Value of Not Knowing: Partisan Cue-Taking and Belief Updating of the Uninformed, the Ambiguous, and the Misinformed," *Journal of Communication* 70, no. 5 (2020): 646–69, https://doi.org/10.1093/joc/jqaa022.

8. Nir Grinberg et al., "Fake News on Twitter During the 2016 U.S. Presidential Election," *Science* 363, no. 6425 (2019): 374–78, https://doi.org/10.1126/science.aau2706; Gordon Pennycook et al., "Shifting Attention to Accuracy Can Reduce Misinformation Online," *Nature* 592, no. 7855 (2021): 590–95, https://doi.org/10.1038/s41586-021-03344-2.

9. Emily K. Vraga and Leticia Bode, "Defining Misinformation and Understanding Its Bounded Nature: Using Expertise and Evidence for Describing Misinformation," *Political Communication* 37, no. 1 (2020): 136–44, https://doi.org/10.1080/10584609.2020.1716500.

10. Frederick N. Rasmussen, "100 Years After the Titanic Disaster," *Baltimore Sun*, April 14, 2012, https://www.baltimoresun.com/2012/04/14/100-years-after-the-titanic-disaster/.

11. Lyneyve Finch, "Psychological Propaganda: The War of Ideas on Ideas During the First Half of the Twentieth Century," *Armed Forces and Society* 26, no. 3 (2000): 367–86, https://doi.org/10.1177/0095327X0002600302. Examples of propaganda leaflets from World War I can be found in the World War I Document Archive, Brigham Young University Library, last edited June 30, 2009, https://wwi.lib.byu.edu/index.php/Propaganda_Leaflets.

12. Gordon Pennycook and David G. Rand, "The Psychology of Fake News," *Trends in Cognitive Sciences* 25, no. 5 (2021): 388–402, https://doi.org/10.1016/j.tics.2021.02.007.

13. "Jon Stewart, Again in the Crossfire," *Washington Post*, October 19, 2004, https://www.washingtonpost.com/archive/lifestyle/2004/10/19/jon-stewart-again-in-the-crossfire/cd6ffdbb-6f06-42cd-9479-21af28ac5b81/.

14. Jeffrey M. Berry and Sarah Sobieraj, *The Outrage Industry: Political Opinion Media and the New Incivility* (Oxford University Press, 2014).

15. Matthew S. Levendusky and Neil Malhotra, "Does Media Coverage of Partisan Polarization Affect Political Attitudes?," *Political Communication* 33, no. 2 (2016): 283–301, https://doi.org/10.1080/10584609.2015.1038455.

16. Eunji Kim et al., "Measuring Dynamic Media Bias," *Proceedings of the National Academy of Sciences* 119, no. 32 (2022): e2202197119, https://doi.org/10.1073/pnas.2202197119.

17. David Rozado and Musa al-Gharbi, "Using Word Embeddings to Probe Sentiment Associations of Politically Loaded Terms in News and Opinion Articles from News Media Outlets," *Journal of Computational Social Science* 5, no. 1 (2022): 427–48, https://doi.org/10.1007/s42001-021-00130-y.

18. "The Color of News: How Different Media Have Covered the General Election," Pew Research Center, October 29, 2008, https://www.pewresearch.org/journalism/2008/10/29/the-color-of-news/.

19. "Prime Time Fox News and WSJ Editorial Climate Coverage Mostly Wrong," *Scientific American*, September 21, 2012, https://www.scientificamerican.com/podcast/episode/primetime-fox-news-and-wsj-editoria-12-09-21/.

20. Daniel de Visé, " 'Hyper-Partisan' Politicians Get Four Times the News Coverage of Bipartisan Colleagues," *The Hill*, March 13, 2023, https://thehill.com/homenews/media/3894486-hyper-partisan-politicians-get-four-times-the-news-coverage-of-bipartisan-colleagues/.

21. *Encyclopaedia Britannica Online*, s.v. "Yellow Journalism," accessed May 2, 2024, https://www.britannica.com/topic/yellow-journalism.

22. Alex Woodward, " 'Fake News': A Guide to Trump's Favourite Phrase—and the Dangers It Obscures," *The Independent*, October 2, 2020, https://www.independent.co.uk/news/world/americas/us-election/trump-fake-news-counter-history-b732873.html.

23. Pennycook and Rand, "The Psychology of Fake News," 389.

24. Mike Sager, "The Fabulist Who Changed Journalism," *Columbia Journalism Review* 54 (2016): 52–60, https://www.cjr.org/the_feature/the_fabulist_who_changed_journalism.php.

25. Sian Lee et al., " 'Fact-Checking' Fact Checkers: A Data-Driven Approach," *Harvard Kennedy School Misinformation Review* 4, no. 5 (2023): 1–22, https://doi.org/10.37016/mr-2020-126.

26. Jennifer Allen, Antonio A. Arechar, Gordon Pennycook, and David G. Rand, "Scaling up Fact-Checking Using the Wisdom of Crowds," *Science Advances* 7, no. 36 (2021): eabf4393.

27. Chloe Lim, "Checking How Fact-Checkers Check," *Research and Politics* 5, no. 3 (2018), https://doi.org/10.1177/2053168018786848.

28. Sakari Nieminen and Valtteri Sankari, "Checking PolitiFact's Fact-Checks," *Journalism Studies* 22, no. 3 (2021): 358–78, https://doi.org/10.1080/1461670X.2021.1873818.

29. Cecilie Steenbuch Traberg, "Misinformation: Broaden Definition to Curb Its Societal Influence," *Nature* 606, no. 7915 (2022): 653, https://doi.org/10.1038/d41586-022-01700-4.

30. Vraga and Bode, "Defining Misinformation and Understanding Its Bounded Nature.

31. "Missing Information, Not Just Misinformation, Is Part of the Problem," Meedan, August 5, 2020, https://meedan.com/post/missing-information-not-just-misinformation-is-part-of-the-problem.

32. John M. Last, ed., *A Dictionary of Public Health* (Oxford University Press, 2007), https://archive.org/details/dictionaryofpubloooolast/page/n439/mode/1up.

33. Naomi Oreskes, *Why Trust Science?* (Princeton University Press, 2019); Ullrich Ecker et al., "Misinformation Poses a Bigger Threat to Democracy Than You Might Think," *Nature* 630, no. 8015 (2024): 29–32, https://doi.org/10.1038/d41586-024-01587-3.

34. Claire Wardle, "Misunderstanding Misinformation," *Issues in Science and Technology* 39, no. 3 (2023): 38–40, https://issues.org/misunderstanding-misinformation-wardle/.

35. Ziva Kunda, "The Case for Motivated Reasoning," *Psychological Bulletin* 108, no. 3 (1990): 480–98, https://doi.org/10.1037/0033-2909.108.3.480; Raymond S. Nickerson, "Confirmation Bias: A Ubiquitous Phenomenon in Many Guises," *Review of General Psychology* 2, no. 2 (1998): 175–220, https://doi.org/10.1037/1089-2680.2.2.175.

36. Carl Sagan, *The Demon-Haunted World: Science as a Candle in the Dark* (Random House, 1995), 25.

37. Craig J. R. Sewall, "Flawed Data Led to Findings of a Connection Between Time Spent on Devices and Mental Health Problems—New Research," *The Conversation*, June 23, 2021, http://theconversation.com/flawed-data-led-to-findings-of-a-connection-between-time-spent-on-devices-and-mental-health-problems-new-research-162585.

38. Amy Mitchell, "Many Americans Say Made-Up News Is a Critical Problem That Needs to Be Fixed," Pew Research Center, June 5, 2019, https://www.pewresearch.org/journalism/2019/06/05/many-americans-say-made-up-news-is-a-critical-problem-that-needs-to-be-fixed/.

39. Isabelle Valdes et al., "KFF Misinformation Poll Snapshot: Public Views Misinformation as a Major Problem, Feels Uncertain About Accuracy of Information on Current Events," KFF, December 15, 2023, https://www.kff.org/coronavirus-covid-19/poll-finding/kff-misinformation-poll-snapshot-public-views-misinformation-as-a-major-problem-feels-uncertain-about-accuracy-of-information/.

40. Jeffrey Gottfried, "Americans' Social Media Use," Pew Research Center, January 31, 2024, https://www.pewresearch.org/internet/2024/01/31/americans-social-media-use/.

41. Sara Atske, "Social Media and News Fact Sheet," Pew Research Center, November 15, 2023, https://www.pewresearch.org/journalism/fact-sheet/social-media-and-news-fact-sheet/.

42. Craig Silverman, "This Analysis Shows How Viral Fake Election News Stories Outperformed Real News on Facebook," *BuzzFeed News*, November 16, 2016, https://www.buzzfeednews.com/article/craigsilverman/viral-fake-election-news-outperformed-real-news-on-facebook.

43. Jennifer Allen et al., "Evaluating the Fake News Problem at the Scale of the Information Ecosystem," *Science Advances* 6, no. 14 (2020), https://doi.org/10.1126/sciadv.aay3539.

44. "Local TV News Fact Sheet," Pew Research Center, September 14, 2023, https://www.pewresearch.org/journalism/fact-sheet/local-tv-news/.

45. Sarah E. Gollust et al., "Television News Coverage of Public Health Issues and Implications for Public Health Policy and Practice," *Annual Review of Public Health* 40 (2019): 167–85, https://doi.org/10.1146/annurev-publhealth-040218-044017.

46. James Tilley, "Why So Many People Believe Conspiracy Theories," BBC, February 12, 2019, https://www.bbc.com/news/world-47144738.

47. Taylor Orth, "Which Conspiracy Theories Do Americans Believe?," YouGov, December 8, 2023, https://today.yougov.com/politics/articles/48113-which-conspiracy-theories-do-americans -believe; J. Eric Oliver and Thomas J. Wood, "Conspiracy Theories and the Paranoid Style(s) of Mass Opinion," *American Journal of Political Science* 58, no. 4 (2014): 952–66, https://doi.org/10.1111 /ajps.12084.

48. Zach Bastick, "Would You Notice If Fake News Changed Your Behavior? An Experiment on the Unconscious Effects of Disinformation," *Computers in Human Behavior* 116 (2021): art. 106633, https://doi.org/10.1016/j.chb.2020.106633.

49. Weimin Hu et al., "Product-Related Emphasis of Skin Disease Information Online," *Archives of Dermatology* 138, no. 6 (2002): 775–80, https://doi.org/10.1001/archderm.138.6.775.

50. "Number of Internet Users Worldwide from 2005 to 2023," Statista, May 22, 2024, https://www .statista.com/statistics/273018/number-of-internet-users-worldwide/.

51. Fang Jin et al., "Misinformation Propagation in the Age of Twitter," *Computer* 47, no. 12 (2014): 90–94, https://doi.ieeecomputersociety.org/10.1109/MC.2014.361.

52. Soroush Vosoughi et al., "The Spread of True and False News Online," *Science* 359, no. 6380 (2018): 1146–51, https://doi.org/10.1126/science.aap9559.

53. Alberto Acerbi, *Cultural Evolution in the Digital Age* (Oxford University Press, 2019), 128.

54. Allen et al., "Evaluating the Fake News Problem."

55. Ceren Budak et al., "Misunderstanding the Harms of Online Misinformation," *Nature* 630, no. 8015 (2024): 45–53, https://doi.org/10.1038/s41586-024-07417-w.

56. "[2020] Presidential Results," NPR, accessed November 2, 2022. https://apps.npr.org/elections20- interactive/; Kate Sullivan and Jennifer Agiesta, "Biden's Popular Vote Margin Over Trump Tops 7 Million," CNN, December 4, 2020, https://www.cnn.com/2020/12/04/politics/biden-popular -vote-margin-7-million/index.html.

57. Reuters, "Fact Check: Courts Have Dismissed Multiple Lawsuits of Alleged Electoral Fraud Presented by Trump Campaign," February 15, 2021, https://www.reuters.com/article/idUSKB- N2AF1FQ/; Christina A. Cassidy, "AP Review Finds Far Too Little Vote Fraud to Tip 2020 Election to Trump," PBS News, December 14, 2021, https://www.pbs.org/newshour/politics/ap-review-finds -far-too-little-vote-fraud-to-tip-2020-election-to-trump.

58. Gordon Pennycook and David G. Rand, "Research Note: Examining False Beliefs About Voter Fraud in the Wake of the 2020 Presidential Election," *Harvard Kennedy School Misinformation Review* 2, no. 1 (2021): 1–10, https://doi.org/10.37016/mr-2020-51.

59. Paul M. Barrett, "Spreading the Big Lie: How Social Media Sites Have Amplified False Claims of U.S. Election Fraud," NYU Stern Center for Business and Human Rights, September 16, 2022, https://bhr.stern.nyu.edu/tech-big-lie.

60. Mark Joyella, "Fox News Hits 23rd Consecutive Month as Most-Watched in Cable News As CNN Sees Gains in January," *Forbes*, February 1, 2023, https://www.forbes.com/sites/markjoyella /2023/02/01/fox-news-hits-23rd-consecutive-month-as-most-watched-in-cable-news-as-cnn-sees -gains-in-january/; Jeremy W. Peters and Katie Robertson, "Fox Stars Privately Expressed Disbe- lief About Election Fraud Claims: 'Crazy Stuff,' " *New York Times*, February 16, 2023, https://www .nytimes.com/2023/02/16/business/media/fox-dominion-lawsuit.html.

61. Chris Cameron, "These Are the People Who Died in Connection with the Capitol Riot," *New York Times*, January 5, 2022, https://www.nytimes.com/2022/01/05/us/politics/jan-6-capitol-deaths.html.

62. Manu Raju and Ted Barrett, "US Capitol Police Chief to Resign After Wednesday's Riots," CNN, January 7, 2021, https://www.cnn.com/2021/01/07/politics/capitol-police-reaction-details/index .html.

63. Dan Mangan, "DOJ Says at Least 1,000 Trump Supporters Arrested for Jan. 6 Capitol Riot," CNBC, March 6, 2023, https://www.cnbc.com/2023/03/06/doj-says-jan-6-capitol-riot-arrests -top-thousand-people.html.

64. Daniel Funke and Susan Benkelman, "Misinformation Is Inciting Violence Around the World: And Tech Platforms Don't Seem to Have a Plan to Stop It," Poynter Institute, April 4, 2019, https://www .poynter.org/fact-checking/2019/misinformation-is-inciting-violence-around-the-world-and-tech -platforms-dont-have-a-plan-to-stop-it/.

65. Eli Meixler, "U.N. Fact Finders Say Facebook Played a 'Determining' Role in Violence Against the Rohingya," *Time*, March 13, 2018, https://time.com/5197039/un-facebook-myanmar-rohingya -violence/.

66. Craig Mod, "The Facebook-Loving Farmers of Myanmar," *The Atlantic*, January 21, 2016, https://www .theatlantic.com/technology/archive/2016/01/the-facebook-loving-farmers-of-myanmar/424812/.

67. Luke Taylor, "Covid-19 Misinformation Sparks Threats and Violence Against Doctors in Latin America," *BMJ* 370 (2020), https://doi.org/10.1136/bmj.m3088.

68. Mathew Ingram, "In India, the Fake News Problem Isn't Facebook, It's WhatsApp," *Columbia Journalism Review*, May 16, 2018, https://www.cjr.org/the_media_today/india-whatsapp.php.

69. Federal Bureau of Investigation, National Center for the Analysis of Violent Crime and Behavioral Assessment Unit, *Lone Offender: A Study of Lone Offender Terrorism in the United States (1972–2015)* (Federal Bureau of Investigation, 2019).

70. Roland Imhoff et al., "Resolving the Puzzle of Conspiracy Worldview and Political Activism: Belief in Secret Plots Decreases Normative but Increases Nonnormative Political Engagement," *Social Psychological and Personality Science* 12, no. 1 (2021): 71–79, https://doi.org/10.1177/1948550619896491.

71. Katherine Ognyanova et al., "Misinformation in Action: Fake News Exposure Is Linked to Lower Trust in Media, Higher Trust in Government When Your Side Is in Power," *Harvard Kennedy School Misinformation Review* 1, no. 4 (2020): 1–19, https://doi.org/10.37016/mr-2020-024.

72. Salman Bin Naeem et al., "An Exploration of How Fake News Is Taking Over Social Media and Putting Public Health at Risk," *Health Information and Libraries Journal* 38, no. 2 (2021): 143–49, https://doi.org/10.1111/hir.12320.

73. Jacob Wallace et al., "Excess Death Rates for Republicans and Democrats During the COVID-19 Pandemic" (Working Paper 30512 National Bureau of Economic Research, 2022).

74. Jenna Sherman, "Gendered Health Misinformation," Meedan, October 2022, https://assets-global.website-files.com/615e270f23c94c3fc683f12c/6360182ce09baba276f9d96d_Gendered%20 Health%20Misinformation%20-%20Meedan.pdf.

75. "The Turnaway Study," Advancing New Standards in Reproductive Health, accessed March 6, 2023, https://www.ansirh.org/research/ongoing/turnaway-study.

76. Hossein Hassanian-Moghaddam et al., "Double Trouble: Methanol Outbreak in the Wake of the COVID-19 Pandemic in Iran—a Cross-Sectional Assessment," *Critical Care* 24, no. 1 (2020): 402, https://doi.org/10.1186/s13054-020-03140-w.

77. U.S. Food and Drug Administration, "FDA Warns Consumers About the Dangerous and Potentially Life-Threatening Side Effects of Miracle Mineral Solution," news release, August 12, 2019, https://www.fda.gov/news-events/press-announcements/fda-warns-consumers-about-dangerous -and-potentially-life-threatening-side-effects-miracle-mineral.

78. Ryan Felton, "Why Did It Take a Pandemic for the FDA to Crack Down on a Bogus Bleach 'Miracle' Cure?," *Consumer Reports*, May 14, 2020, updated July 8, 2020, https://www.consumerreports.org/scams-fraud /bogus-bleach-miracle-cure-fda-crackdown-miracle-mineral-solution-genesis-ii-church/.

79. Tom Porter, "Taking Toxic Bleach MMS Has Killed 7 People in the US, Colombian Prosecutors Say—Far More Than Previously Known," *Business Insider*, August 12, 2020, https://www.business insider.com/mms-bleach-killed-7-americans-new-from-colombia-arrest-2020-8.

80. Paul W. Armstrong and C. David Naylor, "Counteracting Health Misinformation: A Role for Medical Journals?," *JAMA* 321, no. 19 (2019): 1863–64, https://doi.org/10.1001/jama.2019.5168.

81. Informedhealth.org, "Common Colds," updated December 11, 2023, https://www.informedhealth .org/does-vitamin-c-prevent-colds.html.

82. John Heymach et al., "Clinical Cancer Advances 2018: Annual Report on Progress Against Cancer from the American Society of Clinical Oncology," *Journal of Clinical Oncology* 36, no. 10 (2018): 1020–44, https://doi.org/10.1200/JCO.2017.77.0446.

83. Andrew I. Geller et al., "Emergency Department Visits for Adverse Events Related to Dietary Supplements," *New England Journal of Medicine* 373, no. 16 (2015): 1531–40, https://doi.org/10.1056/NEJMsa1504267.

84. Terrence McCoy, "Half of Dr. Oz's Medical Advice Is Baseless or Wrong, Study Says," *Washington Post*, December 19, 2014, https://www.washingtonpost.com/news/morning-mix/wp/2014/12/19/half-of-dr-ozs-medical-advice-is-baseless-or-wrong-study-says/; Christina Korownyk et al., "Televised Medical Talk Shows—What They Recommend and the Evidence to Support Their Recommendations: A Prospective Observational Study," *BMJ* 349 (2014): g7346, https://doi.org/10.1136/bmj.g7346.

85. Mona Hanna-Attisha, *What the Eyes Don't See: A Story of Crisis, Resistance, and Hope in an American City* (Random House, 2018).

86. Vanessa Schipani, "False Claims About Flint Water," FactCheck.org, April 27, 2016, https://www.factcheck.org/2016/04/false-claims-about-flint-water/.

87. Cary Funk et al., "2. Americans' Health Care Behaviors and Use of Conventional and Alternative Medicine," Pew Research Center, February 2, 2017, https://www.pewresearch.org/science/2017/02/02/americans-health-care-behaviors-and-use-of-conventional-and-alternative-medicine/.

88. Edzard Ernst, "Cancer Patients Who Use Alternative Medicine Die Sooner," April 18, 2013, http://edzardernst.com/2013/04/cancer-patients-who-use-alternative-medicine-die-sooner/.

89. Seema Yasmin, *Viral BS: Medical Myths and Why We Fall for Them* (Johns Hopkins University Press, 2021).

90. John P. A. Ioannidis et al., "How to Survive the Medical Misinformation Mess," *European Journal of Clinical Investigation* 47, no. 11 (2017): 795–802, https://doi.org/10.1111/eci.12834.

91. Kristin Myers, "Anti-Vaxxers Are Costing Americans Billions Each Year," *Yahoo Finance*, April 10, 2019, https://finance.yahoo.com/news/antivaxxers-costing-americans-billions-each-year-191839191.html.

92. Nan Zhao et al., "The Impact of Government Interventions on COVID-19 Spread and Consumer Spending," *Management Science* 70, no. 5 (2024): 3302–18, https://doi.org/10.1287/mnsc.2023.4853.

93. Sebnem Kalemli-Ozcan, "The $4 Trillion Economic Cost of Not Vaccinating the Entire World," *The Conversation*, February 12, 2021, http://theconversation.com/the-4-trillion-economic-cost-of-not-vaccinating-the-entire-world-154786.

94. "COVID-19 Vaccine Incentives," National Governors Association, October 19, 2021, https://www.nga.org/center/publications/covid-19-vaccine-incentives/.

95. Victoria Forster, "Ohio Vaccine Lottery Gave Away $5 Million, but Didn't Increase Vaccination Rates, Says New Study," *Forbes*, July 3, 2021, https://www.forbes.com/sites/victoriaforster/2021/07/03/ohio-vaccine-lottery-didnt-increase-vaccination-rates-says-new-study/.

96. "Fake Financial News Is a Real Threat to Majority of Americans: New AICPA Survey," American Institute of CPAs, April 27, 2017, https://www.aicpa.org/press/pressreleases/2017/fake-financial-news-is-a-real-threat-to-majority-of-americans-new-aicpa-survey.html.

97. Heidi Shierholz and Ben Zipperer, "Here Is What's at Stake with the Conflict of Interest ('Fiduciary') Rule," Economic Policy Institute, May 30, 2017, https://www.epi.org/publication/here-is-whats-at-stake-with-the-conflict-of-interest-fiduciary-rule/.

98. Reuters, "Bots Hyped Up GameStop on Major Social Media Platforms, Analysis Finds," February 26, 2021, https://www.reuters.com/article/idUSKBN2AQ2BH/.

99. Dalbar, Inc., "Average Investor Blown Away by Market Turmoil in 2018," PR Newswire, March 25, 2019, https://www.prnewswire.com/news-releases/average-investor-blown-away-by-market-turmoil-in-2018-300817353.html.

100. John Bogle, *The Little Book of Common Sense Investing: The Only Way to Guarantee Your Fair Share of Stock Market Returns* (Wiley, 2017).

101. Oliver Darcy, "ABC News Suspends Brian Ross for 4 Weeks Over Erroneous Flynn Story," CNN Business, December 2, 2017, https://money.cnn.com/2017/12/02/media/abc-news-brian-ross/index.html.

102. Ironman at Political Calculations, "The Cost of Fake News for the S&P 500," Seeking Alpha, December 4, 2017, https://seekingalpha.com/article/4129355-cost-of-fake-news-for-s-and-p-500.

103. Turner Wright, "Fake News: Litecoin Price Surges 35 Percent Following Walmart Adoption Hoax," Cointelegraph, September 13, 2021, https://cointelegraph.com/news/fake-news-litecoin-price-surges-35-following-walmart-adoption-hoax.

104. Vildana Hajric, "Litecoin Foundation 'Screwed Up,' Lee Says of Walmart Snafu," *Bloomberg News*, September 13, 2021, https://www.bloomberg.com/news/articles/2021-09-13/litecoin-foundation-screwed-up-lee-says-about-walmart-snafu.

105. Renée Cho, "How Climate Change Impacts the Economy," *State of the Planet*, Columbia Climate School, Columbia University, June 20, 2019, https://news.climate.columbia.edu/2019/06/20/climate-change-economy-impacts/.

106. Kathie M. d'I. Treen et al., "Online Misinformation About Climate Change," *Wiley Interdisciplinary Reviews: Climate Change* 11, no. 5 (2020), https://doi.org/10.1002/wcc.665.

107. Michael Brüggemann and Sven Engesser, "Beyond False Balance: How Interpretive Journalism Shapes Media Coverage of Climate Change," *Global Environmental Change: Human and Policy Dimensions* 42 (2017): 58–67, https://doi.org/10.1016/j.gloenvcha.2016.11.004.

108. Shaun W. Elsasser and Riley E. Dunlap, "Leading Voices in the Denier Choir: Conservative Columnists' Dismissal of Global Warming and Denigration of Climate Science," *American Behavioral Scientist* 57, no. 6 (2013): 754–76, https://doi.org/10.1177/0002764212469800.

109. Matthew H. Goldberg et al., "Oil and Gas Companies Invest in Legislators That Vote Against the Environment," *Proceedings of the National Academy of Sciences* 117, no. 10 (2020): 5111–12, https://doi.org/10.1073/pnas.1922175117.

110. Ding et al., "Support for Climate Policy and Societal Action Are Linked to Perceptions About Scientific Agreement," *Nature Climate Change* 1, no. 9 (2011): 462–66, https://doi.org/10.1038/nclimate1295.

111. David J. Helfand, *A Survival Guide to the Misinformation Age: Scientific Habits of Mind* (Columbia University Press, 2016); Carl T. Bergstrom and Jevin D. West, *Calling Bullshit: The Art of Skepticism in a Data-Driven World* (Random House, 2021).

2. HOW OUR IDENTITIES CAN MAKE US VULNERABLE TO MISINFORMATION

1. Stephen G. Bloom, "Lesson of a Lifetime," *Smithsonian Magazine*, August 31, 2005, https://www.smithsonianmag.com/science-nature/lesson-of-a-lifetime-72754306/.

2. Peter J. Burke, "Identity," in *The Cambridge Handbook of Social Theory*, ed. Peter Kivisto (Cambridge University Press, 2020), 63.

3. Peter J. Burke and Jan E. Stets, *Identity Theory: Revised and Expanded* (Oxford University Press, 2022).

4. Peter J. Burke, "Identity Control Theory," in *The Blackwell Encyclopedia of Sociology*, ed. George Ritzer (Blackwell, 2007): 2202–7.

5. Daniel Kahneman, *Thinking, Fast and Slow* (Macmillan, 2011).

6. Sheldon Stryker and Richard T. Serpe, "Commitment, Identity Salience, and Role Behavior: Theory and Research Example," in *Personality, Roles, and Social Behavior*, ed. William Ickes and Eric S.

Knowles (Springer New York, 1982), 199–218; Laurie H. Ervin and Sheldon Stryker, "Theorizing the Relationship Between Self-Esteem and Identity," in *Extending Self-Esteem Theory and Research: Sociological and Psychological Currents*, ed. Timothy J. Owens et al. (Cambridge University Press, 2001), 29–55, https://doi.org/10.1017/CBO9780511527739.

7. Peter J. Burke, "Conceptualizing Identity Prominence, Salience, and Commitment," in *Advancing Identity Theory, Measurement, and Research*, ed. Jan E. Stets et al. (Springer International, 2023), 17–33, https://doi.org/10.1007/978-3-031-32986-9_2.

8. Peter J. Burke and Donald C. Reitzes, "An Identity Theory Approach to Commitment," *Social Psychology Quarterly* 54, no. 3 (1991): 239–51, https://doi.org/10.2307/2786653; Richard B. Felson, "Reflected Appraisal and the Development of Self," *Social Psychology Quarterly* 48, no. 1 (1985): 71–78, https://doi.org/10.2307/3033783.

9. Charles Horton Cooley, *Human Nature and the Social Order* (Routledge, 2017).

10. Richard T. Serpe et al., "Multiple Identities and Self-Esteem," in *Identities in Everyday Life*, ed. Jan E. Stets and Richard T. Serpe (Oxford University Press, 2019), 72; Jon W. Hoelter, "The Relationship Between Specific and Global Evaluations of Self: A Comparison of Several Models," *Social Psychology Quarterly* 49, no. 2 (1986): 129–41, https://doi.org/10.2307/2786724.

11. Peter J. Burke, "Social Identities and Psychosocial Stress," in *Psychosocial Stress: Perspectives on Structure, Theory, Life-Course, and Methods*, ed. H. B. Kaplan (Academic Press, 1996), 141–74; David K. Sherman and Geoffrey L. Cohen, "The Psychology of Self-Defense: Self-Affirmation Theory," in *Advances in Experimental Social Psychology*, vol. 38, ed. Mark P. Zanna (Academic Press, 2006), 183–242.

12. Will Kalkhoff et al., "Neural Processing of Identity-Relevant Feedback," in *New Directions in Identity Theory and Research*, ed. Jan E. Stets and Richard T. Serpe (Oxford University Press, 2016): 195–238, https://doi.org/10.1093/acprof:oso/9780190457532.003.0008.

13. Kelly-Ann Allen et al., "The Need to Belong: A Deep Dive Into the Origins, Implications, and Future of a Foundational Construct," *Educational Psychology Review* 34, no. 2 (2022): 1133–56, https://doi.org/10.1007/s10648-021-09633-6.

14. Chun Shen et al., "Associations of Social Isolation and Loneliness with Later Dementia," *Neurology* 99, no. 2 (2022): e164–75, https://doi.org/10.1212/WNL.0000000000200583.

15. "This Former Surgeon General Says There's a 'Loneliness Epidemic' and Work Is Partly to Blame," *Washington Post*, October 4, 2017, https://www.washingtonpost.com/news/on-leadership/wp/2017/10/04/this-former-surgeon-general-says-theres-a-loneliness-epidemic-and-work-is-partly-to-blame/.

16. Ilja Van Beest and Kipling D. Williams, "When Inclusion Costs and Ostracism Pays, Ostracism Still Hurts," *Journal of Personality and Social Psychology* 91, no. 5 (2006): 918–28, https://doi.org/10.1037/0022-3514.91.5.918.

17. Robert Waldinger and Marc Schulz, *The Good Life: Lessons from the World's Longest Scientific Study Of Happiness* (Simon and Schuster, 2023).

18. Solomon E. Asch, "Studies of Independence and Conformity: I. A Minority of One Against a Unanimous Majority," *Psychological Monographs: General and Applied* 70, no. 9 (1956): 1–70, https://doi.org/10.1037/h0093718.

19. Mariola Paruzel-Czachura et al., "Online Moral Conformity: How Powerful Is a Group of Strangers When Influencing an Individual's Moral Judgments During a Video Meeting?," *Current Psychology* 43, no. 7 (2024): 6125–35, https://doi.org/10.1007/s12144-023-04765-0.

20. Daniel Kreiss et al., "Trump Gave Them Hope: Studying the Strangers in Their Own Land," *Political Communication* 34, no. 3 (2017): 470–78, https://doi.org/10.1080/10584609.2017.1330076.

21. Dan M. Kahan, "Misconceptions, Misinformation, and the Logic of Identity-Protective Cognition," Cultural Cognition Project Working Paper No. 164 (Yale University Law School, 2017), http://doi.org/10.2139/SSRN.2973067.

22. Andrew Whalen, "'Behind the Curve' Ending: Flat Earthers Disprove Themselves with Own Experiments in Netflix Documentary," *Newsweek*, February 25, 2019, https://www.newsweek.com/behind-curve-netflix-ending-light-experiment-mark-sargent-documentary-movie-1343362.

23. Asheley R. Landrum and Alex Olshansky, "The Role of Conspiracy Mentality in Denial of Science and Susceptibility to Viral Deception About Science," *Politics and the Life Sciences* 38, no. 2 (2019): 193–209, https://doi.org/10.1017/pls.2019.9.

24. Henri Tajfel and John C. Turner, "An Integrative Theory of Intergroup Conflict," in *Organizational Identity: A Reader*, ed. Mary Jo Hatch and Majken Schultz (Oxford University Press, 2004), 56–65.

25. Jan E. Stets and Peter J. Burke, "Identity Theory and Social Identity Theory," *Social Psychology Quarterly* 63, no. 3 (2000): 224–37, https://doi.org/10.2307/2695870.

26. J. M. Rabbie and M. Horwitz, "Arousal of Ingroup-Outgroup Bias by a Chance Win or Loss," *Journal of Personality and Social Psychology* 13, no. 3 (1969): 269–77, https://doi.org/10.1037/h0028284; Henri Tajfel et al., "Social Categorization and Intergroup Behaviour," *European Journal of Social Psychology* 1, no. 2 (1971): 149–78, https://doi.org/10.1002/ejsp.2420010202.

27. Leslie Ashburn-Nardo et al., "Implicit Associations as the Seeds of Intergroup Bias: How Easily Do They Take Root?," *Journal of Personality and Social Psychology* 81, no. 5 (2001): 789–99.

28. Kirsten G. Volz et al., "In-Group as Part of the Self: In-Group Favoritism Is Mediated by Medial Prefrontal Cortex Activation," *Social Neuroscience* 4, no. 3 (2009): 244–60, https://doi.org/10.1080/17470910802553565.

29. Yarrow Dunham et al., "Consequences of 'Minimal' Group Affiliations in Children," *Child Development* 82, no. 3 (2011): 793–811, https://doi.org/10.1111/j.1467-8624.2011.01577.x.

30. Marilynn B. Brewer, "The Psychology of Prejudice: Ingroup Love and Outgroup Hate?," *Journal of Social Issues* 55, no. 3 (1999): 429–44, https://doi.org/10.1111/0022-4537.00126.

31. Henri Tajfel, ed., *Differentiation Between Social Groups: Studies in the Social Psychology of Intergroup Relations* (Academic Press, 1978).

32. Albert H. Hastorf and Hadley Cantril, "They Saw a Game: A Case Study," *Journal of Abnormal and Social Psychology* 49, no. 1 (1954): 129–34, https://doi.org/10.1037/h0057880.

33. Patrick R. Miller and Pamela Johnston Conover, "Red and Blue States of Mind: Partisan Hostility and Voting in the United States," *Political Research Quarterly* 68, no. 2 (2015): 225–39, https://doi.org/10.1177/1065912915577208.

34. Donald P. Green et al., *Partisan Hearts and Minds: Political Parties and the Social Identities of Voters* (Yale University Press, 2004); Jay J. Van Bavel and Andrea Pereira, "The Partisan Brain: An Identity-Based Model of Political Belief," *Trends in Cognitive Sciences* 22, no. 3 (2018): 213–24, https://doi.org/10.1016/j.tics.2018.01.004.

35. Steven Levitsky and Daniel Ziblatt, *How Democracies Die* (Crown, 2019).

36. "1. Feelings About Partisans and the Parties," Pew Research Center, June 22, 2016, https://www.pewresearch.org/politics/2016/06/22/1-feelings-about-partisans-and-the-parties/.

37. Reem Nadeem, "As Partisan Hostility Grows, Signs of Frustration with the Two-Party System," Pew Research Center, August 9, 2022, https://www.pewresearch.org/politics/2022/08/09/as-partisan-hostility-grows-signs-of-frustration-with-the-two-party-system/.

38. Nathan P. Kalmoe and Lilliana Mason, "Lethal Mass Partisanship: Prevalence, Correlates, and Electoral Contingencies," paper presented at the meeting of the National Capital Area Political Science Association, Washington, DC, August 29–September 1, 2019.

39. William B. Swann Jr. and Michael D. Buhrmester, "Identity Fusion," *Current Directions in Psychological Science* 24, no. 1 (2015): 52–57, https://doi.org/10.1177/0963721414551363.

40. Jay J. Van Bavel et al., "Political Psychology in the Digital (Mis)information Age: A Model of News Belief and Sharing," *Social Issues and Policy Review* 15, no. 1 (2021): 91, https://doi.org/10.1111/sipr.12077.

41. Abigail Geiger, "Key Facts About Americans and Guns," Pew Research Center, September 13, 2023, https://www.pewresearch.org/short-reads/2023/09/13/key-facts-about-americans-and-guns/; Reem Nadeem, "Abortion Rises in Importance as a Voting Issue, Driven by Democrats," Pew Research Center, August 23, 2022, https://www.pewresearch.org/politics/2022/08/23/abortion-rises-in-importance-as-a-voting-issue-driven-by-democrats/.

42. Dan M. Kahan et al., "Motivated Numeracy and Enlightened Self-Government," *Behavioural Public Policy* 1, no. 1 (2017): 54–86, https://doi.org/10.1017/bpp.2016.2; Nicholas Scurich and Adam Shniderman, "The Selective Allure of Neuroscientific Explanations," *PLOS One* 9, no. 9 (2014): e107529, https://doi.org/10.1371/journal.pone.0107529.

43. Adam M. Enders and Joseph E. Uscinski, "Are Misinformation, Antiscientific Claims, and Conspiracy Theories for Political Extremists?," *Group Processes and Intergroup Relations* 24, no. 4 (2021): 583–605, https://doi.org/10.1177/1368430220960805.

44. Shanto Iyengar and Sean J. Westwood, "Fear and Loathing Across Party Lines: New Evidence on Group Polarization," *American Journal of Political Science* 59, no. 3 (2015): 690–707, https://doi.org/10.1111/ajps.12152; Karen Gift and Thomas Gift, "Does Politics Influence Hiring? Evidence from a Randomized Experiment," *Political Behavior* 37 (2015): 653–75, https://doi.org/10.1007/s11109-014-9286-0.

45. Chris Kahn, "Half of Republicans Say Biden Won Because of a 'Rigged' Election: Reuters/Ipsos Poll," Reuters, November 19, 2020, https://www.reuters.com/article/world/india/half-of-republicans-say-biden-won-because-of-a-rigged-election-reutersipsos-idUSKBN27Y1AD/.

46. Jeremy Schulman, "59 Percent of Republicans Say It's Important to Believe Trump Won the Election," *Mother Jones*, September 12, 2021, https://www.motherjones.com/mojo-wire/2021/09/59-percent-of-republicans-say-its-important-to-believe-trump-won-the-election/.

47. Leor Zmigrod and Manos Tsakiris, "Computational and Neurocognitive Approaches to the Political Brain: Key Insights and Future Avenues for Political Neuroscience," *Philosophical Transactions of the Royal Society of London, Series B, Biological Sciences* 376, no. 1822 (2021): 20200130, https://doi.org/10.1098/rstb.2020.0130.

48. Jonas T. Kaplan, Sarah I. Gimbel, and Sam Harris, "Neural Correlates of maintaining One's Political Beliefs in the Face of Counterevidence," *Scientific reports* 6, no. 1 (2016): 39589.

49. Drew Westen et al., "An fMRI Study of Motivated Reasoning: Partisan Political Reasoning in the US Presidential Election," unpublished manuscript, Emory University, Psychology Department, 2006.

50. Adam Moore et al., "Trust in Information, Political Identity and the Brain: An Interdisciplinary fMRI Study," *Philosophical Transactions of the Royal Society of London, Series B, Biological Sciences* 376, no. 1822 (2021), 6, https://doi.org/10.1098/rstb.2020.0140.

51. Kerrie L. Unsworth and Kelly S. Fielding, "It's Political: How the Salience of One's Political Identity Changes Climate Change Beliefs and Policy Support," *Global Environmental Change* 27 (2014): 131–37, https://doi.org/10.1016/j.gloenvcha.2014.05.002.

52. Chris Wang et al., "There Is an 'I' in Truth: How Salient Identities Shape Dynamic Perceptions of Truth," *European Journal of Social Psychology* 53, no. 2 (2023), https://doi.org/10.1002/ejsp.2909.

53. Jamie B. Luguri and Jaime L. Napier, "Of Two Minds: The Interactive Effect of Construal Level and Identity on Political Polarization," *Journal of Experimental Social Psychology* 49, no. 6 (2013): 972–77, https://doi.org/10.1016/j.jesp.2013.06.002.

54. Burke and Stets, *Identity Theory*.

55. Michael J. Mahoney, "Publication Prejudices: An Experimental Study of Confirmatory Bias in the Peer Review System," *Cognitive Therapy and Research* 1 (1977): 161–75, https://doi.org/10.1007/BF01173636.

56. Junghwan Yang et al., "Why Are 'Others' So Polarized? Perceived Political Polarization and Media Use in 10 Countries," *Journal of Computer-Mediated Communication* 21, no. 5 (2016): 349–67, https://doi.org/10.1111/jcc4.12166.

57. Lilliana Mason, "Best Of: The Age of 'Mega-Identity Politics,' " interview by Ezra Klein, *Vox Conversations*, podcast, reaired November 28, 2019, 1 hr., 15 min., 59 sec., https://radiopublic.com/Ezra/s1!e039b. See also Lilliana Mason, *Uncivil Agreement: How Politics Became Our Identity* (University of Chicago Press, 2018).

58. Raymond S. Nickerson, "Confirmation Bias: A Ubiquitous Phenomenon in Many Guises," *Review of General Psychology* 2, no. 2 (1998): 175–220, https://doi.org/10.1037/1089-2680.2.2.175.

59. Natalie Jomini Stroud, "Polarization and Partisan Selective Exposure," *Journal of Communication* 60, no. 3 (2010): 556–76, https://doi.org/10.1111/j.1460-2466.2010.01497.x; R. Kelly Garrett, "Echo Chambers Online? Politically Motivated Selective Exposure Among Internet News Users," *Journal of Computer-Mediated Communication* 14, no. 2 (2009): 265–85, https://doi.org/10.1111/j.1083-6101.2009.01440.x.

60. Desirée Schmuck et al., "Avoiding the Other Side? An Eye-Tracking Study of Selective Exposure and Selective Avoidance Effects in Response to Political Advertising," *Journal of Media Psychology* 32, no. 3 (2020): 158–64, http://doi.org/10.1027/1864-1105/a000265.

61. Lynn Hasher et al., "Frequency and the Conference of Referential Validity," *Journal of Verbal Learning and Verbal Behavior* 16, no. 1 (1977): 107–12, https://doi.org/10.1016/S0022-5371(77)80012-1.

62. Gordon Pennycook et al., "Prior Exposure Increases Perceived Accuracy of Fake News," *Journal of Experimental Psychology: General* 147, no. 12 (2018): 1865–80, https://doi.org/10.1037/xge0000465.

63. Michael D. Slater, "Reinforcing Spirals: The Mutual Influence of Media Selectivity and Media Effects and Their Impact on Individual Behavior and Social Identity," *Communication Theory* 17, no. 3 (2007): 281–303, https://doi.org/10.1111/j.1468-2885.2007.00296.x.

64. John Petrocelli, *The Life-Changing Science of Detecting Bullshit* (St. Martin's Press, 2021). John Petrocelli was actually one of my advisers for my undergraduate thesis. My thesis attempted to improve people's performance on the Monty Hall Problem while also recording their EEG activity. Looking back, my undergraduate thesis was foreshadowing my future career: not only did I have Dr. Petrocelli as an adviser, but also the broader goal of my project was to improve people's abilities to evaluate information and reduce their biases through education.

65. Thomas Strandberg et al., "Depolarizing American Voters: Democrats and Republicans Are Equally Susceptible to False Attitude Feedback," *PLOS ONE* 15, no. 2 (2020): 5, https://doi.org/10.1371/journal.pone.0226799.

66. Leon Festinger, *A Theory of Cognitive Dissonance* (Stanford University Press, 1962).

67. Julia Kneer et al., "Fast and Not Furious?," *Social Psychology* 43, no. 2 (2012), https://doi.org/10.1027/1864-9335/a000086; Daisy Jane C. Orcullo and Teo Hui San, "Understanding Cognitive Dissonance in Smoking Behaviour: A Qualitative Study," *International Journal of Social Science and Humanity* 6, no. 6 (2016): 481–84, https://doi.org/10.7763/IJSSH.2016.V6.695.

68. Leon Festinger and Stanley Schachter, *When Prophecy Fails* (Simon and Schuster, 2013).

69. David McRaney (@davidmcraney), "It's useful to think of confirmation bias," Twitter (now X), August 10, 2021, https://x.com/davidmcraney/status/1425082061782724608.

70. David R. Heise, "Affect Control Theory: Concepts and Model," in *Analyzing Social Interaction*, ed. L. Smith-Lovin and David R. Heise (Routledge, 2016), 1–33.

71. The formal structure of affect control theory also offers promise as a bridge between sociological and neuroscientific literatures because it can capture our feelings toward many different types of social events by applying a mathematical framework. See Neil J. MacKinnon and Jesse Hoey, "Operationalizing the Relation Between Affect and Cognition with the Somatic Transform," *Emotion Review* 13, no. 3 (2021): 245–56.

72. Steven M. Nelson, "Redefining a Bizarre Situation: Relative Concept Stability in Affect Control Theory," *Social Psychology Quarterly* 69, no. 3 (2006): 215–34, https://doi.org/10.1177/019027250606900301.

73. M. B. Fallin Hunzaker, "Cultural Sentiments and Schema-Consistency Bias in Information Transmission," *American Sociological Review* 81, no. 6 (2016): 1223–50, https://doi.org/10.1177/0003122416671742.

74. Matthew Facciani and Cecilie Steenbuch Traberg, "Personal Network Composition and Cognitive Reflection Predict Susceptibility to Different Types of Misinformation," *Connections*, ahead of print, June 27, 2024, https://intapi.sciendo.com/pdf/10.21307/connections-2019.044.

75. Caitlin Drummond and Baruch Fischhoff, "Individuals with Greater Science Literacy and Education Have More Polarized Beliefs on Controversial Science Topics," *Proceedings of the National*

Academy of Sciences 114, no. 36 (2017): 9587–92, https://doi.org/10.1073/pnas.1704882114; Dan M. Kahan, " 'Ordinary Science Intelligence': A Science-Comprehension Measure for Study of Risk and Science Communication, with Notes on Evolution and Climate Change," *Journal of Risk Research* 20, no. 8 (2017): 995–1016, https://doi.org/10.1080/13669877.2016.1148067.

76. Benjamin A. Lyons et al., "Overconfidence in News Judgments Is Associated with False News Susceptibility," *Proceedings of the National Academy of Sciences* 118, no. 23 (2021): e2019527118, https://doi.org/10.1073/pnas.2019527118.

77. Douglas J. Ahler and Gaurav Sood, "The Parties in Our Heads: Misperceptions About Party Composition and Their Consequences," *Journal of Politics* 80, no. 3 (2018): 964–81, https://doi.org/10.1086/697253.

78. Dan M. Kahan, "Climate-Science Communication and the Measurement Problem," *Political Psychology* 36 (2015): 1–43, https://doi.org/10.1111/pops.12244.

79. Marlis Stubenvoll and Jörg Matthes, "Why Retractions of Numerical Misinformation Fail: The Anchoring Effect of Inaccurate Numbers in the News," *Journalism and Mass Communication Quarterly* 99, no. 2 (2022): 368–89, https://doi.org/10.1177/10776990211021800.

80. Michael C. Schwalbe, Katie Joseff, Samuel Woolley, and Geoffrey L. Cohen, "When Politics Trumps Truth: Political Concordance Versus Veracity as a Determinant of Believing, Sharing, and Recalling the News," *Journal of Experimental Psychology. General* 153, no. 10 (2024): 2524–51.

81. Christopher Achen and Larry Bartels, *Democracy for Realists: Why Elections Do Not Produce Responsive Government* (Princeton University Press, 2017), 268.

82. Tyler F. Stillman and Roy F. Baumeister, "Uncertainty, Belongingness, and Four Needs for Meaning," *Psychological Inquiry* 20, no. 4 (2009): 249–51, https://doi.org/10.1080/10478400903333544.

83. Kip Williams, "Ostracism: The Impact of Being Rendered Meaningless," in *Meaning, Mortality, and Choice: The Social Psychology of Existential Concerns*, ed. P. R. Shaver and M. Mikulincer (American Psychological Association, 2012), 309–23.

84. Ian McGregor et al., "Reactive Approach Motivation (RAM) for Religion," *Journal of Personality and Social Psychology* 99, no. 1 (2010): 148–61, https://doi.org/10.1037/a0019702; Michael A. Hogg and Mark J. Rinella, "Social Identities and Shared Realities," *Current Opinion in Psychology* 23 (2018): 6–10, https://doi.org/10.1016/j.copsyc.2017.10.003.

85. Fiona Grant and Michael A. Hogg, "Self-Uncertainty, Social Identity Prominence and Group Identification," *Journal of Experimental Social Psychology* 48, no. 2 (2012): 538–42, https://doi.org/10.1016/j.jesp.2011.11.006.

86. Zachary P. Hohman et al., "Identity and Intergroup Leadership: Asymmetrical Political and National Identification in Response to Uncertainty," *Self and Identity* 9, no. 2 (2010): 113–28, https://doi.org/10.1080/15298860802605937.

87. Ian McGregor et al., "Compensatory Conviction in the Face of Personal Uncertainty: Going to Extremes and Being Oneself," *Journal of Personality and Social Psychology* 80, no. 3 (2001): 472–88, https://doi.org/10.1037/0022-3514.80.3.472.

88. Paul G. Grieve and Michael A. Hogg, "Subjective Uncertainty and Intergroup Discrimination in the Minimal Group Situation," *Personality and Social Psychology Bulletin* 25, no. 8 (1999): 926–40, https://doi.org/10.1177/01461672992511002.

89. Michael A. Hogg, "From Uncertainty to Extremism: Social Categorization and Identity Processes," *Current Directions in Psychological Science* 23, no. 5 (2014): 338–42, https://doi.org/10.1177/0963721414540168.

90. Michael Inzlicht et al., "Neural Markers of Religious Conviction," *Psychological Science* 20, no. 3 (2009): 385–92, https://doi.org/10.1111/j.1467-9280.2009.02305.x.

91. Michael Inzlicht and Alexa M. Tullett, "Reflecting on God: Religious Primes Can Reduce Neurophysiological Response to Errors," *Psychological Science* 21, no. 8 (2010): 1184–90, https://doi.org/10.1177/0956797610375451.

92. John T. Jost et al., "Are Needs to Manage Uncertainty and Threat Associated with Political Conservatism or Ideological Extremity?," *Personality and Social Psychology Bulletin* 33, no. 7 (2007): 989–1007, https://doi.org/10.1177/0146167207301028.

93. Miguel Farias et al., "Scientific Faith: Belief in Science Increases in the Face of Stress and Existential Anxiety," *Journal of Experimental Social Psychology* 49, no. 6 (2013): 1210–13, https://doi.org/10.1016/j.jesp.2013.05.008.

94. Stefan Reiss et al., "Strength of Socio-political Attitudes Moderates Electrophysiological Responses to Perceptual Anomalies," *PLOS One* 14, no. 8 (2019): e0220732, https://doi.org/10.1371/journal.pone.0220732.

95. Jeroen M. van Baar et al., "Intolerance of Uncertainty Modulates Brain-to-Brain Synchrony During Politically Polarized Perception," *Proceedings of the National Academy of Sciences* 118, no. 20 (2021), https://doi.org/10.1073/pnas.2022491118.

96. Isabelle Freiling et al., "Believing and Sharing Misinformation, Fact-Checks, and Accurate Information on Social Media: The Role of Anxiety During COVID-19," *New Media and Society* 25, no. 1 (2023): 141–62, https://doi.org/10.1177/14614448211011451.

97. Geoffrey L. Cohen and David K. Sherman, "The Psychology of Change: Self-Affirmation and Social Psychological Intervention," *Annual Review of Psychology* 65 (2014): 333–71, https://doi.org/10.1146/annurev-psych-010213-115137.

98. David K. Sherman, "Self-Affirmation: Understanding the Effects," *Social and Personality Psychology Compass* 7, no. 11 (2013): 834–45, https://doi.org/10.1111/spc3.12072.

99. William M. P. Klein et al., "Feelings of Vulnerability in Response to Threatening Messages: Effects of Self-Affirmation," *Journal of Experimental Social Psychology* 47, no. 6 (2011): 1237–42, https://doi.org/10.1016/j.jesp.2011.05.005.

100. Kevin R Binning et al., "Seeing the Other Side: Reducing Political Partisanship via Self-Affirmation in the 2008 Presidential Election," *Analyses of Social Issues and Public Policy* 10, no. 1 (2010): 276–92, https://doi.org/10.1111/j.1530-2415.2010.01210.x; Ian McGregor et al., "Can Ingroup Affirmation Relieve Outgroup Derogation?," *Journal of Experimental Social Psychology* 44, no. 5 (2008): 1395–1401, https://doi.org/10.1016/j.jesp.2008.06.001; Geoffrey L. Cohen et al., "Bridging the Partisan Divide: Self-Affirmation Reduces Ideological Closed-Mindedness and Inflexibility in Negotiation," *Journal of Personality and Social Psychology* 93, no. 3 (2007): 415–30, https://doi.org/10.1037/0022-3514.93.3.415.

101. Tracy Epton et al., "The Impact of Self-Affirmation on Health-Behavior Change: A Meta-Analysis," *Health Psychology* 34, no. 3 (2015): 187–96, https://doi.org/10.1037/hea0000116.

102. Emily B. Falk et al., "Self-Affirmation Alters the Brain's Response to Health Messages and Subsequent Behavior Change," *Proceedings of the National Academy of Sciences* 112, no. 7 (2015): 1977–82, https://doi.org/10.1073/pnas.1500247112.

103. Janine M. Dutcher et al., "Neural Mechanisms of Self-Affirmation's Stress Buffering Effects," *Social Cognitive and Affective Neuroscience* 15, no. 10 (2020): 1086–96, https://doi.org/10.1093/scan/nsaa042.

104. Benjamin A. Lyons et al., "Self-Affirmation and Identity-Driven Political Behavior," *Journal of Experimental Political Science* 9, no. 2 (2022): 225–40, https://doi.org/10.1017/XPS.2020.46.

105. Ian McGregor, "Offensive Defensiveness: Toward an Integrative Neuroscience of Compensatory Zeal After Mortality Salience, Personal Uncertainty, and Other Poignant Self-Threats," *Psychological Inquiry* 17, no. 4 (2006): 299–308, https://doi.org/10.1080/10478400701366977.

106. Matthew J. Facciani, "A Novel Approach for Measuring Self-Affirmation," *Current Research in Social Psychology* 30 (2021): 12–20.

107. Mathias Osmundsen et al., "Partisan Polarization Is the Primary Psychological Motivation Behind Political Fake News Sharing on Twitter," *American Political Science Review* 115, no. 3 (2021): 999–1015, https://doi.org/10.1017/S0003055421000290.

108. Louise Lemyre and Philip M. Smith, "Intergroup Discrimination and Self-Esteem in the Minimal Group Paradigm," *Journal of Personality and Social Psychology* 49, no. 3 (1985): 660–70; Steven Fein and Steven J. Spencer, "Prejudice as Self-Image Maintenance: Affirming the Self Through Derogating Others," *Journal of Personality and Social Psychology* 73, no. 1 (1997): 31–44, https://doi.org/10.1037/0022-3514.73.1.31.

109. Sonia Roccas and Marilynn B. Brewer, "Social Identity Complexity," *Personality and Social Psychology Review* 6, no. 2 (2002): 88–106, https://doi.org/10.1207/S15327957PSPR0602_01.

110. Grant and Hogg, "Self-Uncertainty, Social Identity Prominence and Group Identification."

111. Lilliana Mason and Julie Wronski, "One Tribe to Bind Them All: How Our Social Group Attachments Strengthen Partisanship," *Political Psychology* 39 (2018): 257–77, https://doi.org/10.1111/pops.12485.

112. Xun Zhu and Youllee Kim, "Mitigating Identity Threat in Health Messaging: A Social Identity Complexity Perspective," *Health Communication*, May 22, 2024, 1–12, https://doi.org/10.1080/10410236.2024.2358275.

113. Roccas and Brewer, "Social Identity Complexity."

114. Marilynn B. Brewer and Kathleen P. Pierce, "Social Identity Complexity and Outgroup Tolerance," *Personality and Social Psychology Bulletin* 31, no. 3 (2005): 428–37, https://doi.org/10.1177/0146167204271710.

115. Kevin P. Miller et al., "Social Identity Complexity: Its Correlates and Antecedents," *Group Processes and Intergroup Relations* 12, no. 1 (2009): 79–94, https://doi.org/10.1177/1368430208098778.

116. Grant and Hogg, "Self-Uncertainty, Social Identity Prominence and Group Identification."

117. Benjamin A. Lyons, "Unbiasing Information Search and Processing Through Personal and Social Identity Mechanisms" (PhD diss., Southern Illinois University at Carbondale, 2016).

118. Joseph Cesario, "Priming, Replication, and the Hardest Science," *Perspectives on Psychological Science* 9, no. 1 (2014): 40–48, https://doi.org/10.1177/1745691613513470.

119. Peter J. Burke and Christine Cerven, "Identity Accumulation, Verification, and Well-Being," in *Identities in Everyday Life*, ed. Jan E. Stets and Richard T. Serpe (Oxford University Press, 2019), 17–33, https://doi.org/10.1093/oso/9780190873066.003.0002.

120. David B. Yaden et al., "The Overview Effect: Awe and Self-Transcendent Experience in Space Flight," *Psychology of Consciousness: Theory, Research, and Practice* 3, no. 1 (2016): 1–11, http://doi.org/10.1037/cns0000092.

121. Leonard David, "Space Philosopher Frank White on 'The Overview Effect' and Humanity's Connection with Earth," *Space*, August 2, 2022, https://www.space.com/frank-white-overview-effect.

122. Omnia Salah, "First Egyptian, Arab Woman to Go to Space Recounts Her Journey," *Al-Monitor*, November 25, 2022, https://www.al-monitor.com/originals/2022/11/first-egyptian-arab-woman-go-space-recounts-her-journey.

123. Marina Koren, "Seeing Earth from Space Will Change You," *The Atlantic*, December 10, 2022, https://www.theatlantic.com/magazine/archive/2023/01/astronauts-visiting-space-overview-effect-spacex-blue-origin/672226/.

124. Luguri and Napier, "Of Two Minds."

125. Rachel Hartman et al., "Interventions to Reduce Partisan Animosity," *Nature Human Behaviour* 6, no. 9 (2022): 1194–1205, https://doi.org/10.1038/s41562-022-01442-3.

126. Stefano Balietti et al., "Reducing Opinion Polarization: Effects of Exposure to Similar People with Differing Political Views," *Proceedings of the National Academy of Sciences* 118, no. 52 (2021): e2112552118, https://doi.org/10.1073/pnas.2112552118.

127. Emily P. Diamond, "The Influence of Identity Salience on Framing Effectiveness: An Experiment," *Political Psychology* 41, no. 6 (2020): 1133–50, https://doi.org/10.1111/pops.12669.

128. Jan G. Voelkel et al., "Megastudy Identifying Successful Interventions to Strengthen Americans' Democratic Attitudes," Working Paper No. 22-38 (Institute for Policy Research, Northwestern University, 2022).

129. Kahneman, *Thinking, Fast and Slow*.

130. Shane Frederick, "Cognitive Reflection and Decision Making," *Journal of Economic Perspectives* 19, no. 4 (2005): 25–42, https://doi.org/10.1257/089533005775196732.

131. Gordon Pennycook and David G. Rand, "Who Falls for Fake News? The Roles of Bullshit Receptivity, Overclaiming, Familiarity, and Analytic Thinking," *Journal of Personality* 88, no. 2 (2020): 185–200, https://doi.org/10.1111/jopy.12476.

132. Gordon Pennycook and David G. Rand, "Lazy, Not Biased: Susceptibility to Partisan Fake News Is Better Explained by Lack of Reasoning than by Motivated Reasoning," *Cognition* 188 (2019): 39–50, https://doi.org/10.1016/j.cognition.2018.06.011.

133. John T. Cacioppo and Richard E. Petty, "The Need for Cognition," *Journal of Personality and Social Psychology* 42, no. 1 (1982): 116–31, https://doi.org/10.1037/0022-3514.42.1.116.

134. Porismita Borah, "The Moderating Role of Political Ideology: Need for Cognition, Media Locus of Control, Misinformation Efficacy, and Misperceptions About COVID-19," *International Journal of Communication Systems* 16 (2022): 3534–59, https://ijoc.org/index.php/ijoc/article/view/18261.

135. Juliana K. Leding and Lilyeth Antonio, "Need for Cognition and Discrepancy Detection in the Misinformation Effect," *Journal of Cognitive Psychology* 31, no. 4 (2019): 409–15, https://doi.org/10.1080/20445911.2019.1626400.

136. Gordon Pennycook et al., "Shifting Attention to Accuracy Can Reduce Misinformation Online," *Nature* 592, no. 7855 (2021): 590–95, https://doi.org/10.1038/s41586-021-03344-2.

137. Nadia M. Brashier et al., "An Initial Accuracy Focus Prevents Illusory Truth," *Cognition* 194 (2020): art. 104054, https://doi.org/10.1016/j.cognition.2019.104054.

138. Cédric Batailler et al., "A Signal Detection Approach to Understanding the Identification of Fake News," *Perspectives on Psychological Science* 17, no. 1 (2022): 78–98, https://doi.org/10.1177/1745691620986135.

139. Bertram Gawronski, "Partisan Bias in the Identification of Fake News," *Trends in Cognitive Sciences* 25, no. 9 (2021): 723, https://doi.org/10.1016/j.tics.2021.05.001.

140. Jay J. Van Bavel et al., "Updating the Identity-Based Model of Belief: From False Belief to the Spread of Misinformation," *Current Opinion in Psychology* 56 (2024): art. 101787, https://doi.org/10.1016/j.copsyc.2023.101787.

141. Carey K. Morewedge et al., "Debiasing Decisions: Improved Decision Making with a Single Training Intervention," *Policy Insights from the Behavioral and Brain Sciences* 2, no. 1 (2015): 129–40, https://doi.org/10.1177/2372732215600886.

3. POLITICAL IDENTITIES, THE RESPONSE TO COVID-19, AND HOW TO HAVE PRODUCTIVE CONVERSATIONS

1. Hannah Hartig, "Two Decades Later, the Enduring Legacy of 9/11," Pew Research Center, September 2, 2021, https://www.pewresearch.org/politics/2021/09/02/two-decades-later-the-enduring-legacy-of-9-11/.

2. Laura Petrecca, "America's Division: We United in the Wake of 9/11, Then Partisanship Re-emerged," *USA Today*, September 11, 2017, https://www.usatoday.com/story/news/2017/09/11/americas-division-we-united-wake-9-11-then-partisanship-re-emerged/639473001/.

3. Megan Brenan, "Record-Low 38 Percent Extremely Proud to Be American," Gallup, June 29, 2022, https://news.gallup.com/poll/394202/record-low-extremely-proud-american.aspx.

4. Peter Bell, "Public Trust in Government: 1958–2023," Pew Research Center, September 19, 2023, https://www.pewresearch.org/politics/2023/09/19/public-trust-in-government-1958-2023/.

5. Lydia Saad, "Gallup Vault: A Country Unified After Pearl Harbor," Gallup, December 5, 2016, https://news.gallup.com/vault/199049/gallup-vault-country-unified-pearl-harbor.aspx.

6. "How Did Public Opinion About Entering World War II Change Between 1939 and 1941?," U.S. Holocaust Memorial Museum, accessed June 20, 2024, https://exhibitions.ushmm.org/americans -and-the-holocaust/us-public-opinion-world-war-II-1939-1941.

7. "Presidential Job Approval," American Presidency Project, University of California–Santa Barbara, accessed June 20, 2024, https://www.presidency.ucsb.edu/statistics/data/presidential-job-approval.

8. Associated Press, "What to Wear: Feds' Mixed Messages on Masks Sow Confusion," *U.S. News & World Report*, June 27, 2020, https://www.usnews.com/news/health-news/articles/2020-06-27/what-to -wear-feds-mixed-messages-on-masks-sow-confusion; Ashley Yeager, "Government's Mixed Messages on Coronavirus Are Dangerous: Experts," *The Scientist*, February 28, 2020, https://www.the-scientist .com/news-opinion/governments-mixed-messages-on-coronavirus-are-dangerous-experts-67202.

9. Bethany Albertson and Shana Kushner Gadarian, *Anxious Politics: Democratic Citizenship in a Threatening World* (Cambridge University Press, 2015).

10. Jaclyn Kettler et al., "Democratic Governors Are Quicker in Responding to the Coronavirus than Republicans," *The Conversation*, April 6, 2020, http://theconversation.com/democratic-governors -are-quicker-in-responding-to-the-coronavirus-than-republicans-135599.

11. Ciro Indolfi and Carmen Spaccarotella, "The Outbreak of COVID-19 in Italy: Fighting the Pandemic," *JACC: Case Reports* 2, no. 9 (2020): 1414–18, https://doi.org/10.1016/j.jaccas.2020.03.012.

12. "COVID-19 Cases," World Health Organization, accessed June 20, 2024, https://covid19.who.int /region/amro/country/us.

13. Carrie Blazina, "Republicans Remain Far Less Likely than Democrats to View COVID-19 as a Major Threat to Public Health," Pew Research Center, July 22, 2020, https://www.pewresearch.org/fact -tank/2020/07/22/republicans-remain-far-less-likely-than-democrats-to-view-covid-19-as-a-major -threat-to-public-health/; Matthew Facciani, "Video: How Did Mask Wearing Become So Politicized?," *The Conversation*, September 9, 2020, http://theconversation.com/video-how-did-mask-wearing -become-so-politicized-144268.

14. "COVID-19 Was Third Leading Cause of Death in the United States in Both 2020 and 2021," National Institutes of Health, July 5, 2022, https://www.nih.gov/news-events/news-releases/covid-19 -was-third-leading-cause-death-united-states-both-2020-2021.

15. Alyssa Pereira, "What Mask Use Looks Like in 10 Other Countries Compared to the U.S.," SFGATE, July 5, 2020, https://www.sfgate.com/news/article/mask-wearing-japan-korea-brazil-germany-zealand -15383513.php.

16. Kettler et al., "Democratic Governors Are Quicker in Responding to the Coronavirus than Republicans."

17. Juliette Cubanski et al., "What Happens When COVID-19 Emergency Declarations End? Implications for Coverage, Costs, and Access," *KFF*, January 31, 2023, https://www.kff.org/coronavirus -covid-19/issue-brief/what-happens-when-covid-19-emergency-declarations-end-implications-for -coverage-costs-and-access/.

18. Maggie Severns, "From Distraction to Disaster: How Coronavirus Crept Up on Washington," *Politico*, March 30, 2020, https://www.politico.com/news/2020/03/30/how-coronavirus-shook -congress-complacency-155058.

19. Jon Green et al., "Elusive Consensus: Polarization in Elite Communication on the COVID-19 Pandemic," *Science Advances* 6, no. 28 (2020): eabc2717, https://doi.org/10.1126/sciadv.abc2717.

20. Jeremy Padgett et al., "As Seen on TV? How Gatekeeping Makes the U.S. House Seem More Extreme," *Journal of Communication* 69, no. 6 (2019): 696–719, https://doi.org/10.1093/joc/jqz039; Matthew S. Levendusky and Neil Malhotra, "Does Media Coverage of Partisan Polarization Affect Political Attitudes?," *Political Communication* 33, no. 2 (2016): 283–301, https://doi.org/10.1080 /10584609.2015.1038455.

21. Brad Adgate, "Nielsen: How the Pandemic Changed At Home Media Consumption," *Forbes*, August 21, 2020, https://www.forbes.com/sites/bradadgate/2020/08/21/nielsen-how-the-pandemic -changed-at-home-media-consumption/.

22. Katherine Schaeffer, "Those on Ideological Right Favor Fewer COVID-19 Restrictions in Most Advanced Economies," Pew Research Center, July 30, 2021, https://www.pewresearch.org/short -reads/2021/07/30/those-on-ideological-right-favor-fewer-covid-19-restrictions-in-most-advanced -economies/.

23. Liz Goodwin, "Trump's Refusal to Wear a Mask Is Helping Politicize a Crucial Tool for Fighting Virus," *Boston Globe*, May 27, 2020, https://www.bostonglobe.com/2020/05/27/nation/trumps -refusal-wear-mask-is-helping-politicize-crucial-tool-fighting-virus/.

24. "NPR/PBS NewsHour/Marist Poll Results: Coronavirus," Marist Poll, February 4, 2020, https://maristpoll.marist.edu/npr-pbs-newshour-marist-poll-results-15/; Shana Kushner Gadar- ian et al., "Partisanship, Health Behavior, and Policy Attitudes in the Early Stages of the COVID- 19 Pandemic," *PLOS One* 16, no. 4 (2021): e0249596, https://doi.org/10.1371/journal.pone .0249596.

25. James Bisbee and Diana Lee, "Mobility and Elite Cues: Partisan Responses to COVID-19," APSA Preprints, August 28, 2020, https://doi.org/10.33774/apsa-2020-76tv9; Geoffrey Skelley and Amelia Thomson-DeVeaux, "How Americans Are Reacting to Trump's COVID-19 Diagnosis," FiveThirtyEight, October 5, 2020, https://fivethirtyeight.com/features/will-trumps-diagnosis-change-the-way-republicans -think-about-covid-19/.

26. Ben Tappin, email message to author, December 19, 2020.

27. Mert Moral and Robin E. Best, "On the Relationship Between Party Polarization and Citizen Polar- ization," *Party Politics* 29, no. 2 (2023): 229–47, https://doi.org/10.1177/13540688211069544.

28. Gordon Pennycook et al., "Beliefs About COVID-19 in Canada, the United Kingdom, and the United States: A Novel Test of Political Polarization and Motivated Reasoning," *Personality and Social Psychology Bulletin* 48, no. 5 (2022): 750–65, https://doi.org/10.1177/01461672211023652.

29. Jay J. Van Bavel, Aleksandra Cichocka, Valerio Capraro, Hallgeir Sjåstad, John B. Nezlek, Tomislav Pavlović, Mark Alfano et al, "National Identity Predicts Public Health Support During a Global Pandemic," *Nature communications* 13, no. 1 (2022): 517.

30. Jay Van Bavel, "," X, March 10, 2023, https://x.com/jayvanbavel/status/1501540453686075393.

31. Jeremy Page et al., "In Hunt for Covid-19 Origin, Patient Zero Points to Second Wuhan Market," *Wall Street Journal*, February 26, 2021, https://www.wsj.com/articles/in-hunt-for-covid-19-origin -patient-zero-points-to-second-wuhan-market-11614335404.

32. Dan Mangan and Berkeley Lovelace Jr., "Trump Suspects Coronavirus Outbreak Came from China Lab, Doesn't Cite Evidence," CNBC, April 30, 2020, https://www.cnbc.com/2020/04/30 /coronavirus-trump-suspects-covid-19-came-from-china-lab.html.

33. Brian Flood and Nikolas Lanum, "Credibility Crisis: Egg on Media's Face After Dismissing COVID Lab Leak as 'Debunked' Conspiracy Theory," Fox News, February 28, 2023, https://www.foxnews .com/media/credibility-crisis-egg-medias-face-after-dismissing-covid-lab-leak-debunked-conspiracy -theory.

34. Linley Sanders and Kathy Frankovic, "Two-Thirds of Americans Believe That the COVID-19 Virus Originated from a Lab in China," YouGov, March 10, 2023, https://today.yougov.com/topics /politics/articles-reports/2023/03/10/americans-believe-covid-origin-lab.

35. Michaeleen Doucleff, "U.S. Dept of Energy Says with 'Low Confidence' That COVID May Have Leaked from a Lab," NPR, February 28, 2023, https://www.npr.org/2023/02/28/1160157977/u-s -dept-of-energy-says-with-low-confidence-that-covid-may-have-leaked-from-a-la.

36. Ana Faguy, "U.S. Government Divided on Covid Lab Leak Theory—Here's Where Each Agency Stands," *Forbes*, February 27, 2023, https://www.forbes.com/sites/anafaguy/2023/02/27/us-government -divided-on-covid-lab-leak-theory-heres-where-each-agency-stands/.

37. Michael Worobey et al., "The Huanan Seafood Wholesale Market in Wuhan Was the Early Epicenter of the COVID-19 Pandemic," *Science* 377, no. 6609 (2022): 951–59, https://doi.org/10.1126/science.abp8715.

38. William J. Liu et al., "Surveillance of SARS-CoV-2 at the Huanan Seafood Market," *Nature* 631, no. 8020 (2024): 402–8, https://doi.org/10.1038/s41586-023-06043-2; Dyani Lewis et al., "COVID-Origins Data from Wuhan Market Published: What Scientists Think," *Nature* 616, no. 7956 (2023): 225–26, https://doi.org/10.1038/d41586-023-00998-y.

39. Gary Ackerman et al., *The Origin and Implications of the COVID-19 Pandemic: An Expert Survey,"* Technical Report 24–1 (Global Catastrophic Risk Institute, 2024).

40. I considered including more details about the data surrounding the COVID-19 origin, but I didn't want to stray too far from the main points of the chapter. While the expert consensus leaned toward zoonotic origins, the uncertainty in the data wasn't communicated well enough by some media figures and science journalists. As I discuss in chapter 5, admitting uncertainty is good practice for both public health officials and scientists.

41. Andrew Solender, "Trump Said U.S. Was 'Rounding the Final Turn' on Aug. 31—and on 39 of the 57 Days Since," *Forbes*, October 27, 2020, https://www.forbes.com/sites/andrewsolender/2020/10/27/trump-said-us-was-rounding-the-final-turn-on-aug-31-and-on-39-of-the-57-days-since/.

42. Cecelia Smith-Schoenwalder, "Fauci Disagrees with Trump's Claim About Rounding the 'Final Turn' on the COVID-19 Outbreak," *U.S. News & World Report*, September 11, 2020, https://www.usnews.com/news/national-news/articles/2020-09-11/fauci-disagrees-with-trumps-claim-about-rounding-the-corner-on-the-coronavirus-outbreak.

43. Adam Schrader, "Gallup: Chief Justice John Roberts Earns Top Approval Rating Among Federal Leaders," UPI, December 27, 2021, https://www.upi.com/Top_News/US/2021/12/27/chief-justice-john-roberts-highets-approval-rating-federal-leaders/9461640629282/.

44. Michelle M. Mello et al., "Attacks on Public Health Officials During COVID-19," *JAMA* 324, no. 8 (2020): 741–42, https://doi.org/10.1001/jama.2020.14423.

45. Peter J. Hotez, *The Deadly Rise of Anti-Science: A Scientist's Warning* (Johns Hopkins University Press, 2023).

46. As I wrote this book, I wondered how many attacks I would receive for presenting the scientific consensus on COVID-19 and vaccinations. I made an effort to approach these topics with as much nuance as I could, and it may not matter for some individuals. It's so unfortunate that these scientific topics have become so politicized.

47. Blazina, "Republicans Remain Far Less Likely than Democrats to View COVID-19 as a Major Threat to Public Health."

48. Sara Atske, "Republicans, Democrats Move Even Further Apart in Coronavirus Concerns," Pew Research Center, June 25, 2020, https://www.pewresearch.org/politics/2020/06/25/republicans-democrats-move-even-further-apart-in-coronavirus-concerns/.

49. Nattavudh Powdthavee et al., "When Face Masks Signal Social Identity: Explaining the Deep Face-Mask Divide During the COVID-19 Pandemic," *PLOS One* 16, no. 6 (2021): e0253195, https://doi.org/10.1371/journal.pone.0253195; "Research Reveals Why People Refuse to Wear Face Masks," Warwick Business School, August 10, 2021, https://www.wbs.ac.uk/news/research-reveals-why-people-refuse-to-wear-face-masks/.

50. "Republicans See U.S. As Better Off Now than 4 Years Ago Ahead of Convention—Battleground Tracker Poll," CBS News, August 23, 2020, https://www.cbsnews.com/news/republicans-economy-coronavirus-opinion-poll-cbs-news-battleground-tracker/.

51. "400 [U.S.] Cities," World Population Review, accessed June 20, 2024, https://worldpopulation-review.com/countries/cities/united-states.

52. Christine Hauser, "The Mask Slackers of 1918," *New York Times*, August 3, 2020, https://www.nytimes.com/2020/08/03/us/mask-protests-1918.html.

53. Sarah Pruitt, "How the US Pulled Off Midterm Elections Amid the 1918 Flu Pandemic," History, April 22, 2020, https://www.history.com/news/1918-pandemic-midterm-elections.

54. Berkeley Lovelace Jr., "Medical Historian Compares the Coronavirus to the 1918 Flu Pandemic: Both Were Highly Political," CNBC, September 29, 2020, https://www.cnbc.com/2020/09/28/comparing-1918-flu-vs-coronavirus.html.

55. John Sands, "Local News Is More Trusted than National News—but That Could Change," Knight Foundation, October 29, 2019, https://knightfoundation.org/articles/local-news-is-more-trusted-than-national-news-but-that-could-change/.

56. Jay J. Van Bavel et al., "National Identity Predicts Public Health Support During a Global Pandemic," *Nature Communications* 13, no. 1 (2022): 517, https://doi.org/10.1038/s41467-021-27668-9.

57. "Tracking the COVID-19 Economy's Effects on Food, Housing, and Employment Hardships," Center on Budget and Policy Priorities, August 13, 2020, https://www.cbpp.org/research/poverty-and-inequality/tracking-the-covid-19-economys-effects-on-food-housing-and.

58. Monmouth University Polling Institute, "1 in 4 Say 'No Thanks' to Vaccine," February 3, 2021, https://www.monmouth.edu/polling-institute/reports/MonmouthPoll_US_020321/.

59. Sol Hart et al., "Politicization and Polarization in COVID-19 News Coverage," *Science Communication* 42, no. 5 (2020): 679–97, https://doi.org/10.1177/1075547020950735.

60. Robert Hart, "As Fox's Tucker Carlson Stokes Covid-19 Vaccine Fears—Here's What You Really Need to Know About Pfizer's Covid-19 Vaccine," *Forbes*, December 18, 2020, https://www.forbes.com/sites/roberthart/2020/12/18/as-foxs-tucker-carlson-stokes-covid-19-vaccine-fears--heres-what-you-really-need-to-know-about-pfizers-covid-19-vaccine/.

61. Shannon Bond, "Facebook Widens Ban on COVID-19 Vaccine Misinformation in Push to Boost Confidence," NPR, February 8, 2021, https://www.npr.org/2021/02/08/965390755/facebook-widens-ban-on-covid-19-vaccine-misinformation-in-push-to-boost-confiden.

62. Jacob Wallace et al., "Excess Death Rates for Republican and Democratic Registered Voters in Florida and Ohio During the COVID-19 Pandemic," *JAMA Internal Medicine* 183, no. 9 (2023): 916–23.

63. "The COVID-19 Infodemic," editorial, *Lancet Infectious Diseases* 20, no. 8 (2020): 875.

64. Eric Kleefeld and Bobby Lewis, "'Long COVID' and the Ongoing Public Health Dangers That Fox News Hosts Ignore," Media Matters for America, August 7, 2020, https://www.mediamatters.org/coronavirus-covid-19/long-covid-and-ongoing-public-health-dangers-fox-news-hosts-ignore.

65. "Fox's Dr. Marc Siegel Says 'Worse Case Scenario' for Coronavirus Is 'It Could Be the Flu,'" Media Matters for America, March 6, 2020, https://www.mediamatters.org/sean-hannity/foxs-dr-marc-siegel-says-worse-case-scenario-coronavirus-it-could-be-flu.

66. Robert Hornik et al., "Association of COVID-19 Misinformation with Face Mask Wearing and Social Distancing in a Nationally Representative US Sample," *Health Communication* 36, no. 1 (2021): 6–14, https://doi.org/10.1080/10410236.2020.1847437.

67. "New Research Explores How Conservative Media Misinformation May Have Intensified the Severity of the Pandemic," *Washington Post*, June 25, 2020, https://www.washingtonpost.com/business/2020/06/25/fox-news-hannity-coronavirus-misinformation/.

68. Jeremy Howard et al., "An Evidence Review of Face Masks Against COVID-19," *Proceedings of the National Academy of Sciences* 118, no. 4 (2021): e2014564118, https://doi.org/10.1073/pnas.2014564118; Krista Conger, "Surgical Masks Reduce COVID-19 Spread, Large-Scale Study Shows," News Center, Stanford Medicine, September 1, 2021, https://med.stanford.edu/news/all-news/2021/09/surgical-masks-covid-19.html; Hannah E. Clapham and Alex R. Cook, "Face Masks Help Control Transmission of COVID-19," *Lancet Digital Health* 3, no. 3 (2021): e136–37, https://doi.org/10.1016/S2589-7500(21)00003-0.

69. "Why Doctors Wear Masks," Yale Medicine, September 1, 2020, https://www.yalemedicine.org/news/why-doctors-wear-masks.

70. Rhett Allain et al., "The Physics of the N95 Face Mask," *Wired*, January 28, 2022, https://www.wired.com/story/the-physics-of-the-n95-face-mask/; Hiroshi Ueki et al., "Effectiveness of Face Masks in Preventing Airborne Transmission of SARS-CoV-2," *mSphere* 5, no. 5 (2020): e00637-20, https://doi.org/10.1128/mSphere.00637-20; Jianyu Lai et al., "Relative Efficacy of Masks and Respirators as Source Control for Viral Aerosol Shedding from People Infected with SARS-CoV-2: A Controlled Human Exhaled Breath Aerosol Experimental Study," *EBioMedicine* 104 (2024): art. 105157, https://doi.org/10.1016/j.ebiom.2024.105157.

71. Lucky Tran, "Don't Believe Those Who Claim Science Proves Masks Don't Work," *The Guardian*, February 27, 2023, https://www.theguardian.com/commentisfree/2023/feb/27/dont-believe-those-who-claim-science-proves-masks-dont-work.

72. Mary Kekatos, "Masks Are Effective but Here's How a Study from a Respected Group Was Misinterpreted to Say They Weren't," ABC News, March 14, 2023, https://abcnews.go.com/Health/masks-effective-study-respected-group-misinterpreted/story?id=97846561.

73. Jason Abaluck et al., "Impact of Community Masking on COVID-19: A Cluster-Randomized Trial in Bangladesh," *Science* 375, no. 6577 (2022): eabi9069, https://doi.org/10.1126/science.abi9069.

74. Xiaowen Wang et al., "Association Between Universal Masking in a Health Care System and SARS-CoV-2 Positivity Among Health Care Workers," *JAMA* 324, no. 7 (2020): 703–4, https://doi.org/10.1001/jama.2020.12897.

75. "Poll: Majority of Americans Say Key COVID-19 Policies Were a Good Idea—but Views of Individual Policies Vary," Harvard T. H. Chan School of Public Health, June 17, 2024, https://www.hsph.harvard.edu/news/press-releases/poll-majority-of-americans-say-key-covid-19-policies-were-a-good-idea-but-views-of-individual-policies-vary/.

76. Reuters, "Herman Cain, Ex-Presidential Candidate Who Refused to Wear Mask, Dies After COVID-19 Diagnosis," July 31, 2020. https://www.reuters.com/article/world/herman-cain-ex-presidential-candidate-who-refused-to-wear-mask-dies-after-covi-idUSKCN24V2OH/.

77. Nathan Place, "Anti-Mask Maine Lawmaker Resigns After Wife Dies of Covid-19," *The Independent*, November 30, 2021, https://www.independent.co.uk/news/world/americas/us-politics/chris-johansen-covid-death-maine-b1967013.html.

78. Blazina, "Republicans Remain Far Less Likely than Democrats to View COVID-19 as a Major Threat to Public Health"; Atske, "Republicans, Democrats Move Even Further Apart in Coronavirus Concerns."

79. Emma Green, "The Liberals Who Can't Quit Lockdown," *The Atlantic*, May 4, 2021, https://www.theatlantic.com/politics/archive/2021/05/liberals-covid-19-science-denial-lockdown/618780/.

80. Jonathan Rothwell and Sonal Desai, "How Misinformation Is Distorting COVID Policies and Behaviors," *Brookings*, December 22, 2020, https://www.brookings.edu/research/how-misinformation-is-distorting-covid-policies-and-behaviors/.

81. Zeynep Tufekci, "Scolding Beachgoers Isn't Helping," *The Atlantic*, July 4, 2020, https://www.theatlantic.com/health/archive/2020/07/it-okay-go-beach/613849/.

82. Poppy Noor, "The Beach-Going Grim Reaper on His Florida Protest: 'Someone Has to Stand Up,' " *The Guardian*, May 7, 2020, https://www.theguardian.com/us-news/2020/may/07/florida-grim-reaper-beach-interview.

83. Tommaso Celeste Bulfone et al., "Outdoor Transmission of SARS-CoV-2 and Other Respiratory Viruses: A Systematic Review," *Journal of Infectious Diseases* 223, no. 4 (2021): 550–61, https://doi.org/10.1093/infdis/jiaa742; "COVID-19: Is It Safe to Swim?," Cleveland Clinic, September 17, 2021, https://health.clevelandclinic.org/is-the-pool-safe-during-coronavirus-pandemic/.

84. Zeynep Tufekci, "Keep the Parks Open," *The Atlantic*, April 7, 2020, https://www.theatlantic.com/health/archive/2020/04/closing-parks-ineffective-pandemic-theater/609580/.

85. Tufekci, "Scolding Beachgoers Isn't Helping."

86.   "A Year Into the Pandemic, It's Even More Clear That It's Safer to Be Outside," *Washington Post*, April 13, 2021, https://www.washingtonpost.com/health/2021/04/13/covid-outside-safety/.

87.   Yesola Kweon and Byeonghwa Choi, "Fueling Conspiracy Beliefs: Political Conservatism and the Backlash Against COVID-19 Containment Policies," *Governance* 37, no. 3 (2024): 867–86.

88.   Derek Thompson, "Hygiene Theater Is a Huge Waste of Time," *The Atlantic*, July 27, 2020, https://www.theatlantic.com/ideas/archive/2020/07/scourge-hygiene-theater/614599/.

89.   Emanuel Goldman, "Exaggerated Risk of Transmission of COVID-19 by Fomites," *Lancet Infectious Diseases* 20, no. 8 (2020): 892–93, https://doi.org/10.1016/S1473-3099(20)30561-2; National Center for Immunization and Respiratory Diseases, Division of Viral Diseases, "Science Brief: SARS-CoV-2 and Surface (Fomite) Transmission for Indoor Community Environments," updated April 5, 2021, https://archive.cdc.gov/#/details?url=https://www.cdc.gov/coronavirus/2019-ncov/more/science-and-research/surface-transmission.html.

90.   Derek Thompson, "Deep Cleaning Isn't a Victimless Crime," *The Atlantic*, April 13, 2021, https://www.theatlantic.com/ideas/archive/2021/04/end-hygiene-theater/618576/.

91.   Rebecca Leber, "Hygiene Theater at Restaurants Is Creating Endless Plastic Waste," *Mother Jones*, accessed June 24, 2024, https://www.motherjones.com/food/2020/10/hygiene-theater-at-restaurants-is-creating-endless-plastic-waste/.

92.   Green, "The Liberals Who Can't Quit Lockdown."

93.   Alexander W. Bartik et al., "The Impact of COVID-19 on Small Business Outcomes and Expectations," *Proceedings of the National Academy of Sciences* 117, no. 30 (2020): 17656–66. https://doi.org/10.1073/pnas.2006991117.

94.   "Washington Post-ABC News Poll March 22–25, 2020." *Washington Post*, March 29, 2020, https://www.washingtonpost.com/context/washington-post-abc-news-poll-march-22-25-2020/974c3312-5a40-4764-afb1-4bb6b86f1cf4/.

95.   Sara Moniuszko, "Dr. Anthony Fauci Says Keeping Schools Shut Down for So Long Amid COVID 'Was Not a Good Idea,' " CBS News, June 18, 2024, https://www.cbsnews.com/news/fauci-schools-shut-down-covid/.

96.   AFP, "Hoax Circulates Online That an Old Indian Textbook Lists Treatments for COVID-19," April 14, 2020, https://factcheck.afp.com/hoax-circulates-online-old-indian-textbook-lists-treatments-covid-19.

97.   Katrina A. Bramstedt, "Unicorn Poo and Blessed Waters: COVID-19 Quackery and FDA Warning Letters," *Therapeutic Innovation and Regulatory Science* 55, no. 1 (2021): 239–44.

98.   U.S. Food and Drug Administration, "FDA Warns Consumers About the Dangerous and Potentially Life Threatening Side Effects of Miracle Mineral Solution," news release, August 12, 2019, https://www.fda.gov/news-events/press-announcements/fda-warns-consumers-about-dangerous-and-potentially-life-threatening-side-effects-miracle-mineral; Ryan Felton, "Why Did It Take a Pandemic for the FDA to Crack Down on a Bogus Bleach 'Miracle' Cure?," *Consumer Reports*, May 14, 2020, updated July 8, 2020, https://www.consumerreports.org/scams-fraud/bogus-bleach-miracle-cure-fda-crackdown-miracle-mineral-solution-genesis-ii-church/.

99.   Ben Collins and Brandy Zadrozny, "The Far Right Is Struggling to Contain QAnon After Giving It Life," NBC News, August 10, 2018, https://www.nbcnews.com/tech/tech-news/far-right-struggling-contain-qanon-after-giving-it-life-n899741; Will Sommer, "QAnon Conspiracy Theorists' Magic Cure for Coronavirus Is Drinking Lethal Bleach," *Daily Beast*, January 28, 2020, https://www.thedailybeast.com/qanon-conspiracy-theorists-magic-cure-for-coronavirus-is-drinking-lethal-bleach.

100.   Tom Porter, "Taking Toxic Bleach MMS Has Killed 7 People in the US, Colombian Prosecutors Say—Far More Than Previously Known," *Business Insider*, August 12, 2020, https://www.businessinsider.com/mms-bleach-killed-7-americans-new-from-colombia-arrest-2020-8.

101.   U.S. Department of Justice, "Leaders of 'Genesis II Church of Health and Healing,' Who Sold Toxic Bleach as Fake 'Miracle' Cure for COVID-19 and Other Serious Diseases, Sentenced to

More than 12 Years in Federal Prison," October 6, 2023, https://www.justice.gov/usao-sdfl/pr/leaders-genesis-ii-church-health-and-healing-who-sold-toxic-bleach-fake-miracle-cure.

102.   Kacper Niburski and Oskar Niburski, "Impact of Trump's Promotion of Unproven COVID-19 Treatments and Subsequent Internet Trends: Observational Study," *Journal of Medical Internet Research* 22, no. 11 (2020): e20044, https://doi.org/10.2196/20044.

103.   Michael S. Saag, "Misguided Use of Hydroxychloroquine for COVID-19: The Infusion of Politics Into Science," *JAMA* 324, no. 21 (2020): 2161–62, https://doi.org/10.1001/jama.2020.22389; Frits R. Rosendaal, "Review of: 'Hydroxychloroquine and Azithromycin as a Treatment of COVID-19: Results of an Open-Label Non-randomized Clinical Trial Gautret et al 2010, DOI:10.1016/j.ijantimicag.2020.105949,' " *International Journal of Antimicrobial Agents* 56, no. 1 (2020): art. 106063, https://doi.org/10.1016%2Fj.ijantimicag.2020.106063.

104.   Gilmar Reis et al., "Effect of Early Treatment with Hydroxychloroquine or Lopinavir and Ritonavir on Risk of Hospitalization Among Patients with COVID-19: The TOGETHER Randomized Clinical Trial," *JAMA Network Open* 4, no. 4 (2021): e216468, https://doi.org/10.1001/jamanetworkopen.2021.6468; Charles H. Hennekens et al., "Updates on Hydroxychloroquine in Prevention and Treatment of COVID-19," *American Journal of Medicine* 135, no. 1 (2022): 7–9, https://doi.org/10.1016/j.amjmed.2021.07.035; Reuters, "FDA Cautions Against Use of Hydroxychloroquine or Chloroquine for Covid-19 Outside of Hospital Setting Due to Risk of Heart Rhythm Problems," April 24, 2020, https://www.reuters.com/article/business/healthcare-pharmaceuticals/fda-cautions-against-use-of-hydroxychloroquine-or-chloroquine-for-covid-19-outsi-idUSFWN2CC20M/.

105.   "Fox News and Trump Are Still Pushing Hydroxychloroquine: Here's What the Data Actually Shows," *Washington Post*, June 21, 2021, https://www.washingtonpost.com/politics/2021/06/21/hydroxycholoroquine-coronavirus-treatment-trump-allies-cant-quit/.

106.   Steven Chee Loon Lim et al., "Efficacy of Ivermectin Treatment on Disease Progression Among Adults with Mild to Moderate COVID-19 and Comorbidities: The I-TECH Randomized Clinical Trial," *JAMA Internal Medicine* 182, no. 4 (2022): 426–35, https://doi.org/10.1001/jamainternmed.2022.0189.

107.   Maria Popp et al., "Ivermectin for Preventing and Treating COVID-19," *Cochrane Database of Systematic Reviews*, no. 7 (2021): art. CD015017, https://doi.org/10.1002/14651858.CD015017.pub2.

108.   Rachel Schraer and Jack Goodman, "Ivermectin: How False Science Created a Covid 'Miracle' Drug," BBC, October 6, 2021, https://www.bbc.com/news/health-58170809.

109.   "Rand Paul Has a *Very* Wacky Theory About Ivermectin," CNN, August 31, 2021, https://www.cnn.com/2021/08/31/politics/rand-paul-covid-19-ivermectin/index.html;
        Riley Vetterkind, "Sen. Ron Johnson Doubles Down on Unproven Early COVID-19 Treatments Including Ivermectin," *Wisconsin State Journal*, September 2, 2021, https://madison.com/wsj/news/local/govt-and-politics/sen-ron-johnson-doubles-down-on-unproven-early-covid-19-treatments-including-ivermectin/article_22d17b96-3b26-5a6b-a1e7-a778dde40893.html.

110.   "Rand Paul Has a *Very* Wacky Theory About Ivermectin."

111.   Kao-Ping Chua et al., "US Insurer Spending on Ivermectin Prescriptions for COVID-19," *JAMA* 327, no. 6 (2022): 584–87, https://doi.org/10.1001/jama.2021.24352.

112.   Kathy Frankovic, "Most Republicans Who Have Heard of Ivermectin as a COVID-19 Treatment Think It May Be Effective," YouGov, September 2, 2021, https://today.yougov.com/topics/politics/articles-reports/2021/09/02/most-republicans-who-have-heard-ivermectin.

113.   Kathy Katella, "Comparing the COVID-19 Vaccines: How Are They Different?," Yale Medicine, April 24, 2024, https://www.yalemedicine.org/news/covid-19-vaccine-comparison.

114.   "Anti-SARS-CoV-2 Monoclonal Antibodies," COVID-19 Treatment Guidelines, accessed June 24, 2024, https://www.covid19treatmentguidelines.nih.gov/therapies/antivirals-including-antibody-products/anti-sars-cov-2-monoclonal-antibodies/.

115.   Roni Rabin, "Paxlovid Cuts Covid Deaths Among Older People, Israeli Study Finds," *New York Times*, August 30, 2022, https://www.nytimes.com/2022/08/30/health/paxlovid-efficacy-seniors .html; Dylan Scott, "Why People Who Don't Trust Vaccines Are Embracing Unproven Drugs," *Vox*, October 1, 2021, https://www.vox.com/coronavirus-covid19/22686147/covid-19-vaccine -betadine-hydroxychloroquine-ivermectin-trump-conspiracy.

116.   Linqi Lu et al., "Source Trust and COVID-19 Information Sharing: The Mediating Roles of Emotions and Beliefs About Sharing," *Health Education and Behavior* 48, no. 2 (2021): 132–39, https:// doi.org/10.1177/1090198120984760.

117.   Luca Simione et al., "Mistrust and Beliefs in Conspiracy Theories Differently Mediate the Effects of Psychological Factors on Propensity for COVID-19 Vaccine," *Frontiers in Psychology* 12 (2021): art. 683684, https://doi.org/10.3389/fpsyg.2021.683684.

118.   Betsy Broaddus, "Amidst the Pandemic, Confidence in the Scientific Community Becomes Increasingly Polarized," AP-NORC, January 26, 2022, https://apnorc.org/projects/amidst-the-pandemic -confidence-in-the-scientific-community-becomes-increasingly-polarized/.

119.   Robert Wood Johnson Foundation and Harvard T. H. Chan School of Public Health, *The Public's Perspective on the United States Public Health System* (Harvard Opinion Research Program, 2021), https://cdn1.sph.harvard.edu/wp-content/uploads/sites/94/2021/05/RWJF-Harvard-Report _FINAL-051321.pdf.

120.   Abigail Geiger, "How Americans See the Future of Space Exploration, 50 Years After the First Moon Landing," Pew Research Center, July 17, 2019, https://www.pewresearch.org/fact-tank/2019/07/17 /how-americans-see-the-future-of-space-exploration-50-years-after-the-first-moon-landing/.

121.   Simione et al., "Mistrust and Beliefs in Conspiracy Theories Differently Mediate the Effects of Psychological Factors on Propensity for COVID-19 Vaccine."

122.   Liz Hamel et al., "KFF COVID-19 Vaccine Monitor: Media and Misinformation," *KFF*, November 8, 2021, https://www.kff.org/coronavirus-covid-19/poll-finding/kff-covid-19-vaccine-monitor-media- and-misinformation/?utm_campaign=KFF-2021-polling-surveys&utm_medium.

123.   Molly Callahan, "Are You More Likely to Believe Misinformation About Ukraine or COVID-19?," *Northeastern Global News*, April 27, 2022, https://news.northeastern.edu/2022/04/27/misinformation -about-ukraine-or-covid-19/.

124.   Amy Mitchell et al., "5. Republicans' Views on COVID-19 Shifted Over Course of 2020; Democrats' Hardly Budged," Pew Research Center, February 22, 2021, https://www.pewresearch.org /journalism/2021/02/22/republicans-views-on-covid-19-shifted-over-course-of-2020-democrats -hardly-budged/.

125.   Carrie Blazina, "Despite Wide Partisan Gaps in Views of Many Aspects of the Pandemic, Some Common Ground Exists," Pew Research Center, March 24, 2021, https://www.pewresearch.org /fact-tank/2021/03/24/despite-wide-partisan-gaps-in-views-of-many-aspects-of-the-pandemic -some-common-ground-exists/.

126.   Deborah Netburn, "Timeline: CDC Mask Guidelines During the COVID Pandemic," *Los Angeles Times*, July 27, 2021, https://www.latimes.com/science/story/2021-07-27/timeline-cdc-mask -guidance-during-covid-19-pandemic.

127.   American Medical Association, "AMA: CDC Quarantine and Isolation Guidance Is Confusing, Counterproductive," press release, January 5, 2022, https://www.ama-assn.org/press-center /press-releases/ama-cdc-quarantine-and-isolation-guidance-confusing-counterproductive.

128.   Nicole Wetsman, "Masks May Be Good, but the Messaging Around Them Has Been Very Bad," *The Verge*, April 3, 2020, https://www.theverge.com/2020/4/3/21206728/cloth-face-masks-white -house-coronavirus-covid-cdc-messaging.

129.   "'You've Got Bad Blood': The Horror of the Tuskegee Syphilis Experiment," *Washington Post*, May 16, 2017, https://www.washingtonpost.com/news/retropolis/wp/2017/05/16/youve-got-bad -blood-the-horror-of-the-tuskegee-syphilis-experiment/.

130. Zeynep Tufekci, "5 Pandemic Mistakes We Keep Repeating," *The Atlantic*, February 26, 2021, https://www.theatlantic.com/ideas/archive/2021/02/how-public-health-messaging-backfired/618147/.

131. "CDC, Under Fire, Lays Out Plan to Become More Nimble and Accountable," *Washington Post*, August 17, 2022. https://www.washingtonpost.com/health/2022/08/17/walensky-revamp-cdc-culture-covid/.

132. Jackie Flynn Mogensen, "5 Tips for How to Actually Change an Anti-Masker's Mind, According to Experts," *Mother Jones*, December 21, 2020, https://www.motherjones.com/politics/2020/12/how-to-win-an-argument-change-mind-anti-masker-tips/.

133. Claire Hooker, "How to Talk to Someone Who Doesn't Wear a Mask, and Actually Change Their Mind," *The Conversation*, August 14, 2020, http://theconversation.com/how-to-talk-to-someone-who-doesnt-wear-a-mask-and-actually-change-their-mind-143995.

134. Harry Weger Jr. et al., "The Relative Effectiveness of Active Listening in Initial Interactions," *International Journal of Listening* 28, no. 1 (2014): 13–31, https://doi.org/10.1080/10904018.2013.813234.

135. Guy Itzchakov et al., "Listening to Understand: The Role of High-Quality Listening on Speakers' Attitude Depolarization During Disagreements," *Journal of Personality and Social Psychology* 126, no. 2 (2024): 213–39, https://doi.org/10.1037/pspa0000366.

136. Julia Marcus, "The Dudes Who Won't Wear Masks," *The Atlantic*, June 23, 2020, https://www.theatlantic.com/ideas/archive/2020/06/dudes-who-wont-wear-masks/613375/.

137. Emily Stewart, "Anti-Maskers Explain Themselves," *Vox*, August 7, 2020, https://www.vox.com/the-goods/2020/8/7/21357400/anti-mask-protest-rallies-donald-trump-covid-19.

138. Daniel J. O'Keefe, "Persuasion," in *The Handbook of Communication Skills*, ed. Mark A. Bodie (Routledge, 2006), 333–52.

139. Diane Musho Hamilton, "Calming Your Brain During Conflict," *Harvard Business Review*, December 22, 2015, https://hbr.org/2015/12/calming-your-brain-during-conflict.

140. James Fishkin et al., "Is Deliberation an Antidote to Extreme Partisan Polarization? Reflections on 'America in One Room,'" *American Political Science Review* 115, no. 4 (2021): 1464–81, https://doi.org/10.1017/S0003055421000642.

141. Larry Diamond, "Could Deliberative Democracy Depolarize America?," Stanford Report, February 4, 2021, https://news.stanford.edu/2021/02/04/deliberative-democracy-depolarize-america/.

142. Guy Itzchakov and Harry T. Reis, "Perceived Responsiveness Increases Tolerance of Attitude Ambivalence and Enhances Intentions to Behave in an Open-Minded Manner," *Personality and Social Psychology Bulletin* 47, no. 3 (2021): 468–85, https://doi.org/10.1177/0146167220929218.

143. Amy McQueen and William M. P. Klein, "Experimental Manipulations of Self-Affirmation: A Systematic Review," *Self and Identity* 5, no. 4 (2006): 289–354, https://doi.org/10.1080/15298860600805325.

144. Guy Itzchakov et al., "I Am Aware of My Inconsistencies but Can Tolerate Them: The Effect of High Quality Listening on Speakers' Attitude Ambivalence," *Personality and Social Psychology Bulletin* 43, no. 1 (2017): 105–20, https://doi.org/10.1177/0146167216675339.

145. Guy Itzchakov and Kenneth G. DeMarree, "Attitudes in an Interpersonal Context: Psychological Safety as a Route to Attitude Change," *Frontiers in Psychology* 13 (2022): 932413, https://doi.org/10.3389/fpsyg.2022.932413.

146. Brian Resnick, "How to Talk Someone out of Bigotry," *Vox*, January 29, 2020, https://www.vox.com/2020/1/29/21065620/broockman-kalla-deep-canvassing.

147. Joshua L. Kalla and David E. Broockman, "Reducing Exclusionary Attitudes Through Interpersonal Conversation: Evidence from Three Field Experiments," *American Political Science Review* 114, no. 2 (2020): 410–25, https://doi.org/10.1017/S0003055419000923.

148. Katharine Hayhoe (@KHayhoe), "The key ingredient to constructive conversations is mutual respect," Twitter (now X), August 20, 2021, https://x.com/KHayhoe/status/1428812272441470976.

149. Blake M. Riek et al., "Does a Common Ingroup Identity Reduce Intergroup Threat?," *Group Processes and Intergroup Relations* 13, no. 4 (2010): 403–23, https://doi.org/10.1177/1368430209346701.

150. Emily Kubin et al., "Personal Experiences Bridge Moral and Political Divides Better than Facts," *Proceedings of the National Academy of Sciences* 118, no. 6 (2021): e2008389118, https://doi.org/10.1073/pnas.2008389118.

151. Jamie B. Luguri and Jaime L. Napier, "Of Two Minds: The Interactive Effect of Construal Level and Identity on Political Polarization," *Journal of Experimental Social Psychology* 49, no. 6 (2013): 972–77, https://doi.org/10.1016/j.jesp.2013.06.002.

152. "Strengthening Democracy Challenge: Winning Interventions," Stanford University, accessed June 27, 2024, https://www.strengtheningdemocracychallenge.org/winning-interventions.

153. Cecilie Steenbuch Traberg and Sander van der Linden, "Birds of a Feather Are Persuaded Together: Perceived Source Credibility Mediates the Effect of Political Bias on Misinformation Susceptibility," *Personality and Individual Differences* 185 (2022): art. 111269, https://doi.org/10.1016/j.paid.2021.111269; Adam J. Berinsky, "Rumors and Health Care Reform: Experiments in Political Misinformation," *British Journal of Political Science* 47, no. 2 (2017): 241–62, https://doi.org/10.1017/S0007123415000186.

154. Jonah Koetke et al., "Trust in Science Increases Conservative Support for Social Distancing," *Group Processes and Intergroup Relations* 24, no. 4 (2021): 680–97, https://doi.org/10.1177/1368430220985918.

155. Geoffrey L. Cohen, "Party Over Policy: The Dominating Impact of Group Influence on Political Beliefs," *Journal of Personality and Social Psychology* 85, no. 5 (2003): 808–22, https://doi.org/10.1037/0022-3514.85.5.808.

156. Amy Sokolow, "With Science and Scripture, a Baltimore Pastor Is Fighting Covid-19 Vaccine Skepticism," *STAT*, August 31, 2020, https://www.statnews.com/2020/08/31/with-science-and-scripture-a-baltimore-pastor-is-fighting-covid-19-vaccine-skepticism/.

157. Travis Mitchell, "Most Americans Who Go to Religious Services Say They Would Trust Their Clergy's Advice on COVID-19 Vaccines," Pew Research Center, October 15, 2021, https://www.pewforum.org/2021/10/15/most-americans-who-go-to-religious-services-say-they-would-trust-their-clergys-advice-on-covid-19-vaccines/.

158. Dominic Packer and Jay Van Bavel, "Navigating Political Divides at Thanksgiving," The Power of Us, November 22, 2022, https://powerofus.substack.com/p/navigating-political-divides-at-thanksgiving.

159. Jesse Graham and Jonathan Haidt, "Sacred Values and Evil Adversaries: A Moral Foundations Approach," in *The Social Psychology of Morality: Exploring the Causes of Good and Evil*, ed. M. Mikulincer and P. R. Shaver (American Psychological Association, 2012), 11–31, https://doi.org/10.1037/13091-001.

160. Jesse Graham et al., "Liberals and Conservatives Rely on Different Sets of Moral Foundations," *Journal of Personality and Social Psychology* 96, no. 5 (2009): 1029–46, https://doi.org/10.1037/a0015141.

161. Matthew Feinberg and Robb Willer, "The Moral Roots of Environmental Attitudes," *Psychological Science* 24, no. 1 (2013): 56–62, https://doi.org/10.1177/0956797612449177.

162. Matthew Motta et al., "A Call to Arms for Climate Change? How Military Service Member Concern About Climate Change Can Inform Effective Climate Communication," *Environmental Communication* 15, no. 1 (2021): 85–98, https://doi.org/10.1080/17524032.2020.1799836.

163. Reem Nadeem, "5. The U.S. Military," Pew Research Center, February 1, 2024, https://www.pewresearch.org/politics/2024/02/01/the-u-s-military/.

164. Matthew Feinberg and Robb Willer, "From Gulf to Bridge: When Do Moral Arguments Facilitate Political Influence?," *Personality and Social Psychology Bulletin* 41, no. 12 (2015): 1665–81, https://doi.org/10.1177/0146167215607842.

165. Oliver Scott Curry et al., "Mapping Morality with a Compass: Testing the Theory of 'Morality-as-Cooperation' with a New Questionnaire," *Journal of Research in Personality* 78 (February 1, 2019): 106–24, https://doi.org/10.1016/j.jrp.2018.10.008.

166. Edward D. Vargas and Gabriel R. Sanchez, "American Individualism Is an Obstacle to Wider Mask Wearing in the US," *Brookings*, August 31, 2020, https://www.brookings.edu/blog/up-front /2020/08/31/american-individualism-is-an-obstacle-to-wider-mask-wearing-in-the-us/.

167. Steven Taylor and Gordon J. G. Asmundson, "Negative Attitudes About Facemasks During the COVID-19 Pandemic: The Dual Importance of Perceived Ineffectiveness and Psychological Reactance," *PLOS One* 16, no. 2 (2021): e0246317, https://doi.org/10.1371/journal.pone.0246317.

168. Stephanie L. DeMora et al., "Reducing Mask Resistance Among White Evangelical Christians with Value-Consistent Messages," *Proceedings of the National Academy of Sciences* 118, no. 21 (2021): e2101723118, https://doi.org/10.1073/pnas.2101723118.

169. Gordon Pennycook et al., "Shifting Attention to Accuracy Can Reduce Misinformation Online," *Nature* 592, no. 7855 (2021): 590–95, https://doi.org/10.1038/s41586-021-03344-2.

170. Mohsen Mosleh et al., "Cognitive Reflection Correlates with Behavior on Twitter," *Nature Communications* 12, no. 1 (2021): 921, https://doi.org/10.1038/s41467-020-20043-0.

171. Richard E. Petty et al., "Elaboration as a Determinant of Attitude Strength: Creating Attitudes That Are Persistent, Resistant, and Predictive of Behavior," in *Attitude Strength*, ed. Richard E. Petty and Jon A. Krosnick (Psychology Press, 2014), 93–130; Mark W. Susmann et al., "Persuasion Amidst a Pandemic: Insights from the Elaboration Likelihood Model," *European Review of Social Psychology* 33, no. 2 (2022): 323–59, https://doi.org/10.1080/10463283.2021.1964744.

172. Amnon Glassner et al., "Pupils' Evaluation and Generation of Evidence and Explanation in Argumentation," *British Journal of Educational Psychology* 75, no. 1 (2005): 105–18, https://doi .org/10.1348/000709904X22278.

173. Rebecca Lawson, "The Science of Cycology: Failures to Understand How Everyday Objects Work," *Memory and Cognition* 34, no. 8 (2006): 1667–75, https://doi.org/10.3758/BF03195929; Steven Sloman and Philip Fernbach, *The Knowledge Illusion: Why We Never Think Alone* (Penguin, 2018).

174. Jessica Keating et al., "Partisan Underestimation of the Polarizing Influence of Group Discussion," *Journal of Experimental Social Psychology* 65 (2016): 52–58, https://doi.org/10.1016/j.jesp.2016.03.002.

175. Kim Strandberg et al., "Do Discussions in Like-Minded Groups Necessarily Lead to More Extreme Opinions? Deliberative Democracy and Group Polarization," *International Political Science Review* 40, no. 1 (2019): 41–57, https://doi.org/10.1177/0192512117692136.

176. Philip M. Fernbach et al., "Political Extremism Is Supported by an Illusion of Understanding," *Psychological Science* 24, no. 6 (2013): 939–46, https://doi.org/10.1177/0956797612464058.

177. Jarret T. Crawford and John Ruscio, "Asking People to Explain Complex Policies Does Not Increase Political Moderation: Three Preregistered Failures to Closely Replicate Fernbach, Rogers, Fox, and Sloman's (2013) Findings," *Psychological Science* 32, no. 4 (2021): 611–21, https://doi.org /10.1177/0956797620972367.

178. Abraham Tesser et al., "The Impact of Thought on Attitude Extremity and Attitude-Behavior Consistency," in *Attitude Strength*, ed. Richard E. Petty and Jon A. Krosnick (Psychology Press, 2014), 73–92; Kerrie L. Unsworth and Kelly S. Fielding, "It's Political: How the Salience of One's Political Identity Changes Climate Change Beliefs and Policy Support," *Global Environmental Change* 27 (2014): 131–37, https://doi.org/10.1016/j.gloenvcha.2014.05.002.

179. Peter T. Coleman, *The Way Out: How to Overcome Toxic Polarization* (Columbia University Press, 2021), 39.

180. Peter T. Coleman, "Chapter 7 Exercises," The Way Out, accessed August 14, 2024, https://www .thewayoutofpolarization.com/chapter-7-exercises.

181. Briony Swire-Thompson, "Hello! Our new paper "Correction format has a limited role when debunking misinformation" is in press," Twitter (now X), November 18, 2021, https://twitter.com /Briony_Swire/status/1461455196954013700.

182. Karin Tamerius, "Resources," Smart Politics, accessed July 15, 2024, https://www.joinsmart.org /resources/.

183. Karin Tamerius, "The 10 Types of Trust You Need to Persuade a Republican," *Medium*, February 17, 2022, https://medium.com/progressively-speaking/trust-is-everything-when-talking-politics-5cd84140a11f.

184. Eran Halperin et al., eds., *Psychological Intergroup Interventions: Evidence-Based Approaches to Improve Intergroup Relations* (Taylor & Francis, 2023).

185. Emile G. Bruneau et al., "Minding the Gap: Narrative Descriptions About Mental States Attenuate Parochial Empathy," *PLOS One* 10, no. 10 (2015): e0140838, https://doi.org/10.1371/journal.pone.0140838.

186. Matthew S. Levendusky and Dominik A. Stecula, *We Need to Talk: How Cross-Party Dialogue Reduces Affective Polarization* (Cambridge University Press, 2021).

187. Erik Santoro and David E. Broockman, "The Promise and Pitfalls of Cross-Partisan Conversations for Reducing Affective Polarization: Evidence from Randomized Experiments," *Science Advances* 8, no. 25 (2022): eabn5515, https://doi.org/10.1126/sciadv.abn5515.

188. Rakoen Maertens et al., "Long-Term Effectiveness of Inoculation Against Misinformation: Three Longitudinal Experiments," *Journal of Experimental Psychology: Applied* 27, no. 1 (2021): 1–16, https://doi.org/10.1037/xap0000315; Thomas Zerback et al., "The Disconcerting Potential of Online Disinformation: Persuasive Effects of Astroturfing Comments and Three Strategies for Inoculation Against Them," *New Media and Society* 23, no. 5 (2021): 1080–98, https://doi.org/10.1177/1461444820908530.

189. Emily Pronin, "The Introspection Illusion," in *Advances in Experimental Social Psychology*, vol. 41, ed. Mark P. Zanna (Academic Press, 2009), 1–67.

190. Julia A. Minson and Charles A. Dorison, "Why Is Exposure to Opposing Views Aversive? Reconciling Three Theoretical Perspectives," *Current Opinion in Psychology* 47 (2022): art. 101435, https://doi.org/10.1016/j.copsyc.2022.101435.

191. David Dunning, *Self-Insight: Roadblocks and Detours on the Path to Knowing Thyself* (Psychology Press, 2012).

192. Steven A. Sloman, "How Do We Believe?," *Topics in Cognitive Science* 14, no. 1 (2022): 31, https://doi.org/10.1111/tops.12580.

193. Dylan Marron, "Empathy Is Not Endorsement," TED Talk, Vancouver, BC, April 13, 2018, 10 min., 52 sec., https://www.ted.com/talks/dylan_marron_empathy_is_not_endorsement.

194. Robert Jones Jr. (formerly known as Son of Baldwin), "Contact," accessed August 14, 2024, https://www.sonofbaldwin.com/contact/. (On the attribution to Jones, see snopes.com, https://www.snopes.com/fact-check/james-baldwin-disagree-love/.)

195. Santoro and Broockman, "The Promise and Pitfalls of Cross-Partisan Conversations for Reducing Affective Polarization."

196. David E. Broockman et al., "Does Affective Polarization Undermine Democratic Norms or Accountability? Maybe Not," *American Journal of Political Science* 67, no. 3 (2023): 808–28, https://doi.org/10.1111/ajps.12719.

197. Mogensen, "5 Tips for How to Actually Change an Anti-Masker's Mind, According to Experts."

198. Katharine Hayhoe, "Contrary to what many think, I don't spend my time talking to dismissives," Twitter (now X), March 4, 2020, https://twitter.com/KHayhoe/status/1235204287891988481.

4. HOW SOCIAL NETWORKS AND SOCIAL MEDIA SPREAD MISINFORMATION

1. Ben Collins, "YouTube Search Results for A-list Celebrities Hijacked by Conspiracy Theorists," NBC News, July 30, 2018, https://www.nbcnews.com/tech/tech-news/youtube-search-results-list-celebrities-hijacked-conspiracy-theorists-n895926.

2. Joel Rose, "Even If It's 'Bonkers,' Poll Finds Many Believe QAnon and Other Conspiracy Theo-
    ries," NPR, December 30, 2020, https://www.npr.org/2020/12/30/951095644/even-if-its-bonkers
    -poll-finds-many-believe-qanon-and-other-conspiracy-theories; Rachel E. Greenspan, "The QAnon
    Conspiracy Theory and a Stew of Misinformation Fueled the Insurrection at the Capitol,"
    Insider, January 7, 2021, https://www.insider.com/capitol-riots-qanon-protest-conspiracy-theory
    -washington-dc-protests-2021-1.

3. Anastasiia Carrier, "'This Crap Means More to Him than My Life': When QAnon Invades American
    Homes," *Politico*, February 19, 2021, https://www.politico.com/news/magazine/2021/02/19/qanon
    -conspiracy-theory-family-members-reddit-forum-469485.

4. Marc-André Argentino and Sara Aniano, "QAnon and Beyond: Analysing QAnon Trends a Year
    After January 6th," GNET, January 6, 2022, https://gnet-research.org/2022/01/06/qanon-and-beyond
    -analysing-qanon-trends-a-year-after-january-6th/.

5. Mack Lamoureux, "Q Is Dead, Long Live QAnon," *VICE*, November 15, 2022, https://www.vice
    .com/en/article/wxnkzq/qanon-q-drop-midterms.

6. Miller McPherson et al., "Birds of a Feather: Homophily in Social Networks," *Annual Review of
    Sociology* 27, no. 1 (2001): 415–44, https://doi.org/10.1146/annurev.soc.27.1.415.

7. Adam D. I. Kramer et al., "Experimental Evidence of Massive-Scale Emotional Contagion Through
    Social Networks," *Proceedings of the National Academy of Sciences* 111, no. 24 (2014): 8788–90,
    https://doi.org/10.1073/pnas.1320040111; Nicholas A. Christakis and James H. Fowler, *Connected:
    The Surprising Power of Our Social Networks and How They Shape Our Lives* (Little, Brown Spark,
    2009): Stephen Vaisey and Omar Lizardo, "Can Cultural Worldviews Influence Network Composi-
    tion?," *Social Forces* 88, no. 4 (2010): 1595–1618, https://doi.org/10.1353/sof.2010.0009.

8. Carolyn Parkinson et al., "Similar Neural Responses Predict Friendship," *Nature Communications* 9,
    no. 1 (2018): 332, https://doi.org/10.1038/s41467-017-02722-7.

9. Abigail Geiger, "Political Polarization in the American Public," Pew Research Center, June 12, 2014,
    https://www.pewresearch.org/politics/2014/06/12/political-polarization-in-the-american-public/.

10. Hannah Fingerhut, "Partisanship in the U.S. Isn't Just About Politics, but How People See Their Neigh-
    bors," Pew Research Center, June 27, 2016, https://www.pewresearch.org/fact-tank/2016/06/27
    /partisanship-in-u-s-isnt-just-about-politics-but-how-people-see-their-neighbors/.

11. Jacob R. Brown and Ryan D. Enos, "The Measurement of Partisan Sorting for 180 Million Voters,"
    Nature Human Behaviour 5, no. 8 (2021): 998–1008, https://doi.org/10.1038/s41562-021-01066-z.

12. Byungkyu Lee and Peter Bearman, "Political Isolation in America," *Network Science* 8, no. 3 (2020):
    333–55, https://doi.org/10.1017/nws.2020.9.

13. M. Keith Chen and Ryne Rohla, "The Effect of Partisanship and Political Advertising on Close
    Family Ties," *Science* 360, no. 6392 (2018): 1020–24, https://doi.org/10.1126/science.aaq1433.

14. John Burnett, "Americans Are Fleeing to Places Where Political Views Match Their Own," NPR, Feb-
    ruary 18, 2022, https://www.npr.org/2022/02/18/1081295373/the-big-sort-americans-move-to-areas
    -political-alignment.

15. Rhodes Cook, "The 'Big Sort' Continues, with Trump as a Driving Force," Center for Politics:
    Sabato's Crystal Ball, February 17, 2022, https://centerforpolitics.org/crystalball/articles/the-big-sort
    -continues-with-trump-as-a-driving-force/.

16. Matthew S. Levendusky and Neil Malhotra, "(Mis)perceptions of Partisan Polarization in the
    American Public," *Public Opinion Quarterly* 80, no. S1 (2015): 378–91, https://doi.org/10.1093/poq
    /nfv045.

17. Daniel Yudkin et al., *The Perception Gap: How False Impressions Are Pulling Americans Apart* (More
    in Common, 2019).

18. Stefanie Stantcheva et al., "The Polarization of Reality," Discussion Paper No. 14348 (Centre for
    Economic Policy Research, 2020).

19. Kai Ruggeri et al., "The General Fault in Our Fault Lines," *Nature Human Behaviour* 5, no. 10
    (2021): 1369–80, https://doi.org/10.1038/s41562-021-01092-x.

20. Amber Hye-Yon Lee, "Social Trust in Polarized Times: How Perceptions of Political Polarization Affect Americans' Trust in Each Other," *Political Behavior* 44, no. 3 (2022): 1533–54, https://doi.org /10.1007/s11109-022-09787-1.

21. Cass Sunstein, "The Law of Group Polarization," *Journal of Political Philosophy* 10 (2002): 175–95, and "Sunstein on the Internet and Political Polarization," University of Chicago Law School, December 14, 2007, https://www.law.uchicago.edu/news/sunstein-internet-and-political-polarization.

22. Sarah K. Cowan, "Secrets and Misperceptions: The Creation of Self-Fulfilling Illusions," *Sociological Science* 1 (2014): 466–92, https://doi.org/10.15195/v1.a26.

23. Gordon D. A. Brown et al., "Social Sampling and Expressed Attitudes: Authenticity Preference and Social Extremeness Aversion Lead to Social Norm Effects and Polarization," *Psychological Review* 129, no. 1 (2022): 18–48, https://doi.org/10.1037/rev0000342.

24. Jessica Keating et al., "Partisan Underestimation of the Polarizing Influence of Group Discussion," *Journal of Experimental Social Psychology* 65 (2016): 52–58, https://doi.org/10.1016/j.jesp .2016.03.002.

25. John C. Blanchar and Catherine J. Norris, "Political Homophily, Bifurcated Social Reality, and Perceived Legitimacy of the 2020 US Presidential Election Results: A Four-Wave Longitudinal Study," *Analyses of Social Issues and Public Policy* 21, no. 1 (2021): 259–83, https://doi.org/10.1111/asap.12276.

26. Christian Elliott, "Climate Conversations: How to Talk with Friends Who Repeat Misinformation," Medill Reports Chicago, December 10, 2021, https://news.medill.northwestern.edu/chicago /climate-conversations-how-to-talk-with-friends-who-repeat-misinformation/.

27. Saher Khan and Vignesh Ramachandran, "Millions Depend on Private Messaging Apps to Keep in Touch: They're Ripe with Misinformation," PBS News, November 5, 2021, https://www.pbs .org/newshour/world/millions-depend-on-private-messaging-apps-to-keep-in-touch-theyre -ripe-with-misinformation; Ali Breland, "How the Coronavirus Brought Chain Mail Back to the Mainstream," *Mother Jones*, March 27, 2020, https://www.motherjones.com/politics/2020/03 /coronavirus-chain-mail/.

28. Khushbu Shah, "When Your Family Spreads Misinformation," *The Atlantic*, June 16, 2020, https://www.theatlantic.com/family/archive/2020/06/when-family-members-spread-coronavirus -misinformation/613129/.

29. Amy Mitchell, "Many Americans Say Made-Up News Is a Critical Problem That Needs to Be Fixed," Pew Research Center, June 5, 2019, https://www.pewresearch.org/journalism/2019/06/05 /many-americans-say-made-up-news-is-a-critical-problem-that-needs-to-be-fixed/.

30. David Lazer et al., "The Coevolution of Networks and Political Attitudes," *Political Communication* 27, no. 3 (2010): 248–74, https://doi.org/10.1080/10584609.2010.500187.

31. Bryan M. Parsons, "The Social Identity Politics of Peer Networks," *American Politics Research* 43, no. 4 (2015): 680–707, https://doi.org/10.1177/1532673X14546856.

32. Jennifer Wolak, "The Social Foundations of Public Support for Political Compromise," *The Forum* 20, no. 1 (2022): 185–207, https://doi.org/10.1515/for-2022-2050.

33. Peter Berger, *The Sacred Canopy: Elements of a Sociological Theory of Religion* (Doubleday, 1967).

34. Christian Smith et al., *American Evangelicalism: Embattled and Thriving* (University of Chicago Press, 1998).

35. Matthew E. Brashears, "Anomia and the Sacred Canopy: Testing a Network Theory," *Social Networks* 32, no. 3 (2010): 187–96, https://doi.org/10.1016/j.socnet.2009.12.003.

36. Joseph Langston et al., "Toward Faith: A Qualitative Study of How Atheists Convert to Christianity," *Journal of Religion and Society* 21 (2019): 1–23, http://moses.creighton.edu/jrs/toc/2019 .html.

37. Matthew Facciani and Matthew E. Brashears, "Sacred Alters: The Effects of Ego Network Structure on Religious and Political Beliefs," *Socius* 5 (2019): 2378023119873825, https://doi.org/10.1177 /2378023119873825.

38. Matthew E. Brashears, "'Trivial' Topics and Rich Ties: The Relationship Between Discussion Topic, Alter Role, and Resource Availability Using the 'Important Matters' Name Generator," *Sociological Science* 1 (2014): 493–511, https://doi.org/10.15195/v1.a27.

39. Casey A. Klofstad et al., "Disagreeing About Disagreement: How Conflict in Social Networks Affects Political Behavior," *American Journal of Political Science* 57, no. 1 (2013): 120–34, https://doi.org/10.1111/j.1540-5907.2012.00620.x; Joshua Robison et al., "Do Disagreeable Political Discussion Networks Undermine Attitude Strength?," *Political Psychology* 39, no. 2 (2018): 479–94, https://doi.org/10.1111/pops.12374.

40. Matthew Facciani and Cecilie Steenbuch Traberg, "Personal Network Composition and Cognitive Reflection Predict Susceptibility to Different Types of Misinformation," *Connections*, ahead of print, June 27, 2024.

41. Jonas Stein et al., "Network Segregation and the Propagation of Misinformation," *Scientific Reports* 13, no. 1 (2023): 917, https://doi.org/10.1038/s41598-022-26913-5.

42. Matthew S. Levendusky and Dominik A. Stecula, *We Need to Talk: How Cross-Party Dialogue Reduces Affective Polarization* (Cambridge University Press, 2021).

43. Kim Strandberg et al., "Do Discussions in Like-Minded Groups Necessarily Lead to More Extreme Opinions? Deliberative Democracy and Group Polarization," *International Political Science Review* 40, no. 1 (2019): 41–57, https://doi.org/10.1177/0192512117692136.

44. Matthew Facciani and Tara McKay, "Network Loss Following the 2016 Presidential Election Among LGBTQ+ Adults," *Applied Network Science* 7 (2022), https://doi.org/10.1007/s41109-022-00474-y.

45. Jay J. Van Bavel et al., "Political Psychology in the Digital (Mis)information Age: A Model of News Belief and Sharing," *Social Issues and Policy Review* 15, no. 1 (2021): 84–113, https://doi.org/10.1111/sipr.12077.

46. Dannagal G. Young and Amy Bleakley, "Ideological Health Spirals: An Integrated Political and Health Communication Approach to COVID Interventions," *International Journal of Communication Systems* 14, no. 14 (2020): 3508–24.

47. Jan E. Stets et al., "Using Identity Theory to Understand Homophily in Groups," in *Identities in Action: Developments in Identity* Theory, ed. Philip S. Brenner et al. (Springer, 2021), 285–302, https://doi.org/10.1007/978-3-030-76966-6_14.

48. Peter J. Burke and Jan E. Stets, "Trust and Commitment Through Self-Verification," *Social Psychology Quarterly* 62, no. 4 (1999): 347–66, https://doi.org/10.2307/2695833; Peter J. Burke and Michael M. Harrod, "Too Much of a Good Thing?," *Social Psychology Quarterly* 68, no. 4 (2005): 359–74, https://doi.org/10.1177/019027250506800404.

49. Christian Keysers and Valeria Gazzola, "Hebbian Learning and Predictive Mirror Neurons for Actions, Sensations and Emotions," *Philosophical Transactions of the Royal Society of London, Series B, Biological Sciences* 369, no. 1644 (2014): 20130175, https://doi.org/10.1098/rstb.2013.0175.

50. D. O. Hebb, *The Organization of Behavior: A Neuropsychological Theory* (Wiley, 1949; repr., Lawrence Erlbaum Associates, 2002), https://doi.org/10.4324/9781410612403.

 Donald Hebb was invited to teach a semester at my alma mater, Westminster College, toward the end of his career, according to my undergraduate neuroscience professor, Alan Gittis. Sadly, Hebb declined this offer because he was quite old at this point and was dealing with health issues.

51. Anna K. Zinn et al., "Social Identity Switching: How Effective Is It?," *Journal of Experimental Social Psychology* 101 (2022): 104309, https://doi.org/10.1016/j.jesp.2022.104309.

52. "Social Media Fact Sheet," Pew Research Center, January 31, 2024, https://www.pewresearch.org/internet/fact-sheet/social-media/.

53. Elisa Shearer, "Americans Are Wary of the Role Social Media Sites Play in Delivering the News," Pew Research Center, October 2, 2019, https://www.pewresearch.org/journalism/2019/10/02/americans-are-wary-of-the-role-social-media-sites-play-in-delivering-the-news/.

54. "Social Media Fact Sheet."

55. Ciera E. Kirkpatrick and Larissa L. Lawrie, "TikTok as a Source of Health Information and Misinformation for Young Women in the United States: Survey Study," *JMIR Infodemiology* 4 (2024): e54663, https://doi.org/10.2196/54663.

56. Shearer. "Americans Are Wary of the Role Social Media Sites Play in Delivering the News."

57. Rohit Shewale, "Facebook Users Statistics 2024 (Worldwide Data)," DemandSage, April 5, 2024, https://www.demandsage.com/facebook-statistics/.

58. Shubham Singh, "How Many People Use Instagram 2024 [Global Data]," DemandSage, May 6, 2024, https://www.demandsage.com/instagram-statistics/, and "TikTok User Statistics 2024 (Global Data)," DemandSage, accessed July 15, 2024, https://www.demandsage.com/tiktok-user-statistics/.

59. Juan Carlos Medina Serrano et al., "Dancing to the Partisan Beat: A First Analysis of Political Communication on TikTok," in *WebSci '20: Proceedings of the 12th ACM Conference on Web Science* (Association for Computing Machinery, 2020), 257–66, https://doi.org/10.1145/3394231.3397916; John H. Parmelee and Nataliya Roman, "Insta-politicos: Motivations for Following Political Leaders on Instagram," *Social Media + Society* 5, no. 2 (2019): 2056305119837662, https://doi.org/10.1177/2056305119837662.

60. Shearer, "Americans Are Wary of the Role Social Media Sites Play in Delivering the News."

61. Joan Donovan (@BostonJoan), "Donovan's 5 key principles of misinformation," Twitter (now X), March 3, 2022, https://x.com/BostonJoan/status/1499441476526292998.

62. Mark S. Granovetter, "The Strength of Weak Ties," *American Journal of Sociology* 78, no. 6 (1973): 1360–80, https://doi.org/10.1086/225469.

63. Jason Jeffrey Jones and Nick Rogers, "Online in the US, Personal Identity Is Increasingly Political," SocArXiv, August 25, 2021, https://doi.org/10.31235/osf.io/7k8xr.

64. Neil Levy, "Do Your Own Research!," *Synthese* 200, no. 5 (2022): 356, https://doi.org/10.1007/s11229-022-03793-w.

65. Francesca Bolla Tripodi, "Google Search Is Quietly Damaging Democracy," *Wired*, August 16, 2022, https://www.wired.com/story/google-search-quietly-damaging-democracy/.

66. Francesca Bolla Tripodi (@ftripodi), "This leads to what I refer to in my book, as the 'IKEA effect of misinformation,'" Twitter (now X), August 16, 2022, https://twitter.com/ftripodi/status/1559575975335059456.

67. Benjamin A. Lyons et al., "Overconfidence in News Judgments Is Associated with False News Susceptibility," *Proceedings of the National Academy of Sciences* 118, no. 23 (2021): e2019527118, https://doi.org/10.1073/pnas.2019527118.

68. Ryan Prior, "Most Americans Think They Can Spot Fake News: They Can't, Study Finds." CNN, May 31, 2021, https://www.cnn.com/2021/05/31/health/fake-news-study/index.html.

69. Eran Amsalem and Alon Zoizner, "Do People Learn About Politics on Social Media? A Meta-Analysis of 76 Studies," *Journal of Communication* 73, no. 1 (2022): 3–13, https://doi.org/10.1093/joc/jqac034.

70. Adam M. Enders et al., "The Relationship Between Social Media Use and Beliefs in Conspiracy Theories and Misinformation," *Political Behavior* 45, no. 2 (2023): 781–804, https://doi.org/10.1007/s11109-021-09734-6.

71. Nir Grinberg et al., "Fake News on Twitter During the 2016 U.S. Presidential Election," *Science* 363, no. 6425 (2019): 374–78, https://doi.org/10.1126/science.aau2706.

72. Andrew M. Guess et al., "Less Than You Think: Prevalence and Predictors of Fake News Dissemination on Facebook," *Science Advances* 5, no. 1 (2019): eaau4586, https://doi.org/10.1126/sciadv.aau4586.

73. Brittany Shaughnessy et al., "That Is So Mainstream: The Impact of Hyper-Partisan Media Use and Right-, Left-Wing Alternative Media Repertoires on Consumers' Belief in Political Misperceptions in the United States," *International Journal of Communication* 18 (2024): 1561–81,

1932-8036/20240005; Andrew M. Guess et al., "The Consequences of Online Partisan Media," *Proceedings of the National Academy of Sciences* 118, no. 14 (2021): e2013464118, https://doi .org/10.1073/pnas.2013464118.

74. Luxuan Wang, "Many Americans Find Value in Getting News on Social Media, but Concerns About Inaccuracy Have Risen," Pew Research Center, February 7, 2024, https://www.pewresearch.org /short-reads/2024/02/07/many-americans-find-value-in-getting-news-on-social-media-but -concerns-about-inaccuracy-have-risen/.

75. Filippo Menczer. "4 Reasons Why Social Media Make Us Vulnerable to Manipulation," Keynote speech to International Conference on Computational Social Science, 2020, https://www.youtube .com/watch?v=BQYveMPwlNg.

76. Martin Greenberger, "Designing Organizations for an Information-Rich World," in *Computers, Communications, and the Public Interest*, ed. Martin Greenberger (Johns Hopkins University Press, 1971), 40–41.

77. "Paying Attention: The Attention Economy," *Berkeley Economic Review*, March 31, 2020, https:// econreview.berkeley.edu/paying-attention-the-attention-economy/.

78. "Data Growth Worldwide 2010–2025," Statista, accessed June 30, 2024, https://www.statista.com /statistics/871513/worldwide-data-created/.

79. Robert Sheldon et al., "Gigabyte (GB)," TechTarget, October 21, 2021, https://www.techtarget.com /searchstorage/definition/gigabyte.

80. Maksym Gabielkov et al., "Social Clicks: What and Who Gets Read on Twitter?," in *Sigmetrics '16: Proceedings of the 2016 ACM Sigmetrics International Conference on Measurement and Modeling of Computer Science* (Association for Computing Machinery, 2016), 179–92, https://doi.org/10.1145 /2896377.2901462.

81. Tony Haile, "What You Think You Know About the Web Is Wrong," *Time*, March 9, 2014, https:// time.com/12933/what-you-think-you-know-about-the-web-is-wrong/.

82. L. Weng et al., "Competition Among Memes in a World with Limited Attention," *Scientific Reports* 2 (2012): art. 335, https://doi.org/10.1038/srep00335.

83. Xiaoyan Qiu et al., "Limited Individual Attention and Online Virality of Low-Quality Informa- tion," *Nature Human Behaviour* 1 (2017): art. 132, https://doi.org/10.1038/s41562-017-0132.

84. Filippo Menczer and Thomas Hills, "Information Overload Helps Fake News Spread, and Social Media Knows It," *Scientific American*, December 1, 2020, https://www.scientificamerican.com /article/information-overload-helps-fake-news-spread-and-social-media-knows-it/.

85. Kate Starbird, "How a Crisis Researcher Makes Sense of Covid-19 Misinformation," Medium, March 9, 2020, https://onezero.medium.com/reflecting-on-the-covid-19-infodemic-as-a-crisis-informatics -researcher-ce0656fa4d0a.

86. Taher S. Valika et al., "A Second Pandemic? Perspective on Information Overload in the COVID- 19 Era," *Otolaryngology—Head and Neck Surgery* 163, no. 5 (2020): 931–33, https://doi.org/10.1177 /0194599820935850.

87. Javier Abrego, "How Many Tweets About Covid-19 and Coronavirus? 508 MM Tweets So Far," *TweetBinder*, April 14, 2020, https://www.tweetbinder.com/blog/covid-19-coronavirus-twitter/.

88. Gordon Pennycook et al., "Shifting Attention to Accuracy Can Reduce Misinformation Online," *Nature* 592, no. 7855 (2021): 590–95, https://doi.org/10.1038/s41586-021-03344-2; Gordon Penny- cook et al., "Fighting COVID-19 Misinformation on Social Media: Experimental Evidence for a Scalable Accuracy-Nudge Intervention," *Psychological Science* 31, no. 7 (2020): 770–80, https://doi .org/10.1177/0956797620939054.

89. Toby Hopp et al., "Why Do People Share Ideologically Extreme, False, and Misleading Content on Social Media? A Self-Report and Trace Data–Based Analysis of Countermedia Content Dissemina- tion on Facebook and Twitter," *Human Communication Research* 46, no. 4 (2020): 357–84, https:// doi.org/10.1093/hcr/hqz022.

90. Robert M. Ross et al., "Beyond 'Fake News': Analytic Thinking and the Detection of False and Hyperpartisan News Headlines," *Judgment and Decision Making* 16, no. 2 (2021): 484–504, https://doi.org/10.1017/S1930297500008640.

91. Grinberg et al., "Fake News on Twitter During the 2016 U.S. Presidential Election."

92. Ryan Mac, "Who Is Frances Haugen, the Facebook Whistle-Blower?," *New York Times*, October 5, 2021, https://www.nytimes.com/2021/10/05/technology/who-is-frances-haugen.html.

93. Claire Sanford, "Facebook Whistleblower Frances Haugen Testifies Before UK Parliament Transcript," *Rev*, October 26, 2021, https://www.rev.com/blog/transcripts/facebook-whistleblower-frances-haugen-testifies-before-uk-parliament-transcript.

94. Scott Pelley, "Whistleblower: Facebook Is Misleading the Public on Progress Against Hate Speech, Violence, Misinformation," CBS News, October 4, 2021, https://www.cbsnews.com/news/facebook-whistleblower-misinformation-public-60-minutes-2021-10-03/; Cat Zakrzewski and Cristiano Lima-Strong, "Former Facebook Employee Frances Haugen Revealed as 'Whistleblower' Behind Leaked Documents That Plunged the Company Into Scandal," *Washington Post*, October 3, 2021, https://www.washingtonpost.com/technology/2021/10/03/facebook-whistleblower-frances-haugen-revealed/.

95. "Facebook Whistleblower: Another Ex-Employee Speaks Out," October 1, 2021, posted by Now This Impact, 3 min., 47 sec., https://www.youtube.com/watch?v=8XkREyzDgnA.

96. Emily Glazer, "Mark Zuckerberg Breaks Silence on Facebook Whistleblower Testimony, Media Reports," *Wall Street Journal*, October 6, 2021, https://www.wsj.com/articles/mark-zuckerberg-says-facebooks-work-mischaracterized-in-reports-whistleblower-testimony-11633482725; Jeff Horwitz, "Who Is Facebook Whistleblower Frances Haugen? What to Know After Her Senate Testimony," *Wall Street Journal*, October 5, 2021, https://www.wsj.com/articles/who-is-frances-haugen-facebook-whistleblower-11633409993.

97. Jeff Allen, "Misinformation Amplification Analysis and Tracking Dashboard," *Integrity Institute*, October 13, 2022, https://integrityinstitute.org/blog/misinformation-amplification-tracking-dashboard; "A Blueprint for Content Governance and Enforcement," Facebook, May 5, 2021, https://www.facebook.com/notes/751449002072082/.

98. Filippo Menczer, "How 'Engagement' Makes You Vulnerable to Manipulation and Misinformation on Social Media," Nieman Lab, September 13, 2021, https://www.niemanlab.org/2021/09/how-engagement-makes-you-vulnerable-to-manipulation-and-misinformation-on-social-media/.

99. Sinan Aral, *The Hype Machine: How Social Media Disrupts Our Elections, Our Economy, and Our Health—and How We Must Adapt* (Crown Currency, 2021).

100. Nicola Slawson, "'Women Have Been Woefully Neglected': Does Medical Science Have a Gender Problem?," *The Guardian*, December 18, 2019, https://www.theguardian.com/education/2019/dec/18/women-have-been-woefully-neglected-does-medical-science-have-a-gender-problem.

101. Lev Muchnik et al., "Social Influence Bias: A Randomized Experiment," *Science* 341, no. 6146 (2013): 647–51, https://doi.org/10.1126/science.1240466.

102. Menczer, "How 'Engagement' Makes You Vulnerable to Manipulation and Misinformation on Social Media."

103. Soroush Vosoughi et al., "The Spread of True and False News Online," *Science* 359, no. 6380 (2018): 1146–51, https://doi.org/10.1126/science.aap9559.

104. William J. Brady et al., "The MAD Model of Moral Contagion: The Role of Motivation, Attention, and Design in the Spread of Moralized Content Online," *Perspectives on Psychological Science* 15, no. 4 (2020): 978–1010, https://doi.org/10.1177/1745691620917336; Steve Rathje et al., "Out-Group Animosity Drives Engagement on Social Media," *Proceedings of the National Academy of Sciences* 118, no. 26 (2021): e2024292118, https://doi.org/10.1073/pnas.2024292118.

105. "Social Media and Political Polarization in America," *60 Minutes*, November 6, 2022, YouTube, 13 min., 38 sec., https://www.youtube.com/watch?v=WLfr7sU5W2E.

106. Patrick Rafail et al., "Polarizing Feedback Loops on Twitter: Congressional Tweets During the 2022 Midterm Elections," *Socius* 10 (2024): e23780231241228924, https://doi.org/10.1177/23780231241228924; Michael Kowal, "The Value of a Like: Facebook, Viral Posts, and Campaign Finance in US Congressional Elections," *Media and Communication* 11, no. 3 (2023): 153–63, https://doi.org/10.17645/mac.v11i3.6661.

107. Claire Wardle and Hossein Derakhshan, *Information Disorder: Toward an Interdisciplinary Framework for Research and Policymaking* (Council of Europe, 2017).

108. Chase Budnieski, "First Draft Teaches Journalists How to Avoid Amplifying Misinformation," NewsCo/Lab, February 14, 2020, https://newscollab.org/2020/02/14/first-draft-teaches-journalists-how-to-avoid-amplifying-misinformation/.

109. Tom Buchanan, "How to Reduce the Spread of Fake News—by Doing Nothing," Nieman Lab, January 5, 2021, https://www.niemanlab.org/2021/01/how-to-reduce-the-spread-of-fake-news-by-doing-nothing/.

110. Hunt Allcott et al., "Trends in the Diffusion of Misinformation on Social Media," *Research and Politics* 6, no. 2 (2019): e2053168019848554, https://doi.org/10.1177/2053168019848554.

111. Lauren E. Sherman et al., "The Power of the *Like* in Adolescence: Effects of Peer Influence on Neural and Behavioral Responses to Social Media," *Psychological Science* 27, no. 7 (2016): 1027–35, https://doi.org/10.1177/0956797616645673.

112. Mike Allen, "Sean Parker Unloads on Facebook: 'God Only Knows What It's Doing to Our Children's Brains,'" *Axios*, November 9, 2017, https://www.axios.com/2017/12/15/sean-parker-unloads-on-facebook-god-only-knows-what-its-doing-to-our-childrens-brains-1513306792.

114. Elon Musk (@elonmusk), "Trashing accounts that you hate will cause our algorithm to show you more of those accounts," Twitter (now X), January 16, 2023, https://twitter.com/elonmusk/status/1615194151737520128.

115. Joanna Stern, "Social-Media Algorithms Rule How We See the World: Good Luck Trying to Stop Them," *Wall Street Journal*, January 17, 2021, https://www.wsj.com/articles/social-media-algorithms-rule-how-we-see-the-world-good-luck-trying-to-stop-them-11610884800.

113. Renée DiResta, "Free Speech Is Not the Same as Free Reach," *Wired*, August 30, 2018, https://www.wired.com/story/free-speech-is-not-the-same-as-free-reach/.

116. Travis Mitchell, "How the Public Reacted on Facebook," Pew Research Center, February 23, 2017, https://www.pewresearch.org/politics/2017/02/23/how-the-public-reacted-on-facebook/.

117. William J. Brady et al., "Emotion Shapes the Diffusion of Moralized Content in Social Networks," *Proceedings of the National Academy of Sciences* 114, no. 28 (2017): 7313–18, https://doi.org/10.1073/pnas.1618923114.

118. Rathje et al., "Out-Group Animosity Drives Engagement on Social Media."

119. N. Velásquez et al., "Online Hate Network Spreads Malicious COVID-19 Content Outside the Control of Individual Social Media Platforms," *Scientific Reports* 11, no. 1 (2021): art. 11549, https://doi.org/10.1038/s41598-021-89467-y.

120. Carrie Blazina, "70 Percent of U.S. Social Media Users Never or Rarely Post or Share About Political, Social Issues," Pew Research Center, May 4, 2021, https://www.pewresearch.org/fact-tank/2021/05/04/70-of-u-s-social-media-users-never-or-rarely-post-or-share-about-political-social-issues/.

121. Magdalena Wojcieszak et al., "Most Users Do Not Follow Political Elites on Twitter; Those Who Do Show Overwhelming Preferences for Ideological Congruity," *Science Advances* 8, no. 39 (2022): eabn9418, https://doi.org/10.1126/sciadv.abn9418.

122. Abigail Geiger, "A Small Group of Prolific Users Account for a Majority of Political Tweets Sent by U.S. Adults," Pew Research Center, October 23, 2019, https://www.pewresearch.org/fact-tank/2019/10/23/a-small-group-of-prolific-users-account-for-a-majority-of-political-tweets-sent-by-u-s-adults/.

123. William J. Brady et al., "How Social Learning Amplifies Moral Outrage Expression in Online Social Networks," *Science Advances* 7, no. 33 (2021), https://doi.org/10.1126/sciadv.abe5641.

124. Jaeho Cho et al., "Influencing Myself: Self-Reinforcement Through Online Political Expression," *Communication Research* 45, no. 1 (2018): 83–111, https://doi.org/10.1177/0093650216644020.

125. Pierce D. Ekstrom and Calvin K. Lai, "The Selective Communication of Political Information," *Social Psychological and Personality Science* 12, no. 5 (2021): 789–800, https://doi.org/10.1177/1948550620942365.

126. Mathias Osmundsen et al., "Partisan Polarization Is the Primary Psychological Motivation Behind Political Fake News Sharing on Twitter," *American Political Science Review* 115, no. 3 (2021): 999–1015, https://doi.org/10.1017/S0003055421000290.

127. Andrew Duffy et al., "Too Good to Be True, Too Good Not to Share: The Social Utility of Fake News," *Information, Communication and Society* 23, no. 13 (2020): 1965–79, https://doi.org/10.1080/1369118X.2019.1623904.

128. Zhiying Ren et al., "Beyond Belief: How Social Engagement Motives Influence the Spread of Conspiracy Theories," *Journal of Experimental Social Psychology* 104 (2023): art. 104421, https://doi.org/10.1016/j.jesp.2022.104421.

129. Yini Zhang et al., "Network Amplification of Politicized Information and Misinformation About COVID-19 by Conservative Media and Partisan Influencers on Twitter," *Political Communication* 40, no. 1 (2022): 24–47, https://doi.org/10.1080/10584609.2022.2113844.

130. Jana Lasser et al., "Social Media Sharing of Low-Quality News Sources by Political Elites," *PNAS Nexus* 1, no. 4 (2022): pgac186, https://doi.org/10.1093/pnasnexus/pgac186.

131. Todd Spangler, "X/twitter Verified Blue Check-Mark Users Are 'Superspreaders' of Disinformation About Israel-Hamas War, Study Says," *Variety*, October 20, 2023, https://variety.com/2023/digital/news/x-twitter-blue-check-mark-users-superspreaders-disinformation-israel-hamas-war-1235763100/.

132. Michael Bang Petersen et al., "The 'Need for Chaos' and Motivations to Share Hostile Political Rumors," *American Political Science Review* 117, no. 4 (2023): 1486–1505, https://doi.org/10.1017/S0003055422001447.

133. Jin Woo Kim et al., "The Distorting Prism of Social Media: How Self-Selection and Exposure to Incivility Fuel Online Comment Toxicity," *Journal of Communication* 71, no. 6 (2021): 922–46, https://doi.org/10.1093/joc/jqab034.

134. Ashwin Rajadesingan et al., "Quick, Community-Specific Learning: How Distinctive Toxicity Norms Are Maintained in Political Subreddits," *Proceedings of the International AAAI Conference on Web and Social Media* 14, no. 1 (2020): 557–68, https://doi.org/10.1609/icwsm.v14i1.7323.

135. Matthew Barnidge, "Exposure to Political Disagreement in Social Media Versus Face-to-Face and Anonymous Online Settings," *Political Communication* 34, no. 2 (2017): 302–21, https://doi.org/10.1080/10584609.2016.1235639.

136. William J. Brady et al., "Overperception of Moral Outrage in Online Social Networks Inflates Beliefs About Intergroup Hostility," *Nature Human Behaviour* 7, no. 6 (2023): 917–27, https://doi.org/10.1038/s41562-023-01582-0.

137. Elizabeth Suhay et al., "The Polarizing Effects of Online Partisan Criticism: Evidence from Two Experiments," *International Journal of Press/Politics* 23, no. 1 (2018): 95–115, https://doi.org/10.1177/1940161217740697.

138. Hunt Allcott et al., "The Welfare Effects of Social Media," *American Economic Review* 110, no. 3 (2020): 629–76, https://doi.org/10.1257/aer.20190658.

139. Victoria Oldemburgo de Mello et al., "Twitter (X) Use Predicts Substantial Changes in Well-Being, Polarization, Sense of Belonging, and Outrage," *Communications Psychology* 2, no. 1 (2024): 15, https://doi.org/10.1038/s44271-024-00062-z.

140. Steve Rathje, Clara Pretus, James Kunling He, Trisha Harjani, Jon Roozenbeek, Kurt Gray, Sander van der Linden, and Jay Joseph Van Bavel, "Unfollowing Hyperpartisan Social Media Influencers Durably Reduces Out-Party Animosity," *PsyArXiv*, October 3, 2024, https://doi.org/10.31234/osf .io/acbwg.

141. Hilary Andersson, "Social Media Apps Are 'Deliberately' Addictive to Users," BBC, July 3, 2018, https://www.bbc.com/news/technology-44640959.

142. Aza Raskin (@aza), "One of my lessons from infinite scroll," Twitter (now X), June 10, 2019, https:// twitter.com/aza/status/1138268959982022656.

143. Danielle Cohen, "He Created Your Phone's Most Addictive Feature: Now He Wants to Build a Rosetta Stone for Animal Language," *GQ*, June 30, 2021, https://www.gq.com/story/aza-raskin-interview.

144. Eli Pariser, *The Filter Bubble: What the Internet Is Hiding from You* (Penguin UK, 2011).

145. Kathleen Hall Jamieson and Joseph N. Cappella, *Echo Chamber: Rush Limbaugh and the Conservative Media Establishment* (Oxford University Press, 2008), 76.

146. Michael D. Slater, "Reinforcing Spirals Model: Conceptualizing the Relationship Between Media Content Exposure and the Development and Maintenance of Attitudes," *Media Psychology* 18, no. 3 (2015): 370–95, https://doi.org/10.1080/15213269.2014.897236.

147. Christine Abdalla Mikhaeil and Richard L. Baskerville, "Explaining Online Conspiracy Theory Radicalization: A Second-Order Affordance for Identity-Driven Escalation," *Information Systems Journal* 34, no. 3 (2024): 711–35, https://doi.org/10.1111/isj.12427.

148. "How Facebook's Algorithm Amplifies Climate Disinformation," Global Witness, March 10, 2022, https://www.globalwitness.org/en/campaigns/digital-threats/climate-divide-how-facebooks -algorithm-amplifies-climate-disinformation/.

149. "Tackling Climate Change Together," Meta, September 16, 2021, https://about.fb.com/news/2021 /09/tackling-climate-change-together/.

150. Mark Lynas et al., "Greater than 99 Percent Consensus on Human-Caused Climate Change in the Peer-Reviewed Scientific Literature," *Environmental Research Letters* 16, no. 11 (2021): 114005, https://doi.org/10.1088/1748-9326/ac2966.

151. Amy Ross Arguedas et al., "Echo Chambers, Filter Bubbles, and Polarisation: A Literature Review," Reuters Institute for the Study of Journalism, January 19, 2022, https://reutersinstitute.politics .ox.ac.uk/echo-chambers-filter-bubbles-and-polarisation-literature-review.

152. Pablo Barberá, "Birds of the Same Feather Tweet Together: Bayesian Ideal Point Estimation Using Twitter Data," *Political Analysis* 23, no. 1 (2015): 76–91, https://doi.org/10.1093/pan/mpu011.

153. Michael D. Conover et al., "Predicting the Political Alignment of Twitter Users," in *2011 IEEE Third International Conference on Privacy, Security, Risk and Trust and 2011 IEEE Third International Conference on Social Computing* (IEEE, 2011), 192–99, https://doi.org/10.1109/PASSAT/SocialCom.2011.34.

154. Matteo Cinelli et al., "The Echo Chamber Effect on Social Media," *Proceedings of the National Academy of Sciences* 118, no. 9 (2021): e2023301118, https://doi.org/10.1073/pnas.2023301118.

155. Michael D. Conover et al., "Partisan Asymmetries in Online Political Activity," *EPJ Data Science* 1, no. 1 (2012): 1–19, https://doi.org/10.1140/epjds6; Yosh Halberstam and Brian Knight, "Homophily, Group Size, and the Diffusion of Political Information in Social Networks: Evidence from Twitter," *Journal of Public Economics* 143 (2016): 73–88, https://doi.org/10.1016/j.jpubeco.2016.08.011.

156. Chengcheng Shao et al., "Anatomy of an Online Misinformation Network," *PLOS One* 13, no. 4 (2018): e0196087, https://doi.org/10.1371/journal.pone.0196087.

157. Michela Del Vicario et al., "The Spreading of Misinformation Online," *Proceedings of the National Academy of Sciences* 113, no. 3 (2016): 554–59, https://doi.org/10.1073/pnas.1517441113.

158. Anthony Lantian et al., "Stigmatized Beliefs: Conspiracy Theories, Anticipated Negative Evaluation of the Self, and Fear of Social Exclusion," *European Journal of Social Psychology* 48, no. 7 (2018): 939–54, https://doi.org/10.1002/ejsp.2498; Lu Tang et al., "'Down the Rabbit Hole' of Vaccine

Misinformation on YouTube: Network Exposure Study," *Journal of Medical Internet Research* 23, no. 1 (2021): e23262, https://doi.org/10.2196/23262.

159.   M. Asher Lawson et al., "Tribalism and Tribulations: The Social Costs of Not Sharing Fake News," *Journal of Experimental Psychology: General* 152, no. 3 (2023): 611–31, https://doi.org/10.1037 /xge0001374.

160.   Marcella Tambuscio et al., "Fact-Checking Effect on Viral Hoaxes: A Model of Misinformation Spread in Social Networks," in *WWW '15 Companion: Proceedings of the 24th International Conference on World Wide Web* (Association for Computing Machinery, 2015), 977–82, https://doi .org/10.1145/2740908.2742572.

161.   Jennifer Hochschild and Katherine Levine Einstein, "'It Isn't What We Don't Know That Gives Us Trouble, It's What We Know That Ain't So': Misinformation and Democratic Politics," *British Journal of Political Science* 45, no. 3 (2015): 467–75, https://doi.org/10.1017/S000712341400043X.

162.   Sunstein, "Sunstein on the Internet and Political Polarization."

163.   Mohsen Mosleh et al., "Shared Partisanship Dramatically Increases Social Tie Formation in a Twitter Field Experiment," *Proceedings of the National Academy of Sciences* 118, no. 7 (2021): e2022761118, https://doi.org/10.1073/pnas.2022761118.

164.   "Interactive Media Bias Chart," Ad Fontes Media, accessed July 9, 2022, https://adfontesmedia .com/interactive-media-bias-chart/.

165.   Lisa-Maria Neudert et al., "Polarization, Partisanship and Junk News Consumption on Social Media During the 2018 US Midterm Elections," *America* 8 (2017): 10; Guess et al., "Less Than You Think."

166.   Rebecca Heilweil, "Right-Wing Media Thrives on Facebook: Whether It Rules Is More Complicated," *Vox*, September 9, 2020, https://www.vox.com/recode/21419328/facebook-conservative-bias -right-wing-crowdtangle-election.

167.   M. Asher Lawson and Hemant Kakkar, "Of Pandemics, Politics, and Personality: The Role of Conscientiousness and Political Ideology in the Sharing of Fake News," *Journal of Experimental Psychology: General* 151, no. 5 (2022): 1154–77, https://doi.org/10.1037/xge0001120.

168.   Christopher A. Bail et al., "Exposure to Opposing Views on Social Media Can Increase Political Polarization," *Proceedings of the National Academy of Sciences* 115, no. 37 (2018): 9216–21, https://doi .org/10.1073/pnas.1804840115.

169.   Suhay et al., "The Polarizing Effects of Online Partisan Criticism."

170.   Christopher A. Bail, *Breaking the Social Media Prism: How to Make Our Platforms Less Polarizing* (Princeton University Press, 2022), 52–53.

171.   Robert King Merton, *Sociological Ambivalence and Other Essays* (Simon and Schuster, 1976).

172.   Natalia Arugute et al., "Network Activated Frames: Content Sharing and Perceived Polarization in Social Media," *Journal of Communication* 73, no. 1 (2022): 14–24, https://doi.org/10.1093/joc /jqac035.

173.   Abigail Geiger, "46 Percent of U.S. Social Media Users Say They Are 'Worn Out' by Political Posts and Discussions," Pew Research Center, August 8, 2019, https://www.pewresearch.org/fact-tank /2019/08/08/46-of-u-s-social-media-users-say-they-are-worn-out-by-political-posts-and-discussions/.

174.   Maeve Duggan, "The Political Environment on Social Media," Pew Research Center, October 25, 2016, https://www.pewresearch.org/internet/2016/10/25/the-political-environment-on-social-media/.

175.   Jaime E. Settle and Taylor N. Carlson, "Opting Out of Political Discussions," *Political Communication* 36, no. 3 (2019): 476–96, https://doi.org/10.1080/10584609.2018.1561563.

176.   Ro'ee Levy, "Social Media, News Consumption, and Polarization: Evidence from a Field Experiment," *American Economic Review* 111, no. 3 (2021): 831–70, https://doi.org/10.1257/aer.20191777.

177.   "Echo Chambers, Filter Bubbles, and Polarisation: A Literature Review."

178.   Gregory Eady et al., "How Many People Live in Political Bubbles on Social Media? Evidence from Linked Survey and Twitter Data," *Sage Open* 9, no. 1 (2019): 2158244019832705, https://doi .org/10.1177/2158244019832705.

179. Seth Flaxman et al., "Filter Bubbles, Echo Chambers, and Online News Consumption," *Public Opinion Quarterly* 80, no. S1 (2016): 298–320, https://doi.org/10.1093/poq/nfw006.

180. Jin Woo Kim and Eunji Kim, "Temporal Selective Exposure: How Partisans Choose When to Follow Politics," *Political Behavior* 43, no. 4 (2021): 1663–83, https://doi.org/10.1007/s11109-021-09690-1; Matthew Tyler et al., "Partisan Enclaves and Information Bazaars: Mapping Selective Exposure to Online News," *Journal of Politics* 84, no. 2 (2022): 1057–73, https://doi.org/10.1086/716950.

181. Kevin Roose, "The Making of a YouTube Radical," *New York Times*, June 8, 2019, https://www.nytimes.com/interactive/2019/06/08/technology/youtube-radical.html.

182. Manoel Horta Ribeiro et al., "Auditing Radicalization Pathways on YouTube," in *FAT* '20: Proceedings of the 2020 Conference on Fairness, Accountability, and Transparency* (Association for Computing Machinery, 2020), 131–41, https://doi.org/10.1145/3351095.3372879.

183. Kevin Munger and Joseph Phillips, "Right-Wing YouTube: A Supply and Demand Perspective," *International Journal of Press/Politics* 27, no. 1 (2022): 186–219, https://doi.org/10.1177/1940161220964767.

184. Ronald E. Robertson et al., "Users Choose to Engage with More Partisan News Than They Are Exposed to on Google Search," *Nature* 618, no. 7964 (2023): 342–48, https://doi.org/10.1038/s41586-023-06078-5.

185. Annie Y. Chen et al., "Subscriptions and External Links Help Drive Resentful Users to Alternative and Extremist YouTube Channels," *Science Advances* 9, no. 35 (2023): eadd8080, https://doi.org/10.1126/sciadv.add8080.

186. Essam El-Dardiry, "Giving You More Control Over Your Homepage and Up Next Videos," *YouTube Official Blog*, June 26, 2019, https://blog.youtube/news-and-events/giving-you-more-control-over-homepage/.

187. Dean Jackson et al., "Insiders' View of the January 6th Committee's Social Media Investigation," Just Security, January 5, 2023, https://www.justsecurity.org/84658/insiders-view-of-the-january-6th-committees-social-media-investigation/.

188. David Klepper, " 'It's Not Simple': Researchers Tweaked Facebook's Algorithms to See If They Could Fix America's Political Polarization: They Failed," *Fortune*, July 27, 2023, https://fortune.com/2023/07/27/facebook-algorithm-political-polarization/.

189. John Gramlich, "Q&A: How Pew Research Center Identified Bots on Twitter," Pew Research Center, April 19, 2018, https://www.pewresearch.org/fact-tank/2018/04/19/qa-how-pew-research-center-identified-bots-on-twitter/.

190. Bobby Allyn, "Researchers: Nearly Half of Accounts Tweeting About Coronavirus Are Likely Bots," NPR, May 20, 2020, https://www.npr.org/sections/coronavirus-live-updates/2020/05/20/859814085/researchers-nearly-half-of-accounts-tweeting-about-coronavirus-are-likely-bots.

191. Alessandro Bessi and Emilio Ferrara, "Social Bots Distort the 2016 US Presidential Election Online Discussion," *First Monday* 21, no. 11 (2016), https://doi.org/10.5210/fm.v21i11.7090.

192. Onur Varol et al., "Online Human-Bot Interactions: Detection, Estimation, and Characterization," *Eleventh International AAAI Conference on Web and Social Media* 11, no. 1 (2017): 280–89, https://aaai.org/ocs/index.php/ICWSM/ICWSM17/paper/view/15587.

193. Rose Marie Santini et al., "Comparative Approaches to Mis/Disinformation: When Machine Behavior Targets Future Voters—The Use of Social Bots to Test Narratives for Political Campaigns in Brazil," *International Journal of Communication* 15 (2021): 1220–43.

194. Samantha Bradshaw and Philip N. Howard, "The Global Disinformation Order: 2019 Global Inventory of Organised Social Media Manipulation," University of Oxford, 2019, https://digitalcommons.unl.edu/scholcom/207.

195. Chengcheng Shao et al., "The Spread of Low-Credibility Content by Social Bots," *Nature Communications* 9, no. 1 (2018): 4787, https://doi.org/10.1038/s41467-018-06930-7.

196. "As a Conservative Twitter User Sleeps, His Account Is Hard at Work," *Washington Post*, February 5, 2017, https://www.washingtonpost.com/business/economy/as-a-conservative-twitter-user-sleeps-his-account-is-hard-at-work/2017/02/05/18d5a532-df31-11e6-918c-99ede3c8cafa_story.html.

197. *Spot the Troll*, accessed July 3, 2024, https://spotthetroll.org/.

198. Harry Yaojun Yan et al., "Asymmetrical Perceptions of Partisan Political Bots," *New Media and Society* 23, no. 10 (2021): 3016–37, https://doi.org/10.1177/1461444820942744.

199. Adrian Chen, "The Agency," *New York Times*, June 2, 2015, https://www.nytimes.com/2015/06/07/magazine/the-agency.html.

200. Dan Mangan and Mike Calia, "Special Counsel Mueller: Russians Conducted 'Information Warfare' Against US During Election to Help Donald Trump Win," CNBC, February 16, 2018, https://www.cnbc.com/2018/02/16/russians-indicted-in-special-counsel-robert-muellers-probe.html.

201. Philip N. Howard et al., "The IRA, Social Media and Political Polarization in the United States, 2012–2018," University of Oxford, 2019, https://digitalcommons.unl.edu/senatedocs/1.

202. Darren L. Linvill and Patrick L. Warren, "Troll Factories: Manufacturing Specialized Disinformation on Twitter," *Political Communication* 37, no. 4 (2020): 447–67, https://doi.org/10.1080/10584609.2020.1718257.

203. Todd J. Gillman, "Russian Trolls Orchestrated 2016 Clash at Houston Islamic Center, New Senate Intel Report Recalls," *Dallas Morning News*, October 8, 2019, https://www.dallasnews.com/news/politics/2019/10/08/russian-trolls-orchestrated-2016-clash-houston-islamic-center-senate-intel-report-says/.

204. Christopher A. Bail et al., "Assessing the Russian Internet Research Agency's Impact on the Political Attitudes and Behaviors of American Twitter Users in Late 2017," *Proceedings of the National Academy of Sciences* 117, no. 1 (2020): 243–50, https://doi.org/10.1073/pnas.1906420116.

205. Franziska Martini et al., "Bot, or Not? Comparing Three Methods for Detecting Social Bots in Five Political Discourses," *Big Data and Society* 8, no. 2 (2021): 20539517211033566, https://doi.org/10.1177/20539517211033566.

206. Marco T. Bastos and Dan Mercea, "The Brexit Botnet and User-Generated Hyperpartisan News," *Social Science Computer Review* 37, no. 1 (2019): 38–54, https://doi.org/10.1177/0894439317734157.

207. Diogo Pacheco et al., "Uncovering Coordinated Networks on Social Media: Methods and Case Studies," *Proceedings of the International AAAI Conference on Web and Social Media* 15, no. 1 (2021): 455–66, https://doi.org/10.1609/icwsm.v15i1.18075.

208. Bjarke Mønsted et al., "Evidence of Complex Contagion of Information in Social Media: An Experiment Using Twitter Bots," *PLOS One* 12, no. 9 (2017): e0184148, https://doi.org/10.1371/journal.pone.0184148.

209. Cecilia Kang and Sheera Frenkel, "Facebook Says Cambridge Analytica Harvested Data of Up to 87 Million Users," *New York Times*, April 4, 2018, https://www.nytimes.com/2018/04/04/technology/mark-zuckerberg-testify-congress.html.

210. Alex Hern, "Cambridge Analytica: How Did It Turn Clicks Into Votes?," *The Guardian*, May 6, 2018, https://www.theguardian.com/news/2018/may/06/cambridge-analytica-how-turn-clicks-into-votes-christopher-wylie.

211. Michal Kosinski et al., "Private Traits and Attributes Are Predictable from Digital Records of Human Behavior," *Proceedings of the National Academy of Sciences* 110, no. 15 (2013): 5802–5, https://doi.org/10.1073/pnas.1218772110.

212. Kurt Wagner, "Here's How Facebook Allowed Cambridge Analytica to Get Data for 50 Million Users," *Vox*, March 17, 2018, https://www.vox.com/2018/3/17/17134072/facebook-cambridge-analytica-trump-explained-user-data.

213. Jon Greenberg, "Trump Campaign Used Cambridge Analytica in Final Months of Campaign," PolitiFact, March 21, 2018, https://www.politifact.com/factchecks/2018/mar/21/jack-posopiec/trump-campaign-used-cambridge-analytica-final-mont/.

214. S. C. Matz et al., "Psychological Targeting as an Effective Approach to Digital Mass Persuasion," *Proceedings of the National Academy of Sciences* 114, no. 48 (2017): 12714–19, https://doi.org/10.1073/pnas.1710966114.

215. Ben M. Tappin et al., "Quantifying the Potential Persuasive Returns to Political Microtargeting," *Proceedings of the National Academy of Sciences* 120, no. 25 (2023): e2216261120, https://doi.org/10.1073/pnas.2216261120.

216. Nicholas Confessore and Danny Hakim, "Data Firm Says 'Secret Sauce' Aided Trump; Many Scoff," *New York Times*, March 6, 2017, https://www.nytimes.com/2017/03/06/us/politics/cambridge-analytica.html.

217. Brendan Nyhan, "Fake News and Bots May Be Worrisome, but Their Political Power Is Overblown," *New York Times*, February 13, 2018, https://www.nytimes.com/2018/02/13/upshot/fake-news-and-bots-may-be-worrisome-but-their-political-power-is-overblown.html.

218. Gregory Eady et al., "Exposure to the Russian Internet Research Agency Foreign Influence Campaign on Twitter in the 2016 US Election and Its Relationship to Attitudes and Voting Behavior," *Nature Communications* 14, no. 1 (2023): 62, https://doi.org/10.1038/s41467-022-35576-9.

219. Serena Tardelli et al., "Characterizing Social Bots Spreading Financial Disinformation," in *Social Computing and Social Media: Design, Ethics, User Behavior, and Social Network Analysis*, ed. Gabriele Meiselwitz (Springer International, 2020), 376–92, https://doi.org/10.1007/978-3-030-49570-1_26.

220. Kate Starbird et al., "Influence and Improvisation: Participatory Disinformation During the 2020 US Election," *Social Media + Society* 9, no. 2 (2023), https://doi.org/10.1177/20563051231177943.

221. Pam Fessler, "Former Election Security Official Says It Will Take 'Years' to Undo Disinformation," NPR, December 22, 2020, https://www.npr.org/2020/12/22/949157510/former-election-security-official-says-it-will-take-years-to-undo-disinformation.

222. Katherine Ognyanova et al., "Misinformation in Action: Fake News Exposure Is Linked to Lower Trust in Media, Higher Trust in Government When Your Side Is in Power," *Harvard Kennedy School Misinformation Review* 1, no. 4 (2020): 1–19, https://doi.org/10.37016/mr-2020-024.

223. Sangwon Lee and S. Mo Jones-Jang, "Cynical Nonpartisans: The Role of Misinformation in Political Cynicism During the 2020 U.S. Presidential Election," *New Media and Society* 26, no. 7 (2024): 4255–76, https://doi.org/10.1177/14614448221116036.

224. Sarah Kreps, *Social Media and International Relations* (Cambridge University Press, 2020), 3, https://doi.org/10.1017/9781108920377.

225. Sora Park et al., "Global Mistrust in News: The Impact of Social Media on Trust," *International Journal on Media Management* 22, no. 2 (2020): 83–96, https://doi.org/10.1080/14241277.2020.1799794.

226. Andrea Ceron, "Internet, News, and Political Trust: The Difference Between Social Media and Online Media Outlets," *Journal of Computer-Mediated Communication* 20, no. 5 (2015): 487–503, https://doi.org/10.1111/jcc4.12129; Jaime E. Settle, *Frenemies: How Social Media Polarizes America* (Cambridge University Press, 2018).

227. Natalia Aruguete et al., "Trustful Voters, Trustworthy Politicians: A Survey Experiment on the Influence of Social Media in Politics," Working Paper No. IDB-WP-1169 (Inter-American Development Bank, 2021), https://doi.org/10.18235/0003389.

228. Elad Klein and Joshua Robison, "Like, Post, and Distrust? How Social Media Use Affects Trust in Government," *Political Communication* 37, no. 1 (2020): 46–64.

229. Bail, *Breaking the Social Media Prism*, 67.

230. Shruti Phadke et al., "What Makes People Join Conspiracy Communities? Role of Social Factors in Conspiracy Engagement," *Proceedings of the ACM on Human-Computer Interaction* 4, no. CSCW3 (2021): art. 223, https://doi.org/10.1145/3432922.

231. Damaris Graeupner and Alin Coman, "The Dark Side of Meaning-Making: How Social Exclusion Leads to Superstitious Thinking," *Journal of Experimental Social Psychology* 69 (2017): 218–22, https://doi.org/10.1016/j.jesp.2016.10.003.

232. Alice Marwick and Rebecca Lewis, *Media Manipulation and Disinformation Online* (Data & Society Research Institute, 2017).

233. Octavia Ionescu et al., "Political Extremism and Perceived Anomie: New Evidence of Political Extremes' Symmetries and Asymmetries Within French Samples," *International Review of Social Psychology* 34, no. 1 (2021): 1–16, https://doi.org/10.5334/irsp.573; Molly McCarthy et al., "Examining the Relationship Between Conspiracy Theories and COVID-19 Vaccine Hesitancy: A Mediating Role for Perceived Health Threats, Trust, and Anomie?," *Analyses of Social Issues and Public Policy* 22, no. 1 (2022): 106–29, https://doi.org/10.1111/asap.12291.

5. THE SPREAD OF VACCINE MISINFORMATION AND HOW WE CAN REBUILD TRUST IN INSTITUTIONS

1. Steve Mullis, "She Resisted Getting Her Kids the Usual Vaccines: Then the Pandemic Hit," NPR, January 22, 2021, https://www.npr.org/2021/01/22/956935520/she-resisted-getting-her-kids-the-usual-vaccines-then-the-pandemic-hit.

2. Carina C. Mallard et al., "The Myth of the Immature Barrier Systems in the Developing Brain: Role in Perinatal Brain Injury," *Journal of Physiology* 596, no. 23 (2018): 5655–64, https://doi.org/10.1113/JP274938.

3. Max Kozlov, "Introducing Inoculation, 1721," *The Scientist*, January 1, 2021, https://www.the-scientist.com/foundations/introducing-inoculation-1721-68275.

4. Stefan Riedel, "Edward Jenner and the History of Smallpox and Vaccination," *Baylor University Medical Center Proceedings* 18, no. 1 (2005): 21–25, https://doi.org/10.1080/08998280.2005.11928028.

5. "Types of Vaccines," Immunisation Advisory Centre, last updated October 2022, https://www.immune.org.nz/vaccines/vaccine-development; Walter A. Orenstein and Rafi Ahmed, "Simply Put: Vaccination Saves Lives," *Proceedings of the National Academy of Sciences* 114, no. 16 (2017): 4031–33, https://doi.org/10.1073/pnas.1704507114.

6. Jaspreet Toor et al., "Lives Saved with Vaccination for 10 Pathogens Across 112 Countries in a Pre-COVID-19 World," *eLife* 10 (2021): e67635, https://doi.org/10.7554/eLife.67635.

7. "How Does a mRNA Vaccine Compare to a Traditional Vaccine?," Vanderbilt Institute for Infection, Immunology and Inflammation, Vanderbilt University Medical Center, November 16, 2020, https://www.vumc.org/viiii/infographics/how-does-mrna-vaccine-compare-traditional-vaccine.

8. Molly K. Steele et al., "Estimated Number of COVID-19 Infections, Hospitalizations, and Deaths Prevented Among Vaccinated Persons in the US, December 2020 to September 2021," *JAMA Network Open* 5, no. 7 (2022): e2220385, https://doi.org/10.1001/jamanetworkopen.2022.20385.

9. Ari Daniel, "First Malaria Vaccine Hits 1 Million Dose Milestone—Although It Has Its Shortcomings," NPR, May 13, 2022, https://www.npr.org/sections/goatsandsoda/2022/05/13/1098536246/first-malaria-vaccine-hits-1-million-dose-milestone-although-it-has-its-shortcom.

10. Tanya Lewis, "The Benefits of Vaccinating Kids Against COVID Far Outweigh the Risks of Myocarditis," *Scientific American*, December 2, 2021, https://www.scientificamerican.com/article/the-benefits-of-vaccinating-kids-against-covid-far-outweigh-the-risks-of-myocarditis1/; Matthew E. Oster et al., "Myocarditis Cases Reported After mRNA-Based COVID-19 Vaccination in the US from December 2020 to August 2021," *JAMA* 327, no. 4 (2022): 331–40, https://doi.org/10.1001/jama.2021.24110.

11. Isabella Backman, "Q&A: What Causes Rare Instances of Myocarditis After mRNA COVID-19 Vaccines?," Yale School of Medicine, May 16, 2023, https://medicine.yale.edu/news-article/qanda-what-causes-rare-instances-of-myocarditis-after-mrna-covid-19-vaccines/.

12. Berkeley Lovelace Jr., "Myocarditis After Covid Vaccine Low Among Teens and Young Adults, Large Study Finds," NBC News, December 5, 2022, https://www.nbcnews.com/health/health-news/myocarditis-covid-vaccine-teens-study-rcna60118.

13. Maya Bar-Hillel, "The Base-Rate Fallacy in Probability Judgments," *Acta Psychologica* 44, no. 3 (1980): 211–33, https://doi.org/10.1016/0001-6918(80)90046-3; Dan Pilat and Sekoul Krastev, "Base Rate Fallacy," Decision Lab, accessed July 3, 2024, https://thedecisionlab.com/biases/base-rate-fallacy.

14. Hossein Azarpanah et al., "Vaccine Hesitancy: Evidence from an Adverse Events Following Immunization Database, and the Role of Cognitive Biases." *BMC Public Health* 21 (2021): art. 1686, https://doi.org/10.1186/s12889-021-11745-1; Amos Tversky and Daniel Kahneman, "Loss Aversion in Riskless Choice: A Reference-Dependent Model," *Quarterly Journal of Economics* 106, no. 4 (1991): 1039–61, https://doi.org/10.2307/2937956.

15. Jennifer McLenon and Mary A. M. Rogers, "The Fear of Needles: A Systematic Review and Meta-Analysis," *Journal of Advanced Nursing* 75, no. 1 (2019): 30–42, https://doi.org/10.1111/jan.13818.

16. "Ten Threats to Global Health in 2019," World Health Organization, accessed July 4, 2024, https://www.who.int/news-room/spotlight/ten-threats-to-global-health-in-2019.

17. Lauren Gardner et al., "Persistence of US Measles Risk due to Vaccine Hesitancy and Outbreaks Abroad," *Lancet Infectious Diseases* 20, no. 10 (2020): 1114–15, https://doi.org/10.1016/S1473-3099(20)30522-3.

18. Ken Alltucker, "A Quarter of All Kindergartners in This County in Washington Aren't Immunized: Now There's a Measles Crisis," *USA Today*, February 11, 2019, https://www.usatoday.com/story/news/health/2019/02/11/measles-spread-anti-vaccination-communities-new-york-clar-county-washington/2812667002/.

19. Jan Hoffman, "Mistrust of a Coronavirus Vaccine Could Imperil Widespread Immunity," *New York Times*, July 18, 2020, https://www.nytimes.com/2020/07/18/health/coronavirus-anti-vaccine.html.

20. Steve Benen, "Departing NIH Chief Eyes More Research 'on Human Behavior,'" MSNBC, December 21, 2021, https://www.msnbc.com/rachel-maddow-show/departing-nih-chief-eyes-more-research-human-behavior-n1286390.

21. Matthew Facciani, "Video: How Did Mask Wearing Become So Politicized?," *The Conversation*, September 9, 2020, http://theconversation.com/video-how-did-mask-wearing-become-so-politicized-144268.

22. Tara Haelle, "Vaccine Hesitancy Is Nothing New: Here's the Damage It's Done Over Centuries," *ScienceNews*, May 11, 2021, https://www.sciencenews.org/article/vaccine-hesitancy-history-damage-anti-vaccination.

23. Georges Peter, "Vaccine Crisis: An Emerging Societal Problem," *Journal of Infectious Diseases* 151, no. 6 (1985): 981–83, http://www.jstor.org/stable/30130075.

24. James D. Cherry, "'Pertussis Vaccine Encephalopathy': It Is Time to Recognize It as the Myth That It Is," *JAMA* 263, no. 12 (1990): 1679–80, https://doi.org/10.1001/jama.1990.03440120101046.

25. Robert T. Chen and Frank DeStefano, "Vaccine Adverse Events: Causal or Coincidental?," *The Lancet* 351, no. 9103 (1998): 611–12, https://doi.org/10.1016/S0140-6736(05)78423-3.

26. Brian Deer, "How the Case Against the MMR Vaccine Was Fixed," *BMJ* 342 (2011): c5347, https://doi.org/10.1136/bmj.c5347.

27. Sarah Boseley, "How Disgraced Anti-Vaxxer Andrew Wakefield Was Embraced by Trump's America," *The Guardian*, July 18, 2018, https://www.theguardian.com/society/2018/jul/18/how-disgraced-anti-vaxxer-andrew-wakefield-was-embraced-by-trumps-america; Saad B. Omer, "The Discredited Doctor Hailed by the Anti-Vaccine Movement," *Nature* 586, no. 7831 (2020): 668–69, https://doi.org/10.1038/d41586-020-02989-9.

28. Mark Lynas, "Are the Anti-GMO and Anti-Vaccine Movements Merging?," *Alliance for Science*, December 6, 2017, https://allianceforscience.cornell.edu/blog/2017/12/are-the-anti-gmo-and-anti-vaccine-movements-merging/.

29. Kavin Senapathy, "The Anti-Vaccine and Anti-GMO Movements Are Inextricably Linked and Cause Preventable Suffering," *Forbes*, May 18, 2017, https://www.forbes.com/sites/kavinsenapathy

/2017/05/18/the-anti-vaccine-and-anti-gmo-movements-are-inextricably-linked-and-cause-preventable-suffering/.

30. Alan Levinovitz, *Natural: How Faith in Nature's Goodness Leads to Harmful Fads, Unjust Laws, and Flawed Science* (Beacon Press, 2021).

31. Dan Milmo, "Anti-Vaxxers Making 'at Least $2.5m' a Year from Publishing on Substack," *The Guardian*, January 27, 2022, https://www.theguardian.com/technology/2022/jan/27/anti-vaxxers-making-at-least-25m-a-year-from-publishing-on-substack.

32. Neena Satija and Lena H. Sun, "A Major Funder of the Anti-Vaccine Movement Has Made Millions Selling Natural Health Products." *Washington Post*, October 15, 2019, https://www.washingtonpost.com/investigations/2019/10/15/fdc01078-c29c-11e9-b5e4-54aa56d5b7ce_story.html.

33. "Event 201 Didn't Predict the Covid-19 Pandemic," Full Fact, April 27, 2020, https://fullfact.org/health/event-201-coronavirus-pandemic/.https://fullfact.org/health/event-201-coronavirus-pandemic/.

34. Jonathan Jarry, "The Upside-Down Doctor," Office for Science and Society, McGill University, June 4, 2021, https://www.mcgill.ca/oss/article/covid-19-health-pseudoscience/upside-down-doctor.

35. V. A. Parker, E. Kehoe, J. Lees, M. Facciani, and A. E. Wilson, "Alluring or Alarming? The Polarizing Effect of Forbidden Knowledge in Political Discourse," *Personality & Social Psychology Bulletin*, November 6, 2024, 1461672241288332.

36. Bruce Y. Lee, "Anti-Vaxxers Exploit Damar Hamlin's Crisis with Unfounded Covid-19 Vaccine Claims," *Forbes*, January 3, 2023, https://www.forbes.com/sites/brucelee/2023/01/03/anti-vaxxers-exploit-damar-hamlins-crisis-with-unfounded-covid-19-vaccine-claims/.

37. Saranac Hale Spencer, "Grant Wahl Died from Aortic Aneurysm, No Link to COVID-19 Vaccine," FactCheck.org, December 16, 2022, https://www.factcheck.org/2022/12/scicheck-grant-wahl-died-from-aortic-aneurysm-no-link-to-covid-19-vaccine/.

38. Derek Thompson, "The Pandemic's Wrongest Man," *The Atlantic*, April 1, 2021, https://www.theatlantic.com/ideas/archive/2021/04/pandemics-wrongest-man/618475/.

39. Derek Beres, "How Anti-Vaxxers Monetize Misinformation," re:frame, January 19, 2023, https://derekberes.substack.com/p/how-anti-vaxxers-monetize-misinformation.

40. Jake Lahut, "Fox News Dominated Primetime Ratings for COVID Summer—Not Just on Cable, but All of TV," *Business Insider*, September 11, 2020, https://www.businessinsider.com/fox-news-ratings-most-watched-channel-summer-2020-primetime-2020-9; Joe Flint et al., "Fox News Ousts Tucker Carlson," *Wall Street Journal*, April 24, 2023, https://www.wsj.com/articles/tucker-carlson-is-leaving-fox-news-db31f2fa.

41. Bill McCarthy, "Tucker Carlson Falsely Claims COVID-19 Vaccines Might Not Work," PolitiFact, April 15, 2021, https://www.politifact.com/factchecks/2021/apr/15/tucker-carlson/tucker-carlson-falsely-claims-covid-19-vaccines-mi/.

42. Bill McCarthy, "Fox News Host Will Cain Falsely Claims Vaccine More Dangerous for Children than COVID-19," PolitiFact, October 7, 2021, https://www.politifact.com/factchecks/2021/oct/07/will-cain/fox-news-host-will-cain-falsely-claims-vaccine-mor/.

43. Amy Sherman, "Biden Said People Vaccinated for COVID-19 'Do Not Spread the Disease to Anyone Else': That Contradicts a CDC Presentation in Dec That Said It's Likely That Vaccinated People 'Can Spread the Virus to Others.'" PolitiFact, December 22, 2021, https://www.politifact.com/factchecks/2021/dec/22/joe-biden/biden-says-vaccinated-people-cant-spread-covid-19-/.

44. Louis Jacobson, "Joe Biden Overstates How Well Vaccines Prevent Person-to-Person Virus Spread," PolitiFact, October 14, 2021, https://www.politifact.com/factchecks/2021/oct/14/joe-biden/joe-biden-overstates-effectiveness-vaccines-preven/.

45. Cortney O'Brien, "Viewers Demand Apology from MSNBC, Rachel Maddow for Previous COVID Vaccine Comments," Fox News, December 28, 2021, https://www.foxnews.com/media/social-media-users-demand-apology-msnbc-rachel-maddow-vaccines.

46.   Bill McCarthy, "What Trump Said to Encourage COVID-19 Vaccine Use," PolitiFact, March 4, 2021, https://www.politifact.com/factchecks/2021/mar/04/rachel-maddow/what-trump-said -encourage-covid-19-vaccine-use/.

47.   Will Sommer, "QAnon Star Cirsten Weldon Who Said Only 'Idiots' Get Vaccinated Dies of COVID," *Daily Beast*, January 7, 2022, https://www.thedailybeast.com/qanon-star-cirsten-weldon- who-said-only-idiots-get-vaccinated-dies-of-covid; Geoff Brumfiel, "Their Mom Died of COVID: They Say Conspiracy Theories Are What Really Killed Her," NPR, April 24, 2022, https://www .npr.org/sections/health-shots/2022/04/24/1089786147/covid-conspiracy-theories.

48.   Cary Funk, "Vast Majority of Americans Say Benefits of Childhood Vaccines Outweigh Risks," Pew Research Center, February 2, 2017, https://www.pewresearch.org/science/2017/02/02/vast -majority-of-americans-say-benefits-of-childhood-vaccines-outweigh-risks/.

49.   "Wellcome Global Monitor 2018," Wellcome, September 18, 2020, https://wellcome.org/reports /wellcome-global-monitor/2018.

50.   Funk, "Vast Majority of Americans Say Benefits of Childhood Vaccines Outweigh Risks.

51.   Charles McCoy, "Anti-Vaccination Beliefs Don't Follow the Usual Political Polarization," *The Conversation*, August 24, 2017, http://theconversation.com/anti-vaccination-beliefs-dont-follow -the-usual-political-polarization-81001.

52.   Alec Tyson et al., "Majority in U.S. Says Public Health Benefits of COVID-19 Restrictions Worth the Costs, Even as Large Shares Also See Downsides," Pew Research Center, September 15, 2021, https://www.pewresearch.org/science/2021/09/15/majority-in-u-s-says-public-health-benefits-of -covid-19-restrictions-worth-the-costs-even-as-large-shares-also-see-downsides/.

53.   "COVID-19 Vaccines Prevented Nearly 140,000 U.S. Deaths," *NIH News in Health*, October 2021, https://newsinhealth.nih.gov/2021/10/covid-19-vaccines-prevented-nearly-140000-us-deaths.

54.   Liz Hamel et al., "KFF COVID-19 Vaccine Monitor: January 2022," *KFF*, January 28, 2022, https:// www.kff.org/coronavirus-covid-19/poll-finding/kff-covid-19-vaccine-monitor-january-2022/; "Booster Shots in U.S. Have Strongly Protected Against Severe Disease from Omicron Variant, CDC Studies Show," *Washington Post*, January 21, 2022, https://www.washingtonpost.com/health /2022/01/21/cdc-studies-booster-shots-omicron/.

55.   Adam Cancryn, "A Sharp Partisan Divide Remains Over New Covid Boosters," *Politico*, September 15, 2023, https://www.politico.com/news/2023/09/15/poll-covid-booster-democrats-00116123.

56.   Carrie Blazina, "About Half of Recent Online Daters in U.S. Say It's Important to See COVID-19 Vaccination Status on Profiles," Pew Research Center, September 27, 2022, https://www.pewresearch .org/fact-tank/2022/09/27/about-half-of-recent-online-daters-in-u-s-say-its-important-to-see -covid-19-vaccination-status-on-profiles/.

57.   Katherine Ognyanova et al., "The COVID States Project #82: COVID-19 Vaccine Misinformation Trends, Awareness of Expert Consensus, and Trust in Social Institutions," OSF Preprints, February 15, 2022, https://doi.org/10.31219/osf.io/9ua2x.

58.   Eliza Relman, "The Gap Between Republicans and Democrats on Flu Shots Is 20 Percent- age-Points Bigger Than It Was Pre-pandemic," *Business Insider*, November 15, 2021, https://www .businessinsider.com/theres-a-25-point-gap-between-republicans-democrats-flu-shots-2021-11; Carla K. Johnson and Hannah Fingerhut, "AP-NORC Poll: More Americans Worry About Flu than New Virus," AP News, February 20, 2020, https://apnews.com/article/us-news-health-china -virus-outbreak-ap-top-news-c3eddb289d20d279de31a7c1b75f73d2.

59.   Harry Enten, "Flu Shots Uptake Is Now Partisan: It Didn't Use to Be," CNN, November 14, 2021, https://www.cnn.com/2021/11/14/politics/flu-partisan-divide-analysis/index.html.

60.   Alessandro Siani and Amy Tranter, "Is Vaccine Confidence an Unexpected Victim of the COVID- 19 Pandemic?," *Vaccine* 40, no. 50 (2022): 7262–69, https://doi.org/10.1016/j.vaccine.2022.10.061.

61.   Kevin Estep et al., "Partisan Polarization of Childhood Vaccination Policies, 1995-2020," *American Journal of Public Health* 112, no. 10 (2022): 1471–79, https://doi.org/10.2105/AJPH.2022.306964.

62. Matt Motta, "Is Cancer Treatment Immune from Partisan Conflict? How Partisan Communication Motivates Opposition to Preventative Cancer Vaccination in the U.S.," *Journal of Elections, Public Opinion and Parties* 34, no. 2 (2024): 319–43, https://doi.org/10.1080/17457289.2023.2168678.

63. Matt Motta et al., "Sick as a Dog? The Prevalence, Politicization, and Health Policy Consequences of Canine Vaccine Hesitancy (CVH)," *Vaccine* 41 (2023): 5946–50, https://doi.org/10.1016/j.vaccine.2023.08.059.

64. Dominik A. Stecula et al., "Policy Views and Negative Beliefs About Vaccines in the United States, 2019," *American Journal of Public Health* 110, no. 10 (2020): 1561–63, https://doi.org/10.2105/AJPH.2020.305828.

65. Robert Böhm and Cornelia Betsch, "Prosocial Vaccination," *Current Opinion in Psychology* 43 (2022): 307–11, https://doi.org/10.1016/j.copsyc.2021.08.010.

66. Ed Yong, "America Is Getting Unvaccinated People All Wrong," *The Atlantic*, July 22, 2021, https://www.theatlantic.com/health/archive/2021/07/unvaccinated-different-anti-vax/619523/.

67. Timothy B. Gravelle et al., "Estimating the Size of 'Anti-Vax' and Vaccine Hesitant Populations in the US, UK, and Canada: Comparative Latent Class Modeling of Vaccine Attitudes," *Human Vaccines and Immunotherapeutics* 18, no. 1 (2022): 2008214, https://doi.org/10.1080/21645515.2021.2008214.

68. Yong, "America Is Getting Unvaccinated People All Wrong."

69. Lesley Torres, "Unvaccinated Adults Are More Likely to Be Uninsured, Study Finds," *Bloomberg Law*, June 11, 2021, https://news.bloomberglaw.com/pharma-and-life-sciences/unvaccinated-adults-are-more-likely-to-be-uninsured-study-finds.

70. Rebecca Weintraub et al., "We Must 'Boost' COVID Vaccinations to Prevent a Winter Surge," *The Hill*, October 12, 2021, https://thehill.com/opinion/healthcare/575879-we-must-boost-covid-vaccinations-to-prevent-a-winter-surge/.

71. Pinelopi Konstantinou et al., "Transmission of Vaccination Attitudes and Uptake Based on Social Contagion Theory: A Scoping Review," *Vaccines* 9, no. 6 (2021): art. 607, https://doi.org/10.3390/vaccines9060607.

72. Maria Cordina et al., "Attitudes Towards COVID-19 Vaccination, Vaccine Hesitancy and Intention to Take the Vaccine," *Pharmacy Practice* 19, no. 1 (2021): 2317, https://doi.org/10.18549/PharmPract.2021.1.2317; Carl Latkin et al., "A Longitudinal Study of Vaccine Hesitancy Attitudes and Social Influence as Predictors of COVID-19 Vaccine Uptake in the US," *Human Vaccines and Immunotherapeutics* 18, no. 5 (2022): art. 2043102, https://doi.org/10.1080/21645515.2022.2043102.

73. Felix E. Fernández-Penny et al., "COVID-19 Vaccine Hesitancy Among Patients in Two Urban Emergency Departments," *Academic Emergency Medicine* 28, no. 10 (2021): 1100–7, https://doi.org/10.1111/acem.14376.

74. Sarah A. Nowak et al., "Association Among Trust in Health Care Providers, Friends, and Family, and Vaccine Hesitancy," *Vaccine* 39, no. 40 (2021): 5737–40, https://doi.org/10.1016/j.vaccine.2021.08.035.

75. Fridman et al., "COVID-19 and Vaccine Hesitancy."

76. Robert Böhm et al., "Selfish-Rational Non-vaccination: Experimental Evidence from an Interactive Vaccination Game," *Journal of Economic Behavior and Organization* 131 (2016): 183–95, https://doi.org/10.1016/j.jebo.2015.11.008.

77. Seth Masket, "Seth Masket: The Great Vaccine Divide Puts Republican Leaders in a Moral Quandary," *Denver Post*, June 25, 2021, https://www.denverpost.com/2021/06/25/covid-19-vaccine-rates-donald-trump-joe-biden/.

78. Sophia L. Pink et al., "Elite Party Cues Increase Vaccination Intentions Among Republicans," *Proceedings of the National Academy of Sciences* 118, no. 32 (2021): e2106559118, https://doi.org/10.1073/pnas.2106559118.

79. German Lopez, "How Political Polarization Broke America's Vaccine Campaign," *Vox*, July 6, 2021, https://www.vox.com/2021/7/6/22554198/political-polarization-vaccine-covid-19-coronavirus.

80. Matthew Facciani et al., "Political Network Composition Predicts Vaccination Attitudes," *Social Science and Medicine* 328 (2023): art. 116004, https://doi.org/10.1016/j.socscimed.2023 .116004.

81. Feng Fu et al., "Dueling Biological and Social Contagions," *Scientific Reports* 7 (2017): art. 43634, https://doi.org/10.1038/srep43634.

82. Kevin Winter et al., "Pro-Vaccination Subjective Norms Moderate the Relationship Between Conspiracy Mentality and Vaccination Intentions," *British Journal of Health Psychology* 27, no. 2 (2022): 390–405, https://doi.org/10.1111/bjhp.12550.

83. Damon Centola, "An Experimental Study of Homophily in the Adoption of Health Behavior," *Science* 334, no. 6060 (2011): 1269–72, https://doi.org/10.1126/science.1207055.

84. Alex Moehring et al., "Providing Normative Information Increases Intentions to Accept a COVID-19 Vaccine," *Nature Communications* 14, no. 1 (2023): 126, https://doi.org/10.1038 /s41467-022-35052-4.

85. Samantha Sinclair and Jens Agerström, "Do Social Norms Influence Young People's Willingness to Take the COVID-19 Vaccine?," *Health Communication* 38, no. 1 (2023): 152–59, https://doi.org/10 .1080/10410236.2021.1937832.

86. Paweł Waszkiewicz et al., "Public Vaccination Reluctance: What Makes Us Change Our Minds? Results of a Longitudinal Cohort Survey," *Vaccines* 10, no. 7 (2022): art. 1081, https://doi.org /10.3390/vaccines10071081.

87. Summer Lin, "Here's Why Some Vaccine Skeptics Changed Their Minds and Got COVID Shots, Poll Says," *Miami Herald*, July 14, 2021, https://www.miamiherald.com/news/coronavirus/article252777353.html; Ashley Kirzinger et al., "KFF COVID-19 Vaccine Monitor: In Their Own Words, Six Months Later," *KFF*, July 13, 2021, https://www.kff.org/coronavirus-covid-19 /poll-finding/kff-covid-19-vaccine-monitor-in-their-own-words-six-months-later/.

88. Brendan Nyhan et al., "Effective Messages in Vaccine Promotion: A Randomized Trial," *Pediatrics* 133, no. 4 (2014): e835–42, https://doi.org/10.1542/peds.2013-2365; Sara Pluviano et al., "Parents' Beliefs in Misinformation About Vaccines Are Strengthened by Pro-Vaccine Campaigns," *Cognitive Processing* 20, no. 3 (2019): 325–31, https://doi.org/10.1007/s10339-019-00919-w.

89. Maryke S. Steffens et al., "How Organisations Promoting Vaccination Respond to Misinformation on Social Media: A Qualitative Investigation," *BMC Public Health* 19, no. 1 (2019): art, 1348, 1, https:// doi.org/10.1186/s12889-019-7659-3.

90. Ashley Fetters, "How to Talk to an Anti-Vax Relative," *The Atlantic*, April 22, 2019, https://www .theatlantic.com/family/archive/2019/04/when-families-feud-over-vaccines/587629/.

91. Brooke Harrington, "Here's Why Your Efforts to Convince Anti-Vaxxers Aren't Working," *The Guardian*, August 9, 2021, https://www.theguardian.com/commentisfree/2021/aug/09/convince-anti -vaxxers.

92. Brooke Harrington, "How Sociologists Can Battle Covid Denialism," *Chronicle of Higher Education*, September 1, 2021, https://www.chronicle.com/article/how-sociologists-can-battle-covid -denialism?sra=true.

93. Pink et al., "Elite Party Cues Increase Vaccination Intentions Among Republicans."

94. James Chu et al., "Religious Identity Cues Increase Vaccination Intentions and Trust in Medical Experts Among American Christians," *Proceedings of the National Academy of Sciences* 118, no. 49 (2021): e2106481118, https://doi.org/10.1073/pnas.2106481118.

95. Ariel Hart, "Dr. Kimberly Manning: Yes, We Can Reach the Unvaccinated," *Atlanta Journal-Constitution*, October 29, 2021, https://www.ajc.com/news/coronavirus/dr-kimberly-manning-yes-we -can-reach-the-unvaccinated/ZMNQOJIGHBASBDTMODPTVD7QAQ/.

96. Elizabeth Nix, "Tuskegee Experiment: The Infamous Syphilis Study," History, May 16, 2017, https:// www.history.com/news/the-infamous-40-year-tuskegee-study.

97. Matt Motta, email message to author, December 5, 2020.

98. David A. Broniatowski et al., "Weaponized Health Communication: Twitter Bots and Russian Trolls Amplify the Vaccine Debate," *American Journal of Public Health* 108, no. 10 (2018): 1378–84, https://doi.org/10.2105/AJPH.2018.304567.

99. Rachel Gutman-Wei, "Of Course Biden Has Rebound COVID," *The Atlantic*, July 30, 2022, https://www.theatlantic.com/health/archive/2022/07/biden-paxlovid-covid-drug-rebound-infections/671009/.

100. Brenda Goodman and Virginia Langmaid, "Fauci Says His Covid Rebounded After Paxlovid," CNN, June 30, 2022, https://www.cnn.com/2022/06/30/health/covid-paxlovid-fauci-rebound/index.html.

101. Kristin Lunz Trujillo et al., "The COVID States Project #77: Healthcare Workers' Perception of COVID-19 Misinformation," OSF Preprints, January 15, 2022, https://doi.org/10.31219/osf.io/6pzqj.

102. Neil F. Johnson et al., "The Online Competition Between Pro- and Anti-Vaccination Views," *Nature* 582, no. 7811 (2020): 230–33, https://doi.org/10.1038/s41586-020-2281-1.

103. Samira Yousefinaghani et al., "An Analysis of COVID-19 Vaccine Sentiments and Opinions on Twitter," *International Journal of Infectious Diseases* 108 (2021): 256–62, https://doi.org/10.1016/j.ijid.2021.05.059.

104. Erika Bonnevie et al., "Quantifying the Rise of Vaccine Opposition on Twitter During the COVID-19 Pandemic," *Journal of Communication in Healthcare* 14, no. 1 (2021): 12–19, https://doi.org/10.1080/17538068.2020.1858222.

105. Elvira Ortiz-Sánchez et al., "Analysis of the Anti-Vaccine Movement in Social Networks: A Systematic Review," *International Journal of Environmental Research and Public Health* 17, no. 15 (2020): art. 5394, https://doi.org/10.3390/ijerph17155394; Broniatowski et al., "Weaponized Health Communication."

106. Shannon Bond, "Just 12 People Are Behind Most Vaccine Hoaxes on Social Media, Research Shows," NPR, May 13, 2021, https://www.npr.org/2021/05/13/996570855/disinformation-dozen-test-facebooks-twitters-ability-to-curb-vaccine-hoaxes.

107. Francesco Pierri et al., "Online Misinformation Is Linked to Early COVID-19 Vaccination Hesitancy and Refusal," *Scientific Reports* 12, no. 1 (2022): art. 5966, https://doi.org/10.1038/s41598-022-10070-w.

108. Saifuddin Ahmed et al., "Social Media News Use Induces COVID-19 Vaccine Hesitancy Through Skepticism Regarding Its Efficacy: A Longitudinal Study from the United States," *Frontiers in Psychology* 13 (2022): 900386, https://doi.org/10.3389/fpsyg.2022.900386.

109. Matthew J. Hornsey et al., "Donald Trump and Vaccination: The Effect of Political Identity, Conspiracist Ideation and Presidential Tweets on Vaccine Hesitancy," *Journal of Experimental Social Psychology* 88 (2020): art. 103947, https://doi.org/10.1016/j.jesp.2019.103947.

110. Ana Lucía Schmidt et al., "Polarization of the Vaccination Debate on Facebook," *Vaccine* 36, no. 25 (2018): 3606–12, https://doi.org/10.1016/j.vaccine.2018.05.040.

111. Helge Giese et al., "The Echo in Flu-Vaccination Echo Chambers: Selective Attention Trumps Social Influence," *Vaccine* 38, no. 8 (2020): 2070–76, https://doi.org/10.1016/j.vaccine.2019.11.038.

112. Dominik A. Stecuła and Matt Motta, "Unverified Reports of Vaccine Side Effects in VAERS Aren't the Smoking Guns Portrayed by Right-Wing Media Outlets—They Can Offer Insight Into Vaccine Hesitancy," *The Conversation*, August 25, 2021, http://theconversation.com/unverified-reports-of-vaccine-side-effects-in-vaers-arent-the-smoking-guns-portrayed-by-right-wing-media-outlets-they-can-offer-insight-into-vaccine-hesitancy-166401.

113. Saranac Hale Spencer, "Tucker Carlson Misrepresents Vaccine Safety Reporting Data," FactCheck.org, May 14, 2021, https://www.factcheck.org/2021/05/scicheck-tucker-carlson-misrepresents-vaccine-safety-reporting-data/.

114. Constance de Saint Laurent et al., "Measuring the Effects of Misinformation Exposure and Beliefs on Behavioural Intentions: A COVID-19 Vaccination Study," *Cognitive Research: Principles and Implications* 7, no. 1 (2022): 87, https://doi.org/10.1186/s41235-022-00437-y.

115. Ciara M. Greene and Gillian Murphy, "Quantifying the Effects of Fake News on Behavior: Evidence from a Study of COVID-19 Misinformation," *Journal of Experimental Psychology: Applied* 27, no. 4 (2021): 773–84, https://doi.org/10.1037/xap0000371.

116. Sahil Loomba et al., "Measuring the Impact of COVID-19 Vaccine Misinformation on Vaccination Intent in the UK and USA," *Nature Human Behaviour* 5, no. 3 (2021): 337–48, https://doi.org/10.1038/s41562-021-01056-1.

117. Johnson et al., "The Online Competition Between Pro- and Anti-Vaccination Views."

118. Megan N. Imundo and David N. Rapp, "When Fairness Is Flawed: Effects of False Balance Reporting and Weight-of-Evidence Statements on Beliefs and Perceptions of Climate Change," *Journal of Applied Research in Memory and Cognition* 11, no. 2 (2022): 258–71, https://doi.org/10.1016/j.jarmac.2021.10.002.

119. Max Witynski, "False Balance in News Coverage of Climate Change Makes It Harder to Address Crisis," Phys.org, July 22, 2022, https://phys.org/news/2022-07-false-news-coverage-climate-harder.html.

120. Nicholas Light et al., "Knowledge Overconfidence Is Associated with Anti-Consensus Views on Controversial Scientific Issues," *Science Advances* 8, no. 29 (2022): eabo0038, https://doi.org/10.1126/sciadv.abo0038.

121. Adrian F. Ward et al., "I Share, Therefore I Know? Sharing Online Content—Even Without Reading It—Inflates Subjective Knowledge," *Journal of Consumer Psychology* 33, no. 3 (2022): 469–88, https://doi.org/10.1002/jcpy.1321.

122. Timothy S. Rich et al., "Research Note: Does the Public Support Fact-Checking Social Media? It Depends Who and How You Ask," *Harvard Kennedy School Misinformation Review* 1, no. 8 (2020): 1–10, https://doi.org/10.37016/mr-2020-46.

123. Lee Rainie et al., "3. Mixed Views About Social Media Companies Using Algorithms to Find False Information," Pew Research Center, March 17, 2022, https://www.pewresearch.org/internet/2022/03/17/mixed-views-about-social-media-companies-using-algorithms-to-find-false-information/.

124. Nathan Walter et al., "Fact-Checking: A Meta-Analysis of What Works and for Whom," *Political Communication* 37, no. 3 (2020): 350–75, https://doi.org/10.1080/10584609.2019.1668894.

125. John M. Carey et al., "The Ephemeral Effects of Fact-Checks on COVID-19 Misperceptions in the United States, Great Britain and Canada," *Nature Human Behaviour* 6, no. 2 (2022): 236–43, https://doi.org/10.1038/s41562-021-01278-3.

126. Ethan Porter et al., "Correcting COVID-19 Vaccine Misinformation in 10 Countries," *Royal Society Open Science* 10, no. 3 (2023): art. 221097, https://doi.org/10.1098/rsos.221097.

127. Emily K. Vraga and Leticia Bode, "Correction as a Solution for Health Misinformation on Social Media," *American Journal of Public Health* 110, no. S3 (2020): S278–80, https://doi.org/10.2105/AJPH.2020.305916.

128. "A Group of Moms on Facebook Built an Island of Good-Faith Vaccine Debate in a Sea of Misinformation," *Washington Post*, August 23, 2021, https://www.washingtonpost.com/technology/2021/08/23/facebook-vaccine-talk-group/.

129. "'Vaccine Talk' Facebook Group Is a Carefully Moderated Forum for Vaccine Questions," NPR, September 18, 2021, https://www.npr.org/2021/09/18/1038533086/vaccine-talk-facebook-group-is-a-carefully-moderated-forum-for-vaccine-questions.

130. Craig R. McClain, "Practices and Promises of Facebook for Science Outreach: Becoming a 'Nerd of Trust,'" *PLOS Biology* 15, no. 6 (2017): e2002020. https://doi.org/10.1371/journal.pbio.2002020.

131. Gordon Pennycook and David G. Rand, "Fighting Misinformation on Social Media Using Crowdsourced Judgments of News Source Quality," *Proceedings of the National Academy of Sciences* 116, no. 7 (2019): 2521–26, https://doi.org/10.1073/pnas.1806781116.

132. Feng Shi et al., "The Wisdom of Polarized Crowds," *Nature Human Behaviour* 3, no. 4 (2019): 329–36, https://doi.org/10.1038/s41562-019-0541-6.

133. Folco Panizza, Piero Ronzani, Tiffany Morisseau, Simone Mattavelli, and Carlo Martini, "How Do Online Users Respond to Crowdsourced Fact-Checking?," *Humanities & Social Sciences Communications* 10, no. 1 (November 25, 2023): 1–11.

134. Jennifer Allen, Antonio A. Arechar, Gordon Pennycook, and David G. Rand, "Scaling up Fact-Checking Using the Wisdom of Crowds," *Science Advances* 7, no. 36 (2021): eabf4393.

135. William Godel et al., "Moderating with the Mob: Evaluating the Efficacy of Real-Time Crowdsourced Fact-Checking," *Journal of Online Trust and Safety* 1, no. 1 (2021): 1–36, https://doi.org /10.54501/jots.v1i1.15.

136. Tali Sharot, "To Quell Misinformation, Use Carrots—Not Just Sticks," *Nature*, March 17, 2021, https://doi.org/10.1038/d41586-021-00657-0.

137. Jeff Turrentine, "Climate Misinformation on Social Media Is Undermining Climate Action," Natural Resources Defense Council, April 19, 2022, https://www.nrdc.org/stories/climate -misinformation-social-media-undermining-climate-action.

138. Bruna Horvath, Jason Abbruzzese and Ben Goggin, "Meta Is Ending Its Fact-Checking Program in Favor of a 'Community Notes' System Similar to X's," NBC News, January 7, 2025, https://www .nbcnews.com/tech/social-media/meta-ends-fact-checking-program-community-notes-x -rcna186468.

139. Anastasia Kozyreva et al., "Resolving Content Moderation Dilemmas Between Free Speech and Harmful Misinformation," *Proceedings of the National Academy of Sciences* 120, no. 7 (2023): e2210666120, https://doi.org/10.1073/pnas.2210666120.

140. Filippo Menczer and Thomas Hills, "Information Overload Helps Fake News Spread, and Social Media Knows It," *Scientific American*, December 1, 2020, https://www.scientificamerican.com /article/information-overload-helps-fake-news-spread-and-social-media-knows-it/.

141. Filippo Menczer, "How 'Engagement' Makes You Vulnerable to Manipulation and Misinformation on Social Media," *The Conversation*, September 20, 2021, http://theconversation.com/how-engagement -makes-you-vulnerable-to-manipulation-and-misinformation-on-social-media-145375.

142. Lisa Fazio, "Pausing to Consider Why a Headline Is True or False Can Help Reduce the Sharing of False News," *Harvard Kennedy School Misinformation Review* 1, no. 2 (2020): 1–8, https://doi.org /10.37016/mr-2020-009.

143. Renée DiResta and Tobias Rose-Stockwell, "How to Stop Misinformation Before It Gets Shared," *Wired*, March 26, 2021, https://www.wired.com/story/how-to-stop-misinformation-before-it-gets-shared/.

144. Steve Rathje et al., "Letter to the Editors of *Psychological Science*: Meta-Analysis Reveals That Accuracy Nudges Have Little to No Effect for US Conservatives: Regarding Pennycook et al.," preprint, PsyArXiv, April 2, 2022, https://doi.org/10.31234/osf.io/945na.

145. Frances Haugen, "Keynote Facilitated by Brian Klaas," Cambridge Disinformation Summit, University of Cambridge, July 28, 2023, YouTube, 44 min., 49 sec., https://www.youtube.com/watch?v =bSCy6y5wu1g.

146. Makena Kelly, "New Algorithm Bill Could Force Facebook to Change How the News Feed Works," *The Verge*, February 10, 2022, https://www.theverge.com/2022/2/10/22927472/klobuchar-lummis -algorithm-bill-section-230-misinformation-teenager-mental-health.

147. Natasha Lomas, "Twitter Offers More Support to Researchers—to 'Keep Us Accountable,'" TechCrunch, January 6, 2020, https://techcrunch.com/2020/01/06/twitter-offers-more-support-to -researchers-to-keep-us-accountable/.

148. Jess Weatherbed, "Twitter Replaces Its Free API with a Paid Tier in Quest to Make More Money," *The Verge*, February 2, 2023, https://www.theverge.com/2023/2/2/23582615/twitter-removing-free-api -developer-apps-price-announcement.

149. Stuart A. Thompson, "To Fight Election Falsehoods, Social Media Companies Ready a Familiar Playbook," *New York Times*, August 23, 2022, https://www.nytimes.com/2022/08/23/technology /midterms-misinformation-tiktok-facebook.html.

150. Jack Brewster et al., "Beware the 'New Google': TikTok's Search Engine Pumps Toxic Misinformation to Its Young Users," NewsGuard, September 14, 2022. https://www.newsguardtech.com/misinformation-monitor/september-2022/.

151. Joshua A. Tucker and Nathaniel Persily, "How to Fix Social Media? Start with Independent Research," *Brookings*, December 1, 2021, https://www.brookings.edu/articles/how-to-fix-social-media-start-with-independent-research/.

152. Renée DiResta et al., "It's Time to Open the Black Box of Social Media," *Scientific American*, April 28, 2022, https://www.scientificamerican.com/article/its-time-to-open-the-black-box-of-social-media/.

153. Irene Pasquetto et al., "Tackling Misinformation: What Researchers Could Do with Social Media Data," *Harvard Kennedy School Misinformation Review* 1, no. 8 (2020): 1–14 https://doi.org/10.37016/mr-2020-49.

154. "3M State of Science Index: Connecting the 2023 Survey to 3M Forward," 3M, accessed August 14, 2024, https://www.3m.com/3M/en_US/3m-forward-us/about-the-survey/.

155. Cary Funk et al., "1. Scientists Are Among the Most Trusted Groups in Society, Though Many Value Practical Experience Over Expertise," Pew Research Center, September 29, 2020, https://www.pewresearch.org/science/2020/09/29/scientists-are-among-the-most-trusted-groups-in-society-though-many-value-practical-experience-over-expertise/.

156. Brian Kennedy, "Americans' Trust in Scientists, Other Groups Declines," Pew Research Center, February 15, 2022, https://www.pewresearch.org/science/2022/02/15/americans-trust-in-scientists-other-groups-declines/.

157. Annenberg Public Policy Center of the University of Pennsylvania, "Survey Finds Public Perception of Scientists' Credibility Has Slipped," Phys.org, June 26, 2024, https://phys.org/news/2024-06-survey-perception-scientists-credibility.html.

158. Kennedy, "Americans' Trust in Scientists, Other Groups Declines."

159. Betsy Broaddus, "Amidst the Pandemic, Confidence in the Scientific Community Becomes Increasingly Polarized," AP-NORC, January 26, 2022, https://apnorc.org/projects/amidst-the-pandemic-confidence-in-the-scientific-community-becomes-increasingly-polarized/.

160. Robert Wood Johnson Foundation and Harvard T. H. Chan School of Public Health, *The Public's Perspective on the United States Public Health System* (Harvard Opinion Research Program, 2021), https://www.hsph.harvard.edu/wp-content/uploads/sites/94/2021/05/RWJF-Harvard-Report_FINAL-051321.pdf; Abigail Geiger, "How Americans See the Future of Space Exploration, 50 Years After the First Moon Landing," Pew Research Center, July 17, 2019, https://www.pewresearch.org/fact-tank/2019/07/17/how-americans-see-the-future-of-space-exploration-50-years-after-the-first-moon-landing/.

161. Tom Nichols, *The Death of Expertise: The Campaign Against Established Knowledge and Why It Matters* (Oxford University Press, 2017), x. See also Michiko Kakutani, "'The Death of Expertise' Explores How Ignorance Became a Virtue," *New York Times*, March 21, 2017, https://www.nytimes.com/2017/03/21/books/the-death-of-expertise-explores-how-ignorance-became-a-virtue.html.

162. Charlotte Alter, "Nothing, Not Even Hard Facts, Can Make Anti-Vaxxers Change Their Minds," *Time*, March 4, 2014, https://healthland.time.com/2014/03/04/nothing-not-even-hard-facts-can-make-anti-vaxxers-change-their-minds/.

163. Nyhan et al., "Effective Messages in Vaccine Promotion."

164. "Howard Stern Says Anti-Vaxxers Should Be Denied Hospital Care If They Catch COVID-19," *Hollywood Reporter*, September 9, 2021, YouTube, 1 min. 28 sec., https://www.youtube.com/watch?v=OPOfyGOnJtA; Ed Yong, "It's a Terrible Idea to Deny Medical Care to Unvaccinated People," *The Atlantic*, January 20, 2022, https://www.theatlantic.com/health/archive/2022/01/unvaccinated-medical-care-hospitals-omicron/621299/.

165. Maya J. Goldenberg, *Vaccine Hesitancy: Public Trust, Expertise, and the War on Science* (University of Pittsburgh Press, 2021).

166. Torsten Wilholt, "Epistemic Trust in Science," *British Journal for the Philosophy of Science* 64, no. 2 (2013): 233–53, http://www.jstor.org/stable/24563046.

167. John Hardwig, "The Role of Trust in Knowledge," *Journal of Philosophy* 88, no. 12 (1991): 693–708, https://doi.org/10.2307/2027007; Doran Smolkin, "Puzzles About Trust," *Southern Journal of Philosophy* 46, no. 3 (2008): 431–49, https://doi.org/10.1111/j.2041-6962.2008.tb00127.x.

168. Rupali J. Limaye et al., "Patient Decision Making Related to Maternal and Childhood Vaccines: Exploring the Role of Trust in Providers Through a Relational Theory of Power Approach," *Health Education and Behavior* 47, no. 3 (2020): 449–56, https://doi.org/10.1177/1090198120915432.

169. Martin Tanis and Tom Postmes, "A Social Identity Approach to Trust: Interpersonal Perception, Group Membership and Trusting Behaviour," *European Journal of Social Psychology* 35, no. 3 (2005): 413–24, https://doi.org/10.1002/ejsp.256; Sufei Xin et al., "Effects of Trustors' Social Identity Complexity on Interpersonal and Intergroup Trust," *European Journal of Social Psychology* 46, no. 4 (2016): 428–40, https://doi.org/10.1002/ejsp.2156; Ryan E. Carlin and Gregory J. Love, "The Politics of Interpersonal Trust and Reciprocity: An Experimental Approach," *Political Behavior* 35, no. 1 (2013): 43–63, https://doi.org/10.1007/s11109-011-9181-x.

170. Jarry, "The Upside-Down Doctor."

171. David Robert Grimes and David H. Gorski, "Quantifying Public Engagement with Medical Science, Misinformation, and Malinformation," OSF Preprints, June 1, 2022, https://doi.org/10.31219/osf.io/g4jwr.

172. Heidi J. Larson, *Stuck: How Vaccine Rumors Start—and Why They Don't Go Away* (Oxford University Press, 2020), xxxv.

173. Alice Fabbri et al., "The Influence of Industry Sponsorship on the Research Agenda: A Scoping Review," *American Journal of Public Health* 108, no. 11 (2018): e9–16, https://doi.org/10.2105/AJPH.2018.304677.

174. Brian Mann, "4 U.S. Companies Will Pay $26 Billion to Settle Claims They Fueled the Opioid Crisis," NPR, February 25, 2022, https://www.npr.org/2022/02/25/1082901958/opioid-settlement-johnson-26-billion; Reuters, "Big 3 U.S. Drug Distributors, Johnson & Johnson Reach Landmark $26 Billion Opioid Settlement," CNBC, July 21, 2021, https://www.cnbc.com/2021/07/21/drug-distributors-jj-reach-landmark-26-billion-opioid-settlement-.html.

175. Amy Maxmen, "Unseating Big Pharma: The Radical Plan for Vaccine Equity," *Nature* 607, no. 7918 (2022): 226–33, https://doi.org/10.1038/d41586-022-01898-3.

176. Thomas B. Cueni, "Waiving Intellectual Property Rights Is a Flawed Solution to Achieving Covid-19 Vaccine Equity," *STAT*, June 10, 2022, https://www.statnews.com/2022/06/10/waiving-intellectual-property-rights-is-a-flawed-solution-to-achieving-covid-19-vaccine-equity/.

177. Zion Market Research, "Insights on Global Homeopathy Products Market Size & Share Projected to Hit at USD 50,203.3 Million and Rise at a CAGR of 18.7 Percent by 2028: Industry Trends, Demand, Value, Analysis & Forecast Report," PR Newswire, May 17, 2022. https://www.prnewswire.com/news-releases/insights-on-global-homeopathy-products-market-size--share-projected-to-hit-at-usd-50-203-3-million-and-rise-at-a-cagr-of-18-7-by-2028-industry-trends-demand-value-analysis--forecast-report--zion-market-research-301549050.html; Michael Simpson, "Anti-Vaccine Activists Support Big Pharma Profits—My Irony Meter Dies," Skeptical Raptor, March 8, 2022, https://www.skepticalraptor.com/skepticalraptorblog.php/anti-vaccine-activists-support-big-pharma-profits-my-irony-meter-dies/.

178. Liz Hamel et al., "KFF COVID-19 Vaccine Monitor: December 2020," *KFF*, December 15, 2020, https://www.kff.org/coronavirus-covid-19/report/kff-covid-vaccine-monitor-december-2020/.

179. Mark Navin, *Values and Vaccine Refusal: Hard Questions in Ethics, Epistemology, and Health Care* (Routledge, 2015).

180. "Yale Experts Join Campaign to Boost Vaccinations in Communities of Color," YaleNews, December 17, 2021, https://news.yale.edu/2021/12/17/yale-experts-join-campaign-boost-vaccinations-communities-color.

181. Charlotte Dries et al., "When Evidence Changes: Communicating Uncertainty Protects Against a Loss of Trust," *Public Understanding of Science* 33, no. 6 (2024): 777–94, https://doi.org/10.1177 /09636625241228449.

182. Francis Collins, interview by David DeSteno, uploaded May 29, 2024, Vimeo, 1 min., 42 sec., https://vimeo.com/951537027.

183. Alexander A. Kaurov et al., "Trends in American Scientists' Political Donations and Implications for Trust in Science," *Humanities and Social Sciences Communications* 9, no. 1 (2022): 1–8, https://doi.org /10.1057/s41599-022-01382-3.

184. Robert D. Putnam, *The Upswing: How America Came Together a Century Ago and How We Can Do It Again* (Simon and Schuster, 2020).

185. Jonathan M. Berman, *Anti-Vaxxers: How to Challenge a Misinformed Movement* (MIT Press, 2020).

186. Rashawn Ray et al., *Examining and Addressing COVID-19 Racial Disparities in Detroit* (Brookings Institution, 2021), https://www.brookings.edu/wp-content/uploads/2021/02/Detroit_Covid_report _final.pdf.

187. Matthew Motta et al., " 'The CDC Won't Let Me Be': The Opinion Dynamics of Support for CDC Regulatory Authority," *Journal of Health Politics, Policy and Law* 48, no. 6 (2023): 829–57, https:// doi.org/10.1215/03616878-10852592.

188. Ethan Zuckerman, *Mistrust: Why Losing Faith in Institutions Provides the Tools to Transform Them* (Norton, 2021); Caroline Harting, "Can America Restore Its Trust in Government?" Columbia News, February 22, 2021 https://news.columbia.edu/news/mistrust-ethan-zuckerman-knight.

189. Katie Mack (@AstroKatie), "Personally, I think part of the problem," Twitter (now X), February 6, 2022, https://twitter.com/AstroKatie/status/1490431034470584327.

190. Katie Mack (@AstroKatie), "Anyway there are lots of ways," Twitter (now X), February 6, 2022, https://twitter.com/AstroKatie/status/1490439888256352266.

191. Carol Tavris, "Episode 1: Carol Tavris on Mistakes, Justification, and Cognitive Dissonance," interview by Sean Carroll, *Sean Carroll's Mindscape*, podcast, July 9, 2018, https://www.preposterousuniverse.com /podcast/2018/07/09/episode-1-carol-tavris-on-mistakes-justification-and-cognitive-dissonance/.

192. Bianca Manago, "Preregistration and Registered Reports in Sociology: Strengths, Weaknesses, and Other Considerations," *American Sociologist* 54, no. 1 (2023): 193–210, https://doi.org/10.1007/s12108 -023-09563-6.

193. Tamarinde Haven et al., "Promoting Trust in Research and Researchers: How Open Science and Research Integrity Are Intertwined," *BMC Research Notes* 15, no. 1 (2022): 302, https://doi.org /10.1186/s13104-022-06169-y.

194. Vimal Patel, "White House Pushes Journals to Drop Paywalls on Publicly Funded Research," *New York Times*, August 26, 2022, https://www.nytimes.com/2022/08/25/us/white-house-federally-funded -research-access.html.

195. Jürgen Huber et al., "Nobel and Novice: Author Prominence Affects Peer Review," *Proceedings of the National Academy of Sciences* 119, no. 41 (2022): e2205779119, https://doi.org/10.1073/pnas .2205779119.

196. "Reinventing Academic Publishing," Advances.in Reinventing Academic Publishing, accessed August 14, 2024, https://advances.in/.

197. Richard L. Street et al., "Understanding Concordance in Patient-Physician Relationships: Personal and Ethnic Dimensions of Shared Identity," *Annals of Family Medicine* 6, no. 3 (2008): 198–205, https://doi.org/10.1370/afm.821.

198. Douglas L. Medin and Carol D. Lee, "Diversity Makes Better Science," *APS Observer*, May-June 2012, https://www.psychologicalscience.org/observer/diversity-makes-better-science.

199. Kendall Powell, "The Power of Diversity," *Nature* 558, no. 7708 (2018): 19–22, https://doi.org/10.1038 /d41586-018-05316-5.

200. "The Scientific Community: Diversity Makes the Difference," Understanding Science, April 14, 2022, https://undsci.berkeley.edu/article/0_0_0/socialsideofscience_02.

201. Jen Gunter, *The Vagina Bible: The Vulva and the Vagina; Separating the Myth from the Medicine* (Citadel Press, 2019).

202. Yang Yang et al., "Gender-Diverse Teams Produce More Novel and Higher-Impact Scientific Ideas," *Proceedings of the National Academy of Sciences* 119, no. 36 (2022): e2200841119, https://doi .org/10.1073/pnas.2200841119.

203. Michael M. E. Johns et al., "Restoring Balance to Industry-Academia Relationships in an Era of Institutional Financial Conflicts of Interest: Promoting Research While Maintaining Trust," *JAMA* 289, no. 6 (2003): 741–46, https://doi.org/10.1001/jama.289.6.741.

6. HOW EDUCATION, MEDIA LITERACY, AND PREBUNKING CAN COMBAT MISINFORMATION

1. Sarah R. Olutola, "Nicki Minaj's COVID-19 Vaccine Tweet About Swollen Testicles Signals the Dangers of Celebrity Misinformation and Fandom," *The Conversation*, September 20, 2021, http:// theconversation.com/nicki-minajs-covid-19-vaccine-tweet-about-swollen-testicles-signals-the -dangers-of-celebrity-misinformation-and-fandom-168242; Ramishah Maruf, "These Four Words Are Helping Spread Vaccine Misinformation," CNN, September 19, 2021, https://www.cnn.com /2021/09/19/media/reliable-sources-covid-research/index.html.

2. John Herrman, "They Did Their Own 'Research': Now What?" *New York Times*, May 29, 2022, https://www.nytimes.com/2022/05/29/style/do-your-own-research.html.

3. Sedona Chinn and Ariel Hasell, "Support for 'Doing Your Own Research' Is Associated with COVID-19 Misperceptions and Scientific Mistrust," *Harvard Kennedy School Misinformation Review* 4, no. 3 (2023): 1–15, https://doi.org/10.37016/mr-2020-117.

4. Benjamin A. Lyons et al., "Overconfidence in News Judgments Is Associated with False News Susceptibility," *Proceedings of the National Academy of Sciences* 118, no. 23 (2021): e2019527118, https:// doi.org/10.1073/pnas.2019527118.

5. Andrea Vranic et al., "'I Did My Own Research': Overconfidence, (Dis)trust in Science, and Endorsement of Conspiracy Theories," *Frontiers in Psychology* 13 (2022): 1–9, https://doi.org/10.3389 /fpsyg.2022.931865.

6. Nadia M. Brashier and Daniel L. Schacter, "Aging in an Era of Fake News," *Current Directions in Psychological Science* 29, no. 3 (2020): 316–23, https://doi.org/10.1177/0963721420915872.

7. Didem Pehlivanoglu et al., "Aging in an 'Infodemic': The Role of Analytical Reasoning, Affect, and News Consumption Frequency on News Veracity Detection," *Journal of Experimental Psychology: Applied* 28, no. 3 (2022): 468–85, https://doi.org/10.1037/xap0000426.

8. "Students' Civic Online Reasoning," Digital Inquiry Group, November 14, 2019, https://sheg.stan-ford.edu/students-civic-online-reasoning.

9. David Rosenberg et al., "Teens Are Spending the Equivalent of a 40-Hour Work Week on Their Devices: Here's How to Help Them Find the Right Balance," *Fortune*, October 24, 2023, https:// fortune.com/well/2023/10/24/teens-too-much-screen-time-find-balance/.

10. Joe Carr, "A New Index Shows That the US Scores Low on Media Literacy Education," Media Literacy Now, July 27, 2021, https://medialiteracynow.org/a-new-index-shows-that-the-us-scores -low-on-media-literacy-education/.

11. Chris Teale, "7 in 10 Voters Support Teaching Social Media Literacy in Schools," Morning Consult Pro, December 8, 2021, https://morningconsult.com/2021/12/08/social-media-literacy -polling/.

12. "National Survey Finds Most U.S. Adults Have Not Had Media Literacy Education in High School," Media Literacy Now, September 7, 2022, https://medialiteracynow.org/nationalsurvey2022/.

13. Patricia Aufderheide, *Media Literacy: A Report of the National Leadership Conference on Media Literacy, the Aspen Institute Wye Center, Queenstown, Maryland, December 7–9, 1992* (Aspen Institute, 1993), v.

14. Taşkın Inan and Turan Temur, "Examining Media Literacy Levels of Prospective Teachers," *International Electronic Journal of Elementary Education* 4, no. 2 (2012): 269–85, https://www.iejee.com /index.php/IEJEE/article/view/199/195.

15. Erica Weintraub Austin et al., "How Media Literacy and Science Media Literacy Predicted the Adoption of Protective Behaviors Amidst the COVID-19 Pandemic," *Journal of Health Communication* 26, no. 4 (2021): 239–52, https://doi.org/10.1080/10810730.2021.1899345.

16. Momin M. Malik et al., "The Challenges of Defining 'News Literacy,'" Research Publication No. 2013–20 (Berkman Center for Internet and Society, 2013), https://doi.org/10.2139/ssrn.2342313.

17. Seth Ashley et al., "Developing a News Media Literacy Scale," *Journalism and Mass Communication Educator* 68, no. 1 (2013): 7–21, https://doi.org/10.1177/1077695812469802.

18. "Information Literacy Competency Standards for Higher Education," American Library Association, accessed July 15, 2024, http://hdl.handle.net/11213/7668.

19. Sonia Livingstone et al., "Converging Traditions of Research on Media and Information Literacies: Disciplinary, Critical, and Methodological Issues," in *Handbook of Research on New Literacies*, ed. Julie Coiro et al. (Routledge, 2008), 106.

20. Bojana Boh Podgornik et al., "Development, Testing, and Validation of an Information Literacy Test (ILT) for Higher Education," *Journal of the Association for Information Science and Technology* 67, no. 10 (2016): 2420–36, https://doi.org/10.1002/asi.23586.

21. S. Mo Jones-Jang et al., "Does Media Literacy Help Identification of Fake News? Information Literacy Helps, but Other Literacies Don't," *American Behavioral Scientist* 65, no. 2 (2021): 371–88, https://doi.org/10.1177/0002764219869406.

22. Ann Marie Deer Owens, "Former Vice President Al Gore Kicks Off Vanderbilt Project on Unity and American Democracy, Followed by Case Study on PEPFAR with 66th Secretary of State Condoleezza Rice," Vanderbilt University, January 15, 2021, https://news.vanderbilt.edu/2021/01/15 /former-vice-president-al-gore-kicks-off-vanderbilt-project-on-unity-and-american-democracy -followed-by-case-study-on-pepfar-with-66th-secretary-of-state-condoleezza-rice/.

23. Caitlin Drummond and Baruch Fischhoff, "Individuals with Greater Science Literacy and Education Have More Polarized Beliefs on Controversial Science Topics," *Proceedings of the National Academy of Sciences* 114, no. 36 (2017): 9587–92, https://doi.org/10.1073/pnas.1704882114.

24. Stephanie Preston et al., "Detecting Fake News on Facebook: The Role of Emotional Intelligence," *PLOS One* 16, no. 3 (2021): e0246757, https://doi.org/10.1371/journal.pone.0246757; Neophytos Georgiou et al., "COVID-19-Related Conspiracy Beliefs and Their Relationship with Perceived Stress and Pre-existing Conspiracy Beliefs," *Personality and Individual Differences* 166 (2020): art. 110201, https://doi.org/10.1016/j.paid.2020.110201.

25. Asheley R. Landrum and Alex Olshansky, "The Role of Conspiracy Mentality in Denial of Science and Susceptibility to Viral Deception About Science," *Politics and the Life Sciences* 38, no. 2 (2019): 193–209, https://doi.org/10.1017/pls.2019.9.

26. Jan-Willem van Prooijen, "Why Education Predicts Decreased Belief in Conspiracy Theories," *Applied Cognitive Psychology* 31, no. 1 (2017): 50–58, https://doi.org/10.1002/acp.3301.

27. Stefan Stieger and Ulf-Dietrich Reips, "A Limitation of the Cognitive Reflection Test: Familiarity," *PeerJ* 4 (2016): e2395, https://doi.org/10.7717/peerj.2395.

28. Shane Frederick, "Cognitive Reflection and Decision Making," *Journal of Economic Perspectives* 19, no. 4 (2005): 25–42, https://doi.org/10.1257/089533005775196732.

29. Gordon Pennycook et al., "Shifting Attention to Accuracy Can Reduce Misinformation Online," *Nature* 592, no. 7855 (2021): 590–95, https://doi.org/10.1038/s41586-021-03344-2; Gordon Pennycook and David G. Rand, "Who Falls for Fake News? The Roles of Bullshit Receptivity,

Overclaiming, Familiarity, and Analytic Thinking," *Journal of Personality* 88, no. 2 (2020): 185–200, https://doi.org/10.1111/jopy.12476.

30. Frederick, "Cognitive Reflection and Decision Making," 35.

31. Robert Duron et al., "Critical Thinking Framework for Any Discipline," *International Journal of Teaching and Learning in Higher Education* 17, no. 2 (2006): 160.

32. Anthony Lantian et al., "Maybe a Free Thinker but Not a Critical One: High Conspiracy Belief Is Associated with Low Critical Thinking Ability," *Applied Cognitive Psychology* 35, no. 3 (2021): 674–84, https://doi.org/10.1002/acp.3790.

33. Ben Motz et al., "A Scalable, Versatile Approach for Improving Critical Thinking Skills," Reboot Foundation, last modified 2021, https://reboot-foundation.org/wp-content/uploads/_docs/Improving_Critical_Thinking_Skills.pdf.

34. Lauren Lutzke et al., "Priming Critical Thinking: Simple Interventions Limit the Influence of Fake News About Climate Change on Facebook," *Global Environmental Change: Human and Policy Dimensions* 58 (2019): art. 101964, https://doi.org/10.1016/j.gloenvcha.2019.101964.

35. D. Alan Bensley et al., "Skepticism, Cynicism, and Cognitive Style Predictors of the Generality of Unsubstantiated Belief," *Applied Cognitive Psychology* 36, no. 1 (2022): 83–99, https://doi.org/10.1002/acp.3900.

36. Jianing Li, "Not All Skepticism Is 'Healthy' Skepticism: Theorizing Accuracy- and Identity-Motivated Skepticism Toward Social Media Misinformation," *New Media and Society*, June 26, 2023, https://doi.org/10.1177/14614448231179941.

37. Tim Moore, "Knowledge, Disciplinarity and the Teaching of Critical Thinking," in *The Routledge International Handbook of Research on Teaching Thinking*, ed. Rupert Wegerif et al. (Routledge, 2015), 243–53; Emily R. Lai, "Critical Thinking: A Literature Review," *Pearson's Research Reports* 6, no. 1 (2011): 40–41, https://citeseerx.ist.psu.edu/document?repid=rep1&type=pdf&doi=b42cffa5a2ad63a31fcf99869e7c-b8ef72b44374.

38. Corine S. Meppelink et al., "'I Was Right About Vaccination': Confirmation Bias and Health Literacy in Online Health Information Seeking," *Journal of Health Communication* 24, no. 2 (2019): 129–40, https://doi.org/10.1080/10810730.2019.1583701.

39. Tenelle Porter et al., "Predictors and Consequences of Intellectual Humility," *Nature Reviews Psychology* 1, no. 9 (2022): 525, https://doi.org/10.1038/s44159-022-00081-9.

40. Elizabeth J. Krumrei-Mancuso et al., "Links Between Intellectual Humility and Acquiring Knowledge," *Journal of Positive Psychology* 15, no. 2 (2020): 155–70, https://doi.org/10.1080/17439760.2019.1579359.

41. Joshua N. Hook et al., "Intellectual Humility and Religious Tolerance," *Journal of Positive Psychology* 12, no. 1 (2017): 29–35, https://doi.org/10.1080/17439760.2016.1167937; Shauna M. Bowes et al., "Intellectual Humility and Between-Party Animus: Implications for Affective Polarization in Two Community Samples," *Journal of Research in Personality* 88 (2020): art. 103992, https://doi.org/10.1016/j.jrp.2020.103992; Matthew L. Stanley et al., "Intellectual Humility and Perceptions of Political Opponents," *Journal of Personality* 88, no. 6 (December 2020): 1196–1216, https://doi.org/10.1111/jopy.12566.

42. Amy R. Senger and Ho P. Huynh, "Intellectual Humility's Association with Vaccine Attitudes and Intentions," *Psychology, Health and Medicine* 26, no. 9 (2021): 1053–62, https://doi.org/10.1080/13548506.2020.1778753; Jonah Koetke et al., "Intellectual Humility Predicts Scrutiny of COVID-19 Misinformation," *Social Psychological and Personality Science* 13, no. 1 (2022): 277–84, https://doi.org/10.1177/1948550620988242.

43. Ho P. Huynh and Amy R. Senger, "A Little Shot of Humility: Intellectual Humility Predicts Vaccination Attitudes and Intention to Vaccinate Against COVID-19," *Journal of Applied Social Psychology* 51, no. 4 (2021): 449–60, https://doi.org/10.1111/jasp.12747.

44. David J. Anderson et al., "The Development of Intellectual Humility as an Impact of a Week-Long Philosophy Summer Camp for Teens and Tweens: Preliminary Results," *Precollege Philosophy and Public Practice* 3 (2021): 41–65, https://doi.org/10.5840/p4202151418.

45. Benjamin R. Meagher et al., "An Intellectually Humbling Experience: Changes in Interpersonal Perception and Cultural Reasoning Across a Five-Week Course," *Journal of Psychology and Theology* 47, no. 3 (2019): 217–29, https://doi.org/10.1177/0091647119837010.

46. Ethan Kross and Igor Grossmann, "Boosting Wisdom: Distance from the Self Enhances Wise Reasoning, Attitudes, and Behavior," *Journal of Experimental Psychology: General* 141, no. 1 (2012): 43–48, https://doi.org/10.1037/a0024158; Tenelle Porter et al., "Intellectual Humility Predicts Mastery Behaviors When Learning," *Learning and Individual Differences* 80 (2020): art. 101888, https://doi.org/10.1016/j.lindif.2020.101888.

47. Tenelle Porter et al., "Predictors and Consequences of Intellectual Humility," *Nature Reviews Psychology* 1, no. 9 (2022): 524–36, https://doi.org/10.1038/s44159-022-00081-9.

48. Anastasia Kozyreva et al., "Critical Ignoring as a Core Competence for Digital Citizens," *Current Directions in Psychological Science* 32, no. 1 (2023): 82, https://doi.org/10.1177/09637214221121570.

49. Hunt Allcott et al., "The Welfare Effects of Social Media," *American Economic Review* 110, no. 3 (2020): 629–76, https://doi.org/10.1257/aer.20190658.

50. Sam Wineburg and Sarah McGrew, "Lateral Reading and the Nature of Expertise: Reading Less and Learning More When Evaluating Digital Information," *Teachers College Record* 121, no. 11 (2019): 1–40, https://doi.org/10.1177/016146811912101102.

51. Folco Panizza et al., "Lateral Reading and Monetary Incentives to Spot Disinformation About Science," *Scientific Reports* 12, no. 1 (2022): 5678, https://doi.org/10.1038/s41598-022-09168-y; Sam Wineburg et al., "Lateral Reading on the Open Internet: A District-Wide Field Study in High School Government Classes," *Journal of Educational Psychology* 114, no. 5 (2022): 893–909, https://doi.org/10.1037/edu0000740.

52. Mike Caulfield, "Introducing SIFT, a Four Moves Acronym," Hapgood, May 12, 2019, https://hapgood.us/2019/05/12/sift-and-a-check-please-preview/; Mike Caulfield, "Information Literacy for Mortals," Project Information Literacy, December 14, 2021, https://projectinfolit.org/pubs/provocation-series/essays/information-literacy-for-mortals.html.

53. Jevin D. West and Carl T. Bergstrom, "Misinformation in and About Science," *Proceedings of the National Academy of Sciences* 118, no. 15 (2021): e1912444117, https://doi.org/10.1073/pnas.1912444117; Carl T. Bergstrom et al., "To Fight Misinformation, We Need to Teach That Science Is Dynamic," *Scientific American*, October 26, 2022, https://www.scientificamerican.com/article/to-fight-misinformation-we-need-to-teach-that-science-is-dynamic/.

54. "Is It Legit? Five Steps for Vetting a News Source," News Literacy Project, last modified November 5, 2021, https://newslit.org/educators/resources/is-it-legit/.

55. David Yarrow, "From Fact-Checking to Value-Checking: Normative Reasoning in the New Public Sphere," *Political Quarterly* 92, no. 4 (2021): 621–28, https://doi.org/10.1111/1467-923X.12999.

56. "The Commitments of the Code of Principles," International Fact Checking Network, last modified 2024, https://www.ifcncodeofprinciples.poynter.org/know-more/the-commitments-of-the-code-of-principles.

57. Sources of free information and media literacy resources include

    News Literacy Project, https://newslit.org, and its e-learning platform Checkology, https://get.checkology.org.

    Reuters, https://reuters.com/manipulatedmedia.

    Center for News Literacy, Stony Brook University School of Journalism, https://centerfornewsliteracy.org/.

Karla Lassonde and Melissa Birkett, eds., *Psychological Myths, Mistruths and Misconceptions: Curriculum-Based Strategies for Knowledge Change* (Society for the Teaching of Psychology, 2021), http://teachpsych.org/ebooks/mythsmistruths.

"Determine Credibility (Evaluating)," Illinois State University, last updated July 15, 2024, https://guides.library.illinoisstate.edu/evaluating/craap.

Civic Online Reasoning, accessed August 14, 2024, https://cor.stanford.edu/.

"Do Your Own Research? A Reference Librarian's Recommendations," Denver Public Library, November 2, 2021, https://www.denverlibrary.org/blog/research/ross/do-your-own-research-reference -librarians-recommendations.

Jonathan Jarry, "Doing Your Own Research a Little Bit Better," McGill University, April 14, 2022, https://www.mcgill.ca/oss/article/critical-thinking/doing-your-own-research-little-bit-better.

58. "How AllSides Rates Media Bias," AllSides, August 10, 2016, https://www.allsides.com/media-bias /media-bias-rating-methods.

59. "How Ad Fontes Ranks News Sources," Ad Fontes Media, accessed July 9, 2022, https://adfontes-media.com/how-ad-fontes-ranks-news-sources/. See also "Interactive Media Bias Chart," Ad Fontes Media, accessed July 9, 2022, https://adfontesmedia.com/interactive-media-bias-chart/.

60. Maruf, "These Four Words Are Helping Spread Vaccine Misinformation."

61. Melanie Trecek-King, "Is What You Believe True? Use These 6 Questions to Find Out," Thinking Is Power, December 17, 2020, https://thinkingispower.com/the-power-of-questioning-our -beliefs/.

62. Melanie Trecek-King, "Teach Skills, Not Facts," Thinking Is Power, accessed February 23, 2021, https:// thinkingispower.com/from-non-majors-biology-to-critical-thinking-an-educators-journey/.

63. William J. McGuire, "Some Contemporary Approaches," *Advances in Experimental Social Psychology* 1 (1964): 191–229, https://doi.org/10.1016/S0065-2601(08)60052-0.

64. Josh Compton et al., "Inoculation Theory in the Post-Truth Era: Extant Findings and New Frontiers for Contested Science, Misinformation, and Conspiracy Theories," *Social and Personality Psychology Compass* 15, no. 6 (2021): e12602, https://doi.org/10.1111/spc3.12602.

65. W. J. McGuire and D. Papageorgis, "The Relative Efficacy of Various Types of Prior Belief-Defense in Producing Immunity Against Persuasion," *Journal of Abnormal and Social Psychology* 62, no. 2 (1961): 327–37, https://doi.org/10.1037/h0042026.

66. Sander Van der Linden et al., "Inoculating the Public Against Misinformation About Climate Change," *Global Challenges* 1, no. 2 (2017): 1600008, https://doi.org/10.1002/gch2.201600008.

67. Sunny Jung Kim et al., "Countering Antivax Misinformation via Social Media: Message-Testing Randomized Experiment for Human Papillomavirus Vaccination Uptake," *Journal of Medical Internet Research* 24, no. 11 (2022): e37559, https://doi.org/10.2196/37559; Robin L. Nabi, "'Feeling' Resistance: Exploring the Role of Emotionally Evocative Visuals in Inducing Inoculation," *Media Psychology* 5, no. 2 (2003): 199–223, https://doi.org/10.1207/S1532785XMEP0502_4.

68. Michelle L. M. Wood, "Rethinking the Inoculation Analogy: Effects on Subjects with Differing Preexisting Attitudes," *Human Communication Research* 33, no. 3 (2007): 357–78, https://doi.org /10.1111/j.1468-2958.2007.00303.x.

69. Daniel Jolley and Karen M. Douglas, "Prevention Is Better than Cure: Addressing Anti-Vaccine Conspiracy Theories," *Journal of Applied Social Psychology* 47, no. 8 (2017): 459–69, https://doi .org/10.1111/jasp.12453; Norman C. H. Wong, "'Vaccinations Are Safe and Effective': Inoculating Positive HPV Vaccine Attitudes Against Antivaccination Attack Messages," *Communication Reports* 29, no. 3 (2016): 127–38, https://doi.org/10.1080/08934215.2015.1083599.

70. John A. Banas et al., "Inoculating Against Anti-Vaccination Conspiracies," *Health Communication* 39, no. 9 (2023): 1760–68, https://doi.org/10.1080/10410236.2023.2235733.

71. Brendan Nyhan et al., "Effective Messages in Vaccine Promotion: A Randomized Trial," *Pediatrics* 133, no. 4 (2014): e835–42, https://doi.org/10.1542/peds.2013-2365.

72. John Cook, "Teaching About Our Climate Crisis: Combining Games and Critical Thinking to Fight Misinformation," *American Educator* 45, no. 4 (2022): 12, https://files.eric.ed.gov/fulltext /EJ1321545.pdf.

73. John L. McCuin et al., "Comparing the Effects of Traditional vs. Misconceptions-Based Instruction on Student Understanding of the Greenhouse Effect," *Journal of Geoscience Education* 62, no. 3 (2014): 445–59, https://doi.org/10.5408/13-068.1.

74. Rodney Schmaltz and Scott O. Lilienfeld, "Hauntings, Homeopathy, and the Hopkinsville Goblins: Using Pseudoscience to Teach Scientific Thinking," *Frontiers in Psychology* 5 (2014): art. 336, https:// doi.org/10.3389/fpsyg.2014.00336.

75. Jon Roozenbeek and Sander Van der Linden, "Fake News Game Confers Psychological Resistance Against Online Misinformation," *Palgrave Communications* 5 (2019): art. 65, https://doi.org /10.1057/s41599-019-0279-9.

76. Melisa Basol et al., "Good News About *Bad News*: Gamified Inoculation Boosts Confidence and Cognitive Immunity Against Fake News," *Journal of Cognition* 3, no. 1 (2020): 2, https://doi.org /10.5334%2Fjoc.91.

77. Jon Roozenbeek et al., "Technique-Based Inoculation Against Real-World Misinformation," *Royal Society Open Science* 9, no. 5 (2022): art. 211719, https://doi.org/10.1098/rsos.211719.

78. Jon Roozenbeek and Sander van der Linden, "Breaking *Harmony Square*: A Game That 'Inoculates' Against Political Misinformation," *Harvard Kennedy School Misinformation Review* 1, no. 8 (2020): 1–26, https://doi.org/10.37016/mr-2020-47.

79. Daniele Campello, "The Fake News Game Received the Brouwer Trust Prize from the Royal Holland Society of Sciences," University of Cambridge, Department of Psychology, January 16, 2020, https://www.psychol.cam.ac.uk/news/fake-news-game-received-brouwer-trust-prize-royal -holland-society-sciences.

80. Gillaume Klossa, *Towards European Media Sovereignty: An Industrial Media Strategy to Leverage Data, Algorithms, and Artificial Intelligence* (European Commission, 2019), 23, https://ec.europa.eu /commission/sites/beta-political/files/gk_report_final.pdf,

81. Jon Roozenbeek et al., "Psychological Inoculation Improves Resilience Against Misinformation on Social Media," *Science Advances* 8, no. 34 (2022): eabo6254, https://doi.org/10.1126/sciadv.abo6254.

82. Michael V. Bronstein and Sophia Vinogradov, "Education Alone Is Insufficient to Combat Online Medical Misinformation," *EMBO Reports* 22, no. 3 (2021): e52282, https://doi.org/10.15252/embr .202052282.

83. Rakoen Maertens et al., "Psychological Booster Shots Targeting Memory Increase Long-Term Resistance Against Misinformation," preprint, PsyArXiv, April 17, 2023, https://doi.org/10.31234/osf .io/6r9as.

84. Ariana Modirrousta-Galian and Philip A. Higham, "Gamified Inoculation Interventions Do Not Improve Discrimination Between True and Fake News: Reanalyzing Existing Research with Receiver Operating Characteristic Analysis," *Journal of Experimental Psychology: General* 152, no. 9 (2023): 2411–37, https://doi.org/10.1037/xge0001395.

85. Chasu An and Michael Pfau, "The Efficacy of Inoculation in Televised Political Debates," *Journal of Communication* 54, no. 3 (2004): 421–36, https://doi.org/10.1111/j.1460-2466.2004.tb02637.x; Josh Compton, "Inoculation Against/as Character Assassination," in *Routledge Handbook of Character Assassination and Reputation Management*, ed. Sergei A. Samoilenko et al. (Routledge, 2021), 25–35.

86. Jon Roozenbeek et al., "Susceptibility to Misinformation Is Consistent Across Question Framings and Response Modes and Better Explained by Myside Bias and Partisanship than Analytical Thinking," *Judgment and Decision Making* 17, no. 3 (2022): 547–73, https://doi.org/10.1017 /S1930297500003570; David Borukhson et al., "When Does an Individual Accept Misinformation? An Extended Investigation Through Cognitive Modeling," *Computational Brain and Behavior* 5, no. 2 (2022): 244–60, https://doi.org/10.1007/s42113-022-00136-3.

87. Cecilie Steenbuch Traberg and Sander van der Linden, "Birds of a Feather Are Persuaded Together: Perceived Source Credibility Mediates the Effect of Political Bias on Misinformation Susceptibility," *Personality and Individual Differences* 185 (2022): art. 111269, https://doi.org/10.1016/j .paid.2021.111269.

88. Tatiana Celadin et al., "Displaying News Source Trustworthiness Ratings Reduces Sharing Intentions for False News Posts," *Journal of Online Trust and Safety* 1, no. 5 (2023): 1–20, https://doi .org/10.54501/jots.v1i5.100.

89. Alexander Agadjanian et al., "Counting the Pinocchios: The Effect of Summary Fact-Checking Data on Perceived Accuracy and Favorability of Politicians," *Research and Politics* 6, no. 3 (2019): 2053168019870351, https://doi.org/10.1177/2053168019870351.

90. Brendan Nyhan and Jason Reifler, "The Effect of Fact-Checking on Elites: A Field Experiment on U.S. State Legislators," *American Journal of Political Science* 59, no. 3 (2015): 628–40, https://doi .org/10.1111/ajps.12162.

91. Nathaniel Sirlin et al., "Digital Literacy Is Associated with More Discerning Accuracy Judgments but Not Sharing Intentions," *Harvard Kennedy School Misinformation Review* 2, no. 6 (2021): 1–13, https://doi.org/10.37016/mr-2020-83.

92. Sumitra Badrinathan and Simon Chauchard, "Researching and Countering Misinformation in the Global South," *Current Opinion in Psychology* 55 (2024): art. 101733, https://doi.org/10.1016/j .copsyc.2023.101733.

93. Joseph Henrich et al., "Most People Are Not WEIRD," *Nature* 466, no. 7302 (2010): 29, https://doi .org/10.1038/466029a.

94. Andrew M. Guess and Kevin Munger, "Digital Literacy and Online Political Behavior," *Political Science Research and Methods* 11, no. 1 (2023): 110–28, https://doi.org/10.1017/psrm.2022.17.

95. "Internet and Social Media Users in the World 2024," Statista, accessed July 6, 2024, https://www .statista.com/statistics/617136/digital-population-worldwide/.

96. Brooke Auxier, "Social Media Use in 2021," Pew Research Center, April 7, 2021, https://www .pewresearch.org/internet/2021/04/07/social-media-use-in-2021/.

97. "Facebook Users by Country 2024," World Population Review, accessed July 6, 2024, https:// worldpopulationreview.com/country-rankings/facebook-users-by-country.

98. "WhatsApp Market Share Among Messaging App Users Worldwide 2022, by Country," Statista, accessed July 6, 2024, https://www.statista.com/statistics/1311229/whatsapp-usage-messaging-app -users-by-country/.

99. Thomas Nygren et al., "Teachers' Views on Disinformation and Media Literacy Supported by a Tool Designed for Professional Fact-Checkers: Perspectives from France, Romania, Spain and Sweden," *SN Social Sciences* 2, no. 4 (2022): 40, https://doi.org/10.1007/s43545-022-00340-9.

100. Nuurrianti Jalli, "Lack of Internet Access in Southeast Asia Poses Challenges for Students to Study Online Amid COVID-19 Pandemic," *The Conversation*, March 17, 2020, http://theconversation .com/lack-of-internet-access-in-southeast-asia-poses-challenges-for-students-to-study-online-amid -covid-19-pandemic-133787.

101. Angela Fan, "Introducing the First AI Model That Translates 100 Languages Without Relying on English," Meta, October 19, 2020, https://about.fb.com/news/2020/10/first-multilingual-machine -translation-model/; "Live: Facebook Whistleblower Testifies at Senate Hearing," NBC News, October 5, 2021, YouTube, 3 hr., 16 min., 27 sec., https://www.youtube.com/watch?v=_IhWeVHxdXg&t =2836s&ab_channel=NBCNews.

102. "How Facebook Can Flatten the Curve of the Coronavirus Infodemic," Avaaz, April 15, 2020, https://secure.avaaz.org/campaign/en/facebook_coronavirus_misinformation/.

103. Keith Zubrow, "Facebook Whistleblower Says Company Incentivizes 'Angry, Polarizing, Divisive Content,'" *60 Minutes Overtime*, CBS News, October 4, 2021, https://www.cbsnews.com/news /facebook-whistleblower-frances-haugen-60-minutes-polarizing-divisive-content/; "Live: Facebook Whistleblower Testifies at Senate Hearing."

104. Jenny Gross, "How Finland Is Teaching a Generation to Spot Misinformation," *New York Times*, January 10, 2023, https://www.nytimes.com/2023/01/10/world/europe/finland-misinformation -classes.html.

105. Esa Reunanen, "Finland," Reuters Institute for the Study of Journalism, June 14, 2023, https:// reutersinstitute.politics.ox.ac.uk/digital-news-report/2023/finland; "Digital News Report 2023," Reuters Institute for the Study of Journalism, accessed July 6, 2024, https://reutersinstitute.poli tics.ox.ac.uk/digital-news-report/2023; Eliza Mackintosh, "Finland Is Winning the War on Fake News. What's It Learning?," CNN, May 2019, https://edition.cnn.com/interactive/2019/05/europe /finland-fake-news-intl/.

106. Zubrow, "Facebook Whistleblower Says Company Incentivizes 'Angry, Polarizing, Divisive Content.'"

107. "Project Look Sharp," Ithaca College, accessed July 6, 2024, https://www.projectlooksharp.org/.

108. "Fact Checks," Africa Check, accessed August 14, 2024, https://africacheck.org/fact-checks.

109. Jeremy Bowles et al., "Sustaining Exposure to Fact-Checks: Misinformation Discernment, Media Consumption, and Its Political Implications," SSRN, September 25, 2023, https://doi.org/10.2139 /ssrn.4582703.

110. Jon Roozenbeek et al., "Prebunking Interventions Based on the Psychological Theory of 'Inocula- tion' Can Reduce Susceptibility to Misinformation Across Cultures," *Harvard Kennedy School Mis- information Review* 1, no. 2 (2020): 1–23, https://doi.org/10.37016//mr-2020-008.

111. Andrew M. Guess et al., "A Digital Media Literacy Intervention Increases Discernment Between Mainstream and False News in the United States and India," *Proceedings of the National Academy of Sciences* 117, no. 27 (2020): 15536–45, https://doi.org/10.1073/pnas.1920498117.

112. Statista, "WhatsApp Market Share Among Messaging App Users Worldwide 2022, by Country."

113. "Advancing Media Literacy in Indonesia (Part II)," Moonshot, November 29, 2022, https:// moonshotteam.com/resource/advancing-media-literacy-in-indonesia-part-ii/.

114. "We Won Bronze in Webby's Anthem Awards!," Brave Factor, February 20, 2023, https://bravefactor .com/we-won-bronze-in-webbys-anthem-awards/.

115. Trenton W. Ford et al., "Online Media Literacy Intervention in Indonesia Reduces Misinforma- tion Sharing Intention," *Journal of Media Literacy Education* 15, no. 2 (2023): 99–123, https://doi .org/10.23860/JMLE-2023-15-2-8.

116. Matthew Facciani et al., "Playing *Gali Fakta* Inoculates Indonesian Participants Against False Information," *Harvard Kennedy School Misinformation Review* 5, no. 4 (2024): 1–17, https://doi .org/10.37016/mr-2020-152.

117. Olivia Riggio, "Not All Media Literacy Programs Are Created Equal—and Most Have Yet to Be Created," FAIR, December 15, 2020, https://fair.org/home/not-all-media-literacy-programs -are-created-equal-and-most-have-yet-to-be-created/.

118. Nolan Higdon, *The Anatomy of Fake News: A Critical News Literacy Education* (University of Cali- fornia Press, 2020).

119. Joseph B. Bak-Coleman et al., "Stewardship of Global Collective Behavior," *Proceedings of the National Academy of Sciences* 118, no. 27 (2021): e2025764118, https://doi.org/10.1073/pnas.2025764118.

120. Carl Bergstrom (@CT_Bergstrom), "I want to correct a misperception," Twitter (now X), June 22, 2021, https://twitter.com/CT_Bergstrom/status/1407218435507986476.

121. Taylor Lorenz, "Extremist Influencers Are Generating Millions for Twitter, Report Says," *Wash- ington Post*, February 9, 2023, https://www.washingtonpost.com/technology/2023/02/09/twitter -ads-revenue-suspended-account/.

122. Victor Pickard, *Democracy Without Journalism? Confronting the Misinformation Society* (Oxford University Press, 2019).

123. Jon Roozenbeek et al., "Countering Misinformation: Evidence, Knowledge Gaps, and Implica- tions of Current Interventions," *European Psychologist* 28, no. 3 (2023): 189–205, https://doi.org /10.1027/1016-9040/a000492.

124. Jon Roozenbeek, email message to author, October 27, 2022.

125. Nick Chater and George Loewenstein, "The i-Frame and the s-Frame: How Focusing on Individ-ual-Level Solutions Has Led Behavioral Public Policy Astray," *Behavioral and Brain Sciences* 46 (2023): e147, https://doi.org/10.1017/S0140525X22002023.

126. Michael E Mann, *The New Climate War: The Fight to Take Back Our Planet* (PublicAffairs, 2021).

127. Jason Breslow, "20 Companies Are Behind Half of the World's Single-Use Plastic Waste, Study Finds," NPR, May 18, 2021, https://www.npr.org/2021/05/18/997937090/half-of-the-worlds-single-use-plastic-waste-is-from-just-20-companies-says-a-stu.

128. Finis Dunaway, "The 'Crying Indian' Ad That Fooled the Environmental Movement," *Chicago Tribune*, November 21, 2017, http://www.chicagotribune.com/2017/11/21/the-crying-indian-ad-that-fooled-the-environmental-movement/.

129. Tik Root, "Inside the Long War to Protect Plastic," Center for Public Integrity, May 16, 2019, https://publicintegrity.org/environment/pollution/pushing-plastic/inside-the-long-war-to-protect-plastic/.

130. Matt Wilkins, "More Recycling Won't Solve Plastic Pollution," *Scientific American*, July 6, 2018, https://blogs.scientificamerican.com/observations/more-recycling-wont-solve-plastic-pollution/.

131. iNudgeyou, "Green Nudge: Nudging Litter Into the Bin," February 16, 2012, https://inudgeyou.com/en/green-nudge-nudging-litter-into-the-bin/.

132. Ben Adler, "Banning Plastic Bags Is Great for the World, Right? Not So Fast," *Wired*, June 10, 2016, https://www.wired.com/2016/06/banning-plastic-bags-great-world-right-not-fast/.

133. Margaret Walls, "Extended Producer Responsibility and Product Design: Economic Theory and Selected Case Studies," Discussion Paper No. 06–08 (Resources for the Future, March 1, 2006), https://doi.org/10.2139/ssrn.901661.

134. Gregg Sparkman et al., "Americans Experience a False Social Reality by Underestimating Popular Climate Policy Support by Nearly Half," *Nature Communications* 13, no. 1 (2022): art. 4779, https://doi.org/10.1038/s41467-022-32412-y.

135. Lina Koppel et al., "Individual-Level Solutions May Support System-Level Change â If They Are Internalized as Part of One's Social Identity," *Behavioral and Brain Sciences* 46 (2023): e165, https://doi.org/10.1017/s0140525x2300105x.

136. Tim Weninger, email message to author, October 10, 2022.

137. Marcella Tambuscio et al., "Fact-Checking Effect on Viral Hoaxes: A Model of Misinformation Spread in Social Networks." In *WWW '15 Companion: Proceedings of the 24th International Con-ference on World Wide Web* (Association for Computing Machinery 2015), 977–82, https://doi.org/10.1145/2740908.2742572.

138. Melisa Basol et al., "Towards Psychological Herd Immunity: Cross-Cultural Evidence for Two Pre-bunking Interventions Against COVID-19 Misinformation," *Big Data and Society* 8, no. 1 (2021): 205395172110138,ttps://doi.org/10.1177/20539517211013868.

139. Bobi Ivanov et al., "The General Content of Postinoculation Talk: Recalled Issue-Specific Con-versations Following Inoculation Treatments," *Western Journal of Communication* 79, no. 2 (2015): 218–38, https://doi.org/10.1080/10570314.2014.943423.

140. Bobi Ivanov et al., "Effects of Postinoculation Talk on Resistance to Influence," *Journal of Commu-nication* 62, no. 4 (2012): 701–18, https://doi.org/10.1111/j.1460-2466.2012.01658.x.

141. Lindsay L. Dillingham and Bobi Ivanov, "Using Postinoculation Talk to Strengthen Generated Resistance," *Communication Research Reports* 33, no. 4 (2016): 295–302, https://doi.org/10.1080/08824096.2016.1224161.

142. John Bogle, *The Little Book of Common Sense Investing: The Only Way to Guarantee Your Fair Share of Stock Market Returns* (Wiley, 2007).

143. Compton et al., "Inoculation Theory in the Post-Truth Era," 10.

144. Preliminary results of an ongoing study by Matthew Facciani, Qian Huang, and Tim Weninger.

145. Irene V. Pasquetto et al., "Social Debunking of Misinformation on WhatsApp: The Case for Strong and In-Group Ties," *Proceedings of the ACM on Human-Computer Interaction* 6, no. CSCW1 (2022): art. 117, https://doi.org/10.1145/3512964.

146. Adam Grant (@AdamMGrant), "Great minds don't think alike," Twitter (now X), November 18, 2021, https://twitter.com/adammgrant/status/1461337368611463175?lang=en.

147. Bas Verplanken and Jie Sui, "Habit and Identity: Behavioral, Cognitive, Affective, and Motivational Facets of an Integrated Self," *Frontiers in Psychology* 10 (2019): art. 1504, https://doi.org/10.3389/fpsyg.2019.01504.

7. THE FUTURE OF MISINFORMATION

1. Leticia Bode and Emily Vraga, "The Swiss Cheese Model for Mitigating Online Misinformation," *Bulletin of the Atomic Scientists* 77, no. 3 (2021): 129–33, https://doi.org/10.1080/00963402.2021.1912170.

2. Jamelle Bouie, "Opinion: Glenn Youngkin Knows Exactly What He's Doing," *New York Times*, September 20, 2022, https://www.nytimes.com/2022/09/20/opinion/glenn-youngkin-midterms-trump.html.

3. Ron Johnston, "Manipulating Maps and Winning Elections: Measuring the Impact of Malapportionment and Gerrymandering," *Political Geography* 21, no. 1 (2002): 1–31, https://doi.org/10.1016/S0962-6298(01)00070-1; Steven Levitsky and Daniel Ziblatt, *Tyranny of the Minority: Why American Democracy Reached the Breaking Point* (Crown, 2023).

4. Daniel Kreiss and Shannon C. McGregor, "A Review and Provocation: On Polarization and Platforms," *New Media and Society* 26, no. 1 (2024): 556–79, https://doi.org/10.1177/14614448231161880.

5. Sam Levine, "Partisan Gerrymandering Has Empowered a Hard-Right Turn in Texas," *The Guardian*, September 5, 2021, https://www.theguardian.com/us-news/2021/sep/05/gerrymandering-empowered-hard-right-texas.

6. Cristian Vaccari and Andrew Chadwick, "Deepfakes and Disinformation: Exploring the Impact of Synthetic Political Video on Deception, Uncertainty, and Trust in News," *Social Media + Society* 6, no. 1 (2020), https://doi.org/10.1177/2056305120903408.

7. "AI-Generated or Real?," Detect Fakes, Northwestern Kellogg, accessed August 14, 2024, https://detectfakes.media.mit.edu/.

8. John Ternovski at al., "Deepfake Warnings for Political Videos Increase Disbelief but Do Not Improve Discernment: Evidence from Two Experiments," OSF Preprints, January 14, 2021, https://doi.org/10.31219/osf.io/dta97; Vaccari and Chadwick, "Deepfakes and Disinformation."

9. Robert Chesney and Danielle Keats Citron, "Deep Fakes: A Looming Challenge for Privacy, Democracy, and National Security," *California Law Review* 107 (2019): 1753, https://doi.org/10.2139/ssrn.3213954.

10. Shannon Bond, "AI-Generated Fake Faces Have Become a Hallmark of Online Influence Operations," NPR, December 15, 2022, https://www.npr.org/2022/12/15/1143114122/ai-generated-fake-faces-have-become-a-hallmark-of-online-influence-operations.

11. Jeffrey Lees et al., "The *Spot the Troll* Quiz Game Increases Accuracy in Discerning Between Real and Inauthentic Social Media Accounts," *PNAS Nexus* 2, no. 4 (2023): pgad094, https://doi.org/10.1093/pnasnexus/pgad094.

12. Adam Satariano and Paul Mozur, "The People Onscreen Are Fake: The Disinformation Is Real," *New York Times*, February 7, 2023, https://www.nytimes.com/2023/02/07/technology/artificial-intelligence-training-deepfake.html.

13. Kate Conger, "Hackers' Fake Claims of Ukrainian Surrender Aren't Fooling Anyone: So What's Their Goal?," *New York Times*, April 5, 2022, https://www.nytimes.com/2022/04/05/us/politics/ukraine-russia-hackers.html.

14. Graphika Team, "Deepfake It till You Make It," Graphika, February 7, 2023, https://graphika.com/reports/deepfake-it-till-you-make-it.

15. Preliminary results of an ongoing study by Adnan Hoq, Matthew Facciani, and Tim Weninger: "Empowering Perception: Assessing The Impact of Feedback and Educational Intervention On Image Manipulation Detection," https://osf.io/b9hze.

16. Emma Needleman, "Would Chat GPT Get a Wharton MBA? New White Paper by Christian Terwiesch," Mack Institute for Innovation Management, January 17, 2023, https://mackinstitute.wharton.upenn.edu/2023/would-chat-gpt3-get-a-wharton-mba-new-white-paper-by-christian-terwiesch/; Chris Stokel-Walker, "ChatGPT Listed as Author on Research Papers: Many Scientists Disapprove," *Nature* 613, no. 7945 (2023): 620–21, https://doi.org/10.1038/d41586-023-00107-z.

17. Harry McCracken, "If ChatGPT Doesn't Get a Better Grasp of Facts, Nothing Else Matters," *Fast Company*, January 11, 2023, https://www.fastcompany.com/90833017/openai-chatgpt-accuracy-gpt-4.

18. Melissa Heikkilä, "Why You Shouldn't Trust AI Search Engines," *MIT Technology Review*, February 14, 2023, https://www.technologyreview.com/2023/02/14/1068498/why-you-shouldnt-trust-ai-search-engines/; Cade Metz, "The New Chatbots Could Change the World: Can You Trust Them?," *New York Times*, December 10, 2022, https://www.nytimes.com/2022/12/10/technology/ai-chat-bot-chatgpt.html.

19. Nikhil Sharma et al., "Generative Echo Chamber? Effects of LLM-Powered Search Systems on Diverse Information Seeking," preprint, arXiv:2402.05880v2 [cs.CL], February 8, 2024, https://doi.org/10.48550/arXiv.2402.05880.

20. Michael Gerlich, "AI Tools in Society: Impacts on Cognitive Offloading and the Future of Critical Thinking," *Societies* (Basel, Switzerland) 15, no. 1 (2025): 6.

21. Alex Mahadevan, "This Newspaper Doesn't Exist: How ChatGPT Can Launch Fake News Sites in Minutes," Poynter Institute, February 3, 2023, https://www.poynter.org/fact-checking/2023/chatgpt-build-fake-news-organization-website/.

22. Bai Hui et al., "Artificial Intelligence Can Persuade Humans on Political Issues," OSF Preprints, February 5, 2023, https://doi.org/10.31219/osf.io/stakv.

23. Maurice Jakesch et al., "Human Heuristics for AI-Generated Language Are Flawed," *Proceedings of the National Academy of Sciences* 120, no. 11 (2023): e2208839120, https://doi.org/10.1073/pnas.2208839120.

24. Sayash Kapoor and Arvind Narayanan, "A Misleading Open Letter About Sci-Fi AI Dangers Ignores the Real Risks," AI Snake Oil, March 29, 2023, https://aisnakeoil.substack.com/p/a-misleading-open-letter-about-sci.

25. Yifan Ding, Matthew Facciani, Amrit Poudel, Ellen Joyce, Salvador Aguinaga, Balaji Veeramani, et al., "Citations and Trust in LLM Generated Responses," *arXiv [Cs.CL]*, January 2, 2025, http://arxiv.org/abs/2501.01303.

26. Felix M. Simon et al., "Misinformation Reloaded? Fears About the Impact of Generative AI on Misinformation Are Overblown," *Harvard Kennedy School Misinformation Review* 4, no. 5 (2023): 1–11, https://doi.org/10.37016/mr-2020-127.

27. David Gilbert, "The Israel-Hamas War Is Drowning X in Disinformation," *Wired*, October 9, 2023, https://www.wired.com/story/x-israel-hamas-war-disinformation/.

28. Lauren Leffer, "What Does Artificial General Intelligence Actually Mean?," *Scientific American*, June 25, 2024, https://www.scientificamerican.com/article/what-does-artificial-general-intelligence-actually-mean/.

29. Lisa P. Argyle et al., "Leveraging AI for Democratic Discourse: Chat Interventions Can Improve Online Political Conversations at Scale," *Proceedings of the National Academy of Sciences* 120, no. 41 (2023): e2311627120, https://doi.org/10.1073/pnas.2311627120.

30. Thomas H. Costello et al., "Durably Reducing Conspiracy Beliefs Through Dialogues with AI," preprint, PsyArXiv, April 3, 2024, https://doi.org/10.31234/osf.io/xcwdn.

31. Emma Hoes et al., "Leveraging ChatGPT for Efficient Fact-Checking," preprint, PsyArXiv, May 29, 2023, https://doi.org/10.31234/osf.io/qnjkf.

32. Richard Nieva, "Google Wants to Fix Its Search Engine's Misinformation Problem," *Forbes*, August 11, 2022, https://www.forbes.com/sites/richardnieva/2022/08/11/google-wants-to-fix-its-search-engines-misinformation-problem/; Youness Madani et al., "Using Artificial Intelligence Techniques for Detecting Covid-19 Epidemic Fake News in Moroccan Tweets," *Results in Physics* 25 (2021): art. 104266, https://doi.org/10.1016/j.rinp.2021.104266.

33. Michael Yankoski et al., "An AI Early Warning System to Monitor Online Disinformation, Stop Violence, and Protect Elections," *Bulletin of the Atomic Scientists* 76, no. 2 (2020): 85–90, https://doi.org/10.1080/00963402.2020.1728976.

34. Dave Bayer et al., "Improving the Efficiency and Reliability of Digital Time-Stamping," in *Sequences II: Methods in Communication, Security, and Computer Science*, ed. Renato Capocelli et al. (Springer, 1993), 329–34, https://doi.org/10.1007/978-1-4613-9323-8_24.

35. "What Is Blockchain?," IBM, accessed August 14, 2024, https://www.ibm.com/topics/blockchain.

36. Matt A. V. Chaban, "Can Blockchain Block Fake News and Deep Fakes?," IBM, November 30, 2020, https://www.ibm.com/blog/blockchain-protection-fake-news-deep-fakes-safe-press/.

37. Ed Caesar, "How a Young Couple Failed to Launder Billions of Dollars in Stolen Bitcoin," *New Yorker*, February 14, 2022, https://www.newyorker.com/business/currency/how-a-young-couple-failed-to-launder-billions-of-dollars-in-stolen-bitcoin.

38. Clara Ferreira Marques, "Can Brazil Find an Answer for Fake News?," *Bloomberg*, April 24, 2022, https://www.bloomberg.com/opinion/articles/2022-04-24/can-brazil-find-an-answer-for-fake-news-in-time-for-this-fall-s-election.

39. Nathanael Fast, "AI Is Taking Over Our Workplaces: Here's How It Could Impact Human Psychology—and Vice Versa," *Fast Company*, February 8, 2023, https://www.fastcompany.com/90846471/ai-is-taking-over-our-workplaces-heres-how-it-could-impact-human-psychology.

40. Caroline Friedman Levy and Matthew Facciani, "Human-Shaped Hole in AI Oversight," Vanderbilt Project on Unity & American Democracy, Vanderbilt University, October 4, 2022, https://www.vanderbilt.edu/unity/2022/10/04/human-shaped-hole-in-ai-oversight/.

REFERENCES

Abaluck, Jason, Laura H. Kwong, Ashley Styczynski et al. "Impact of Community Masking on COVID-19: A Cluster-Randomized Trial in Bangladesh." *Science* 375, no. 6577 (2022): eabi9069. https://doi.org /10.1126/science.abi9069.

Abrego, Javier. "How Many Tweets About Covid-19 and Coronavirus? 508 MM Tweets So Far." *Tweet Binder* (blog), April 14, 2020. https://www.tweetbinder.com/blog/covid-19-coronavirus-twitter/.

Acerbi, Alberto. *Cultural Evolution in the Digital Age.* Oxford University Press, 2019.

Achen, Christopher, and Larry Bartels. *Democracy for Realists: Why Elections Do Not Produce Responsive Government.* Princeton University Press, 2017.

Ackerman, Gary, Brandon Behlendorf, Seth Baum, Hayley Peterson, Anna Wetzel, and John Halstead. *The Origin and Implications of the COVID-19 Pandemic: An Expert Survey.* Technical Report 24–1. Global Catastrophic Risk Institute, 2024. https://gcrinstitute.org/covid-origin/.

Ad Fontes Media. "How Ad Fontes Ranks News Sources." Accessed July 9, 2022. https://adfontesmedia.com /how-ad-fontes-ranks-news-sources/.

Ad Fontes Media. "Interactive Media Bias Chart." Accessed July 9, 2022. https://adfontesmedia.com/interactive -media-bias-chart/.

Adgate, Brad. "Nielsen: How the Pandemic Changed at Home Media Consumption." *Forbes*, August 21, 2020. https://www.forbes.com/sites/bradadgate/2020/08/21/nielsen-how-the-pandemic-changed-at-home -media-consumption/.

Adler, Ben. "Banning Plastic Bags Is Great for the World, Right? Not So Fast." *Wired*, June 10, 2016. https:// www.wired.com/2016/06/banning-plastic-bags-great-world-right-not-fast/.

Advances.in Reinventing Academic Publishing. "Reinventing Academic Publishing." Accessed August 14, 2024. https://advances.in/.

Advancing New Standards in Reproductive Health. "The Turnaway Study." Accessed March 6, 2023. https://www.ansirh.org/research/ongoing/turnaway-study.

AFP. "Hoax Circulates Online That an Old Indian Textbook Lists Treatments for COVID-19." April 14, 2020. https://factcheck.afp.com/hoax-circulates-online-old-indian-textbook-lists-treatments-covid-19.

Africa Check. "Fact Checks." Accessed August 14, 2024. https://africacheck.org/fact-checks.

Agadjanian, Alexander, Nikita Bakhru, Victoria Chi et al. "Counting the Pinocchios: The Effect of Summary Fact-Checking Data on Perceived Accuracy and Favorability of Politicians." *Research and Politics* 6, no. 3 (2019): 2053168019870351. https://doi.org/10.1177/2053168019870351.

Ahler, Douglas J., and Gaurav Sood. "The Parties in Our Heads: Misperceptions About Party Composition and Their Consequences." *Journal of Politics* 80, no. 3 (2018): 964–81. https://doi.org/10.1086/697253.

Ahmed, Saifuddin, Muhammad Ehab Rasul, and Jaeho Cho. "Social Media News Use Induces COVID-19 Vaccine Hesitancy Through Skepticism Regarding Its Efficacy: A Longitudinal Study from the United States." *Frontiers in Psychology* 13 (2022): 900386. https://doi.org/10.3389/fpsyg.2022.900386.

Albertson, Bethany, and Shana Kushner Gadarian. *Anxious Politics: Democratic Citizenship in a Threatening World*. Cambridge University Press, 2015.

Allain, Rhett, Brenda Stolyar, and Eric Ravenscraft. "The Physics of the N95 Face Mask." *Wired*, January 28, 2022. https://www.wired.com/story/the-physics-of-the-n95-face-mask/.

Allcott, Hunt, Luca Braghieri, Sarah Eichmeyer, and Matthew Gentzkow. "The Welfare Effects of Social Media." *American Economic Review* 110, no. 3 (2020): 629–76. https://doi.org/10.1257/aer.20190658.

Allcott, Hunt, Matthew Gentzkow, and Chuan Yu. "Trends in the Diffusion of Misinformation on Social Media." *Research and Politics* 6, no. 2 (2019): e2053168019848554. https://doi.org/10.1177/2053168019848554.

Allen, Jeff. "Misinformation Amplification Analysis and Tracking Dashboard." Integrity Institute, October 13, 2022. https://integrityinstitute.org/blog/misinformation-amplification-tracking-dashboard.

Allen, Jennifer, Antonio A. Arechar, Gordon Pennycook, and David G. Rand. "Scaling up Fact-Checking Using the Wisdom of Crowds." *Science Advances* 7, no. 36 (2021): eabf4393.

Allen, Jennifer, Baird Howland, Markus Mobius, David Rothschild, and Duncan J. Watts. "Evaluating the Fake News Problem at the Scale of the Information Ecosystem." *Science Advances* 6, no. 14 (2020): eaay3539. https://doi.org/10.1126/sciadv.aay3539.

Allen, Kelly-Ann, DeLeon L. Gray, Roy F. Baumeister, and Mark R. Leary. "The Need to Belong: A Deep Dive Into the Origins, Implications, and Future of a Foundational Construct." *Educational Psychology Review* 34, no. 2 (2022): 1133–56. https://doi.org/10.1007/s10648-021-09633-6.

Allen, Mike. "Sean Parker Unloads on Facebook: 'God Only Knows What It's Doing to Our Children's Brains.'" *Axios*, November 9, 2017. https://www.axios.com/2017/12/15/sean-parker-unloads-on-facebook-god-only-knows-what-its-doing-to-our-childrens-brains-1513306792.

AllSides. "How AllSides Rates Media Bias." August 10, 2016. https://www.allsides.com/media-bias/media-bias-rating-methods.

Alltucker, Ken. "A Quarter of All Kindergartners in This County in Washington Aren't Immunized: Now There's a Measles Crisis." *USA Today*, February 11, 2019. https://www.usatoday.com/story/news/health/2019/02/11/measles-spread-anti-vaccination-communities-new-york-clar-county-washington/2812667002/.

Allyn, Bobby. "Researchers: Nearly Half of Accounts Tweeting About Coronavirus Are Likely Bots." NPR, May 20, 2020. https://www.npr.org/sections/coronavirus-live-updates/2020/05/20/859814085/researchers-nearly-half-of-accounts-tweeting-about-coronavirus-are-likely-bots.

Alter, Charlotte. "Nothing, Not Even Hard Facts, Can Make Anti-Vaxxers Change Their Minds." *Time*, March 4, 2014. https://healthland.time.com/2014/03/04/nothing-not-even-hard-facts-can-make-anti-vaxxers-change-their-minds/.

American Institute of CPAs. "Fake Financial News Is a Real Threat to Majority of Americans: New AICPA Survey." April 27, 2017. https://www.aicpa.org/press/pressreleases/2017/fake-financial-news-is-a-real-threat-to-majority-of-americans-new-aicpa-survey.html.

American Library Association. "Information Literacy Competency Standards for Higher Education." x. http://hdl.handle.net/11213/7668.

American Medical Association. "AMA: CDC Quarantine and Isolation Guidance Is Confusing, Counterproductive." Press release, January 5, 2022. https://www.ama-assn.org/press-center/press-releases/ama-cdc-quarantine-and-isolation-guidance-confusing-counterproductive.

American Presidency Project. "Presidential Job Approval." University of California–Santa Barbara. Accessed June 20, 2024. https://www.presidency.ucsb.edu/statistics/data/presidential-job-approval.

Amsalem, Eran, and Alon Zoizner. "Do People Learn About Politics on Social Media? A Meta-Analysis of 76 Studies." *Journal of Communication* 73, no. 1 (2022): 3–13. https://doi.org/10.1093/joc/jqac034.

An, Chasu, and Michael Pfau. "The Efficacy of Inoculation in Televised Political Debates." *Journal of Communication* 54, no. 3 (2004): 421–36. https://doi.org/10.1111/j.1460-2466.2004.tb02637.x.

Anderson, David J., Patricia N. Holte, Joseph Maffly-Kipp, Daniel Conway, Claire Elise Katz, and Rebecca J. Schlegel. "The Development of Intellectual Humility as an Impact of a Week-Long Philosophy Summer Camp for Teens and Tweens: Preliminary Results." *Precollege Philosophy and Public Practice* 3 (2021): 41–65. https://doi.org/10.5840/p4202151418.

Andersson, Hilary. "Social Media Apps Are 'Deliberately' Addictive to Users." BBC, July 3, 2018. https://www.bbc.com/news/technology-44640959.

Annenberg Public Policy Center of the University of Pennsylvania. "Survey Finds Public Perception of Scientists' Credibility Has Slipped." Phys.org, June 26, 2024. https://phys.org/news/2024-06-survey-perception-scientists-credibility.html.

Aral, Sinan. *The Hype Machine: How Social Media Disrupts Our Elections, Our Economy, and Our Health—and How We Must Adapt.* Crown Currency, 2021.

Argentino, Marc-André, and Sara Aniano. "QAnon and Beyond: Analysing QAnon Trends a Year After January 6th." GNET, January 6, 2022. https://gnet-research.org/2022/01/06/qanon-and-beyond-analysing-qanon-trends-a-year-after-january-6th/.

Arguedas, Amy Ross, Craig T. Robertson, Richard Fletcher, and Rasmus Kleis Nielsen. "Echo Chambers, Filter Bubbles, and Polarisation: A Literature Review." Reuters Institute for the Study of Journalism, January 19, 2022. https://reutersinstitute.politics.ox.ac.uk/echo-chambers-filter-bubbles-and-polarisation-literature-review.

Argyle, Lisa P., Christopher A. Bail, Ethan C. Busby et al. "Leveraging AI for Democratic Discourse: Chat Interventions Can Improve Online Political Conversations at Scale." *Proceedings of the National Academy of Sciences* 120, no. 41 (2023): e2311627120. https://doi.org/10.1073/pnas.2311627120.

Armstrong, Paul W., and C. David Naylor. "Counteracting Health Misinformation: A Role for Medical Journals?" *JAMA* 321, no. 19 (2019): 1863–64. doi:10.1001/jama.2019.5168.

Aruguete, Natalia, Ernesto Calvo, Carlos Scartascini, and Tiago Ventura. "Trustful Voters, Trustworthy Politicians: A Survey Experiment on the Influence of Social Media in Politics." Working Paper No. IDB-WP-1169. Inter-American Development Bank, 2021. https://doi.org/10.18235/0003389.

Arugu[e]te, Natalia, Ernesto Calvo, and Tiago Ventura. "Network Activated Frames: Content Sharing and Perceived Polarization in Social Media." *Journal of Communication* 73, no. 1 (2022): 14–24. https://doi.org/10.1093/joc/jqac035.

Asch, Solomon E. "Studies of Independence and Conformity: I. A Minority of One Against a Unanimous Majority." *Psychological Monographs: General and Applied* 70, no. 9 (1956): 1–70. https://doi.org/10.1037/h0093718.

Ashburn-Nardo, Leslie, Corrine I. Voils, and Margo J. Monteith. "Implicit Associations as the Seeds of Intergroup Bias: How Easily Do They Take Root?" *Journal of Personality and Social Psychology* 81, no. 5 (2001): 789–99.

Ashley, Seth, Adam Maksl, and Stephanie Craft. "Developing a News Media Literacy Scale." *Journalism and Mass Communication Educator* 68, no. 1 (2013): 7–21. https://doi.org/10.1177/1077695812469802.

Associated Press. "Dictionary.com Chooses 'Misinformation' as Word of the Year." *VOA News*, December 30, 2018. https://www.voanews.com/a/dictionary-com-chooses-misinformation-as-word-of-the-year/4674053.html.

Associated Press. "What to Wear: Feds' Mixed Messages on Masks Sow Confusion." *U.S. News & World Report*, June 27, 2020. https://www.usnews.com/news/health-news/articles/2020-06-27/what-to-wear-feds-mixed-messages-on-masks-sow-confusion.

Atske, Sara. "Republicans, Democrats Move Even Further Apart in Coronavirus Concerns." Pew Research Center, June 25, 2020. https://www.pewresearch.org/politics/2020/06/25/republicans-democrats-move-even-further-apart-in-coronavirus-concerns/.

Atske, Sara. "Social Media and News Fact Sheet." Pew Research Center, November 15, 2023. https://www
.pewresearch.org/journalism/fact-sheet/social-media-and-news-fact-sheet/.

Aufderheide, Patricia. *Media Literacy: A Report of the National Leadership Conference on Media Literacy, the
Aspen Institute Wye Center, Queenstown, Maryland, December 7–9, 1992.* Aspen Institute, 1993.

Austin, Erica Weintraub, Bruce W. Austin, Jessica Fitts Willoughby, Ofer Amram, and Shawn Domgaard.
"How Media Literacy and Science Media Literacy Predicted the Adoption of Protective Behaviors
Amidst the COVID-19 Pandemic." *Journal of Health Communication* 26, no. 4 (2021): 239–52. https://
doi.org/10.1080/10810730.2021.1899345.

Auxier, Brooke. "Social Media Use in 2021." Pew Research Center, April 7, 2021. https://www.pewresearch
.org/internet/2021/04/07/social-media-use-in-2021/.

Avaaz. "How Facebook Can Flatten the Curve of the Coronavirus Infodemic." April 15, 2020. https://secure
.avaaz.org/campaign/en/facebook_coronavirus_misinformation/.

Azarpanah, Hossein, Mohsen Farhadloo, Rustam Vahidov, and Louise Pilote. "Vaccine Hesitancy: Evidence
from an Adverse Events Following Immunization Database, and the Role of Cognitive Biases." *BMC
Public Health* 21, no. 1 (2021): art. 1686. https://doi.org/10.1186/s12889-021-11745-1.

Backman, Isabella. "Q&A: What Causes Rare Instances of Myocarditis After mRNA COVID-19 Vaccines?"
Yale School of Medicine, May 16, 2023. https://medicine.yale.edu/news-article/qanda-what-causes-rare
-instances-of-myocarditis-after-mrna-covid-19-vaccines/.

Badrinathan, Sumitra, and Simon Chauchard. "Researching and Countering Misinformation in the
Global South." *Current Opinion in Psychology* 55 (2024): art. 101733. https://doi.org/10.1016/j.copsyc
.2023.101733.

Bai, Hui, Jan G. Voelkel, Johannes Christopher Eichstaedt, and Robb Willer. "Artificial Intelligence Can
Persuade Humans on Political Issues." OSF Preprints, February 5, 2023. https://doi.org/10.31219/osf
.io/stakv.

Bail, Christopher A. *Breaking the Social Media Prism: How to Make Our Platforms Less Polarizing.* Princeton
University Press, 2022.

Bail, Christopher A., Lisa P. Argyle, Taylor W. Brown et al. "Exposure to Opposing Views on Social Media
Can Increase Political Polarization." *Proceedings of the National Academy of Sciences* 115, no. 37 (2018):
9216–21. https://doi.org/10.1073/pnas.1804840115.

Bail, Christopher A., Brian Guay, Emily Maloney et al. "Assessing the Russian Internet Research Agency's
Impact on the Political Attitudes and Behaviors of American Twitter Users in Late 2017." *Proceedings of
the National Academy of Sciences* 117, no. 1 (2020): 243–50. https://doi.org/10.1073/pnas.1906420116.

Bak-Coleman, Joseph B., Mark Alfano, Wolfram Barfuss et al. "Stewardship of Global Collective Behavior."
Proceedings of the National Academy of Sciences 118, no. 27 (2021): e2025764118. https://doi.org/10.1073
/pnas.2025764118.

Balietti, Stefano, Lise Getoor, Daniel G. Goldstein, and Duncan J. Watts. "Reducing Opinion Polarization:
Effects of Exposure to Similar People with Differing Political Views." *Proceedings of the National Academy of Sciences* 118, no. 52 (2021): e2112552118. https://doi.org/10.1073/pnas.2112552118.

Balliet, Daniel, Junhui Wu, and Carsten K. W. De Dreu. "Ingroup Favoritism in Cooperation: A Meta-
Analysis." *Psychological Bulletin* 140, no. 6 (2014): 1556–81. https://doi.org/10.1037/a0037737.

Banas, John A., Elena Bessarabova, Marisa C. Penkauskas, and Neil Talbert. "Inoculating Against Anti-
Vaccination Conspiracies." *Health Communication* 39, no. 9 (2023): 1760–68. https://doi.org/10.1080
/10410236.2023.2235733.

Bar-Hillel, Maya. "The Base-Rate Fallacy in Probability Judgments." *Acta Psychologica* 44, no. 3 (1980):
211–33. https://doi.org/10.1016/0001-6918(80)90046-3.

Barberá, Pablo. "Birds of the Same Feather Tweet Together: Bayesian Ideal Point Estimation Using Twitter
Data." *Political Analysis* 23, no. 1 (2015): 76–91. https://doi.org/10.1093/pan/mpu011.

Barlev, Michael, and Steven L. Neuberg. "Rational Reasons for Irrational Beliefs." *American Psychologist*,
April 15, 2024. https://doi.org/10.1037/amp0001321.

Barnidge, Matthew. "Exposure to Political Disagreement in Social Media Versus Face-to-Face and Anonymous Online Settings." *Political Communication* 34, no. 2 (2017): 302–21. https://doi.org/10.1080/105 84609.2016.1235639.

Barrett, Paul M. "Spreading the Big Lie: How Social Media Sites Have Amplified False Claims of U.S. Election Fraud." NYU Stern Center for Business & Human Rights, September 16, 2022. https://bhr.stern. nyu.edu/tech-big-lie.

Bartik, Alexander W., Marianne Bertrand, Zoe Cullen, Edward L. Glaeser, Michael Luca, and Christopher Stanton. "The Impact of COVID-19 on Small Business Outcomes and Expectations." *Proceedings of the National Academy of Sciences* 117, no. 30 (2020): 17656–66. https://doi.org/10.1073/pnas.2006991117.

Basol, Melisa, Jon Roozenbeek, Manon Berriche, Fatih Uenal, William P. McClanahan, and Sander van der Linden. "Towards Psychological Herd Immunity: Cross-Cultural Evidence for Two Prebunking Interventions Against COVID-19 Misinformation." *Big Data and Society* 8, no. 1 (2021): 205395172110138. https://doi.org/10.1177/20539517211013868.

Basol, Melisa, Jon Roozenbeek, and Sander van der Linden. "Good News About *Bad News*: Gamified Inoculation Boosts Confidence and Cognitive Immunity Against Fake News." *Journal of Cognition* 3, no. 1 (2020): 2. https://doi.org/10.5334%2Fjoc.91.

Bastick, Zach. "Would You Notice If Fake News Changed Your Behavior? An Experiment on the Unconscious Effects of Disinformation." *Computers in Human Behavior* 116 (2021): art. 106633. https://doi .org/10.1016/j.chb.2020.106633.

Bastos, Marco T., and Dan Mercea. "The Brexit Botnet and User-Generated Hyperpartisan News." *Social Science Computer Review* 37, no. 1 (2019): 38–54. https://doi.org/10.1177/0894439317734157.

Batailler, Cédric, Skylar M. Brannon, Paul E. Teas, and Bertram Gawronski. "A Signal Detection Approach to Understanding the Identification of Fake News." *Perspectives on Psychological Science* 17, no. 1 (2022): 78–98. https://doi.org/10.1177/1745691620986135.

Bayer, Dave, Stuart Haber, and W. Scott Stornetta. "Improving the Efficiency and Reliability of Digital Time-Stamping." In *Sequences II: Methods in Communication, Security, and Computer Science*, ed. Renato Capocelli, Alfredo Santis, and Ugo Vaccaro, 329–34. Springer, 1993. https://doi.org/10.1007/978-1-4613 -9323-8_24.

BBC News. "Oxford Dictionaries' Word of the Year Is 'Post-Truth.' " November 15, 2016. https://www.bbc .com/news/uk-37995600.

Bell, Peter. "Public Trust in Government: 1958–2023." Pew Research Center, September 19, 2023. https:// www.pewresearch.org/politics/2023/09/19/public-trust-in-government-1958-2023/.

Benen, Steve. "Departing NIH Chief Eyes More Research 'on Human Behavior.' " MSNBC, December 21, 2021. https://www.msnbc.com/rachel-maddow-show/departing-nih-chief-eyes-more-research-human -behavior-n1286390.

Bensley, D. Alan, Cody Watkins, Scott O. Lilienfeld, Christopher Masciocchi, Michael P. Murtagh, and Krystal Rowan. "Skepticism, Cynicism, and Cognitive Style Predictors of the Generality of Unsubstantiated Belief." *Applied Cognitive Psychology* 36, no. 1 (2022): 83–99. https://doi.org/10.1002/acp .3900.

Beres, Derek. "How Anti-Vaxxers Monetize Misinformation." re-frame, January 19, 2023. https://derekberes .substack.com/p/how-anti-vaxxers-monetize-misinformation.

Berger, Peter. *The Sacred Canopy: Elements of a Sociological Theory of Religion*. Doubleday, 1967.

Bergstrom, Carl T., Daniel R. Pimentel, and Jonathan Osborne. "To Fight Misinformation, We Need to Teach That Science Is Dynamic." *Scientific American*, October 26, 2022. https://www.scientificamerican .com/article/to-fight-misinformation-we-need-to-teach-that-science-is-dynamic/.

Bergstrom, Carl T., and Jevin D. West. *Calling Bullshit: The Art of Skepticism in a Data-Driven World*. Random House, 2021.

Berinsky, Adam J. "Rumors and Health Care Reform: Experiments in Political Misinformation." *British Journal of Political Science* 47, no. 2 (2017): 241–62. https://doi.org/10.1017/S0007123415000186.

Berkeley Economic Review. "Paying Attention: The Attention Economy." March 31, 2020. https://econreview
.berkeley.edu/paying-attention-the-attention-economy/.

Berman, Jonathan M. *Anti-Vaxxers: How to Challenge a Misinformed Movement.* MIT Press, 2020.

Berry, Jeffrey M., and Sarah Sobieraj. *The Outrage Industry: Political Opinion Media and the New Incivility.*
Oxford University Press, 2014.

Bessi, Alessandro, and Emilio Ferrara. "Social Bots Distort the 2016 US Presidential Election Online
Discussion." *First Monday* 21, no. 11 (2016). https://doi.org/10.5210/fm.v21i11.7090.

Binning, Kevin R., David K. Sherman, Geoffrey L. Cohen, and Kirsten Heitland. "Seeing the Other Side:
Reducing Political Partisanship via Self-Affirmation in the 2008 Presidential Election." *Analyses of Social
Issues and Public Policy* 10, no. 1 (2010): 276–92. https://doi.org/10.1111/j.1530-2415.2010.01210.x.

Bisbee, James, and Diana Lee. "Mobility and Elite Cues: Partisan Responses to COVID-19." APSA Preprints, August 28, 2020. https://doi.org/10.33774/apsa-2020-76tv9.

Blanchar, John C., and Catherine J. Norris. "Political Homophily, Bifurcated Social Reality, and Perceived
Legitimacy of the 2020 US Presidential Election Results: A Four-Wave Longitudinal Study." *Analyses of
Social Issues and Public Policy* 21, no. 1 (2021): 259–83. https://doi.org/10.1111/asap.12276.

Blazina, Carrie. "About Half of Recent Online Daters in U.S. Say It's Important to See COVID-19 Vaccination Status on Profiles." Pew Research Center, September 27, 2022. https://www.pewresearch.org/fact
-tank/2022/09/27/about-half-of-recent-online-daters-in-u-s-say-its-important-to-see-covid-19-vaccination
-status-on-profiles/.

Blazina, Carrie. "Despite Wide Partisan Gaps in Views of Many Aspects of the Pandemic, Some Common
Ground Exists." Pew Research Center, March 24, 2021. https://www.pewresearch.org/fact-tank/2021/03
/24/despite-wide-partisan-gaps-in-views-of-many-aspects-of-the-pandemic-some-common-ground-exists/.

Blazina, Carrie. "Republicans Remain Far Less Likely than Democrats to View COVID-19 as a Major
Threat to Public Health." Pew Research Center, July 22, 2020. https://www.pewresearch.org/fact
-tank/2020/07/22/republicans-remain-far-less-likely-than-democrats-to-view-covid-19-as-a-major
-threat-to-public-health/.

Blazina, Carrie. "70 Percent of U.S. Social Media Users Never or Rarely Post or Share About Political, Social
Issues." Pew Research Center, May 4, 2021. https://www.pewresearch.org/fact-tank/2021/05/04/70-of
-u-s-social-media-users-never-or-rarely-post-or-share-about-political-social-issues/.

Bloom, Stephen G. "Lesson of a Lifetime." *Smithsonian Magazine,* August 31, 2005. https://www.smithso-
nianmag.com/science-nature/lesson-of-a-lifetime-72754306/.

Bode, Leticia, and Emily Vraga. "The Swiss Cheese Model for Mitigating Online Misinformation." *Bulletin
of the Atomic Scientists* 77, no. 3 (2021): 129–33. https://doi.org/10.1080/00963402.2021.1912170.

Bogle, John. *The Little Book of Common Sense Investing: The Only Way to Guarantee Your Fair Share of Stock
Market Returns.* Wiley, 2017.

Boh Podgornik, Bojana, Danica Dolničar, Andrej Šorgo, and Tomaž Bartol. "Development, Testing, and
Validation of an Information Literacy Test (ILT) for Higher Education." *Journal of the Association for
Information Science and Technology* 67, no. 10 (2016): 2420–36. https://doi.org/10.1002/asi.23586.

Böhm, Robert, and Cornelia Betsch. "Prosocial Vaccination." *Current Opinion in Psychology* 43 (2022):
307–11. https://doi.org/10.1016/j.copsyc.2021.08.010.

Böhm, Robert, Cornelia Betsch, and Lars Korn. "Selfish-Rational Non-vaccination: Experimental Evidence
from an Interactive Vaccination Game." *Journal of Economic Behavior and Organization* 131 (2016): 183–
95. https://doi.org/10.1016/j.jebo.2015.11.008.

Bond, Shannon. "AI-Generated Fake Faces Have Become a Hallmark of Online Influence Operations."
NPR, December 15, 2022. https://www.npr.org/2022/12/15/1143114122/ai-generated-fake-faces-have
-become-a-hallmark-of-online-influence-operations.

Bond, Shannon. "'Disinformation' Is the Word of the Year, and a Sign of What's to Come." NPR,
December 30, 2019. https://www.npr.org/2019/12/30/790144099/disinformation-is-the-word-of-the
-year-and-a-sign-of-what-s-to-come.

Bond, Shannon. "Facebook Widens Ban on COVID-19 Vaccine Misinformation in Push to Boost Confidence." NPR, February 8, 2021. https://www.npr.org/2021/02/08/965390755/facebook-widens-ban -on-covid-19-vaccine-misinformation-in-push-to-boost-confiden.

Bond, Shannon. "Just 12 People Are Behind Most Vaccine Hoaxes on Social Media, Research Shows." NPR, May 13, 2021. https://www.npr.org/2021/05/13/996570855/disinformation-dozen-test-facebooks -twitters-ability-to-curb-vaccine-hoaxes.

Bonnevie, Erika, Allison Gallegos-Jeffrey, Jaclyn Goldbarg, Brian Byrd, and Joseph Smyser. "Quantifying the Rise of Vaccine Opposition on Twitter During the COVID-19 Pandemic." *Journal of Communication in Healthcare* 14, no. 1 (2021): 12–19. https://doi.org/10.1080/17538068.2020.1858222.

Borah, Porismita. "The Moderating Role of Political Ideology: Need for Cognition, Media Locus of Control, Misinformation Efficacy, and Misperceptions About COVID-19." *International Journal of Communication Systems* 16 (2022): 3534–59. https://ijoc.org/index.php/ijoc/article/view/18261.

Borukhson, David, Philipp Lorenz-Spreen, and Marco Ragni. "When Does an Individual Accept Misinformation? An Extended Investigation Through Cognitive Modeling." *Computational Brain and Behavior* 5, no. 2 (2022): 244–60. https://doi.org/10.1007/s42113-022-00136-3.

Boseley, Sarah. "How Disgraced Anti-Vaxxer Andrew Wakefield Was Embraced by Trump's America." *The Guardian*, July 18, 2018. https://www.theguardian.com/society/2018/jul/18/how-disgraced-anti-vaxxer -andrew-wakefield-was-embraced-by-trumps-america.

Bouie, Jamelle. "Opinion: Glenn Youngkin Knows Exactly What He's Doing." *New York Times*, September 20, 2022. https://www.nytimes.com/2022/09/20/opinion/glenn-youngkin-midterms-trump.html.

Bowes, Shauna M., Madeline C. Blanchard, Thomas H. Costello, Alan I. Abramowitz, and Scott O. Lilienfeld. "Intellectual Humility and Between-Party Animus: Implications for Affective Polarization in Two Community Samples." *Journal of Research in Personality* 88 (2020): art. 103992. https://doi .org/10.1016/j.jrp.2020.103992.

Bowles, Jeremy, Kevin Croke, Horacio Larreguy, John Marshall, and Shelley Liu. "Sustaining Exposure to Fact-Checks: Misinformation Discernment, Media Consumption, and Its Political Implications." SSRN, September 25, 2023. https://doi.org/10.2139/ssrn.4582703.

Bradshaw, Samantha, and Philip N. Howard. "The Global Disinformation Order: 2019 Global Inventory of Organised Social Media Manipulation." University of Oxford, 2019. https://digitalcommons.unl.edu /scholcom/207.

Brady, William J., Molly J. Crockett, and Jay J. Van Bavel. "The MAD Model of Moral Contagion: The Role of Motivation, Attention, and Design in the Spread of Moralized Content Online." *Perspectives on Psychological Science* 15, no. 4 (2020): 978–1010. https://doi.org/10.1177/1745691620917336.

Brady, William J., Killian McLoughlin, Tuan N. Doan, and Molly J. Crockett. "How Social Learning Amplifies Moral Outrage Expression in Online Social Networks." *Science Advances* 7, no. 33 (2021): eabe5641. https://doi.org/10.1126/sciadv.abe5641.

Brady, William J., Killian L. McLoughlin, Mark P. Torres, Kara F. Luo, Maria Gendron, and M. J. Crockett. "Overperception of Moral Outrage in Online Social Networks Inflates Beliefs About Intergroup Hostility." *Nature Human Behaviour* 7, no. 6 (2023): 917–27. https://doi.org/10.1038/s41562-023-01582-0.

Brady, William J., Julian A. Wills, John T. Jost, Joshua A. Tucker, and Jay J. Van Bavel. "Emotion Shapes the Diffusion of Moralized Content in Social Networks." *Proceedings of the National Academy of Sciences* 114, no. 28 (2017): 7313–18. https://doi.org/10.1073/pnas.1618923114.

Bramstedt, Katrina A. "Unicorn Poo and Blessed Waters: COVID-19 Quackery and FDA Warning Letters." *Therapeutic Innovation and Regulatory Science* 55, no. 1 (2021): 239–44.

Brashears, Matthew E. "Anomia and the Sacred Canopy: Testing a Network Theory." *Social Networks* 32, no. 3 (2010): 187–96. https://doi.org/10.1016/j.socnet.2009.12.003.

Brashears, Matthew E. "'Trivial' Topics and Rich Ties: The Relationship Between Discussion Topic, Alter Role, and Resource Availability Using the 'Important Matters' Name Generator." *Sociological Science* 1 (2014): 493–511. https://doi.org/10.15195/v1.a27.

Brashier, Nadia M., Emmaline Drew Eliseev, and Elizabeth J. Marsh. "An Initial Accuracy Focus Prevents Illusory Truth." *Cognition* 194 (2020): art. 104054. https://doi.org/10.1016/j.cognition.2019.104054.

Brashier, Nadia M., and Daniel L. Schacter. "Aging in an Era of Fake News." *Current Directions in Psychological Science* 29, no. 3 (2020): 316–23. https://doi.org/10.1177/0963721420915872.

Brave Factor. "We Won Bronze in Webby's Anthem Awards!" February 20, 2023. https://bravefactor.com/we-won-bronze-in-webbys-anthem-awards/.

Breland, Ali. "How the Coronavirus Brought Chain Mail Back to the Mainstream." *Mother Jones*, March 27, 2020. https://www.motherjones.com/politics/2020/03/coronavirus-chain-mail/.

Brenan, Megan. "Record-Low 38 Percent Extremely Proud to Be American." Gallup, June 29, 2022. https://news.gallup.com/poll/394202/record-low-extremely-proud-american.aspx.

Breslow, Jason. "20 Companies Are Behind Half of the World's Single-Use Plastic Waste, Study Finds." NPR, May 18, 2021. https://www.npr.org/2021/05/18/997937090/half-of-the-worlds-single-use-plastic-waste-is-from-just-20-companies-says-a-stu.

Brewer, Marilynn B. "The Psychology of Prejudice: Ingroup Love and Outgroup Hate?" *Journal of Social Issues* 55, no. 3 (1999): 429–44. https://doi.org/10.1111/0022-4537.00126.

Brewer, Marilynn B., and Kathleen P. Pierce. "Social Identity Complexity and Outgroup Tolerance." *Personality and Social Psychology Bulletin* 31, no. 3 (2005): 428–37. https://doi.org/10.1177/0146167204271710.

Brewster, Jack, Lorenzo Arvanitis, Valerie Pavilonis, and Macrina Wang. "Beware the 'New Google': TikTok's Search Engine Pumps Toxic Misinformation to Its Young Users." NewsGuard, September 14, 2022. https://www.newsguardtech.com/misinformation-monitor/september-2022.

Broaddus, Betsy. "Amidst the Pandemic, Confidence in the Scientific Community Becomes Increasingly Polarized." AP-NORC, January 26, 2022. https://apnorc.org/projects/amidst-the-pandemic-confidence-in-the-scientific-community-becomes-increasingly-polarized/.

Broniatowski, David A., Amelia M. Jamison, SiHua Qi et al. "Weaponized Health Communication: Twitter Bots and Russian Trolls Amplify the Vaccine Debate." *American Journal of Public Health* 108, no. 10 (2018): 1378–84. https://doi.org/10.2105/AJPH.2018.304567.

Bronstein, Michael V., and Sophia Vinogradov. "Education Alone Is Insufficient to Combat Online Medical Misinformation." *EMBO Reports* 22, no. 3 (2021): e52282. https://doi.org/10.15252/embr.202052282.

Broockman, David E., Joshua L. Kalla, and Sean J. Westwood. "Does Affective Polarization Undermine Democratic Norms or Accountability? Maybe Not." *American Journal of Political Science* 67, no. 3 (2023): 808–28. https://doi.org/10.1111/ajps.12719.

Brown, Gordon D. A., Stephan Lewandowsky, and Zhihong Huang. "Social Sampling and Expressed Attitudes: Authenticity Preference and Social Extremeness Aversion Lead to Social Norm Effects and Polarization." *Psychological Review* 129, no. 1 (2022): 18–48. https://doi.org/10.1037/rev0000342.

Brown, Jacob R., and Ryan D. Enos. "The Measurement of Partisan Sorting for 180 Million Voters." *Nature Human Behaviour* 5, no. 8 (2021): 998–1008. https://doi.org/10.1038/s41562-021-01066-z.

Brüggemann, Michael, and Sven Engesser. "Beyond False Balance: How Interpretive Journalism Shapes Media Coverage of Climate Change." *Global Environmental Change: Human and Policy Dimensions* 42 (2017): 58–67. https://doi.org/10.1016/j.gloenvcha.2016.11.004.

Brumfiel, Geoff. "Their Mom Died of COVID: They Say Conspiracy Theories Are What Really Killed Her." NPR, April 24, 2022. https://www.npr.org/sections/health-shots/2022/04/24/1089786147/covid-conspiracy-theories.

Bruneau, Emile G., Mina Cikara, and Rebecca Saxe. "Minding the Gap: Narrative Descriptions About Mental States Attenuate Parochial Empathy." *PLOS One* 10, no. 10 (2015): e0140838. https://doi.org/10.1371/journal.pone.0140838.

Buchanan, Tom. "How to Reduce the Spread of Fake News—by Doing Nothing." Nieman Lab, January 5, 2021. https://www.niemanlab.org/2021/01/how-to-reduce-the-spread-of-fake-news-by-doing-nothing/.

Budak, Ceren, Brendan Nyhan, David M. Rothschild, Emily Thorson, and Duncan J. Watts. "Misunderstanding the Harms of Online Misinformation." *Nature* 630, no. 8015 (2024): 45–53. https://doi.org/10.1038/s41586-024-07417-w.

Budnieski, Chase. "First Draft Teaches Journalists How to Avoid Amplifying Misinformation." News Co/Lab, February 14, 2020. https://newscollab.org/2020/02/14/first-draft-teaches-journalists-how-to-avoid-amplifying-misinformation/.

Bulfone, Tommaso Celeste, Mohsen Malekinejad, George W. Rutherford, and Nooshin Razani. "Outdoor Transmission of SARS-CoV-2 and Other Respiratory Viruses: A Systematic Review." *Journal of Infectious Diseases* 223, no. 4 (2021): 550–61. https://doi.org/10.1093/infdis/jiaa742.

Burke, Peter J. "Conceptualizing Identity Prominence, Salience, and Commitment." In *Advancing Identity Theory, Measurement, and Research*, ed. Jan E. Stets, Ashley V. Reichelmann, and K. Jill Kiecolt, 17–33. Springer International, 2023. https://doi.org/10.1007/978-3-031-32986-9_2.

Burke, Peter J. "Identity." In *The Cambridge Handbook of Social Theory*, ed. Peter Kivisto, 63–78. Cambridge University Press, 2020.

Burke, Peter J. "Identity Control Theory." In *The Blackwell Encyclopedia of Sociology*, ed. George Ritzer, 2202–7. Blackwell, 2007.

Burke, Peter J. "Social Identities and Psychosocial Stress." In *Psychosocial Stress: Perspectives on Structure, Theory, Life-Course, and Methods*, ed. H. B. Kaplan, 141–74. Academic Press, 1996.

Burke, Peter J., and Christine Cerven. "Identity Accumulation, Verification, and Well-Being." In *Identities in Everyday Life*, ed. Jan E. Stets and Richard T. Serpe, 17–33. Oxford University Press, 2019. https://doi.org/10.1093/oso/9780190873066.003.0002.

Burke, Peter J., and Michael M. Harrod. "Too Much of a Good Thing?" *Social Psychology Quarterly* 68, no. 4 (2005): 359–74. https://doi.org/10.1177/019027250506800404.

Burke, Peter J., and Donald C. Reitzes. "An Identity Theory Approach to Commitment." *Social Psychology Quarterly* 54, no. 3 (1991): 239–51. https://doi.org/10.2307/2786653.

Burke, Peter J., and Jan E. Stets. *Identity Theory: Revised and Expanded*. Oxford University Press, 2022.

Burke, Peter J., and Jan E. Stets. "Trust and Commitment Through Self-Verification." *Social Psychology Quarterly* 62, no. 4 (1999): 347–66. https://doi.org/10.2307/2695833.

Burnett, John. "Americans Are Fleeing to Places Where Political Views Match Their Own." NPR, February 18, 2022. https://www.npr.org/2022/02/18/1081295373/the-big-sort-americans-move-to-areas-political-alignment.

Cacioppo, John T., and Richard E. Petty. "The Need for Cognition." *Journal of Personality and Social Psychology* 42, no. 1 (1982): 116–31. https://doi.org/10.1037/0022-3514.42.1.116.

Caesar, Ed. "How a Young Couple Failed to Launder Billions of Dollars in Stolen Bitcoin." *New Yorker*, February 14, 2022. https://www.newyorker.com/business/currency/how-a-young-couple-failed-to-launder-billions-of-dollars-in-stolen-bitcoin.

Callahan, Molly. "Are You More Likely to Believe Misinformation About Ukraine or COVID-19?" *Northeastern Global News*, April 27, 2022. https://news.northeastern.edu/2022/04/27/misinformation-about-ukraine-or-covid-19/.

Cameron, Chris. "These Are the People Who Died in Connection with the Capitol Riot." *New York Times*, January 5, 2022. https://www.nytimes.com/2022/01/05/us/politics/jan-6-capitol-deaths.html.

Campello, Daniele. "The Fake News Game Received the Brouwer Trust Prize from the Royal Holland Society of Sciences." University of Cambridge, Department of Psychology, January 16, 2020. https://www.psychol.cam.ac.uk/news/fake-news-game-received-brouwer-trust-prize-royal-holland-society-sciences.

Cancryn, Adam. "A Sharp Partisan Divide Remains Over New Covid Boosters." *Politico*, September 15, 2023. https://www.politico.com/news/2023/09/15/poll-covid-booster-democrats-00116123.

Carey, John M., Andrew M. Guess, Peter J. Loewen et al. "The Ephemeral Effects of Fact-Checks on COVID-19 Misperceptions in the United States, Great Britain and Canada." *Nature Human Behaviour* 6, no. 2 (2022): 236–43. https://doi.org/10.1038/s41562-021-01278-3.

Carlin, Ryan E., and Gregory J. Love. "The Politics of Interpersonal Trust and Reciprocity: An Experimental Approach." *Political Behavior* 35, no. 1 (2013): 43–63. https://doi.org/10.1007/s11109-011-9181-x.

Carr, Joe. "A New Index Shows That the US Scores Low on Media Literacy Education." Media Literacy Now, July 27, 2021. https://medialiteracynow.org/a-new-index-shows-that-the-us-scores-low-on-media -literacy-education/.

Carrier, Anastasiia. "'This Crap Means More to Him than My Life': When QAnon Invades American Homes." *Politico*, February 19, 2021. https://www.politico.com/news/magazine/2021/02/19/qanon -conspiracy-theory-family-members-reddit-forum-469485.

Cassidy, Christina A. "AP Review Finds Far Too Little Vote Fraud to Tip 2020 Election to Trump." PBS News, December 14, 2021. https://www.pbs.org/newshour/politics/ap-review-finds-far-too-little-vote -fraud-to-tip-2020-election-to-trump.

Caulfield, Mike. "Information Literacy for Mortals." Project Information Literacy, December 14, 2021. https://projectinfolit.org/pubs/provocation-series/essays/information-literacy-for-mortals.html.

Caulfield, Mike. "Introducing SIFT, a Four Moves Acronym." Hapgood, May 12, 2019. https://hapgood .us/2019/05/12/sift-and-a-check-please-preview/.

CBS News. "Republicans See U.S. as Better Off Now than 4 Years Ago Ahead of Convention—Battleground Tracker Poll." August 23, 2020. https://www.cbsnews.com/news/republicans-economy-coronavirus-opinion -poll-cbs-news-battleground-tracker/.

Celadin, Tatiana, Valerio Capraro, Gordon Pennycook, and David G. Rand. "Displaying News Source Trust-worthiness Ratings Reduces Sharing Intentions for False News Posts." *Journal of Online Trust and Safety* 1, no. 5 (2023): 1–20. https://doi.org/10.54501/jots.v1i5.100.

Center on Budget and Policy Priorities. "Tracking the COVID-19 Economy's Effects on Food, Housing, and Employment Hardships." August 13, 2020. https://www.cbpp.org/research/poverty-and-inequality /tracking-the-covid-19-economys-effects-on-food-housing-and.

Centola, Damon. "An Experimental Study of Homophily in the Adoption of Health Behavior." *Science* 334, no. 6060 (2011): 1269–72. https://doi.org/10.1126/science.1207055.

Ceron, Andrea. "Internet, News, and Political Trust: The Difference Between Social Media and Online Media Outlets." *Journal of Computer-Mediated Communication* 20, no. 5 (2015): 487–503. https://doi .org/10.1111/jcc4.12129.

Cesario, Joseph. "Priming, Replication, and the Hardest Science." *Perspectives on Psychological Science* 9, no. 1 (2014): 40–48. https://doi.org/10.1177/1745691613513470.

Chaban, Matt A. V. "Can Blockchain Block Fake News and Deep Fakes?" IBM, November 30, 2020. https:// www.ibm.com/blog/blockchain-protection-fake-news-deep-fakes-safe-press/.

Chater, Nick, and George Loewenstein. "The i-Frame and the s-Frame: How Focusing on Individual-Level Solutions Has Led Behavioral Public Policy Astray." *Behavioral and Brain Sciences* 46 (2023): e147. https://doi.org/10.1017/S0140525X22002023.

Chen, Adrian. "The Agency." *New York Times*, June 2, 2015. https://www.nytimes.com/2015/06/07/magazine /the-agency.html.

Chen, Annie Y., Brendan Nyhan, Jason Reifler, Ronald E. Robertson, and Christo Wilson. "Subscriptions and External Links Help Drive Resentful Users to Alternative and Extremist YouTube Channels." *Science Advances* 9, no. 35 (2023): eadd8080. https://doi.org/10.1126/sciadv.add8080.

Chen, M. Keith, and Ryne Rohla. "The Effect of Partisanship and Political Advertising on Close Family Ties." *Science* 360, no. 6392 (2018): 1020–24. https://doi.org/10.1126/science.aaq1433.

Chen, Robert T., and Frank DeStefano. "Vaccine Adverse Events: Causal or Coincidental?" *The Lancet* 351, no. 9103 (1998): 611–12. https://doi.org/10.1016/S0140-6736(05)78423-3.

Cherry, James D. "'Pertussis Vaccine Encephalopathy': It Is Time to Recognize It as the Myth That It Is." *JAMA* 263, no. 12 (1990): 1679–80. https://doi.org/10.1001/jama.1990.03440120101046.

Chesney, Robert, and Danielle Keats Citron. "Deep Fakes: A Looming Challenge for Privacy, Democracy, and National Security." *California Law Review* 107 (2019): 1753. https://doi.org/10.2139/ssrn .3213954.

Chinn, Sedona, and Ariel Hasell. "Support for 'Doing Your Own Research' Is Associated with COVID-19 Misperceptions and Scientific Mistrust." *Harvard Kennedy School Misinformation Review* 4, no. 3 (2023): 1–15. https://doi.org/10.37016/mr-2020-117.

Cho, Jaeho, Saifuddin Ahmed, Heejo Keum, Yun Jung Choi, and Jong Hyuk Lee. "Influencing Myself: Self-Reinforcement Through Online Political Expression." *Communication Research* 45, no. 1 (2018): 83–111. https://doi.org/10.1177/0093650216644020.

Cho, Renée. "How Climate Change Impacts the Economy." *State of the Planet*, Columbia Climate School, Columbia University, June 20, 2019. https://news.climate.columbia.edu/2019/06/20/climate-change-economy-impacts/.

Christakis, Nicholas A., and James H. Fowler. *Connected: The Surprising Power of Our Social Networks and How They Shape Our Lives*. Little, Brown Spark, 2009.

Chu, James, Sophia L. Pink, and Robb Willer. "Religious Identity Cues Increase Vaccination Intentions and Trust in Medical Experts Among American Christians." *Proceedings of the National Academy of Sciences* 118, no. 49 (2021): e2106481118. https://doi.org/10.1073/pnas.2106481118.

Chua, Kao-Ping, Rena M. Conti, and Nora V. Becker. "US Insurer Spending on Ivermectin Prescriptions for COVID-19." *JAMA* 327, no. 6 (2022): 584–87. https://doi.org/10.1001/jama.2021.24352.

Cialdini, Robert B. *Influence, New and Expanded: The Psychology of Persuasion*. HarperCollins, 2021.

Cinelli, Matteo, Gianmarco De Francisci Morales, Alessandro Galeazzi, Walter Quattrociocchi, and Michele Starnini. "The Echo Chamber Effect on Social Media." *Proceedings of the National Academy of Sciences* 118, no. 9 (2021): e2023301118. https://doi.org/10.1073/pnas.2023301118.

Clapham, Hannah E., and Alex R. Cook. "Face Masks Help Control Transmission of COVID-19." *Lancet Digital Health* 3, no. 3 (2021): e136–37. https://doi.org/10.1016/S2589-7500(21)00003-0.

Clark, Daniel, dir. *Behind the Curve*. Delta-v Productions, 2018.

Cleveland Clinic. "COVID-19: Is It Safe to Swim?" September 17, 2021. https://health.clevelandclinic.org/is-the-pool-safe-during-coronavirus-pandemic/.

CNN. "Rand Paul Has a *Very* Wacky Theory About Ivermectin." August 31, 2021. https://www.cnn.com/2021/08/31/politics/rand-paul-covid-19-ivermectin/index.html.

Cohen, Danielle. "He Created Your Phone's Most Addictive Feature: Now He Wants to Build a Rosetta Stone for Animal Language." *GQ*, June 30, 2021. https://www.gq.com/story/aza-raskin-interview.

Cohen, Geoffrey L. "Party Over Policy: The Dominating Impact of Group Influence on Political Beliefs." *Journal of Personality and Social Psychology* 85, no. 5 (2003): 808–22. https://doi.org/10.1037/0022-3514.85.5.808.

Cohen, Geoffrey L., and David K. Sherman. "The Psychology of Change: Self-Affirmation and Social Psychological Intervention." *Annual Review of Psychology* 65 (2014): 333–71. https://doi.org/10.1146/annurev-psych-010213-115137.

Cohen, Geoffrey L., David K. Sherman, Anthony Bastardi, Lillian Hsu, Michelle McGoey, and Lee Ross. "Bridging the Partisan Divide: Self-Affirmation Reduces Ideological Closed-Mindedness and Inflexibility in Negotiation." *Journal of Personality and Social Psychology* 93, no. 3 (2007): 415–30. https://doi.org/10.1037/0022-3514.93.3.415.

Coleman, Peter T. "Chapter 7 Exercises." The Way Out. Accessed August 14, 2024. https://www.thewayoutofpolarization.com/chapter-7-exercises.

Coleman, Peter T. *The Way Out: How to Overcome Toxic Polarization*. Columbia University Press, 2021.

Collins, Ben, and Brandy Zadrozny. "The Far Right Is Struggling to Contain QAnon After Giving It Life." NBC News, August 10, 2018. https://www.nbcnews.com/tech/tech-news/far-right-struggling-contain-qanon-after-giving-it-life-n899741.

Collins, Francis. Interview by David DeSteno. Uploaded May 29, 2024. Vimeo, 1 min., 42 sec. https://vimeo.com/951537027.

Compton, Josh. "Inoculation Against/as Character Assassination." In *Routledge Handbook of Character Assassination and Reputation Management*, ed. Sergei A. Samoilenko, Martijn Icks, Jennifer Keohane, and Eric Shiraev, 25–35. Routledge, 2021.

Compton, Josh, Sander van der Linden, John Cook, and Melisa Basol. "Inoculation Theory in the Post-Truth Era: Extant Findings and New Frontiers for Contested Science, Misinformation, and Conspiracy Theories." *Social and Personality Psychology Compass* 15, no. 6 (2021): e12602. https://doi.org/10.1111/spc3.12602.

Confessore, Nicholas, and Danny Hakim. "Data Firm Says 'Secret Sauce' Aided Trump; Many Scoff." *New York Times*, March 6, 2017. https://www.nytimes.com/2017/03/06/us/politics/cambridge-analytica.html.

Conger, Kate. "Hackers' Fake Claims of Ukrainian Surrender Aren't Fooling Anyone: So What's Their Goal?" *New York Times*, April 5, 2022, https://www.nytimes.com/2022/04/05/us/politics/ukraine-russia-hackers.html.

Conger, Krista. "Surgical Masks Reduce COVID-19 Spread, Large-Scale Study Shows." News Center, Stanford Medicine, September 1, 2021. https://med.stanford.edu/news/all-news/2021/09/surgical-masks-covid-19.html.

Conover, Michael D., Bruno Gonçalves, Alessandro Flammini, and Filippo Menczer. "Predicting the Political Alignment of Twitter Users." In *2011 IEEE Third International Conference on Privacy, Security, Risk and Trust and 2011 IEEE Third International Conference on Social Computing*, 192–99. IEEE, 2011. https://doi.org/10.1109/PASSAT/SocialCom.2011.34.

Conover, Michael D., Jacob Ratkiewicz, Matthew Francisco, Bruno Gonçalves, Alessandro Flammini, and Filippo Menczer. "Partisan Asymmetries in Online Political Activity." *EPJ Data Science* 1, no. 1 (2012): 1–19. https://doi.org/10.1140/epjds6.

Cook, John. "Teaching About Our Climate Crisis: Combining Games and Critical Thinking to Fight Misinformation." *American Educator* 45, no. 4 (2022): 12. https://files.eric.ed.gov/fulltext/EJ1321545.pdf.

Cook, Rhodes. "The 'Big Sort' Continues, with Trump as a Driving Force." Center for Politics: Sabato's Crystal Ball, February 17, 2022. https://centerforpolitics.org/crystalball/articles/the-big-sort-continues-with-trump-as-a-driving-force/.

Cooley, Charles Horton. *Human Nature and the Social Order*. Routledge, 2017.

Cordina, Maria, Mary A. Lauri, and Josef Lauri. "Attitudes Towards COVID-19 Vaccination, Vaccine Hesitancy and Intention to Take the Vaccine." *Pharmacy Practice* 19, no. 1 (2021): 2317. https://doi.org/10.18549/PharmPract.2021.1.2317.

Costello, Thomas H., Gordon Pennycook, and David Gertler Rand. "Durably Reducing Conspiracy Beliefs Through Dialogues with AI." Preprint, PsyArXiv, April 3, 2024. https://doi.org/10.31234/osf.io/xcwdn.

COVID-19 Treatment Guidelines. "Anti-SARS-CoV-2 Monoclonal Antibodies." Accessed June 24, 2024. https://www.covid19treatmentguidelines.nih.gov/therapies/antivirals-including-antibody-products/anti-sars-cov-2-monoclonal-antibodies/.

Cowan, Sarah K. "Secrets and Misperceptions: The Creation of Self-Fulfilling Illusions." *Sociological Science* 1 (2014): 466–92. https://doi.org/10.15195/v1.a26.

Crawford, Jarret T., and John Ruscio. "Asking People to Explain Complex Policies Does Not Increase Political Moderation: Three Preregistered Failures to Closely Replicate Fernbach, Rogers, Fox, and Sloman's (2013) Findings." *Psychological Science* 32, no. 4 (2021): 611–21. https://doi.org/10.1177/0956797620972367.

Csikszentmihalyi, Mihaly. *The Evolving Self: A Psychology for the Third Millennium*. HarperCollins, 2009.

Cubanski, Juliette, Jennifer Kates, Jennifer Tolbert, Madeline Guth, Karen Pollitz, and Meredith Freed. "What Happens When COVID-19 Emergency Declarations End? Implications for Coverage, Costs, and Access." *KFF*, January 31, 2023. https://www.kff.org/coronavirus-covid-19/issue-brief/what-happens-when-covid-19-emergency-declarations-end-implications-for-coverage-costs-and-access/.

Cueni, Thomas B. "Waiving Intellectual Property Rights Is a Flawed Solution to Achieving Covid-19 Vaccine Equity." *STAT*, June 10, 2022. https://www.statnews.com/2022/06/10/waiving-intellectual-property-rights-is-a-flawed-solution-to-achieving-covid-19-vaccine-equity/.

Curry, Oliver Scott, Matthew Jones Chesters, and Caspar J. Van Lissa. "Mapping Morality with a Compass: Testing the Theory of 'Morality-as-Cooperation' with a New Questionnaire." *Journal of Research in Personality* 78 (2019): 106–24. https://doi.org/10.1016/j.jrp.2018.10.008.

Dalbar, Inc. "Average Investor Blown Away by Market Turmoil in 2018." PR Newswire, March 25, 2019. https://www.prnewswire.com/news-releases/average-investor-blown-away-by-market-turmoil-in-2018-300817353.html.

Daniel, Ari. "First Malaria Vaccine Hits 1 Million Dose Milestone—Although It Has Its Shortcomings." NPR, May 13, 2022. https://www.npr.org/sections/goatsandsoda/2022/05/13/1098536246/first-malaria-vaccine-hits-1-million-dose-milestone-although-it-has-its-shortcom.

Darcy, Oliver. "ABC News Suspends Brian Ross for 4 Weeks Over Erroneous Flynn Story." CNN Business, December 2, 2017. https://money.cnn.com/2017/12/02/media/abc-news-brian-ross/index.html.

David, Leonard. "Space Philosopher Frank White on 'The Overview Effect' and Humanity's Connection with Earth." *Space*, August 2, 2022. https://www.space.com/frank-white-overview-effect.

de Mello, Victoria Oldemburgo, Felix Cheung, and Michael Inzlicht. "Twitter (X) Use Predicts Substantial Changes in Well-Being, Polarization, Sense of Belonging, and Outrage." *Communications Psychology* 2, no. 1 (2024): 15. https://doi.org/10.1038/s44271-024-00062-z.

de Visé, Daniel. " 'Hyper-Partisan' Politicians Get Four Times the News Coverage of Bipartisan Colleagues." *The Hill*, March 13, 2023. https://thehill.com/homenews/media/3894486-hyper-partisan-politicians-get-four-times-the-news-coverage-of-bipartisan-colleagues/.

Deer, Brian. "How the Case Against the MMR Vaccine Was Fixed." *BMJ* 342 (2011): c5347. https://doi.org/10.1136/bmj.c5347.

Deer Owens, Ann Marie. "Former Vice President Al Gore Kicks Off Vanderbilt Project on Unity and American Democracy, Followed by Case Study on PEPFAR with 66th Secretary of State Condoleezza Rice." Vanderbilt University, January 15, 2021. https://news.vanderbilt.edu/2021/01/15/former-vice-president-al-gore-kicks-off-vanderbilt-project-on-unity-and-american-democracy-followed-by-case-study-on-pepfar-with-66th-secretary-of-state-condoleezza-rice/.

Del Vicario, Michela, Alessandro Bessi, Fabiana Zollo et al. "The Spreading of Misinformation Online." *Proceedings of the National Academy of Sciences* 113, no. 3 (2016): 554–59. https://doi.org/10.1073/pnas.1517441113.

DeMora, Stephanie L., Jennifer L. Merolla, Brian Newman, and Elizabeth J. Zechmeister. "Reducing Mask Resistance Among White Evangelical Christians with Value-Consistent Messages." *Proceedings of the National Academy of Sciences* 118, no. 21 (2021): e2101723118. https://doi.org/10.1073/pnas.2101723118.

Detect Fakes. "AI-Generated or Real?" Northwestern Kellogg. Accessed August 14, 2024. https://detectfakes.media.mit.edu/.

Diamond, Emily P. "The Influence of Identity Salience on Framing Effectiveness: An Experiment." *Political Psychology* 41, no. 6 (2020): 1133–50. https://doi.org/10.1111/pops.12669.

Diamond, Larry. "Could Deliberative Democracy Depolarize America?" Stanford Report, February 4, 2021. https://news.stanford.edu/2021/02/04/deliberative-democracy-depolarize-america/.

Digital Inquiry Group. "Students' Civic Online Reasoning." November 14, 2019. https://sheg.stanford.edu/students-civic-online-reasoning.

Dillingham, Lindsay L., and Bobi Ivanov. "Using Postinoculation Talk to Strengthen Generated Resistance." *Communication Research Reports* 33, no. 4 (2016): 295–302. https://doi.org/10.1080/08824096.2016.1224161.

Ding, Ding, Edward W. Maibach, Xiaoquan Zhao, Connie Roser-Renouf, and Anthony Leiserowitz. "Support for Climate Policy and Societal Action Are Linked to Perceptions About Scientific Agreement." *Nature Climate Change* 1, no. 9 (2011): 462–66. https://doi.org/10.1038/nclimate1295.

Ding, Yifan, Matthew Facciani, Amrit Poudel, Ellen Joyce, Salvador Aguinaga, Balaji Veeramani, Sanmitra Bhattacharya, and Tim Weninger. "Citations and Trust in LLM Generated Responses." *arXiv [Cs.CL]*, January 2, 2025. arXiv. http://arxiv.org/abs/2501.01303.

DiResta, Renée. "Free Speech Is Not the Same as Free Reach." *Wired*, August 30, 2018. https://www.wired.com/story/free-speech-is-not-the-same-as-free-reach/.

DiResta, Renée, Laura Edelson, Brendan Nyhan, and Ethan Zuckerman. "It's Time to Open the Black Box of Social Media." *Scientific American*, April 28, 2022. https://www.scientificamerican.com/article/its-time-to-open-the-black-box-of-social-media/.

DiResta, Renée, and Tobias Rose-Stockwell. "How to Stop Misinformation Before It Gets Shared." *Wired*, March 26, 2021. https://www.wired.com/story/how-to-stop-misinformation-before-it-gets-shared/.

Doucleff, Michaeleen. "U.S. Dept of Energy Says with 'Low Confidence' That COVID May Have Leaked from a Lab." NPR, February 28, 2023. https://www.npr.org/2023/02/28/1160157977/u-s-dept-of-energy -says-with-low-confidence-that-covid-may-have-leaked-from-a-la.

Dries, Charlotte, Michelle McDowell, Felix G. Rebitschek, and Christina Leuker. "When Evidence Changes: Communicating Uncertainty Protects Against a Loss of Trust." *Public Understanding of Science* 33, no. 6 (2024): 777–94. https://doi.org/10.1177/09636625241228449.

Drummond, Caitlin, and Baruch Fischhoff. "Individuals with Greater Science Literacy and Education Have More Polarized Beliefs on Controversial Science Topics." *Proceedings of the National Academy of Sciences* 114, no. 36 (2017): 9587–92. https://doi.org/10.1073/pnas.1704882114.

Duffy, Andrew, Edson Tandoc, and Rich Ling. "Too Good to Be True, Too Good Not to Share: The Social Utility of Fake News." *Information, Communication and Society* 23, no. 13 (2020): 1965–79. https://doi .org/10.1080/1369118X.2019.1623904.

Duggan, Maeve. "The Political Environment on Social Media." Pew Research Center, October 25, 2016. https://www.pewresearch.org/internet/2016/10/25/the-political-environment-on-social-media/.

Dunaway, Finis. "The 'Crying Indian' Ad That Fooled the Environmental Movement." *Chicago Tribune*, November 21, 2017. http://www.chicagotribune.com/2017/11/21/the-crying-indian-ad-that-fooled-the -environmental-movement/.

Dunham, Yarrow, Andrew Scott Baron, and Susan Carey. "Consequences of 'Minimal' Group Affiliations in Children." *Child Development* 82, no. 3 (2011): 793–811. https://doi.org/10.1111/j.1467-8624.2011.01577.x.

Dunning, David. *Self-Insight: Roadblocks and Detours on the Path to Knowing Thyself.* Psychology Press, 2012.

Duron, Robert, Barbara Limbach, and Wendy Waugh. "Critical Thinking Framework for Any Discipline." *International Journal of Teaching and Learning in Higher Education* 17, no. 2 (2006): 160–66.

Dutcher, Janine M., Naomi I. Eisenberger, Hayoung Woo et al. "Neural Mechanisms of Self-Affirmation's Stress Buffering Effects." *Social Cognitive and Affective Neuroscience* 15, no. 10 (2020): 1086–96. https:// doi.org/10.1093/scan/nsaa042.

Eady, Gregory, Jonathan Nagler, Andy Guess, Jan Zilinsky, and Joshua A. Tucker. "How Many People Live in Political Bubbles on Social Media? Evidence from Linked Survey and Twitter Data." *SAGE Open* 9, no. 1 (2019): 215824401983270. https://doi.org/10.1177/2158244019832705.

Eady, Gregory, Tom Paskhalis, Jan Zilinsky, Richard Bonneau, Jonathan Nagler, and Joshua A. Tucker. "Exposure to the Russian Internet Research Agency Foreign Influence Campaign on Twitter in the 2016 US Election and Its Relationship to Attitudes and Voting Behavior." *Nature Communications* 14, no. 1 (2023): 62. https://doi.org/10.1038/s41467-022-35576-9.

Ecker, Ullrich, Jon Roozenbeek, Sander van der Linden et al. "Misinformation Poses a Bigger Threat to Democracy Than You Might Think." *Nature* 630, no. 8015 (2024): 29–32. https://doi.org/10.1038/d41586 -024-01587-3.

Ekstrom, Pierce D., and Calvin K. Lai. "The Selective Communication of Political Information." *Social Psychological and Personality Science* 12, no. 5 (2021): 789–800. https://doi.org/10.1177/1948550620942365.

El-Dardiry, Essam. "Giving You More Control Over Your Homepage and Up Next Videos." *YouTube Official Blog*, June 26, 2019. https://blog.youtube/news-and-events/giving-you-more-control-over-homepage/.

Elliott, Christian. "Climate Conversations: How to Talk with Friends Who Repeat Misinformation." Medill Reports Chicago, December 10, 2021. https://news.medill.northwestern.edu/chicago/climate-conversations -how-to-talk-with-friends-who-repeat-misinformation/.

Elsasser, Shaun W., and Riley E. Dunlap. "Leading Voices in the Denier Choir: Conservative Columnists' Dismissal of Global Warming and Denigration of Climate Science." *American Behavioral Scientist* 57, no. 6 (2013): 754–76. https://doi.org/10.1177/0002764212469800.

Encyclopaedia Britannica Online. s.v. "Yellow Journalism." Accessed May 2, 2024. https://www.britannica .com/topic/yellow-journalism.

Enders, Adam M., and Joseph E. Uscinski. "Are Misinformation, Antiscientific Claims, and Conspiracy Theories for Political Extremists?" *Group Processes and Intergroup Relations* 24, no. 4 (2021): 583–605. https://doi.org/10.1177/1368430220960805.

Enders, Adam M., Joseph E. Uscinski, Michelle I. Seelig et al. "The Relationship Between Social Media Use and Beliefs in Conspiracy Theories and Misinformation." *Political Behavior* 45, no. 2 (2023): 781–804. https://doi.org/10.1007/s11109-021-09734-6.

Enten, Harry. "Flu Shots Uptake Is Now Partisan: It Didn't Use to Be." CNN, November 14, 2021. https://www.cnn.com/2021/11/14/politics/flu-partisan-divide-analysis/index.html.

Epton, Tracy, Peter R. Harris, Rachel Kane, Guido M. van Koningsbruggen, and Paschal Sheeran. "The Impact of Self-Affirmation on Health-Behavior Change: A Meta-Analysis." *Health Psychology* 34, no. 3 (2015): 187–96. https://doi.org/10.1037/hea0000116.

Ernst, Edzard. "Cancer Patients Who Use Alternative Medicine Die Sooner." April 18, 2013. http://edzardernst.com/2013/04/cancer-patients-who-use-alternative-medicine-die-sooner/.

Ervin, Laurie H., and Sheldon Stryker. "Theorizing the Relationship Between Self-Esteem and Identity." In *Extending Self-Esteem Theory and Research: Sociological and Psychological Currents*, ed. Timothy J. Owens, Sheldon Stryker, and Norman Goodman, 29–55. Cambridge University Press, 2001. https://doi.org/10.1017/CBO9780511527739.

Estep, Kevin, Annika Muse, Shannon Sweeney, and Neal D. Goldstein. "Partisan Polarization of Childhood Vaccination Policies, 1995-2020." *American Journal of Public Health* 112, no. 10 (2022): 1471–79. https://doi.org/10.2105/AJPH.2022.306964.

Fabbri, Alice, Alexandra Lai, Quinn Grundy, and Lisa Anne Bero. "The Influence of Industry Sponsorship on the Research Agenda: A Scoping Review." *American Journal of Public Health* 108, no. 11 (2018): e9–16. https://doi.org/10.2105/AJPH.2018.304677.

Facciani, Matthew J. "A Novel Approach for Measuring Self-Affirmation." *Current Research in Social Psychology* 30 (2021): 12–20. https://doi.org/2021-88514-001.

Facciani, Matthew. "Video: How Did Mask Wearing Become So Politicized?" *The Conversation*, September 9, 2020. http://theconversation.com/video-how-did-mask-wearing-become-so-politicized-144268.

Facciani, Matthew, Denisa Apriliawati, and Tim Weninger. "Playing *Gali Fakta* Inoculates Indonesian Participants Against False Information." *Harvard Kennedy School Misinformation Review* 5, no. 4 (2024): 1–17. https://doi.org/10.37016/mr-2020-152.

Facciani, Matthew, and Matthew E. Brashears. "Sacred Alters: The Effects of Ego Network Structure on Religious and Political Beliefs." *Socius* 5 (2019): 2378023119873825. https://doi.org/10.1177/2378023119873825.

Facciani, Matthew, Aleksandra Lazić, Gracemarie Viggiano, and Tara McKay. "Political Network Composition Predicts Vaccination Attitudes." *Social Science and Medicine* 328 (2023): art. 116004. https://doi.org/10.1016/j.socscimed.2023.116004.

Facciani, Matthew, and Tara McKay. "Network Loss Following the 2016 Presidential Election Among LGBTQ+ Adults." *Applied Network Science* 7 (2022). https://doi.org/10.1007/s41109-022-00474-y.

Facciani, Matthew, and Cecilie Steenbuch Traberg. "Personal Network Composition and Cognitive Reflection Predict Susceptibility to Different Types of Misinformation." *Connections*, ahead of print, June 27, 2024. https://doi.org/10.21307/connections-2019.044.

Facebook. "A Blueprint for Content Governance and Enforcement." May 5, 2021. https://www.facebook.com/notes/751449002072082/.

Faguy, Ana. "U.S. Government Divided on Covid Lab Leak Theory—Here's Where Each Agency Stands." *Forbes*, February 27, 2023. https://www.forbes.com/sites/anafaguy/2023/02/27/us-government-divided-on-covid-lab-leak-theory-heres-where-each-agency-stands/.

Falk, Emily B., Matthew Brook O'Donnell, Christopher N. Cascio et al. "Self-Affirmation Alters the Brain's Response to Health Messages and Subsequent Behavior Change." *Proceedings of the National Academy of Sciences* 112, no. 7 (2015): 1977–82. https://doi.org/10.1073/pnas.1500247112.

Fallin Hunzaker, M. B. "Cultural Sentiments and Schema-Consistency Bias in Information Transmission." *American Sociological Review* 81, no. 6 (2016): 1223–50. https://doi.org/10.1177/0003122416671742.

Fan, Angela. "Introducing the First AI Model That Translates 100 Languages Without Relying on English." Meta, October 19, 2020. https://about.fb.com/news/2020/10/first-multilingual-machine-translation-model/.

Farias, Miguel, Anna-Kaisa Newheiser, Guy Kahane, and Zoe de Toledo. "Scientific Faith: Belief in Science Increases in the Face of Stress and Existential Anxiety." *Journal of Experimental Social Psychology* 49, no. 6 (2013): 1210–13. https://doi.org/10.1016/j.jesp.2013.05.008.

Fast, Nathanael. "AI Is Taking Over Our Workplaces: Here's How It Could Impact Human Psychology—and Vice Versa." *Fast Company*, February 8, 2023. https://www.fastcompany.com/90846471/ai-is-taking-over-our-workplaces-heres-how-it-could-impact-human-psychology.

Fazio, Lisa. "Pausing to Consider Why a Headline Is True or False Can Help Reduce the Sharing of False News." *Harvard Kennedy School Misinformation Review* 1, no. 2 (2020): 1–8. https://doi.org/10.37016/mr-2020-009.

Federal Bureau of Investigation, National Center for the Analysis of Violent Crime and Behavioral Assessment Unit. *Lone Offender: A Study of Lone Offender Terrorism in the United States (1972–2015)*. Federal Bureau of Investigation, 2019.

Fein, Steven, and Steven J. Spencer. "Prejudice as Self-Image Maintenance: Affirming the Self Through Derogating Others." *Journal of Personality and Social Psychology* 73, no. 1 (1997): 31–44. https://doi.org/10.1037/0022-3514.73.1.31.

Feinberg, Matthew, and Robb Willer. "From Gulf to Bridge: When Do Moral Arguments Facilitate Political Influence?" *Personality and Social Psychology Bulletin* 41, no. 12 (2015): 1665–81. https://doi.org/10.1177/0146167215607842.

Feinberg, Matthew, and Robb Willer. "The Moral Roots of Environmental Attitudes." *Psychological Science* 24, no. 1 (2013): 56–62. https://doi.org/10.1177/0956797612449177.

Felson, Richard B. "Reflected Appraisal and the Development of Self." *Social Psychology Quarterly* 48, no. 1 (1985): 71–78. https://doi.org/10.2307/3033783.

Felton, Ryan. "Why Did It Take a Pandemic for the FDA to Crack Down on a Bogus Bleach 'Miracle' Cure?" *Consumer Reports*, May 14, 2020, updated July 8, 2020. https://www.consumerreports.org/scams-fraud/bogus-bleach-miracle-cure-fda-crackdown-miracle-mineral-solution-genesis-ii-church/.

Fernández-Penny, Felix E., Eliana L. Jolkovsky, Frances S. Shofer et al. "COVID-19 Vaccine Hesitancy Among Patients in Two Urban Emergency Departments." *Academic Emergency Medicine* 28, no. 10 (2021): 1100–7. https://doi.org/10.1111/acem.14376.

Fernbach, Philip M., Todd Rogers, Craig R. Fox, and Steven A. Sloman. "Political Extremism Is Supported by an Illusion of Understanding." *Psychological Science* 24, no. 6 (2013): 939–46. https://doi.org/10.1177/0956797612464058.

Ferreira Marques, Clara. "Can Brazil Find an Answer for Fake News?" *Bloomberg*, April 24, 2022. https://www.bloomberg.com/opinion/articles/2022-04-24/can-brazil-find-an-answer-for-fake-news-in-time-for-this-fall-s-election.

Fessler, Pam. "Former Election Security Official Says It Will Take 'Years' to Undo Disinformation." NPR, December 22, 2020. https://www.npr.org/2020/12/22/949157510/former-election-security-official-says-it-will-take-years-to-undo-disinformation.

Festinger, Leon. *A Theory of Cognitive Dissonance*. Stanford University Press, 1962.

Festinger, Leon, and Stanley Schachter. *When Prophecy Fails*. Simon and Schuster, 2013.

Fetters, Ashley. "How to Talk to an Anti-Vax Relative." *The Atlantic*, April 22, 2019. https://www.theatlantic.com/family/archive/2019/04/when-families-feud-over-vaccines/587629/.

Finch, Lyneyve. "Psychological Propaganda: The War of Ideas on Ideas During the First Half of the Twentieth Century." *Armed Forces and Society* 26, no. 3 (2000): 367–86. https://doi.org/10.1177/0095327X0002600302.

Fingerhut, Hannah. "Partisanship in the U.S. Isn't Just About Politics, but How People See Their Neighbors." Pew Research Center, June 27, 2016. https://www.pewresearch.org/fact-tank/2016/06/27/partisanship-in-u-s-isnt-just-about-politics-but-how-people-see-their-neighbors/.

Fishkin, James, Alice Siu, Larry Diamond, and Norman Bradburn. "Is Deliberation an Antidote to Extreme Partisan Polarization? Reflections on 'America in One Room.'" *American Political Science Review* 115, no. 4 (2021): 1464–81. https://doi.org/10.1017/S0003055421000642.

Flaxman, Seth, Sharad Goel, and Justin M. Rao. "Filter Bubbles, Echo Chambers, and Online News Consumption." *Public Opinion Quarterly* 80, no. S1 (2016): 298–320. https://doi.org/10.1093/poq/nfw006.

Flint, Joe, Isabella Simonetti, and Keach Hagey. "Fox News Ousts Tucker Carlson." *Wall Street Journal*, April 24, 2023. https://www.wsj.com/articles/tucker-carlson-is-leaving-fox-news-db31f2fa.

Flood, Brian, and Nikolas Lanum. "Credibility Crisis: Egg on Media's Face After Dismissing COVID Lab Leak as 'Debunked' Conspiracy Theory." Fox News, February 28, 2023. https://www.foxnews.com/media/credibility-crisis-egg-medias-face-after-dismissing-covid-lab-leak-debunked-conspiracy-theory.

Ford, Trenton W., Michael Yankoski, Matthew Facciani, and Tim Weninger. "Online Media Literacy Intervention in Indonesia Reduces Misinformation Sharing Intention." *Journal of Media Literacy Education* 15, no. 2 (2023): 99–123. https://doi.org/10.23860/JMLE-2023-15-2-8.

Forster, Victoria. "Ohio Vaccine Lottery Gave Away $5 Million, but Didn't Increase Vaccination Rates, Says New Study." *Forbes*, July 3, 2021. https://www.forbes.com/sites/victoriaforster/2021/07/03/ohio-vaccine-lottery-didnt-increase-vaccination-rates-says-new-study/.

Foster, Diana Greene, and Katrina Kimport. "Who Seeks Abortions at or After 20 Weeks?" *Perspectives on Sexual and Reproductive Health* 45, no. 4 (2013): 210–18. https://doi.org/10.1363/4521013.

Frankovic, Kathy. "Most Republicans Who Have Heard of Ivermectin as a COVID-19 Treatment Think It May Be Effective." YouGov, September 2, 2021. https://today.yougov.com/topics/politics/articles-reports/2021/09/02/most-republicans-who-have-heard-of-ivermectin.

Frederick, Shane. "Cognitive Reflection and Decision Making." *Journal of Economic Perspectives* 19, no. 4 (2005): 25–42, https://doi.org/10.1257/089533005775196732.

Freiling, Isabelle, Nicole M. Krause, Dietram A. Scheufele, and Dominique Brossard. "Believing and Sharing Misinformation, Fact-Checks, and Accurate Information on Social Media: The Role of Anxiety During COVID-19." *New Media and Society* 25, no. 1 (2023): 141–62. https://doi.org/10.1177/14614448211011451.

Fridman, Ariel, Rachel Gershon, and Ayelet Gneezy. "COVID-19 and Vaccine Hesitancy: A Longitudinal Study." *PLOS One* 16, no. 4 (2021): e0250123. https://doi.org/10.1371/journal.pone.0250123.

Friedman Levy, Caroline, and Matthew Facciani. "Human-Shaped Hole in AI Oversight." Vanderbilt Project on Unity & American Democracy, Vanderbilt University, October 4, 2022. https://www.vanderbilt.edu/unity/2022/10/04/human-shaped-hole-in-ai-oversight/.

Fu, Feng, Nicholas A. Christakis, and James H. Fowler. "Dueling Biological and Social Contagions." *Scientific Reports* 7 (2017): art. 43634. https://doi.org/10.1038/srep43634.

Full Fact. "Event 201 Didn't Predict the Covid-19 Pandemic." April 27, 2020. https://fullfact.org/health/event-201-coronavirus-pandemic/.

Funk, Cary. "Vast Majority of Americans Say Benefits of Childhood Vaccines Outweigh Risks." Pew Research Center, February 2, 2017. https://www.pewresearch.org/science/2017/02/02/vast-majority-of-americans-say-benefits-of-childhood-vaccines-outweigh-risks/.

Funk, Cary, Brian Kennedy, and Meg Hafferon. "2. Americans' Health Care Behaviors and Use of Conventional and Alternative Medicine." Pew Research Center, February 2, 2017. https://www.pewresearch.org/science/2017/02/02/americans-health-care-behaviors-and-use-of-conventional-and-alternative-medicine/.

Funk, Cary, Alec Tyson, Brian Kennedy, and Courtney Johnson. "1. Scientists Are Among the Most Trusted Groups in Society, Though Many Value Practical Experience Over Expertise." Pew Research Center, September 29, 2020. https://www.pewresearch.org/science/2020/09/29/scientists-are-among-the-most-trusted-groups-in-society-though-many-value-practical-experience-over-expertise/.

Funke, Daniel, and Susan Benkelman. "Misinformation Is Inciting Violence Around the World: And Tech Platforms Don't Seem to Have a Plan to Stop It." Poynter Institute, April 4, 2019. https://www.poynter.org/fact-checking/2019/misinformation-is-inciting-violence-around-the-world-and-tech-platforms-dont-have-a-plan-to-stop-it/.

Gabielkov, Maksym, Arthi Ramachandran, Augustin Chaintreau, and Arnaud Legout. "Social Clicks: What and Who Gets Read on Twitter?" In *Sigmetrics '16: Proceedings of the 2016 ACM Sigmetrics International Conference on Measurement and Modeling of Computer Science*, 179–92. Association for Computing Machinery, 2016. https://doi.org/10.1145/2896377.2901462.

Gadarian, Shana Kushner, Sara Wallace Goodman, and Thomas B. Pepinsky. "Partisanship, Health Behavior, and Policy Attitudes in the Early Stages of the COVID-19 Pandemic." *PLOS One* 16, no. 4 (2021): e0249596. https://doi.org/10.1371/journal.pone.0249596.

Gardner, Lauren, Ensheng Dong, Kamran Khan, and Sahotra Sarkar. "Persistence of US Measles Risk Due to Vaccine Hesitancy and Outbreaks Abroad." *Lancet Infectious Diseases* 20, no. 10 (2020): 1114–15. https://doi.org/10.1016/S1473-3099(20)30522-3.

Garrett, R. Kelly. "Echo Chambers Online? Politically Motivated Selective Exposure Among Internet News Users." *Journal of Computer-Mediated Communication* 14, no. 2 (2009): 265–85. https://doi.org/10.1111/j.1083-6101.2009.01440.x.

Gawronski, Bertram. "Partisan Bias in the Identification of Fake News." *Trends in Cognitive Sciences* 25, no. 9 (2021): 723–24. https://doi.org/10.1016/j.tics.2021.05.001.

Geiger, Abigail. "46 Percent of U.S. Social Media Users Say They Are 'Worn Out' by Political Posts and Discussions." Pew Research Center, August 8, 2019. https://www.pewresearch.org/fact-tank/2019/08/08/46-of-u-s-social-media-users-say-they-are-worn-out-by-political-posts-and-discussions/.

Geiger, Abigail. "How Americans See the Future of Space Exploration, 50 Years After the First Moon Landing." Pew Research Center, July 17, 2019. https://www.pewresearch.org/fact-tank/2019/07/17/how-americans-see-the-future-of-space-exploration-50-years-after-the-first-moon-landing/.

Geiger, Abigail. "Key Facts About Americans and Guns." Pew Research Center, September 13, 2023. https://www.pewresearch.org/short-reads/2023/09/13/key-facts-about-americans-and-guns/.

Geiger, Abigail. "Political Polarization in the American Public." Pew Research Center, June 12, 2014. https://www.pewresearch.org/politics/2014/06/12/political-polarization-in-the-american-public/.

Geiger, Abigail. "A Small Group of Prolific Users Account for a Majority of Political Tweets Sent by U.S. Adults." Pew Research Center, October 23, 2019. https://www.pewresearch.org/fact-tank/2019/10/23/a-small-group-of-prolific-users-account-for-a-majority-of-political-tweets-sent-by-u-s-adults/.

Geller, Andrew I., Nadine Shehab, Nina J. Weidle et al. "Emergency Department Visits for Adverse Events Related to Dietary Supplements." *New England Journal of Medicine* 373, no. 16 (2015): 1531–40. https://doi.org/10.1056/NEJMsa1504267.

Georgiou, Neophytos, Paul Delfabbro, and Ryan Balzan. "COVID-19-Related Conspiracy Beliefs and Their Relationship with Perceived Stress and Pre-existing Conspiracy Beliefs." *Personality and Individual Differences* 166 (2020): art. 110201. https://doi.org/10.1016/j.paid.2020.110201.

Gerlich, Michael. "AI Tools in Society: Impacts on Cognitive Offloading and the Future of Critical Thinking." *Societies (Basel, Switzerland)* 15, no. 1 (January 3, 2025): 6.

Giese, Helge, Hansjörg Neth, Mehdi Moussaïd, Cornelia Betsch, and Wolfgang Gaissmaier. "The Echo in Flu-Vaccination Echo Chambers: Selective Attention Trumps Social Influence." *Vaccine* 38, no. 8 (2020): 2070–76. https://doi.org/10.1016/j.vaccine.2019.11.038.

Gift, Karen, and Thomas Gift. "Does Politics Influence Hiring? Evidence from a Randomized Experiment." *Political Behavior* 37 (2015): 653–75. https://doi.org/10.1007/s11109-014-9286-0.

Gilbert, David. "The Israel-Hamas War Is Drowning X in Disinformation." *Wired*, October 9, 2023. https://www.wired.com/story/x-israel-hamas-war-disinformation/.

Gillman, Todd J. "Russian Trolls Orchestrated 2016 Clash at Houston Islamic Center, New Senate Intel Report Recalls." *Dallas Morning News*, October 8, 2019. https://www.dallasnews.com/news/politics/2019/10/08/russian-trolls-orchestrated-2016-clash-houston-islamic-center-senate-intel-report-says/.

Glassner, Amnon, Michael Weinstock, and Yair Neuman. "Pupils' Evaluation and Generation of Evidence and Explanation in Argumentation." *British Journal of Educational Psychology* 75, no. 1 (2005): 105–18. https://doi.org/10.1348/000709904X22278.

Glazer, Emily. "Mark Zuckerberg Breaks Silence on Facebook Whistleblower Testimony, Media Reports." *Wall Street Journal*, October 6, 2021. https://www.wsj.com/articles/mark-zuckerberg-says-facebooks-work-mischaracterized-in-reports-whistleblower-testimony-11633482725.

Global Witness. "How Facebook's Algorithm Amplifies Climate Disinformation." March 10, 2022. https://www.globalwitness.org/en/campaigns/digital-threats/climate-divide-how-facebooks-algorithm-amplifies-climate-disinformation/.

Godel, William, Zeve Sanderson, Kevin Aslett et al. "Moderating with the Mob: Evaluating the Efficacy of Real-Time Crowdsourced Fact-Checking." *Journal of Online Trust and Safety* 1, no. 1 (2021): 1–36. https://doi.org/10.54501/jots.v1i1.15.

Goldberg, Matthew H., Jennifer R. Marlon, Xinran Wang, Sander van der Linden, and Anthony Leiserowitz. "Oil and Gas Companies Invest in Legislators That Vote Against the Environment." *Proceedings of the National Academy of Sciences* 117, no. 10 (2020): 5111–12. https://doi.org/10.1073/pnas.1922175117.

Goldenberg, Maya J. *Vaccine Hesitancy: Public Trust, Expertise, and the War on Science.* University of Pittsburgh Press, 2021.

Goldman, Emanuel. "Exaggerated Risk of Transmission of COVID-19 by Fomites." *Lancet Infectious Diseases* 20, no. 8 (2020): 892–93. https://www.cdc.gov/coronavirus/2019-ncov/more/science-and-research/surface-transmission.html.

Gollust, Sarah E., Erika Franklin Fowler, and Jeff Niederdeppe. "Television News Coverage of Public Health Issues and Implications for Public Health Policy and Practice." *Annual Review of Public Health* 40 (2019): 167–85. https://doi.org/10.1146/annurev-publhealth-040218-044017.

Goodman, Brenda, and Virginia Langmaid. "Fauci Says His Covid Rebounded After Paxlovid." CNN, June 30, 2022. https://www.cnn.com/2022/06/30/health/covid-paxlovid-fauci-rebound/index.html.

Goodwin, Liz. "Trump's Refusal to Wear a Mask Is Helping Politicize a Crucial Tool for Fighting Virus." *Boston Globe*, May 27, 2020. https://www.bostonglobe.com/2020/05/27/nation/trumps-refusal-wear-mask-is-helping-politicize-crucial-tool-fighting-virus/.

Gottfried, Jeffrey. "Americans' Social Media Use." Pew Research Center, January 31, 2024. https://www.pewresearch.org/internet/2024/01/31/americans-social-media-use/.

Graeupner, Damaris, and Alin Coman. "The Dark Side of Meaning-Making: How Social Exclusion Leads to Superstitious Thinking," *Journal of Experimental Social Psychology* 69 (2017): 218–22. https://doi.org/10.1016/j.jesp.2016.10.003.

Graham, Jesse, and Jonathan Haidt. "Sacred Values and Evil Adversaries: A Moral Foundations Approach." In *The Social Psychology of Morality: Exploring the Causes of Good and Evil*, ed. M. Mikulincer and P. R. Shaver, 11–31. American Psychological Association, 2012. https://doi.org/10.1037/13091-001.

Graham, Jesse, Jonathan Haidt, and Brian A. Nosek. "Liberals and Conservatives Rely on Different Sets of Moral Foundations." *Journal of Personality and Social Psychology* 96, no. 5 (2009): 1029–46. https://doi.org/10.1037/a0015141.

Gramlich, John. "Q&A: How Pew Research Center Identified Bots on Twitter." Pew Research Center, April 19, 2018. https://www.pewresearch.org/fact-tank/2018/04/19/qa-how-pew-research-center-identified-bots-on-twitter/.

Granovetter, Mark S. "The Strength of Weak Ties." *American Journal of Sociology* 78, no. 6 (1973): 1360–80. https://doi.org/10.1086/225469.

Grant, Fiona, and Michael A. Hogg. "Self-Uncertainty, Social Identity Prominence and Group Identification." *Journal of Experimental Social Psychology* 48, no. 2 (2012): 538–42. https://doi.org/10.1016/j.jesp.2011.11.006.

Graphika Team. "Deepfake It till You Make It." Graphika, February 7, 2023. https://graphika.com/reports/deepfake-it-till-you-make-it.

Gravelle, Timothy B., Joseph B. Phillips, Jason Reifler, and Thomas J. Scotto. "Estimating the Size of 'Anti-Vax' and Vaccine Hesitant Populations in the US, UK, and Canada: Comparative Latent Class

Modeling of Vaccine Attitudes." *Human Vaccines and Immunotherapeutics* 18, no. 1 (2022): 2008214. https://doi.org/10.1080/21645515.2021.2008214.

Green, Donald P., Bradley Palmquist, and Eric Schickler. *Partisan Hearts and Minds: Political Parties and the Social Identities of Voters.* Yale University Press, 2004.

Green, Emma. "The Liberals Who Can't Quit Lockdown." *The Atlantic*, May 4, 2021. https://www.theatlantic .com/politics/archive/2021/05/liberals-covid-19-science-denial-lockdown/618780/.

Green, Jon, Jared Edgerton, Daniel Naftel, Kelsey Shoub, and Skyler J. Cranmer. "Elusive Consensus: Polarization in Elite Communication on the COVID-19 Pandemic." *Science Advances* 6, no. 28 (2020): eabc2717. https://doi.org/10.1126/sciadv.abc2717.

Greenberg, Jon. "Trump Campaign Used Cambridge Analytica in Final Months of Campaign." PolitiFact, March 21, 2018. https://www.politifact.com/factchecks/2018/mar/21/jack-posopiec/trump-campaign -used-cambridge-analytica-final-mont/.

Greenberger, Martin. "Designing Organizations for an Information-Rich World." In *Computers, Communications, and the Public Interest*, ed. Martin Greenberger, 40–41. Johns Hopkins University Press, 1971.

Greene, Ciara M., and Gillian Murphy. "Quantifying the Effects of Fake News on Behavior: Evidence from a Study of COVID-19 Misinformation." *Journal of Experimental Psychology. Applied* 27, no. 4 (2021): 773–84. https://doi.org/10.1037/xap0000371.

Greenspan, Rachel E. "The QAnon Conspiracy Theory and a Stew of Misinformation Fueled the Insurrection at the Capitol." *Insider*, January 7, 2021. https://www.insider.com/capitol-riots-qanon-protest-conspiracy -theory-washington-dc-protests-2021-1.

Grieve, Paul G., and Michael A. Hogg. "Subjective Uncertainty and Intergroup Discrimination in the Minimal Group Situation." *Personality and Social Psychology Bulletin* 25, no. 8 (1999): 926–40. https://doi.org /10.1177/01461672992511002.

Grimes, David Robert, and David H. Gorski. "Quantifying Public Engagement with Medical Science, Misinformation, and Malinformation." OSF Preprints, June 1, 2022. https://doi.org/10.31219/osf.io/g4jwr.

Grinberg, Nir, Kenneth Joseph, Lisa Friedland, Briony Swire-Thompson, and David Lazer. "Fake News on Twitter During the 2016 U.S. Presidential Election." *Science* 363, no. 6425 (2019): 374–78. https://doi .org/10.1126/science.aau2706.

Gross, Jenny. "How Finland Is Teaching a Generation to Spot Misinformation." *New York Times*, January 10, 2023. https://www.nytimes.com/2023/01/10/world/europe/finland-misinformation-classes.html.

Guess, Andrew M., Pablo Barberá, Simon Munzert, and JungHwan Yang. "The Consequences of Online Partisan Media." *Proceedings of the National Academy of Sciences* 118, no. 14 (2021): e2013464118. https:// doi.org/10.1073/pnas.2013464118.

Guess, Andrew M., Michael Lerner, Benjamin Lyons et al. "A Digital Media Literacy Intervention Increases Discernment Between Mainstream and False News in the United States and India." *Proceedings of the National Academy of Sciences* 117, no. 27 (2020): 15536–45. https://doi.org/10.1073/pnas.1920498117.

Guess, Andrew M., and Kevin Munger. "Digital Literacy and Online Political Behavior." *Political Science Research and Methods* 11, no. 1 (2023): 110–28. https://doi.org/10.1017/psrm.2022.17.

Guess, Andrew M., Jonathan Nagler, and Joshua Tucker. "Less Than You Think: Prevalence and Predictors of Fake News Dissemination on Facebook." *Science Advances* 5, no. 1 (2019): eaau4586. https://doi .org/10.1126/sciadv.aau4586.

Gunter, Jen. *The Vagina Bible: The Vulva and the Vagina: Separating the Myth from the Medicine.* Citadel Press, 2019.

Gutman-Wei, Rachel. "Of Course Biden Has Rebound COVID." *The Atlantic*, July 30, 2022. https://www .theatlantic.com/health/archive/2022/07/biden-paxlovid-covid-drug-rebound-infections/671009/.

Haelle, Tara. "Vaccine Hesitancy Is Nothing New: Here's the Damage It's Done Over Centuries." *ScienceNews*, May 11, 2021. https://www.sciencenews.org/article/vaccine-hesitancy-history-damage-anti-vaccination.

Haile, Tony. "What You Think You Know About the Web Is Wrong." *Time*, March 9, 2014. https://time .com/12933/what-you-think-you-know-about-the-web-is-wrong/.

Hajric, Vildana. "Litecoin Foundation 'Screwed Up,' Lee Says of Walmart Snafu." *Bloomberg*, September 13, 2021. https://www.bloomberg.com/news/articles/2021-09-13/litecoin-foundation-screwed-up-lee-says-about-walmart-snafu.

Halberstam, Yosh, and Brian Knight. "Homophily, Group Size, and the Diffusion of Political Information in Social Networks: Evidence from Twitter." *Journal of Public Economics* 143 (2016): 73–88. https://doi.org/10.1016/j.jpubeco.2016.08.011.

Halperin, Eran, Boaz Hameiri, and Rebecca Littman, eds. *Psychological Intergroup Interventions: Evidence-Based Approaches to Improve Intergroup Relations.* Taylor & Francis, 2023.

Hamel, Liz, Ashley Kirzinger, Cailey Muñana, and Mollyann Brodie. "KFF COVID-19 Vaccine Monitor: December 2020." *KFF*, December 15, 2020. https://www.kff.org/coronavirus-covid-19/report/kff-covid-19-vaccine-monitor-december-2020/.

Hamel Liz, Lunna Lopes, Ashley Kirzinger, Grace Sparks, Mellisha Stokes, and Mollyann Brodie. "KFF COVID-19 Vaccine Monitor: Media and Misinformation." *KFF*, November 8, 2021. https://www.kff.org/coronavirus-covid-19/poll-finding/kff-covid-19-vaccine-monitor-media-and-misinformation/?utm_campaign=KFF-2021-polling-surveys&utm_medium.

Hamel, Liz, Lunna Lopes, Grace Sparks et al. "KFF COVID-19 Vaccine Monitor: January 2022." *KFF*, January 28, 2022. https://www.kff.org/coronavirus-covid-19/poll-finding/kff-covid-19-vaccine-monitor-january-2022/.

Hamilton, Diane Musho. "Calming Your Brain During Conflict." *Harvard Business Review*, December 22, 2015. https://hbr.org/2015/12/calming-your-brain-during-conflict.

Hanna-Attisha, Mona. *What the Eyes Don't See: A Story of Crisis, Resistance, and Hope in an American City.* Random House, 2018.

Hardwig, John. "The Role of Trust in Knowledge." *Journal of Philosophy* 88, no. 12 (1991): 693–708. https://doi.org/10.2307/2027007.

Harrington, Brooke. "Here's Why Your Efforts to Convince Anti-Vaxxers Aren't Working." *The Guardian*, August 9, 2021. https://www.theguardian.com/commentisfree/2021/aug/09/convince-anti-vaxxers.

Harrington, Brooke. "How Sociologists Can Battle Covid Denialism." *Chronicle of Higher Education*, September 1, 2021. https://www.chronicle.com/article/how-sociologists-can-battle-covid-denialism?sra=true.

Hart, Ariel. "Dr. Kimberly Manning: Yes, We Can Reach the Unvaccinated." *Atlanta Journal-Constitution*, October 29, 2021. https://www.ajc.com/news/coronavirus/dr-kimberly-manning-yes-we-can-reach-the-unvaccinated/ZMNQOJIGHBASBDTMODPTVD7QAQ/.

Hart, Robert. "As Fox's Tucker Carlson Stokes Covid-19 Vaccine Fears—Here's What You Really Need to Know About Pfizer's Covid-19 Vaccine." *Forbes*, December 18, 2020. https://www.forbes.com/sites/roberthart/2020/12/18/as-foxs-tucker-carlson-stokes-covid-19-vaccine-fears--heres-what-you-really-need-to-know-about-pfizers-covid-19-vaccine/.

Hart, Sol, Sedona Chinn, and Stuart Soroka. "Politicization and Polarization in COVID-19 News Coverage." *Science Communication* 42, no. 5 (2020): 679–97. https://doi.org/10.1177/1075547020950735.

Hartig, Hannah. "Two Decades Later, the Enduring Legacy of 9/11." Pew Research Center, September 2, 2021. https://www.pewresearch.org/politics/2021/09/02/two-decades-later-the-enduring-legacy-of-9-11/.

Hartman, Rachel, Will Blakey, Jake Womick et al. "Interventions to Reduce Partisan Animosity." *Nature Human Behaviour* 6, no. 9 (2022): 1194–1205. https://doi.org/10.1038/s41562-022-01442-3.

Harvard T. H. Chan School of Public Health. "Poll: Majority of Americans Say Key COVID-19 Policies Were a Good Idea—but Views of Individual Policies Vary." June 17, 2024. https://www.hsph.harvard.edu/news/press-releases/poll-majority-of-americans-say-key-covid-19-policies-were-a-good-idea-but-views-of-individual-policies-vary/.

Hasher, Lynn, David Goldstein, and Thomas Toppino. "Frequency and the Conference of Referential Validity." *Journal of Verbal Learning and Verbal Behavior* 16, no. 1 (1977): 107–12. https://doi.org/10.1016/S0022-5371(77)80012-1.

Hassanian-Moghaddam, Hossein, Nasim Zamani, Ali-Asghar Kolahi, Rebecca McDonald, and Knut Erik Hovda. "Double Trouble: Methanol Outbreak in the Wake of the COVID-19 Pandemic in Iran—a Cross-Sectional Assessment." *Critical Care* 24, no. 1 (2020): 402. https://doi.org/10.1186/s13054-020-03140-w.

Hastorf, Albert H., and Hadley Cantril. "They Saw a Game: A Case Study." *Journal of Abnormal and Social Psychology* 49, no. 1 (1954): 129–34. https://doi.org/10.1037/h0057880.

Haugen, Frances. "Keynote Facilitated by Brian Klaas." Cambridge Disinformation Summit, University of Cambridge, July 28, 2023. YouTube, 44 min., 49 sec. https://www.youtube.com/watch?v=bSCy6y5wu1g.

Hauser, Christine. "The Mask Slackers of 1918." *New York Times*, August 3, 2020. https://www.nytimes.com/2020/08/03/us/mask-protests-1918.html.

Haven, Tamarinde, Gowri Gopalakrishna, Joeri Tijdink, Dorien van der Schot, and Lex Bouter. "Promoting Trust in Research and Researchers: How Open Science and Research Integrity Are Intertwined." *BMC Research Notes* 15, no. 1 (2022): 302. https://doi.org/10.1186/s13104-022-06169-y.

Heaney, Christopher Z. "Manipulative Silent Scream." *Harvard Crimson*, March 11, 1985, https://www.thecrimson.com/article/1985/3/11/manipulative-silent-scream-pbto-the-editors/.

Hebb, D. O. *The Organization of Behavior: A Neuropsychological Theory*. Wiley, 1949. Reprint, Lawrence Erlbaum Associates, 2002. https://doi.org/10.4324/9781410612403.

Heikkilä, Melissa. "Why You Shouldn't Trust AI Search Engines." *MIT Technology Review*, February 14, 2023. https://www.technologyreview.com/2023/02/14/1068498/why-you-shouldnt-trust-ai-search-engines/.

Heilweil, Rebecca. "Right-Wing Media Thrives on Facebook: Whether It Rules Is More Complicated." *Vox*, September 9, 2020. https://www.vox.com/recode/21419328/facebook-conservative-bias-right-wing-crowdtangle-election.

Heise, David R. "Affect Control Theory: Concepts and Model." In *Analyzing Social Interaction*, ed. L. Smith-Lovin and David R. Heise, 1–33. Routledge, 2016.

Helfand, David J. *A Survival Guide to the Misinformation Age: Scientific Habits of Mind*. Columbia University Press, 2016.

Hennekens, Charles H., Manas Rane, Joshua Solano at al. "Updates on Hydroxychloroquine in Prevention and Treatment of COVID-19." *American Journal of Medicine* 135, no. 1 (2022): 7–9. https://doi.org/10.1016/j.amjmed.2021.07.035.

Henrich, Joseph, Steven J. Heine, and Ara Norenzayan. "Most People Are Not WEIRD." *Nature* 466, no. 7302 (2010): 29. https://doi.org/10.1038/466029a.

Hern, Alex. "Cambridge Analytica: How Did It Turn Clicks Into Votes?" *The Guardian*, May 6, 2018. https://www.theguardian.com/news/2018/may/06/cambridge-analytica-how-turn-clicks-into-votes-christopher-wylie.

Herrman, John. "They Did Their Own 'Research': Now What?" *New York Times*, May 29, 2022. https://www.nytimes.com/2022/05/29/style/do-your-own-research.html.

Heymach, John, Lada Krilov, Anthony Alberg et al. "Clinical Cancer Advances 2018: Annual Report on Progress Against Cancer from the American Society of Clinical Oncology." *Journal of Clinical Oncology* 36, no. 10 (2018): 1020–44. https://doi.org/10.1200/JCO.2017.77.0446.

Higdon, Nolan. *The Anatomy of Fake News: A Critical News Literacy Education*. University of California Press, 2020.

Hochschild, Jennifer, and Katherine Levine Einstein. "'It Isn't What We Don't Know That Gives Us Trouble, It's What We Know That Ain't So': Misinformation and Democratic Politics." *British Journal of Political Science* 45, no. 3 (2015): 467–75. https://doi.org/10.1017/S000712341400043X.

Hoelter, Jon W. "The Relationship Between Specific and Global Evaluations of Self: A Comparison of Several Models." *Social Psychology Quarterly* 49, no. 2 (1986): 129–41. https://doi.org/10.2307/2786724.

Hoes, Emma, Sacha Altay, and Juan Bermeo. "Leveraging ChatGPT for Efficient Fact-Checking." Preprint, PsyArXiv, May 29, 2023. https://doi.org/10.31234/osf.io/qnjkf.

Hoffman, Jan. "Mistrust of a Coronavirus Vaccine Could Imperil Widespread Immunity." *New York Times*, July 18, 2020. https://www.nytimes.com/2020/07/18/health/coronavirus-anti-vaccine.html.

Hogg, Michael A. "From Uncertainty to Extremism: Social Categorization and Identity Processes." *Current Directions in Psychological Science* 23, no. 5 (2014): 338–42. https://doi.org/10.1177/0963721414540168.

Hogg, Michael A., and Mark J. Rinella. "Social Identities and Shared Realities." *Current Opinion in Psychology* 23 (2018): 6–10. https://doi.org/10.1016/j.copsyc.2017.10.003.

Hohman, Zachary P., Michael A. Hogg, and Michelle C. Bligh. "Identity and Intergroup Leadership: Asymmetrical Political and National Identification in Response to Uncertainty." *Self and Identity* 9, no. 2 (2010): 113–28. https://doi.org/10.1080/15298860802605937.

Hollywood Reporter. "Howard Stern Says Anti-Vaxxers Should Be Denied Hospital Care If They Catch COVID-19." September 9, 2021. YouTube, 1 min., 28 sec. https://www.youtube.com/watch?v=OPOfyGOnJtA.

Hook, Joshua N., Jennifer E. Farrell, Kathryn A. Johnson, Daryl R. Van Tongeren, Don E. Davis, and Jamie D. Aten. "Intellectual Humility and Religious Tolerance." *Journal of Positive Psychology* 12, no. 1 (2017): 29–35. https://doi.org/10.1080/17439760.2016.1167937.

Hooker, Claire. "How to Talk to Someone Who Doesn't Wear a Mask, and Actually Change Their Mind." *The Conversation*, August 14, 2020. http://theconversation.com/how-to-talk-to-someone-who-doesnt-wear-a-mask-and-actually-change-their-mind-143995.

Hopp, Toby, Patrick Ferrucci, and Chris J. Vargo. "Why Do People Share Ideologically Extreme, False, and Misleading Content on Social Media? A Self-Report and Trace Data–Based Analysis of Countermedia Content Dissemination on Facebook and Twitter." *Human Communication Research* 46, no. 4 (2020): 357–84. https://doi.org/10.1093/hcr/hqz022.

Hornik, Robert, Ava Kikut, Emma Jesch, Chioma Woko, Leeann Siegel, and Kwanho Kim. "Association of COVID-19 Misinformation with Face Mask Wearing and Social Distancing in a Nationally Representative US Sample." *Health Communication* 36, no. 1 (2021): 6–14. https://doi.org/10.1080/10410236.2020.1847437.

Hornsey, Matthew J., Matthew Finlayson, Gabrielle Chatwood, and Christopher T. Begeny. "Donald Trump and Vaccination: The Effect of Political Identity, Conspiracist Ideation and Presidential Tweets on Vaccine Hesitancy." *Journal of Experimental Social Psychology* 88 (2020): art. 103947. https://doi.org/10.1016/j.jesp.2019.103947.

Horwitz, Jeff. "Who Is Facebook Whistleblower Frances Haugen? What to Know After Her Senate Testimony." *Wall Street Journal*, October 5, 2021. https://www.wsj.com/articles/who-is-frances-haugen-facebook-whistleblower-11633409993.

Hotez, Peter J. *The Deadly Rise of Anti-Science: A Scientist's Warning*. Johns Hopkins University Press, 2023.

Howard, Jeremy, Austin Huang, Zhiyuan Li et al. "An Evidence Review of Face Masks Against COVID-19." *Proceedings of the National Academy of Sciences* 118, no. 4 (2021): e2014564118. https://doi.org/10.1073/pnas.2014564118.

Howard, Philip N., Bharath Ganesh, Dimitra Liotsiou, John Kelly, and Camille François. "The IRA, Social Media and Political Polarization in the United States, 2012–2018." University of Oxford, 2019. https://digitalcommons.unl.edu/senatedocs/1.

Hu, Weimin, Elaine C. Siegfried, and Daniel Mark Siegel. "Product-Related Emphasis of Skin Disease Information Online." *Archives of Dermatology* 138, no. 6 (2002): 775–80. https://doi.org/10.1001/archderm.138.6.775.

Huber, Jürgen, Sabiou Inoua, Rudolf Kerschbamer, Christian König-Kersting, Stefan Palan, and Vernon L. Smith. "Nobel and Novice: Author Prominence Affects Peer Review." *Proceedings of the National Academy of Sciences* 119, no. 41 (2022): e2205779119. https://doi.org/10.1073/pnas.2205779119.

Huynh, Ho P., and Amy R. Senger, "A Little Shot of Humility: Intellectual Humility Predicts Vaccination Attitudes and Intention to Vaccinate Against COVID-19." *Journal of Applied Social Psychology* 51, no. 4 (2021): 449–60. https://doi.org/10.1111/jasp.12747.

IBM. "What Is Blockchain?" Accessed August 14, 2024. https://www.ibm.com/topics/blockchain.

Imhoff, Roland, Lea Dieterle, and Pia Lamberty. "Resolving the Puzzle of Conspiracy Worldview and Political Activism: Belief in Secret Plots Decreases Normative but Increases Nonnormative Political Engagement." *Social Psychological and Personality Science* 12, no. 1 (2021): 71–79. https://doi.org/10.1177/1948550619896491.

Immunisation Advisory Centre. "Types of Vaccines." Last updated October 2022. https://www.immune.org.nz/vaccines/vaccine-development.

Imundo, Megan N., and David N. Rapp. "When Fairness Is Flawed: Effects of False Balance Reporting and Weight-of-Evidence Statements on Beliefs and Perceptions of Climate Change." *Journal of Applied Research in Memory and Cognition* 11, no. 2 (2022): 258–71. https://doi.org/10.1016/j.jarmac.2021.10.002.

Inan, Taşkın, and Turan Temur. "Examining Media Literacy Levels of Prospective Teachers." *International Electronic Journal of Elementary Education* 4, no. 2 (2012): 269–85. https://www.iejee.com/index.php/IEJEE/article/view/199/195.

Indolfi, Ciro, and Carmen Spaccarotella. "The Outbreak of COVID-19 in Italy: Fighting the Pandemic." *JACC: Case Reports* 2, no. 9 (2020): 1414–18. https://doi.org/10.1016/j.jaccas.2020.03.012.

Ingram, Mathew. "In India, the Fake News Problem Isn't Facebook, It's WhatsApp." *Columbia Journalism Review*, May 16, 2018. https://www.cjr.org/the_media_today/india-whatsapp.php.

Institute for Quality and Efficiency in Health Care. "Common Colds: Does Vitamin C Keep You Healthy?" 2020.

International Fact Checking Network. "The Commitments of the Code of Principles." Last modified 2024. https://www.ifcncodeofprinciples.poynter.org/know-more/the-commitments-of-the-code-of-principles.

iNudgeyou. "Green Nudge: Nudging Litter Into the Bin." February 16, 2012. https://inudgeyou.com/en/green-nudge-nudging-litter-into-the-bin/.

Inzlicht, Michael, Ian McGregor, Jacob B. Hirsh, and Kyle Nash. "Neural Markers of Religious Conviction." *Psychological Science* 20, no. 3 (2009): 385–92. https://doi.org/10.1111/j.1467-9280.2009.02305.x.

Inzlicht, Michael, and Alexa M. Tullett. "Reflecting on God: Religious Primes Can Reduce Neurophysiological Response to Errors." *Psychological Science* 21, no. 8 (2010): 1184–90. https://doi.org/10.1177/0956797610375451.

Ioannidis, John P. A., Michael E. Stuart, Shannon Brownlee, and Sheri A. Strite. "How to Survive the Medical Misinformation Mess." *European Journal of Clinical Investigation* 47, no. 11 (2017): 795–802. https://doi.org/10.1111/eci.12834.

Ionescu, Octavia, Jean Louis Tavani, and Julie Collange. "Political Extremism and Perceived Anomie: New Evidence of Political Extremes' Symmetries and Asymmetries Within French Samples." *International Review of Social Psychology* 34, no. 1 (2021): 1–16. https://doi.org/10.5334/irsp.573.

Ironman at Political Calculations. "The Cost of Fake News for the S&P 500." Seeking Alpha, December 4, 2017. https://seekingalpha.com/article/4129355-cost-of-fake-news-for-s-and-p-500.

Ithaca College. "Project Look Sharp." Accessed July 6, 2024. https://www.projectlooksharp.org/.

Itzchakov, Guy, and Kenneth G. DeMarree. "Attitudes in an Interpersonal Context: Psychological Safety as a Route to Attitude Change." *Frontiers in Psychology* 13 (2022): 932413. https://doi.org/10.3389/fpsyg.2022.932413.

Itzchakov, Guy, Avraham N. Kluger, and Dotan R. Castro. "I Am Aware of My Inconsistencies but Can Tolerate Them: The Effect of High Quality Listening on Speakers' Attitude Ambivalence." *Personality and Social Psychology Bulletin* 43, no. 1 (2017): 105–20. https://doi.org/10.1177/0146167216675339.

Itzchakov, Guy, and Harry T. Reis. "Perceived Responsiveness Increases Tolerance of Attitude Ambivalence and Enhances Intentions to Behave in an Open-Minded Manner." *Personality and Social Psychology Bulletin* 47, no. 3 (2021): 468–85. https://doi.org/10.1177/0146167220929218.

Itzchakov, Guy, Netta Weinstein, Mark Leary, Dvori Saluk, and Moty Amar. "Listening to Understand: The Role of High-Quality Listening on Speakers' Attitude Depolarization During Disagreements." *Journal of Personality and Social Psychology* 126, no. 2 (2024): 213–39. https://doi.org/10.1037/pspa0000366.

Ivanov, Bobi, Claude H. Miller, Josh Compton et al. "Effects of Postinoculation Talk on Resistance to Influence." *Journal of Communication* 62, no. 4 (2012): 701–18. https://doi.org/10.1111/j.1460-2466.2012.01658.x.

Ivanov, Bobi, Jeanetta D. Sims, Josh Compton et al. "The General Content of Postinoculation Talk: Recalled Issue-Specific Conversations Following Inoculation Treatments." *Western Journal of Communication* 79, no. 2 (2015): 218–38. https://doi.org/10.1080/10570314.2014.943423.

Iyengar, Shanto, and Sean J. Westwood. "Fear and Loathing Across Party Lines: New Evidence on Group Polarization." *American Journal of Political Science* 59, no. 3 (2015): 690–707. https://doi.org/10.1111/ajps.12152.

Jackson, Dean, Meghan Conroy, and Alex Newhouse. "Insiders' View of the January 6th Committee's Social Media Investigation." Just Security, January 5, 2023. https://www.justsecurity.org/84658/insiders-view-of-the-january-6th-committees-social-media-investigation/.

Jacobson, Louis. "Joe Biden Overstates How Well Vaccines Prevent Person-to-Person Virus Spread." PolitiFact, October 14, 2021. https://www.politifact.com/factchecks/2021/oct/14/joe-biden/joe-biden-overstates-effectiveness-vaccines-preven/.

Jakesch, Maurice, Jeffrey T. Hancock, and Mor Naaman. "Human Heuristics for AI-Generated Language Are Flawed." *Proceedings of the National Academy of Sciences* 120, no. 11 (2023): e2208839120. https://doi.org/10.1073/pnas.2208839120.

Jalli, Nuurrianti. "Lack of Internet Access in Southeast Asia Poses Challenges for Students to Study Online Amid COVID-19 Pandemic." *The Conversation*, March 17, 2020. http://theconversation.com/lack-of-internet-access-in-southeast-asia-poses-challenges-for-students-to-study-online-amid-covid-19-pandemic-133787.

Jamieson, Kathleen Hall, and Joseph N. Cappella. *Echo Chamber: Rush Limbaugh and the Conservative Media Establishment*. Oxford University Press, 2008.

Jarry, Jonathan. "The Upside-Down Doctor." Office for Science and Society, McGill University, June 4, 2021. https://www.mcgill.ca/oss/article/covid-19-health-pseudoscience/upside-down-doctor.

Jin, Fang, Wei Wang, Liang Zhao et al. "Misinformation Propagation in the Age of Twitter." *Computer* 47, no. 12 (2014): 90–94. https://doi.ieeecomputersociety.org/10.1109/MC.2014.361.

Johns, Michael M. E., Mark Barnes, and Patrik S. Florencio. "Restoring Balance to Industry-Academia Relationships in an Era of Institutional Financial Conflicts of Interest: Promoting Research While Maintaining Trust." *JAMA* 289, no. 6 (2003): 741–46. https://doi.org/10.1001/jama.289.6.741.

Johnson, Carla K., and Hannah Fingerhut. "AP-NORC Poll: More Americans Worry About Flu than New Virus." AP News, February 20, 2020. https://apnews.com/article/us-news-health-china-virus-outbreak-ap-top-news-c3eddb289d20d279de31a7c1b75f73d2.

Johnson, Neil F., Nicolas Velásquez, Nicholas Johnson Restrepo et al. "The Online Competition Between Pro- and Anti-Vaccination Views." *Nature* 582, no. 7811 (2020): 230–33. https://doi.org/10.1038/s41586-020-2281-1.

Johnston, Ron. "Manipulating Maps and Winning Elections: Measuring the Impact of Malapportionment and Gerrymandering." *Political Geography* 21, no. 1 (2002): 1–31. https://doi.org/10.1016/S0962-6298(01)00070-1.

Jolley, Daniel, and Karen M. Douglas. "Prevention Is Better than Cure: Addressing Anti-Vaccine Conspiracy Theories." *Journal of Applied Social Psychology* 47, no. 8 (2017): 459–69. https://doi.org/10.1111/jasp.12453.

Jones, Jason Jeffrey, and Nick Rogers. "Online in the US, Personal Identity is Increasingly Political." SocArXiv, August 25, 2021. https://doi.org/10.31235/osf.io/7k8xr.

Jones, Robert, Jr. (formerly known as Son of Baldwin). "Contact." Accessed August 14, 2024. https://www.sonofbaldwin.com/contact/.

Jones-Jang, S. Mo, Tara Mortensen, and Jingjing Liu. "Does Media Literacy Help Identification of Fake News? Information Literacy Helps, but Other Literacies Don't." *American Behavioral Scientist* 65, no. 2 (2021): 371–88. https://doi.org/10.1177/0002764219869406.

Jost, John T., Jaime L. Napier, Hulda Thorisdottir, Samuel D. Gosling, Tibor P. Palfai, and Brian Osta-fin. "Are Needs to Manage Uncertainty and Threat Associated with Political Conservatism or Ideological Extremity?" *Personality and Social Psychology Bulletin* 33, no. 7 (2007): 989–1007. https://doi.org/10.1177/0146167207301028.

Joyella, Mark. "Fox News Hits 23rd Consecutive Month As Most-Watched in Cable News As CNN Sees Gains in January." *Forbes*, February 1, 2023. https://www.forbes.com/sites/markjoyella/2023/02/01/fox-news-hits-23rd-consecutive-month-as-most-watched-in-cable-news-as-cnn-sees-gains-in-january/.

Kahan, Dan M. "Climate-Science Communication and the Measurement Problem." *Political Psychology* 36 (2015): 1–43. https://doi.org/10.1111/pops.12244.

Kahan, Dan M. "Misconceptions, Misinformation, and the Logic of Identity-Protective Cognition." Cultural Cognition Project Working Paper No. 164. Yale University Law School, 2017. https://doi.org/10.2139/ssrn.2973067.

Kahan, Dan M. "'Ordinary Science Intelligence': A Science-Comprehension Measure for Study of Risk and Science Communication, with Notes on Evolution and Climate Change." *Journal of Risk Research* 20, no. 8 (2017): 995–1016. https://doi.org/10.1080/13669877.2016.1148067.

Kahan, Dan M., Ellen Peters, Erica Cantrell Dawson, and Paul Slovic. "Motivated Numeracy and Enlightened Self-Government." *Behavioural Public Policy* 1, no. 1 (2017): 54–86. https://doi.org/10.1017/bpp.2016.2.

Kahn, Chris. "Half of Republicans Say Biden Won Because of a 'Rigged' Election: Reuters/Ipsos Poll." Reuters, November 19, 2020. https://www.reuters.com/article/world/india/half-of-republicans-say-biden-won-because-of-a-rigged-election-reutersipsos-idUSKBN27Y1AD/.

Kahneman, Daniel. *Thinking, Fast and Slow*. Macmillan, 2011.

Kakutani, Michiko. "'The Death of Expertise' Explores How Ignorance Became a Virtue." *New York Times*, March 21, 2017. https://www.nytimes.com/2017/03/21/books/the-death-of-expertise-explores-how-ignorance-became-a-virtue.html.

Kalemli-Ozcan, Sebnem. "The $4 Trillion Economic Cost of Not Vaccinating the Entire World." *The Conversation*, February 12, 2021. http://theconversation.com/the-4-trillion-economic-cost-of-not-vaccinating-the-entire-world-154786.

Kalkhoff, Will, Richard T. Serpe, Joshua Pollock, Brennan Miller, and Matthew Pfeiffer. "Neural Processing of Identity-Relevant Feedback." In *New Directions in Identity Theory and Research*, ed. Jan E. Stets and Richard T. Serpe, 195–238. Oxford University Press, 2016. https://doi.org/10.1093/acprof:oso/9780190457532.003.0008.

Kalla, Joshua L., and David E. Broockman. "Reducing Exclusionary Attitudes Through Interpersonal Conversation: Evidence from Three Field Experiments." *American Political Science Review* 114, no. 2 (2020): 410–25. https://doi.org/10.1017/S0003055419000923.

Kalmoe, Nathan P., and Lilliana Mason. "Lethal Mass Partisanship: Prevalence, Correlates, and Electoral Contingencies." Paper presented at a meeting of the National Capital Area Political Science Association, Washington, DC, 2019.

Kang, Cecilia, and Sheera Frenkel. "Facebook Says Cambridge Analytica Harvested Data of Up to 87 Million Users." *New York Times*, April 4, 2018. https://www.nytimes.com/2018/04/04/technology/mark-zuckerberg-testify-congress.html.

Kapoor, Sayash, and Arvind Narayanan. "A Misleading Open Letter About Sci-Fi AI Dangers Ignores the Real Risks." AI Snake Oil, March 29, 2023. https://aisnakeoil.substack.com/p/a-misleading-open-letter-about-sci.

Katella, Kathy. "Comparing the COVID-19 Vaccines: How Are They Different?" Yale Medicine, April 24, 2024. https://www.yalemedicine.org/news/covid-19-vaccine-comparison.

Kaurov, Alexander A., Viktoria Cologna, Charlie Tyson, and Naomi Oreskes. "Trends in American Scientists' Political Donations and Implications for Trust in Science." *Humanities and Social Sciences Communications* 9, no. 1 (2022): 1–8. https://doi.org/10.1057/s41599-022-01382-3.

Keating, Jessica, Leaf Van Boven, and Charles M. Judd. "Partisan Underestimation of the Polarizing Influence of Group Discussion." *Journal of Experimental Social Psychology* 65 (2016): 52–58. https://doi.org/10.1016/j.jesp.2016.03.002.

Kekatos, Mary. "Masks Are Effective but Here's How a Study from a Respected Group Was Misinterpreted to Say They Weren't." ABC News, March 14, 2023. https://abcnews.go.com/Health/masks-effective-study-respected-group-misinterpreted/story?id=97846561.

Kelly, Makena. "New Algorithm Bill Could Force Facebook to Change How the News Feed Works." *The Verge*, February 10, 2022. https://www.theverge.com/2022/2/10/22927472/klobuchar-lummis-algorithm-bill-section-230-misinformation-teenager-mental-health.

Kennedy, Brian. "Americans' Trust in Scientists, Other Groups Declines." Pew Research Center, February 15, 2022. https://www.pewresearch.org/science/2022/02/15/americans-trust-in-scientists-other-groups-declines/.

Kettler, Jaclyn, Luke Fowler, and Stephanie Witt. "Democratic Governors Are Quicker in Responding to the Coronavirus than Republicans." *The Conversation*, April 6, 2020. http://theconversation.com/democratic-governors-are-quicker-in-responding-to-the-coronavirus-than-republicans-135599.

Keysers, Christian, and Valeria Gazzola. "Hebbian Learning and Predictive Mirror Neurons for Actions, Sensations and Emotions." *Philosophical Transactions of the Royal Society of London, Series B, Biological Sciences* 369, no. 1644 (2014): 20130175. https://doi.org/10.1098/rstb.2013.0175.

Khan, Saher, and Vignesh Ramachandran. "Millions Depend on Private Messaging Apps to Keep in Touch: They're Ripe with Misinformation." PBS News, November 5, 2021. https://www.pbs.org/newshour/world/millions-depend-on-private-messaging-apps-to-keep-in-touch-theyre-ripe-with-misinformation.

Kim, Eunji, Yphtach Lelkes, and Joshua McCrain. "Measuring Dynamic Media Bias." *Proceedings of the National Academy of Sciences* 119, no. 32 (2022): e2202197119. https://doi.org/10.1073/pnas.2202197119.

Kim, Jin Woo, Andrew Guess, Brendan Nyhan, and Jason Reifler. "The Distorting Prism of Social Media: How Self-Selection and Exposure to Incivility Fuel Online Comment Toxicity." *Journal of Communication* 71, no. 6 (2021): 922–46. https://doi.org/10.1093/joc/jqab034.

Kim, Jin Woo, and Eunji Kim. "Temporal Selective Exposure: How Partisans Choose When to Follow Politics." *Political Behavior* 43, no. 4 (2021): 1663–83. https://doi.org/10.1007/s11109-021-09690-1.

Kim, Sunny Jung, Jenna E. Schiffelbein, Inger Imset, and Ardis L. Olson. "Countering Antivax Misinformation via Social Media: Message-Testing Randomized Experiment for Human Papillomavirus Vaccination Uptake." *Journal of Medical Internet Research* 24, no. 11 (2022): e37559. https://doi.org/10.2196/37559.

Kirkpatrick, Ciera E., and Larissa L. Lawrie. "TikTok as a Source of Health Information and Misinformation for Young Women in the United States: Survey Study." *JMIR Infodemiology* 4 (2024): e54663. https://doi.org/10.2196/54663.

Kirzinger, Ashley, Grace Sparks, and Mollyann Brodie. "KFF COVID-19 Vaccine Monitor: In Their Own Words, Six Months Later." *KFF*, July 13, 2021. https://www.kff.org/coronavirus-covid-19/poll-finding/kff-covid-19-vaccine-monitor-in-their-own-words-six-months-later/.

Kleefeld, Eric, and Bobby Lewis. "'Long COVID' and the Ongoing Public Health Dangers That Fox News Hosts Ignore." Media Matters for America, August 7, 2020. https://www.mediamatters.org/coronavirus-covid-19/long-covid-and-ongoing-public-health-dangers-fox-news-hosts-ignore.

Klein, Elad, and Joshua Robison. "Like, Post, and Distrust? How Social Media Use Affects Trust in Government." *Political Communication* 37, no. 1 (2020): 46–64.

Klein, William M. P., Peter R. Harris, Rebecca A. Ferrer, and Laura E. Zajac. "Feelings of Vulnerability in Response to Threatening Messages: Effects of Self-Affirmation." *Journal of Experimental Social Psychology* 47, no. 6 (2011): 1237–42. https://doi.org/10.1016/j.jesp.2011.05.005.

Klepper, David. "'It's Not Simple': Researchers Tweaked Facebook's Algorithms to See If They Could Fix America's Political Polarization: They Failed." *Fortune*, July 27, 2023. https://fortune.com/2023/07/27/facebook-algorithm-political-polarization/.

Klofstad, Casey A., Anand Edward Sokhey, and Scott D. McClurg. "Disagreeing About Disagreement: How Conflict in Social Networks Affects Political Behavior." *American Journal of Political Science* 57, no. 1 (2013): 120–34. https://doi.org/10.1111/j.1540-5907.2012.00620.x.

Klossa, Gillaume. *Towards European Media Sovereignty: An Industrial Media Strategy to Leverage Data, Algorithms, and Artificial Intelligence.* European Commission, 2019. https://ec.europa.eu/commission /sites/beta-political/files/gk_report_final.pdf.

Kneer, Julia, Sabine Glock, and Diana Rieger. "Fast and Not Furious?" *Social Psychology* 43, no. 2 (2012). https://doi.org/10.1027/1864-9335/a000086.

Koetke, Jonah, Karina Schumann, and Tenelle Porter. "Intellectual Humility Predicts Scrutiny of COVID-19 Misinformation." *Social Psychological and Personality Science* 13, no. 1 (2022): 277–84. https://doi .org/10.1177/1948550620988242.

Koetke, Jonah, Karina Schumann, and Tenelle Porter. "Trust in Science Increases Conservative Support for Social Distancing." *Group Processes and Intergroup Relations* 24, no. 4 (2021): 680–97. https://doi .org/10.1177/1368430220985918.

Konnikova, Maria. *The Confidence Game: Why We Fall for It . . . Every Time.* Penguin, 2017.

Konnikova, Maria. "The Conman Who Pulled off History's Most Audacious Scam." BBC, January 27, 2016. https://www.bbc.com/future/article/20160127-the-conman-who-pulled-off-historys-most-audacious -scam.

Konstantinou, Pinelopi, Katerina Georgiou, Navin Kumar et al. "Transmission of Vaccination Attitudes and Uptake Based on Social Contagion Theory: A Scoping Review." *Vaccines* 9, no. 6 (2021): art. 607. https://doi.org/10.3390/vaccines9060607.

Koppel, Lina, Claire E. Robertson, Kimberly C. Doell et al. "Individual-Level Solutions May Support System-Level Change â If They Are Internalized as Part of One's Social Identity." *Behavioral and Brain Sciences* 46 (2023): e165. https://doi.org/10.1017/S0140525X2300105X.

Koren, Marina. "Seeing Earth from Space Will Change You." *The Atlantic,* December 10, 2022. https:// www.theatlantic.com/magazine/archive/2023/01/astronauts-visiting-space-overview-effect-spacex -blue-origin/672226/.

Korownyk, Christina, Michael R. Kolber, James McCormack et al. "Televised Medical Talk Shows–What They Recommend and the Evidence to Support Their Recommendations: A Prospective Observational Study." *BMJ* 349 (2014): g7346. https://doi.org/10.1136/bmj.g7346.

Kosinski, Michal, David Stillwell, and Thore Graepel. "Private Traits and Attributes Are Predictable from Digital Records of Human Behavior." *Proceedings of the National Academy of Sciences* 110, no. 15 (2013): 5802–5. https://doi.org/10.1073/pnas.1218772110.

Kowal, Michael. "The Value of a Like: Facebook, Viral Posts, and Campaign Finance in US Congressional Elections." *Media and Communication* 11, no. 3 (2023): 153–63. https://doi.org/10.17645/mac.v11i3.6661.

Kozlov, Max. "Introducing Inoculation, 1721." *The Scientist,* January 1, 2021. https://www.the-scientist.com /foundations/introducing-inoculation-1721-68275.

Kozyreva, Anastasia, Stefan M. Herzog, Stephan Lewandowsky et al. "Resolving Content Moderation Dilemmas Between Free Speech and Harmful Misinformation." *Proceedings of the National Academy of Sciences* 120, no. 7 (2023): e2210666120. https://doi.org/10.1073/pnas.2210666120.

Kozyreva, Anastasia, Sam Wineburg, Stephan Lewandowsky, and Ralph Hertwig. "Critical Ignoring as a Core Competence for Digital Citizens." *Current Directions in Psychological Science* 32, no. 1 (2023): 81–88. https://doi.org/10.1177/09637214221121570.

Kramer, Adam D. I., Jamie E. Guillory, and Jeffrey T. Hancock. "Experimental Evidence of Massive-Scale Emotional Contagion Through Social Networks." *Proceedings of the National Academy of Sciences* 111, no. 24 (2014): 8788–90. https://doi.org/10.1073/pnas.1320040111.

Kreiss, Daniel, Joshua O. Barker, and Shannon Zenner. "Trump Gave Them Hope: Studying the Strangers in Their Own Land." *Political Communication* 34, no. 3 (2017): 470–78. https://doi.org/10.1080 /10584609.2017.1330076.

Kreiss, Daniel, and Shannon C. McGregor. "A Review and Provocation: On Polarization and Platforms." *New Media and Society* 26, no. 1 (2024): 556–79. https://doi.org/10.1177/14614448231161880.

Kreps, Sarah. *Social Media and International Relations*. Cambridge University Press, 2020. https://doi.org/10.1017/9781108920377.

Kross, Ethan, and Igor Grossmann, "Boosting Wisdom: Distance from the Self Enhances Wise Reasoning, Attitudes, and Behavior," *Journal of Experimental Psychology. General* 141, no. 1 (2012): 43–48, https://doi.org/10.1037/a0024158.

Krumrei-Mancuso, Elizabeth J., Megan C. Haggard, Jordan P. LaBouff, and Wade C. Rowatt. "Links Between Intellectual Humility and Acquiring Knowledge." *Journal of Positive Psychology* 15, no. 2 (2020): 155–70. https://doi.org/10.1080/17439760.2019.1579359.

Kubin, Emily, Curtis Puryear, Chelsea Schein, and Kurt Gray. "Personal Experiences Bridge Moral and Political Divides Better than Facts." *Proceedings of the National Academy of Sciences* 118, no. 6 (2021): e2008389118. https://doi.org/10.1073/pnas.2008389118.

Kunda, Ziva. "The Case for Motivated Reasoning." *Psychological Bulletin* 108, no. 3 (1990): 480–98. https://doi.org/10.1037/0033-2909.108.3.480.

Kweon, Yesola, and Byeonghwa Choi. "Fueling Conspiracy Beliefs: Political Conservatism and the Backlash Against COVID-19 Containment Policies." *Governance* 37, no. 3 (2024): 867–86.

Lahut, Jake. "Fox News Dominated Primetime Ratings for COVID Summer—Not Just on Cable, but All of TV." *Business Insider*, September 11, 2020. https://www.businessinsider.com/fox-news-ratings-most-watched-channel-summer-2020-primetime-2020-9.

Lai, Emily R. "Critical Thinking: A Literature Review." *Pearson's Research Reports* 6, no. 1 (2011): 40–41. https://citeseerx.ist.psu.edu/document?repid=rep1&type=pdf&doi=b42cffa5a2ad63a31fcf99869e7c-b8ef72b44374.

Lai, Jianyu, Kristen K. Coleman, S.-H. Sheldon Tai et al. "Relative Efficacy of Masks and Respirators as Source Control for Viral Aerosol Shedding from People Infected with SARS-CoV-2: A Controlled Human Exhaled Breath Aerosol Experimental Study." *EBioMedicine* 104 (2024): art. 105157. https://doi.org/10.1016/j.ebiom.2024.105157.

Lamoureux, Mack. "Q Is Dead, Long Live QAnon." *VICE*, November 15, 2022. https://www.vice.com/en/article/wxnkzq/qanon-q-drop-midterms.

Lancet Infectious Diseases. "The COVID-19 Infodemic." Editorial. *Lancet Infectious Diseases* 20, no. 8 (2020): 875.

Landrum, Asheley R., and Alex Olshansky. "The Role of Conspiracy Mentality in Denial of Science and Susceptibility to Viral Deception About Science." *Politics and the Life Sciences* 38, no. 2 (2019): 193–209. https://doi.org/10.1017/pls.2019.9.

Langston, Joseph, Heather Albanesi, and Matthew Facciani. "Toward Faith: A Qualitative Study of How Atheists Convert to Christianity." *Journal of Religion and Society* 21 (2019): 1–23. https://doi.org/10.32873/unl.dc.jrs.21.12.

Lantian, Anthony, Virginie Bagneux, Sylvain Delouvée, and Nicolas Gauvrit. "Maybe a Free Thinker but Not a Critical One: High Conspiracy Belief Is Associated with Low Critical Thinking Ability." *Applied Cognitive Psychology* 35, no. 3 (2021): 674–84. https://doi.org/10.1002/acp.3790.

Lantian, Anthony, Dominique Muller, Cécile Nurra, Olivier Klein, Sophie Berjot, and Myrto Pantazi. "Stigmatized Beliefs: Conspiracy Theories, Anticipated Negative Evaluation of the Self, and Fear of Social Exclusion." *European Journal of Social Psychology* 48, no. 7 (2018): 939–54. https://doi.org/10.1002/ejsp.2498.

Larson, Heidi J. *Stuck: How Vaccine Rumors Start–and Why They Don't Go Away*. Oxford University Press, 2020.

Lasser, Jana, Segun Taofeek Aroyehun, Almog Simchon, Fabio Carrella, David Garcia, and Stephan Lewandowsky. "Social Media Sharing of Low-Quality News Sources by Political Elites." *PNAS Nexus* 1, no. 4 (2022): pgac186. https://doi.org/10.1093/pnasnexus/pgac186.

Last, John M., ed. *A Dictionary of Public Health*. Oxford University Press, 2007. https://archive.org/details
  /dictionaryofpubloooolast/page/n439/mode/1up.

Latkin, Carl, Lauren Dayton, Jacob Miller et al. "A Longitudinal Study of Vaccine Hesitancy Attitudes and
  Social Influence as Predictors of COVID-19 Vaccine Uptake in the US." *Human Vaccines and Immunotherapeutics* 18, no. 5 (2022): 2043102. https://doi.org/10.1080/21645515.2022.2043102.

Lawson, M. Asher, Shikhar Anand, and Hemant Kakkar. "Tribalism and Tribulations: The Social Costs of
  Not Sharing Fake News." *Journal of Experimental Psychology: General* 152, no. 3 (2023): 611–31. https://
  doi.org/10.1037/xge0001374.

Lawson, M. Asher, and Hemant Kakkar. "Of Pandemics, Politics, and Personality: The Role of Conscientiousness and Political Ideology in the Sharing of Fake News." *Journal of Experimental Psychology:
  General* 151, no. 5 (2022): 1154–77. https://oi.org/10.1037/xge0001120.

Lawson, Rebecca. "The Science of Cycology: Failures to Understand How Everyday Objects Work." *Memory and Cognition* 34, no. 8 (2006): 1667–75. https://doi.org/10.3758/BF03195929.

Lazer, David, Brian Rubineau, Carol Chetkovich, Nancy Katz, and Michael Neblo. "The Coevolution of
  Networks and Political Attitudes." *Political Communication* 27, no. 3 (2010): 248–74. https://doi.org
  /10.1080/10584609.2010.500187.

Leber, Rebecca. "Hygiene Theater at Restaurants Is Creating Endless Plastic Waste." *Mother Jones*.
  Accessed June 24, 2024. https://www.motherjones.com/food/2020/10/hygiene-theater-at-restaurants
  -is-creating-endless-plastic-waste/.

Leding, Juliana K., and Lilyeth Antonio. "Need for Cognition and Discrepancy Detection in the Misinformation Effect." *Journal of Cognitive Psychology* 31, no. 4 (2019): 409–15. https://doi.org/10.1080
  /20445911.2019.1626400.

Lee, Amber Hye-Yon. "Social Trust in Polarized Times: How Perceptions of Political Polarization Affect
  Americans' Trust in Each Other." *Political Behavior* 44, no. 3 (2022): 1533–54. https://doi.org/10.1007
  /s11109-022-09787-1.

Lee, Bruce Y. "Anti-Vaxxers Exploit Damar Hamlin's Crisis with Unfounded Covid-19 Vaccine Claims."
  Forbes, January 3, 2023. https://www.forbes.com/sites/brucelee/2023/01/03/anti-vaxxers-exploit-damar
  -hamlins-crisis-with-unfounded-covid-19-vaccine-claims/.

Lee, Byungkyu, and Peter Bearman. "Political Isolation in America." *Network Science* 8, no. 3 (2020): 333–55.
  https://doi.org/10.1017/nws.2020.9.

Lee, Sangwon, and S. Mo Jones-Jang. "Cynical Nonpartisans: The Role of Misinformation in Political Cynicism During the 2020 U.S. Presidential Election." *New Media and Society* 26, no. 7 (2024): 4255–76.
  https://doi.org/10.1177/14614448221116036.

Lee, Sian, Aiping Xiong, Haeseung Seo, and Dongwon Lee. "'Fact-Checking' Fact Checkers: A Data-
  Driven Approach." *Harvard Kennedy School Misinformation Review* 4, no, 5 (2023): 1–22. https://doi
  .org/10.37016/mr-2020-126.

Lee, Susan J., Henry J. Peter Ralston, Eleanor A. Drey, John Colin Partridge, and Mark A. Rosen. "Fetal Pain:
  A Systematic Multidisciplinary Review of the Evidence." *JAMA* 294, no. 8 (2005): 947–54. https://doi
  .org/10.1001/jama.294.8.947.

Lees, Jeffrey, John A. Banas, Darren Linvill, Patrick C. Meirick, and Patrick Warren. "The *Spot the Troll* Quiz
  Game Increases Accuracy in Discerning Between Real and Inauthentic Social Media Accounts." *PNAS
  Nexus* 2, no. 4 (2023): pgado94. https://doi.org/10.1093/pnasnexus/pgado94.

Leffer, Lauren. "What Does Artificial General Intelligence Actually Mean?" *Scientific American*, June 25, 2024.
  https://www.scientificamerican.com/article/what-does-artificial-general-intelligence-actually-mean/.

Lemyre, Louise, and Philip M. Smith. "Intergroup Discrimination and Self-Esteem in the Minimal Group
  Paradigm." *Journal of Personality and Social Psychology* 49, no. 3 (1985): 660–670.

Levendusky, Matthew S., and Neil Malhotra. "Does Media Coverage of Partisan Polarization Affect Political
  Attitudes?" *Political Communication* 33, no. 2 (2016): 283–301. https://doi.org/10.1080/10584609.2015
  .1038455.

Levendusky, Matthew S., and Neil Malhotra. "(Mis)perceptions of Partisan Polarization in the American Public." *Public Opinion Quarterly* 80, no. S1 (2015): 378–91. https://doi.org/10.1093/poq/nfv045.

Levendusky, Matthew S., and Dominik A. Stecula. *We Need to Talk: How Cross-Party Dialogue Reduces Affective Polarization.* Cambridge University Press, 2021.

Levine, Sam. "Partisan Gerrymandering Has Empowered a Hard-Right Turn in Texas." *The Guardian*, September 5, 2021. https://www.theguardian.com/us-news/2021/sep/05/gerrymandering-empowered -hard-right-texas.

Levinovitz, Alan. *Natural: How Faith in Nature's Goodness Leads to Harmful Fads, Unjust Laws, and Flawed Science.* Beacon Press, 2021.

Levitan, Dave. "Does a Fetus Feel Pain at 20 Weeks?" FactCheck.org, May 18, 2015. https://www.factcheck .org/2015/05/does-a-fetus-feel-pain-at-20-weeks/.

Levitsky, Steven, and Daniel Ziblatt. *How Democracies Die.* Crown, 2019.

Levitsky, Steven, and Daniel Ziblatt. *Tyranny of the Minority: Why American Democracy Reached the Breaking Point.* Crown, 2023.

Levy, Neil. "Do Your Own Research!" *Synthese* 20, no. 5 (2022): 356. https://doi.org/10.1007/s11229-022 -03793-w.

Levy, Ro'ee. "Social Media, News Consumption, and Polarization: Evidence from a Field Experiment." *American Economic Review* 111, no. 3 (2021): 831–70. https://doi.org/10.1257/aer.20191777.

Lewis, Dyani, Max Kozlov, and Mariana Lenharo. "COVID-Origins Data from Wuhan Market Published: What Scientists Think." *Nature* 616, no. 7956 (2023): 225–26. https://doi.org/10.1038/d41586 -023-00998-y.

Lewis, Tanya. "The Benefits of Vaccinating Kids Against COVID Far Outweigh the Risks of Myocarditis." *Scientific American*, December 2, 2021. https://www.scientificamerican.com/article/the-benefits-of -vaccinating-kids-against-covid-far-outweigh-the-risks-of-myocarditis1/.

Li, Jianing. "Not All Skepticism Is 'Healthy' Skepticism: Theorizing Accuracy- and Identity-Motivated Skepticism Toward Social Media Misinformation." *New Media and Society*, June 26, 2023. https://doi .org/10.1177/14614448231179941.

Li, Jianing, and Michael W. Wagner. "The Value of Not Knowing: Partisan Cue-Taking and Belief Updating of the Uninformed, the Ambiguous, and the Misinformed." *Journal of Communication* 70, no. 5 (2020): 646–69. https://doi.org/10.1093/joc/jqaa022.

Light, Nicholas, Philip M. Fernbach, Nathaniel Rabb, Mugur V. Geana, and Steven A. Sloman. "Knowledge Overconfidence Is Associated with Anti-Consensus Views on Controversial Scientific Issues." *Science Advances* 8, no. 29 (2022): eabo0038. https://doi.org/10.1126/sciadv.abo0038.

Lim, Chloe. "Checking How Fact-Checkers Check." *Research and Politics* 5, no. 3 (2018). https://doi.org /10.1177/2053168018786848.

Lim, Steven Chee Loon, Chee Peng Hor, Kim Heng Tay et al. "Efficacy of Ivermectin Treatment on Disease Progression Among Adults with Mild to Moderate COVID-19 and Comorbidities: The I-TECH Randomized Clinical Trial." *JAMA Internal Medicine* 182, no. 4 (2022): 426–35. https://doi.org/10.1001 /jamainternmed.2022.0189.

Limaye, Rupali J., Fauzia Malik, Paula M. Frew et al. "Patient Decision Making Related to Maternal and Childhood Vaccines: Exploring the Role of Trust in Providers Through a Relational Theory of Power Approach." *Health Education and Behavior* 47, no. 3 (2020): 449–56. https://doi.org/10.1177 /1090198120915432.

Lin, Summer. "Here's Why Some Vaccine Skeptics Changed Their Minds and Got COVID Shots, Poll Says." *Miami Herald*, July 14, 2021. https://www.miamiherald.com/news/coronavirus/article252777353 .html.

Linvill, Darren L., and Patrick L. Warren. "Troll Factories: Manufacturing Specialized Disinformation on Twitter." *Political Communication* 37, no. 4 (2020): 447–67. https://doi.org/10.1080/10584609.2020 .1718257.

Liu, William J., Peipei Liu, Wenwen Lei et al. "Surveillance of SARS-CoV-2 at the Huanan Seafood Market." *Nature* 631, no. 8020 (2024): 402–8. https://doi.org/10.1038/s41586-023-06043-2.

Livingstone, Sonia, Elizabeth Van Couvering, and Nancy Thumim. "Converging Traditions of Research on Media and Information Literacies: Disciplinary, Critical, and Methodological Issues." In *Handbook of Research on New Literacies*, ed. Julie Coiro, Michele Knobel, Colin Lankshear, and Donald J. Leu, 103–32. Routledge, 2008.

Lomas, Natasha. "Twitter Offers More Support to Researchers—to 'Keep Us Accountable.'" TechCrunch, January 6, 2020. https://techcrunch.com/2020/01/06/twitter-offers-more-support-to-researchers-to -keep-us-accountable/.

Loomba, Sahil, Alexandre de Figueiredo, Simon J. Piatek, Kristen de Graaf, and Heidi J. Larson. "Measuring the Impact of COVID-19 Vaccine Misinformation on Vaccination Intent in the UK and USA." *Nature Human Behaviour* 5, no. 3 (2021): 337–48. https://doi.org/10.1038/s41562-021-01056-1.

Lopez, German. "How Political Polarization Broke America's Vaccine Campaign." *Vox*, July 6, 2021. https:// www.vox.com/2021/7/6/22554198/political-polarization-vaccine-covid-19-coronavirus.

Lorenz, Taylor. "Extremist Influencers Are Generating Millions for Twitter, Report Says." *Washington Post*, February 9, 2023. https://www.washingtonpost.com/technology/2023/02/09/twitter-ads-revenue -suspended-account/.

Lovelace, Berkeley, Jr. "Medical Historian Compares the Coronavirus to the 1918 Flu Pandemic: Both Were Highly Political." CNBC, September 29, 2020. https://www.cnbc.com/2020/09/28/comparing -1918-flu-vs-coronavirus.html.

Lovelace, Berkeley, Jr. "Myocarditis After Covid Vaccine Low Among Teens and Young Adults, Large Study Finds." NBC News, December 5, 2022. https://www.nbcnews.com/health/health-news/myocarditis -covid-vaccine-teens-study-rcna60118.

Lu, Linqi, Jiawei Liu, Y. Connie Yuan, Kelli S. Burns, Enze Lu, and Dongxiao Li. "Source Trust and COVID-19 Information Sharing: The Mediating Roles of Emotions and Beliefs About Sharing." *Health Education and Behavior* 48, no. 2 (2021): 132–39. https://doi.org/10.1177/1090198120984760.

Luguri, Jamie B., and Jaime L. Napier. "Of Two Minds: The Interactive Effect of Construal Level and Identity on Political Polarization." *Journal of Experimental Social Psychology* 49, no. 6 (2013): 972–77. https://doi.org/10.1016/j.jesp.2013.06.002.

Lunz Trujillo, Kristin, Roy H. Perlis, Matthew A. Baum, and David Lazer. "The COVID States Project #77: Healthcare Workers' Perception of COVID-19 Misinformation." OSF Preprints, January 15, 2022. https://doi.org/10.31219/osf.io/6pzqj.

Lutzke, Lauren, Caitlin Drummond, Paul Slovic, and Joseph Árvai. "Priming Critical Thinking: Simple Interventions Limit the Influence of Fake News About Climate Change on Facebook." *Global Environmental Change* 58 (): art. 101964. https://doi.org/10.1016/j.gloenvcha.2019.101964.

Lynas, Mark. "Are the Anti-GMO and Anti-Vaccine Movements Merging?" *Alliance for Science*, December 6, 2017. https://allianceforscience.cornell.edu/blog/2017/12/are-the-anti-gmo-and-anti-vaccine -movements-merging/.

Lynas, Mark, Benjamin Z. Houlton, and Simon Perry. "Greater than 99 Percent Consensus on Human-Caused Climate Change in the Peer-Reviewed Scientific Literature." *Environmental Research Letters* 16, no. 11 (2021): 114005. https://doi.org/10.1088/1748-9326/ac2966.

Lyons, Benjamin A. "Unbiasing Information Search and Processing Through Personal and Social Identity Mechanisms." PhD diss., Southern Illinois University at Carbondale, 2016.

Lyons, Benjamin A., Christina E. Farhart, Michael P. Hall et al. "Self-Affirmation and Identity-Driven Political Behavior." *Journal of Experimental Political Science* 9, no. 2 (2022): 225–40. https://doi.org/10.1017 /XPS.2020.46.

Lyons, Benjamin A., Jacob M. Montgomery, Andrew M. Guess, Brendan Nyhan, and Jason Reifler. "Overconfidence in News Judgments Is Associated with False News Susceptibility." *Proceedings of the National Academy of Sciences* 118, no. 23 (2021): e2019527118. https://doi.org/10.1073/pnas.2019527118.

Mac, Ryan. "Who Is Frances Haugen, the Facebook Whistle-Blower?" *New York Times*, October 5, 2021. https://www.nytimes.com/2021/10/05/technology/who-is-frances-haugen.html.

Madani, Youness, Mohammed Erritali, and Belaid Bouikhalene. "Using Artificial Intelligence Techniques for Detecting Covid-19 Epidemic Fake News in Moroccan Tweets." *Results in Physics* 25 (2021): art. 104266. https://doi.org/10.1016/j.rinp.2021.104266.

Maertens, Rakoen, Jon Roozenbeek, Melisa Basol, and Sander van der Linden. "Long-Term Effectiveness of Inoculation Against Misinformation: Three Longitudinal Experiments." *Journal of Experimental Psychology: Applied* 27, no. 1 (2021): 1–16. https://doi.org/10.1037/xap0000315.

Maertens, Rakoen, Jon Roozenbeek, Jon Simons et al. "Psychological Booster Shots Targeting Memory Increase Long-Term Resistance Against Misinformation." Preprint, PsyArXiv, April 17, 2023. https://doi.org/10.31234/osf.io/6r9as.

Mahadevan, Alex. "This Newspaper Doesn't Exist: How ChatGPT Can Launch Fake News Sites in Minutes." Poynter Institute, February 3, 2023. https://www.poynter.org/fact-checking/2023/chatgpt-build-fake-news-organization-website/.

Mahoney, Michael J. "Publication Prejudices: An Experimental Study of Confirmatory Bias in the Peer Review System." *Cognitive Therapy and Research* 1 (1977): 161–75. https://doi.org/10.1007/BF01173636.

Malik, Momin M., Sandra Cortesi, and Urs Gasser. "The Challenges of Defining 'News Literacy.'" Research Publication No. 2013–20. Berkman Center, October 2013. https://doi.org/10.2139/ssrn.2342313.

Mallard, Carina C., Joakim Ek, and Zinaida S. Vexler. "The Myth of the Immature Barrier Systems in the Developing Brain: Role in Perinatal Brain Injury." *Journal of Physiology* 596, no. 23 (2018): 5655–64. https://doi.org/10.1113/JP274938.

Manago, Bianca. "Preregistration and Registered Reports in Sociology: Strengths, Weaknesses, and Other Considerations." *American Sociologist* 54, no. 1 (2023): 193–210. https://doi.org/10.1007/s12108-023-09563-6.

Mangan, Dan. "DOJ Says at Least 1,000 Trump Supporters Arrested for Jan. 6 Capitol Riot." CNBC, March 6, 2023. https://www.cnbc.com/2023/03/06/doj-says-jan-6-capitol-riot-arrests-top-thousand-people.html.

Mangan, Dan, and Mike Calia. "Special Counsel Mueller: Russians Conducted 'Information Warfare' Against US During Election to Help Donald Trump Win." CNBC, February 16, 2018. https://www.cnbc.com/2018/02/16/russians-indicted-in-special-counsel-robert-muellers-probe.html.

Mangan, Dan, and Berkeley Lovelace Jr. "Trump Suspects Coronavirus Outbreak Came from China Lab, Doesn't Cite Evidence." CNBC, April 30, 2020. https://www.cnbc.com/2020/04/30/coronavirus-trump-suspects-covid-19-came-from-china-lab.html.

Mann, Brian. "4 U.S. Companies Will Pay $26 Billion to Settle Claims They Fueled the Opioid Crisis." NPR, February 25, 2022. https://www.npr.org/2022/02/25/1082901958/opioid-settlement-johnson-26-billion.

Mann, Michael E. *The New Climate War: The Fight to Take Back Our Planet*. PublicAffairs, 2021.

Marcus, Julia. "The Dudes Who Won't Wear Masks." *The Atlantic*, June 23, 2020. https://www.theatlantic.com/ideas/archive/2020/06/dudes-who-wont-wear-masks/613375/.

Marist Poll. "NPR/PBS NewsHour/Marist Poll Results: Coronavirus." February 4, 2020. https://maristpoll.marist.edu/npr-pbs-newshour-marist-poll-results-15/.

Marron, Dylan. "Empathy Is Not Endorsement." TED Talk, Vancouver, BC, April 13, 2018. 10 min., 52 sec. https://www.ted.com/talks/dylan_marron_empathy_is_not_endorsement.

Martini, Franziska, Paul Samula, Tobias R. Keller, and Ulrike Klinger. "Bot, or Not? Comparing Three Methods for Detecting Social Bots in Five Political Discourses." *Big Data and Society* 8, no. 2 (2021): 205395172110335. https://doi.org/10.1177/20539517211033566.

Maruf, Ramishah. "These Four Words Are Helping Spread Vaccine Misinformation." CNN, September 19, 2021. https://www.cnn.com/2021/09/19/media/reliable-sources-covid-research/index.html.

Marwick, Alice, and Rebecca Lewis. *Media Manipulation and Disinformation Online*. Data & Society Research Institute, 2017.

Masket, Seth. "Seth Masket: The Great Vaccine Divide Puts Republican Leaders in a Moral Quandary." *Denver Post*, June 25, 2021. https://www.denverpost.com/2021/06/25/covid-19-vaccine-rates-donald-trump-joe-biden/.

Mason, Lilliana. "Best Of: The Age of 'Mega-Identity Politics.'" Interview by Ezra Klein. *Vox Conversations*, podcast, reaired November 28, 2019. 1 hr., 15 min., 59 sec. https://radiopublic.com/Ezra/s1!e039b.

Mason, Lilliana. *Uncivil Agreement: How Politics Became Our Identity*. University of Chicago Press, 2018.

Mason, Lilliana, and Julie Wronski. "One Tribe to Bind Them All: How Our Social Group Attachments Strengthen Partisanship." *Political Psychology* 39 (2018): 257–77. https://doi.org/10.1111/pops.12485.

Matz, S. C., M. Kosinski, G. Nave, and D. J. Stillwell. "Psychological Targeting as an Effective Approach to Digital Mass Persuasion." *Proceedings of the National Academy of Sciences* 114, no. 48 (2017): 12714–19. https://doi.org/10.1073/pnas.1710966114.

Maxmen, Amy. "Unseating Big Pharma: The Radical Plan for Vaccine Equity." *Nature* 607, no. 7918 (2022): 226–33. https://doi.org/10.1038/d41586-022-01898-3.

McCarthy, Bill. "Fox News Host Will Cain Falsely Claims Vaccine More Dangerous for Children than COVID-19." PolitiFact, October 7, 2021. https://www.politifact.com/factchecks/2021/oct/07/will-cain/fox-news-host-will-cain-falsely-claims-vaccine-mor/.

McCarthy, Bill. "Tucker Carlson Falsely Claims COVID-19 Vaccines Might Not Work." PolitiFact, April 15, 2021. https://www.politifact.com/factchecks/2021/apr/15/tucker-carlson/tucker-carlson-falsely-claims-covid-19-vaccines-mi/.

McCarthy, Bill. "What Trump Said to Encourage COVID-19 Vaccine Use." PolitiFact, March 4, 2021. https://www.politifact.com/factchecks/2021/mar/04/rachel-maddow/what-trump-said-encourage-covid-19-vaccine-use/.

McCarthy, Molly, Kristina Murphy, Elise Sargeant, and Harley Williamson. "Examining the Relationship Between Conspiracy Theories and COVID-19 Vaccine Hesitancy: A Mediating Role for Perceived Health Threats, Trust, and Anomie?" *Analyses of Social Issues and Public Policy* 22, no. 1 (2022): 106–29. https://doi.org/10.1111/asap.12291.

McClain, Craig R. "Practices and Promises of Facebook for Science Outreach: Becoming a 'Nerd of Trust.'" *PLOS Biology* 15, no. 6 (2017): e2002020. https://doi.org/10.1371/journal.pbio.2002020.

McCoy, Charles. "Anti-Vaccination Beliefs Don't Follow the Usual Political Polarization." *The Conversation*, August 24, 2017. http://theconversation.com/anti-vaccination-beliefs-dont-follow-the-usual-political-polarization-81001.

McCoy, Terrence. "Half of Dr. Oz's Medical Advice Is Baseless or Wrong, Study Says." *Washington Post*, December 19, 2014. https://www.washingtonpost.com/news/morning-mix/wp/2014/12/19/half-of-dr-ozs-medical-advice-is-baseless-or-wrong-study-says/.

McCracken, Harry. "If ChatGPT Doesn't Get a Better Grasp of Facts, Nothing Else Matters." *Fast Company*, January 11, 2023. https://www.fastcompany.com/90833017/openai-chatgpt-accuracy-gpt-4.

McCuin, John L., Katharine Hayhoe, and Douglas Hayhoe. "Comparing the Effects of Traditional vs. Misconceptions-Based Instruction on Student Understanding of the Greenhouse Effect." *Journal of Geoscience Education* 62, no. 3 (2014): 445–59. https://doi.org/10.5408/13-068.1.

McGregor, Ian. "Offensive Defensiveness: Toward an Integrative Neuroscience of Compensatory Zeal After Mortality Salience, Personal Uncertainty, and Other Poignant Self-Threats." *Psychological Inquiry* 17, no. 4 (2006): 299–308. https://doi.org/10.1080/10478400701366977.

McGregor, Ian, Reeshma Haji, and So-Jin Kang. "Can Ingroup Affirmation Relieve Outgroup Derogation?" *Journal of Experimental Social Psychology* 44, no. 5 (2008): 1395–1401. https://doi.org/10.1016/j.jesp.2008.06.001.

McGregor, Ian, Kyle Nash, and Mike Prentice. "Reactive Approach Motivation (RAM) for Religion." *Journal of Personality and Social Psychology* 99, no. 1 (2010): 148–61. https://doi.org/10.1037/a0019702.

McGregor, Ian, Mark P. Zanna, John G. Holmes, and Steven J. Spencer. "Compensatory Conviction in the Face of Personal Uncertainty: Going to Extremes and Being Oneself." *Journal of Personality and Social Psychology* 80, no. 3 (2001): 472–88. https://doi.org/10.1037/0022-3514.80.3.472.

McGuire, William J. "Some Contemporary Approaches." *Advances in Experimental Social Psychology* 1 (1964): 191–229. https://doi.org/10.1016/S0065-2601(08)60052-0.

McGuire, William J., and D. Papageorgis. "The Relative Efficacy of Various Types of Prior Belief-Defense in Producing Immunity Against Persuasion." *Journal of Abnormal and Social Psychology* 62, no. 2 (1961): 327–37. https://doi.org/10.1037/h0042026.

McLenon, Jennifer, and Mary A. M. Rogers. "The Fear of Needles: A Systematic Review and Meta-Analysis." *Journal of Advanced Nursing* 75, no. 1 (2019): 30–42. https://doi.org/10.1111/jan.13818.

McPherson, Miller, Lynn Smith-Lovin, and James M. Cook. "Birds of a Feather: Homophily in Social Networks." *Annual Review of Sociology* 27, no. 1 (2001): 415–44. https://doi.org/10.1146/annurev.soc .27.1.415.

McQueen, Amy, and William M. P. Klein. "Experimental Manipulations of Self-Affirmation: A Systematic Review." *Self and Identity* 5, no. 4 (2006): 289–354. https://doi.org/10.1080/15298860600805325.

Meagher, Benjamin R., Hanna Gunn, Nathan Sheff, and Daryl R. Van Tongeren. "An Intellectually Humbling Experience: Changes in Interpersonal Perception and Cultural Reasoning Across a Five-Week Course." *Journal of Psychology and Theology* 47, no. 3 (2019): 217–29. https://doi.org/10.1177 /0091647119837010.

Media Literacy Now. "National Survey Finds Most U.S. Adults Have Not Had Media Literacy Education in High School." September 7, 2022. https://medialiteracynow.org/nationalsurvey2022/.

Media Matters for America. "Fox's Dr. Marc Siegel Says 'Worse Case Scenario' for Coronavirus Is 'It Could Be the Flu.'" March 6, 2020. https://www.mediamatters.org/sean-hannity/foxs-dr-marc-siegel -says-worse-case-scenario-coronavirus-it-could-be-flu.

Medin, Douglas L., and Carol D. Lee. "Diversity Makes Better Science." *APS Observer*, May-June 2012. https://www.psychologicalscience.org/observer/diversity-makes-better-science.

Medina Serrano, Juan Carlos, Orestis Papakyriakopoulos, and Simon Hegelich. "Dancing to the Partisan Beat: A First Analysis of Political Communication on TikTok." In *WebSci '20: Proceedings of the 12th ACM Conference on Web Science*, 257–266. Association for Computing Machinery, 2020. https://doi .org/10.1145/3394231.3397916.

Meedan. "Missing Information, Not Just Misinformation, Is Part of the Problem." August 5, 2020. https:// meedan.com/post/missing-information-not-just-misinformation-is-part-of-the-problem.

Meixler, Eli. "U.N. Fact Finders Say Facebook Played a 'Determining' Role in Violence Against the Rohingya." *Time*, March 13, 2018. https://time.com/5197039/un-facebook-myanmar-rohingya-violence/.

Mello, Michelle M., Jeremy A. Greene, and Joshua M. Sharfstein. "Attacks on Public Health Officials During COVID-19." *JAMA* 324, no. 8 (2020): 741–42. https://doi.org/10.1001/jama.2020.14423.

Menczer, Filippo. "4 Reasons Why Social Media Make Us Vulnerable to Manipulation." Keynote speech to International Conference on Computational Social Science. 2020. https://www.youtube.com/watch?v =BQYveMPwlNg.

Menczer, Filippo. "How 'Engagement' Makes You Vulnerable to Manipulation and Misinformation on Social Media." *The Conversation*, September 20, 2021. http://theconversation.com/how-engagement -makes-you-vulnerable-to-manipulation-and-misinformation-on-social-media-145375.

Menczer, Filippo, and Thomas Hills. "Information Overload Helps Fake News Spread, and Social Media Knows It." *Scientific American*, December 1, 2020. https://www.scientificamerican.com/article/information -overload-helps-fake-news-spread-and-social-media-knows-it/.

Meppelink, Corine S., Edith G. Smit, Marieke L. Fransen, and Nicola Diviani. "'I Was Right About Vaccina-tion': Confirmation Bias and Health Literacy in Online Health Information Seeking." *Journal of Health Communication* 24, no. 2 (2019): 129–40. https://doi.org/10.1080/10810730.2019.1583701.

Merton, Robert King. *Sociological Ambivalence and Other Essays*. Simon and Schuster, 1976.

Meta. "Tackling Climate Change Together." September 16, 2021. https://about.fb.com/news/2021/09/tackling -climate-change-together/.

Metz, Cade. "The New Chatbots Could Change the World: Can You Trust Them?" *New York Times*, December 10, 2022. https://www.nytimes.com/2022/12/10/technology/ai-chat-bot-chatgpt.html.

Mikhaeil, Christine Abdalla, and Richard L. Baskerville. "Explaining Online Conspiracy Theory Radicalization: A Second-Order Affordance for Identity-Driven Escalation." *Information Systems Journal* 34, no. 3 (2024): 711–35. https://doi.org/10.1111/isj.12427.

Miller, Kevin P., Marilynn B. Brewer, and Nathan L. Arbuckle. "Social Identity Complexity: Its Correlates and Antecedents." *Group Processes and Intergroup Relations* 12, no. 1 (2009): 79–94. https://doi.org/10.1177/1368430208098778.

Miller, Patrick R., and Pamela Johnston Conover. "Red and Blue States of Mind: Partisan Hostility and Voting in the United States." *Political Research Quarterly* 68, no. 2 (2015): 225–39. https://doi.org/10.1177/1065912915577208.

Milmo, Dan. "Anti-Vaxxers Making 'at Least $2.5m' a Year from Publishing on Substack." *The Guardian*, January 27, 2022. https://www.theguardian.com/technology/2022/jan/27/anti-vaxxers-making-at-least-25m-a-year-from-publishing-on-substack.

Minson, Julia A., and Charles A. Dorison. "Why Is Exposure to Opposing Views Aversive? Reconciling Three Theoretical Perspectives." *Current Opinion in Psychology* 47 (2022): art. 101435. https://doi.org/10.1016/j.copsyc.2022.101435.

Mitchell, Amy. "Many Americans Say Made-Up News Is a Critical Problem That Needs to Be Fixed." Pew Research Center, June 5, 2019. https://www.pewresearch.org/journalism/2019/06/05/many-americans-say-made-up-news-is-a-critical-problem-that-needs-to-be-fixed/.

Mitchell, Amy, Mark Jurkowitz, J. Baxter Oliphant, and Elisa Shearer. "5. Republicans' Views on COVID-19 Shifted Over Course of 2020; Democrats' Hardly Budged." Pew Research Center, February 22, 2021. https://www.pewresearch.org/journalism/2021/02/22/republicans-views-on-covid-19-shifted-over-course-of-2020-democrats-hardly-budged/.

Mitchell, Travis. "How the Public Reacted on Facebook." Pew Research Center, February 23, 2017. https://www.pewresearch.org/politics/2017/02/23/how-the-public-reacted-on-facebook/.

Mitchell, Travis. "Most Americans Who Go to Religious Services Say They Would Trust Their Clergy's Advice on COVID-19 Vaccines." Pew Research Center, October 15, 2021. https://www.pewforum.org/2021/10/15/most-americans-who-go-to-religious-services-say-they-would-trust-their-clergys-advice-on-covid-19-vaccines/.

Mod, Craig. "The Facebook-Loving Farmers of Myanmar." *The Atlantic*, January 21, 2016. https://www.theatlantic.com/technology/archive/2016/01/the-facebook-loving-farmers-of-myanmar/424812/.

Modirrousta-Galian, Ariana, and Philip A. Higham. "Gamified Inoculation Interventions Do Not Improve Discrimination Between True and Fake News: Reanalyzing Existing Research with Receiver Operating Characteristic Analysis." *Journal of Experimental Psychology: General* 152, no. 9 (2023): 2411–37. https://doi.org/10.1037/xge0001395.

Moehring, Alex, Avinash Collis, Kiran Garimella, M. Amin Rahimian, Sinan Aral, and Dean Eckles. "Providing Normative Information Increases Intentions to Accept a COVID-19 Vaccine." *Nature Communications* 14, no. 1 (2023): 126. https://doi.org/10.1038/s41467-022-35052-4.

Mogensen, Jackie Flynn. "5 Tips for How to Actually Change an Anti-Masker's Mind, According to Experts." *Mother Jones*, December 21, 2020. https://www.motherjones.com/politics/2020/12/how-to-win-an-argument-change-mind-anti-masker-tips/.

Moniuszko, Sara. "Dr. Anthony Fauci Says Keeping Schools Shut Down for So Long Amid COVID 'Was Not a Good Idea.'" CBS News, June 18, 2024. https://www.cbsnews.com/news/fauci-schools-shut-down-covid/.

Monmouth University Polling Institute. "1 in 4 Say 'No Thanks' to Vaccine." February 3, 2021. https://www.monmouth.edu/polling-institute/reports/MonmouthPoll_US_020321/.

Mønsted, Bjarke, Piotr Sapieżyński, Emilio Ferrara, and Sune Lehmann. "Evidence of Complex Contagion of Information in Social Media: An Experiment Using Twitter Bots." *PLOS One* 12, no. 9 (2017): e0184148. https://doi.org/10.1371/journal.pone.0184148.

Moonshot. "Advancing Media Literacy in Indonesia Part II: Resilience and Knowledge Change." November 29, 2022. https://moonshotteam.com/resource/advancing-media-literacy-in-indonesia-part-ii/.

Moore, Adam, Sujin Hong, and Laura Cram. "Trust in Information, Political Identity and the Brain: An Interdisciplinary fMRI Study." *Philosophical Transactions of the Royal Society of London, Series B, Biological Sciences* 376, no. 1822 (2021): 20200140. https://doi.org/10.1098/rstb.2020.0140.

Moore, Tim. "Knowledge, Disciplinarity and the Teaching of Critical Thinking." In *The Routledge International Handbook of Research on Teaching Thinking*, ed. Rupert Wegerif, Li Li, and James C. Kaufman, 243–53. Routledge, 2015

Moral, Mert, and Robin E. Best. "On the Relationship Between Party Polarization and Citizen Polarization." *Party Politics* 29, no. 2 (2023): 229–47. https://doi.org/10.1177/13540688211069544.

Morewedge, Carey K., Haewon Yoon, Irene Scopelliti, Carl W. Symborski, James H. Korris, and Karim S. Kassam. "Debiasing Decisions: Improved Decision Making with a Single Training Intervention." *Policy Insights from the Behavioral and Brain Sciences* 2, no. 1 (2015): 129–40. https://doi.org/10.1177/2372732215600886.

Mosleh, Mohsen, Cameron Martel, Dean Eckles, and David G. Rand. "Shared Partisanship Dramatically Increases Social Tie Formation in a Twitter Field Experiment." *Proceedings of the National Academy of Sciences* 118, no. 7 (2021): e2022761118. https://doi.org/10.1073/pnas.2022761118.

Mosleh, Mohsen, Gordon Pennycook, Antonio A. Arechar, and David G. Rand. "Cognitive Reflection Correlates with Behavior on Twitter." *Nature Communications* 12, no. 1 (2021): 921. https://doi.org/10.1038/s41467-020-20043-0.

Motta, Matthew. "Is Cancer Treatment Immune from Partisan Conflict? How Partisan Communication Motivates Opposition to Preventative Cancer Vaccination in the U.S." *Journal of Elections, Public Opinion and Parties* 34, no. 2 (2024): 319–43. https://doi.org/10.1080/17457289.2023.2168678.

Motta, Matthew, Timothy Callaghan, and Kristin Lunz Trujillo. "'The CDC Won't Let Me Be': The Opinion Dynamics of Support for CDC Regulatory Authority." *Journal of Health Politics, Policy and Law* 48, no. 6 (2023): 829–57. https://doi.org/10.1215/03616878-10852592.

Motta, Matthew, Gabriella Motta, and Dominik Stecula. "Sick as a Dog? The Prevalence, Politicization, and Health Policy Consequences of Canine Vaccine Hesitancy (CVH)." *Vaccine* 41 (2023): 5946–50. https://doi.org/10.1016/j.vaccine.2023.08.059.

Motta, Matthew, Robert Ralston, and Jennifer Spindel. "A Call to Arms for Climate Change? How Military Service Member Concern About Climate Change Can Inform Effective Climate Communication." *Environmental Communication* 15, no. 1 (2021): 85–98. https://doi.org/10.1080/17524032.2020.1799836.

Motz, Ben, Emily Fyfe, Helen Lee Bouygues, and Taylor Guba. "A Scalable, Versatile Approach for Improving Critical Thinking Skills." Reboot Foundation, last modified 2021. https://reboot-foundation.org/wp-content/uploads/_docs/Improving_Critical_Thinking_Skills.pdf.

Muchnik, Lev, Sinan Aral, and Sean J. Taylor. "Social Influence Bias: A Randomized Experiment." *Science* 341, no. 6146 (2013): 647–51. https://doi.org/10.1126/science.1240466.

Mullis, Steve. "She Resisted Getting Her Kids the Usual Vaccines: Then the Pandemic Hit." NPR, January 22, 2021. https://www.npr.org/2021/01/22/956935520/she-resisted-getting-her-kids-the-usual-vaccines-then-the-pandemic-hit.

Munger, Kevin, and Joseph Phillips. "Right-Wing YouTube: A Supply and Demand Perspective." *International Journal of Press/Politics* 27, no. 1 (2022): 186–219. https://doi.org/10.1177/1940161220964767.

Myers, Kristin. "Anti-Vaxxers Are Costing Americans Billions Each Year." *Yahoo Finance*, April 10, 2019. https://finance.yahoo.com/news/antivaxxers-costing-americans-billions-each-year-191839191.html.

Nabi, Robin L. "'Feeling' Resistance: Exploring the Role of Emotionally Evocative Visuals in Inducing Inoculation." *Media Psychology* 5, no. 2 (2003): 199–223. https://doi.org/10.1207/S1532785XMEP0502_4.

Nadeem, Reem. "Abortion Rises in Importance as a Voting Issue, Driven by Democrats." Pew Research Center, August 23, 2022. https://www.pewresearch.org/politics/2022/08/23/abortion-rises-in-importance-as-a-voting-issue-driven-by-democrats/.

Nadeem, Reem. "As Partisan Hostility Grows, Signs of Frustration with the Two-Party System." Pew Research Center, August 9, 2022. https://www.pewresearch.org/politics/2022/08/09/as-partisan-hostility-grows-signs-of-frustration-with-the-two-party-system/.

Nadeem, Reem. "5. The U.S. Military." Pew Research Center, February 1, 2024. https://www.pewresearch.org/politics/2024/02/01/the-u-s-military/.

Naeem, Salman Bin, Rubina Bhatti, and Aqsa Khan. "An Exploration of How Fake News Is Taking Over Social Media and Putting Public Health at Risk." Health Information and Libraries Journal 38, no. 2 (2021): 143–49. https://doi.org/10.1111/hir.12320.

National Center for Immunization and Respiratory Diseases, Division of Viral Diseases. "Science Brief: SARS-CoV-2 and Surface (Fomite) Transmission for Indoor Community Environments." Updated April 5, 2021. https://archive.cdc.gov/#/details?url=https://www.cdc.gov/coronavirus/2019-ncov/more/science-and-research/surface-transmission.html.

National Governors Association. "COVID-19 Vaccine Incentives." October 19, 2021. https://www.nga.org/center/publications/covid-19-vaccine-incentives/.

National Institutes of Health. "COVID-19 Was Third Leading Cause of Death in the United States in Both 2020 and 2021." July 5, 2022. https://www.nih.gov/news-events/news-releases/covid-19-was-third-leading-cause-death-united-states-both-2020-2021.

Navin, Mark. Values and Vaccine Refusal: Hard Questions in Ethics, Epistemology, and Health Care. Routledge, 2015.

NBC News. "Live: Facebook Whistleblower Testifies at Senate Hearing." October 5, 2021. YouTube, 3 hr., 16 min., 27 sec. https://www.youtube.com/watch?v=_IhWeVHxdXg&t=2836s&ab_channel=NBCNews.

Needleman, Emma. "Would Chat GPT Get a Wharton MBA? New White Paper by Christian Terwiesch." Mack Institute for Innovation Management, January 17, 2023. https://mackinstitute.wharton.upenn.edu/2023/would-chat-gpt3-get-a-wharton-mba-new-white-paper-by-christian-terwiesch/.

Nelson, Steven M. "Redefining a Bizarre Situation: Relative Concept Stability in Affect Control Theory." Social Psychology Quarterly 69, no. 3 (2006): 215–34. https://doi.org/10.1177/019027250606900301.

Netburn, Deborah. "Timeline: CDC Mask Guidelines During the COVID Pandemic." Los Angeles Times, July 27, 2021. https://www.latimes.com/science/story/2021-07-27/timeline-cdc-mask-guidance-during-covid-19-pandemic.

Neudert, Lisa-Maria, Bence Kollanyi, and Philip N. Howard. "Polarization, Partisanship and Junk News Consumption on Social Media During the 2018 US Midterm Elections." America 8 (2017): 10.

News Literacy Project. "Is It Legit? Five Steps for Vetting a News Source." Last modified November 5, 2021. https://newslit.org/educators/resources/is-it-legit/.

Niburski, Kacper, and Oskar Niburski. "Impact of Trump's Promotion of Unproven COVID-19 Treatments and Subsequent Internet Trends: Observational Study." Journal of Medical Internet Research 22, no. 11 (2020): e20044. https://doi.org/10.2196/20044.

Nichols, Tom. The Death of Expertise: The Campaign Against Established Knowledge and Why It Matters. Oxford University Press, 2017.

Nickerson, Raymond S. "Confirmation Bias: A Ubiquitous Phenomenon in Many Guises." Review of General Psychology 2, no. 2 (1998): 175–220. https://doi.org/10.1037/1089-2680.2.2.175.

Nieminen, Sakari, and Valtteri Sankari. "Checking PolitiFact's Fact-Checks." Journalism Studies 22, no. 3 (2021): 358–78. https://doi.org/10.1080/1461670X.2021.1873818.

Nieva, Richard. "Google Wants to Fix Its Search Engine's Misinformation Problem." Forbes, August 11, 2022. https://www.forbes.com/sites/richardnieva/2022/08/11/google-wants-to-fix-its-search-engines-misinformation-problem/.

NIH News in Health. "COVID-19 Vaccines Prevented Nearly 140,000 U.S. Deaths." October 2021. https://newsinhealth.nih.gov/2021/10/covid-19-vaccines-prevented-nearly-140000-us-deaths.

Nix, Elizabeth. "Tuskegee Experiment: The Infamous Syphilis Study." History, May 16, 2017. https://www.history.com/news/the-infamous-40-year-tuskegee-study.

Noor, Poppy. "The Beach-Going Grim Reaper on His Florida Protest: 'Someone Has to Stand Up.'" *The Guardian*, May 7, 2020. https://www.theguardian.com/us-news/2020/may/07/florida-grim-reaper -beach-interview.

Nowak, Sarah A., Courtney A. Gidengil, Andrew M. Parker, and Luke J. Matthews. "Association Among Trust in Health Care Providers, Friends, and Family, and Vaccine Hesitancy." *Vaccine* 39, no. 40 (2021): 5737–40. https://doi.org/10.1016/j.vaccine.2021.08.035.

Now This Impact. "Facebook Whistleblower: Another Ex-Employee Speaks Out." YouTube, October 8, 2021. Video, 3 min., 47 sec. https://www.youtube.com/watch?v=8XkREyzDgnA.

NPR. "[2020] Election Results." November 2, 2022. https://apps.npr.org/elections20-interactive/.

NPR. "'Vaccine Talk' Facebook Group Is a Carefully Moderated Forum for Vaccine Questions." September 18, 2021. https://www.npr.org/2021/09/18/1038533086/vaccine-talk-facebook-group-is-a-carefully -moderated-forum-for-vaccine-questions.

Nygren, Thomas, Divina Frau-Meigs, Nicoleta Corbu, and Sonia Santoveña-Casal. "Teachers' Views on Disinformation and Media Literacy Supported by a Tool Designed for Professional Fact-Checkers: Perspectives from France, Romania, Spain and Sweden." *SN Social Sciences* 2, no. 4 (2022): 40. https://doi .org/10.1007/s43545-022-00340-9.

Nyhan, Brendan. "Fake News and Bots May Be Worrisome, but Their Political Power Is Overblown." *New York Times*, February 13, 2018. https://www.nytimes.com/2018/02/13/upshot/fake-news-and-bots-may -be-worrisome-but-their-political-power-is-overblown.html.

Nyhan, Brendan, and Jason Reifler. "The Effect of Fact-Checking on Elites: A Field Experiment on U.S. State Legislators." *American Journal of Political Science* 59, no. 3 (2015): 628–40. https://doi.org/10.1111 /ajps.12162.

Nyhan, Brendan, and Jason Reifler. "When Corrections Fail: The Persistence of Political Misperceptions." *Political Behavior* 32, no. 2 (2010): 303–30. https://doi.org/10.1007/s11109-010-9112-2.

Nyhan, Brendan, Jason Reifler, Sean Richey, and Gary L. Freed. "Effective Messages in Vaccine Promotion: A Randomized Trial." *Pediatrics* 133, no. 4 (2014): e835–42. https://doi.org/10.1542/peds.2013-2365.

O'Brien, Cortney. "Viewers Demand Apology from MSNBC, Rachel Maddow for Previous COVID Vaccine Comments." Fox News, December 28, 2021. https://www.foxnews.com/media/social-media-users -demand-apology-msnbc-rachel-maddow-vaccines.

O'Keefe, Daniel J. "Persuasion." In *The Handbook of Communication Skills*, ed. Mark A. Bodie, 333–52. Routledge, 2006.

Ognyanova, Katherine, David Lazer, Matthew Baum et al. "The COVID States Project #82: COVID-19 Vaccine Misinformation Trends, Awareness of Expert Consensus, and Trust in Social Institutions." OSF Preprints, February 15, 2022. https://doi.org/10.31219/osf.io/9ua2x.

Ognyanova, Katherine, David Lazer, Ronald E. Robertson, and Christo Wilson. "Misinformation in Action: Fake News Exposure Is Linked to Lower Trust in Media, Higher Trust in Government When Your Side Is in Power." *Harvard Kennedy School Misinformation Review* 1, no. 4 (2020): 1–19. https:// doi.org/10.37016/mr-2020-024.

Oliver, J. Eric, and Thomas J. Wood. "Conspiracy Theories and the Paranoid Style(s) of Mass Opinion." *American Journal of Political Science* 58, no. 4 (2014): 952–66. https://doi.org/10.1111/ajps.12084.

Olutola, Sarah R. "Nicki Minaj's COVID-19 Vaccine Tweet About Swollen Testicles Signals the Dangers of Celebrity Misinformation and Fandom." *The Conversation*, September 20, 2021. http://the conversation.com/nicki-minajs-covid-19-vaccine-tweet-about-swollen-testicles-signals-the-dangers-of -celebrity-misinformation-and-fandom-168242.

Omer, Saad B. "The Discredited Doctor Hailed by the Anti-Vaccine Movement." *Nature* 586, no. 7831 (2020): 668–69. https://doi.org/10.1038/d41586-020-02989-9.

Orcullo, Daisy Jane C., and Teo Hui San. "Understanding Cognitive Dissonance in Smoking Behaviour: A Qualitative Study." *International Journal of Social Science and Humanity* 6, no. 6 (2016): 481–84. https://doi.org/10.7763/IJSSH.2016.V6.695.

Orenstein, Walter A., and Rafi Ahmed. "Simply Put: Vaccination Saves Lives." *Proceedings of the National Academy of Sciences* 114, no. 16 (2017): 4031–33. https://doi.org/10.1073/pnas.1704507114.

Oreskes, Naomi. *Why Trust Science?* Princeton University Press, 2019.

Orth, Taylor. "Which Conspiracy Theories Do Americans Believe?" YouGov, December 8, 2023. https://today.yougov.com/politics/articles/48113-which-conspiracy-theories-do-americans-believe.

Ortiz-Sánchez, Elvira, Almudena Velando-Soriano, Laura Pradas-Hernández et al. "Analysis of the Anti-Vaccine Movement in Social Networks: A Systematic Review." *International Journal of Environmental Research and Public Health* 17, no. 15 (2020): art. 5394. https://doi.org/10.3390/ijerph17155394.

Osmundsen, Mathias, Alexander Bor, Peter Bjerregaard Vahlstrup, Anja Bechmann, and Michael Bang Petersen. "Partisan Polarization Is the Primary Psychological Motivation Behind Political Fake News Sharing on Twitter." *American Political Science Review* 115, no. 3 (2021): 999–1015. https://doi.org/10.1017/S0003055421000290.

Oster, Matthew E., David K. Shay, John R. Su et al. "Myocarditis Cases Reported After mRNA-Based COVID-19 Vaccination in the US from December 2020 to August 2021." *JAMA* 327, no. 4 (2022): 331–40. https://doi.org/10.1001/jama.2021.24110.

Pacheco, Diogo, Pik-Mai Hui, Christopher Torres-Lugo, Bao Tran Truong, Alessandro Flammini, and Filippo Menczer. "Uncovering Coordinated Networks on Social Media: Methods and Case Studies." *Proceedings of the International AAAI Conference on Web and Social Media* 15, no. 1 (22, 2021): 455–66. https://doi.org/10.1609/icwsm.v15i1.18075.

Packer, Dominic, and Jay Van Bavel. "Navigating Political Divides at Thanksgiving." The Power of Us, November 22, 2022. https://powerofus.substack.com/p/navigating-political-divides-at-thanksgiving.

Padgett, Jeremy, Johanna L. Dunaway, and Joshua P. Darr. "As Seen on TV? How Gatekeeping Makes the U.S. House Seem More Extreme." *Journal of Communication* 69, no. 6 (2019): 696–719. https://doi.org/10.1093/joc/jqz039.

Page, Jeremy, Drew Hinshaw, and Betsy McKay. "In Hunt for Covid-19 Origin, Patient Zero Points to Second Wuhan Market." *Wall Street Journal*, February 26, 2021. https://www.wsj.com/articles/in-hunt-for-covid-19-origin-patient-zero-points-to-second-wuhan-market-11614335404.

Panizza, Folco, Piero Ronzani, Carlo Martini, Simone Mattavelli, Tiffany Morisseau, and Matteo Motterlini. "Lateral Reading and Monetary Incentives to Spot Disinformation About Science." *Scientific Reports* 12, no. 1 (2022): 5678. https://doi.org/10.1038/s41598-022-09168-y.

Panizza, Folco, Piero Ronzani, Tiffany Morisseau, Simone Mattavelli, and Carlo Martini. "How Do Online Users Respond to Crowdsourced Fact-Checking?" *Humanities & Social Sciences Communications* 10, no. 1 (November 25, 2023): 1–11.

Pariser, Eli. *The Filter Bubble: What the Internet Is Hiding from You.* Penguin UK, 2011.

Park, Sora, Caroline Fisher, Terry Flew, and Uwe Dulleck. "Global Mistrust in News: The Impact of Social Media on Trust." *International Journal on Media Management* 22, no. 2 (2020): 83–96. https://doi.org/10.1080/14241277.2020.1799794.

Parker, V. A., E. Kehoe, J. Lees, M. Facciani, and A. E. Wilson. "Alluring or Alarming? The Polarizing Effect of Forbidden Knowledge in Political Discourse." *Personality & Social Psychology Bulletin* (2024), 1461672241288332.

Parkinson, Carolyn, Adam M. Kleinbaum, and Thalia Wheatley. "Similar Neural Responses Predict Friendship." *Nature Communications* 9, no. 1 (2018): 332. https://doi.org/10.1038/s41467-017-02722-7.

Parmelee, John H., and Nataliya Roman. "Insta-politicos: Motivations for Following Political Leaders on Instagram." *Social Media + Society* 5, no. 2 (2019): 2056305119837662. https://doi.org/10.1177/2056305119837662.

Parsons, Bryan M. "The Social Identity Politics of Peer Networks." *American Politics Research* 43, no. 4 (2015): 680–707. https://doi.org/10.1177/1532673X14546856.

Paruzel-Czachura, Mariola, Dominika Wojciechowska, and Dries Bostyn. "Online Moral Conformity: How Powerful Is a Group of Strangers When Influencing an Individual's Moral Judgments During a Video Meeting?" *Current Psychology* 43, no. 7 (2024): 6125–35. https://doi.org/10.1007/s12144-023-04765-0.

Pasquetto, Irene V., Eaman Jahani, Shubham Atreja, and Matthew Baum. "Social Debunking of Misinformation on WhatsApp: The Case for Strong and In-Group Ties." *Proceedings of the ACM on Human -Computer Interaction* 6, no. CSCW1 (2022): art. 117. https://doi.org/10.1145/3512964.

Pasquetto, Irene, Briony Swire-Thompson, Michelle A. Amazeen et al. "Tackling Misinformation: What Researchers Could Do with Social Media Data." *Harvard Kennedy School Misinformation Review* 1, no. 8 (2020): 1–14. https://doi.org/10.37016/mr-2020-49.

Patel, Vimal. "White House Pushes Journals to Drop Paywalls on Publicly Funded Research." *New York Times*, August 26, 2022. https://www.nytimes.com/2022/08/25/us/white-house-federally-funded-research -access.html.

Pehlivanoglu, Didem, Nichole R. Lighthall, Tian Lin et al. "Aging in an 'Infodemic': The Role of Analytical Reasoning, Affect, and News Consumption Frequency on News Veracity Detection." *Journal of Experimental Psychology: Applied* 28, no. 3 (2022): 468–85. https://doi.org/10.1037/xap0000426.

Pelley, Scott. "Whistleblower: Facebook Is Misleading the Public on Progress Against Hate Speech, Violence, Misinformation." CBS News, October 4, 2021. https://www.cbsnews.com/news/facebook-whistleblower -misinformation-public-60-minutes-2021-10-03/.

Pennycook, Gordon, Jabin Binnendyk, Christie Newton, and David G. Rand. "A Practical Guide to Doing Behavioral Research on Fake News and Misinformation." *Collabra: Psychology* 7, no. 1 (2021). https:// doi.org/10.1525/collabra.25293.

Pennycook, Gordon, Tyrone D. Cannon, and David G. Rand. "Prior Exposure Increases Perceived Accuracy of Fake News." *Journal of Experimental Psychology: General* 147, no. 12 (2018): 1865–80. https://doi.org /10.1037/xge0000465.

Pennycook, Gordon, Ziv Epstein, Mohsen Mosleh, Antonio A. Arechar, Dean Eckles, and David G. Rand. "Shifting Attention to Accuracy Can Reduce Misinformation Online." *Nature* 592, no. 7855 (2021): 590–95. https://doi.org/10.1038/s41586-021-03344-2.

Pennycook, Gordon, Jonathon McPhetres, Bence Bago, and David G. Rand. "Beliefs About COVID-19 in Canada, the United Kingdom, and the United States: A Novel Test of Political Polarization and Motivated Reasoning." *Personality and Social Psychology Bulletin* 48, no. 5 (2022): 750–65. https://doi .org/10.1177/01461672211023652.

Pennycook, Gordon, Jonathon McPhetres, Yunhao Zhang, Jackson G. Lu, and David G. Rand. "Fighting COVID-19 Misinformation on Social Media: Experimental Evidence for a Scalable Accuracy-Nudge Intervention." *Psychological Science* 31, no. 7 (2020): 770–80. https://doi.org/10.1177/0956797620939054.

Pennycook, Gordon, and David G. Rand. "Fighting Misinformation on Social Media Using Crowdsourced Judgments of News Source Quality." *Proceedings of the National Academy of Sciences* 116, no. 7 (2019): 2521–26. https://doi.org/10.1073/pnas.1806781116.

Pennycook, Gordon, and David G. Rand. "Lazy, Not Biased: Susceptibility to Partisan Fake News Is Better Explained by Lack of Reasoning than by Motivated Reasoning." *Cognition* 188 (2019): 39–50. https:// doi.org/10.1016/j.cognition.2018.06.011.

Pennycook, Gordon, and David G. Rand. "The Psychology of Fake News." *Trends in Cognitive Sciences* 25, no. 5 (2021): 388–402. https://doi.org/10.1016/j.tics.2021.02.007.

Pennycook, Gordon, and David G. Rand. "Research Note: Examining False Beliefs About Voter Fraud in the Wake of the 2020 Presidential Election." *Harvard Kennedy School Misinformation Review* 1, no. 8 (2021): 1–10. https://doi.org/10.37016/mr-2020-51.

Pennycook, Gordon, and David G. Rand, "Who Falls for Fake News? The Roles of Bullshit Receptivity, Overclaiming, Familiarity, and Analytic Thinking." *Journal of Personality* 88, no. 2 (2020): 185–200, https://doi.org/10.1111/jopy.12476.

Pereira, Alyssa. "What Mask Use Looks Like in 10 Other Countries Compared to the U.S." SFGATE, July 5, 2020. https://www.sfgate.com/news/article/mask-wearing-japan-korea-brazil-germany-zealand-15383513.php.

Peter, Georges. "Vaccine Crisis: An Emerging Societal Problem." *Journal of Infectious Diseases* 151, no. 6 (1985): 981–83. http://www.jstor.org/stable/30130075.

Peters, Jeremy W., and Katie Robertson. "Fox Stars Privately Expressed Disbelief About Election Fraud Claims: 'Crazy Stuff.'" *New York Times*, February 16, 2023. https://www.nytimes.com/2023/02/16/business/media/fox-dominion-lawsuit.html.

Petersen, Michael Bang, Mathias Osmundsen, and Kevin Arceneaux. "The 'Need for Chaos' and Motivations to Share Hostile Political Rumors." *American Political Science Review* 117, no. 4 (2023): 1486–1505. https://doi.org/10.1017/S0003055422001447.

Petrecca, Laura. "America's Division: We United in the Wake of 9/11, Then Partisanship Re-emerged." *USA Today*, September 11, 2017. https://www.usatoday.com/story/news/2017/09/11/americas-division-we-united-wake-9-11-then-partisanship-re-emerged/639473001/.

Petrocelli, John. *The Life-Changing Science of Detecting Bullshit*. St. Martin's Press, 2021.

Petty, Richard E., Curtis P. Haugtvedt, and Stephen M. Smith. "Elaboration as a Determinant of Attitude Strength: Creating Attitudes That Are Persistent, Resistant, and Predictive of Behavior." In *Attitude Strength*, ed. Richard E. Petty and Jon A. Krosnick, 93–130. Psychology Press, 2014.

Pew Research Center. "The Color of News: How Different Media Have Covered the General Election." October 29, 2008. https://www.pewresearch.org/journalism/2008/10/29/the-color-of-news/.

Pew Research Center. "Journalists Highly Concerned About Misinformation, Future of Press Freedoms." June 14, 2022. https://www.pewresearch.org/journalism/2022/06/14/journalists-highly-concerned-about-misinformation-future-of-press-freedoms/.

Pew Research Center. "Local TV News Fact Sheet." September 14, 2023. https://www.pewresearch.org/journalism/fact-sheet/local-tv-news/.

Pew Research Center. "1. Feelings About Partisans and the Parties." June 22, 2016. https://www.pewresearch.org/politics/2016/06/22/1-feelings-about-partisans-and-the-parties/.

Pew Research Center. "Social Media Fact Sheet." January 31, 2024. https://www.pewresearch.org/internet/fact-sheet/social-media/.

Pew Research Center. "Social Media Seen as Mostly Good for Democracy Across Many Nations, but U.S. is a Major Outlier." December 6, 2022. https://www.pewresearch.org/global/2022/12/06/social-media-seen-as-mostly-good-for-democracy-across-many-nations-but-u-s-is-a-major-outlier/.

Phadke, Shruti, Mattia Samory, and Tanushree Mitra. "What Makes People Join Conspiracy Communities? Role of Social Factors in Conspiracy Engagement." *Proceedings of the ACM on Human-Computer Interaction* 4, no. CSCW3 (2021): art. 223. https://doi.org/10.1145/3432922.

Pickard, Victor. *Democracy Without Journalism? Confronting the Misinformation Society*. Oxford University Press, 2019.

Pierri, Francesco, Brea L. Perry, Matthew R. DeVerna et al. "Online Misinformation Is Linked to Early COVID-19 Vaccination Hesitancy and Refusal." *Scientific Reports* 12, no. 1 (2022): art. 5966. https://doi.org/10.1038/s41598-022-10070-w.

Pilat, Dan, and Sekoul Krastev. "Base Rate Fallacy." Decision Lab. Accessed July 3, 2024. https://thedecisionlab.com/biases/base-rate-fallacy.

Pink, Sophia L., James Chu, James N. Druckman, David G. Rand, and Robb Willer. "Elite Party Cues Increase Vaccination Intentions Among Republicans." *Proceedings of the National Academy of Sciences* 118, no. 32 (2021): e2106559118. https://doi.org/10.1073/pnas.2106559118.

Place, Nathan. "Anti-Mask Maine Lawmaker Resigns After Wife Dies of Covid-19." *The Independent*, November 30, 2021. https://www.independent.co.uk/news/world/americas/us-politics/chris-johansen-covid-death-maine-b1967013.html.

Pluviano, Sara, Caroline Watt, Giovanni Ragazzini, and Sergio Della Sala. "Parents' Beliefs in Misinformation About Vaccines Are Strengthened by Pro-Vaccine Campaigns." *Cognitive Processing* 20, no. 3 (2019): 325–31. https://doi.org/10.1007/s10339-019-00919-w.

Popp, Maria, Miriam Stegemann, Maria-Inti Metzendorf et al. "Ivermectin for Preventing and Treating COVID-19." *Cochrane Database of Systematic Reviews*, no. 7 (2021): art. CD015017. https://doi.org/10.1002/14651858.CD015017.pub2.

Porter, Ethan, Yamil Velez, and Thomas J. Wood. "Correcting COVID-19 Vaccine Misinformation in 10 Countries." *Royal Society Open Science* 10, no. 3 (2023): art. 221097. https://doi.org/10.1098/rsos.221097.

Porter, Tenelle, Abdo Elnakouri, Ethan A. Meyers, Takuya Shibayama, Eranda Jayawickreme, and Igor Grossmann. "Predictors and Consequences of Intellectual Humility." *Nature Reviews Psychology* 1, no. 9 (2022): 524–36. https://doi.org/10.1038/s44159-022-00081-9.

Porter, Tenelle, Karina Schumann, Diana Selmeczy, and Kali Trzesniewski. "Intellectual Humility Predicts Mastery Behaviors When Learning." *Learning and Individual Differences* 80 (2020): art. 101888. https://doi.org/10.1016/j.lindif.2020.101888.

Porter, Tom. "Taking Toxic Bleach MMS Has Killed 7 People in the US, Colombian Prosecutors Say—Far More Than Previously Known." *Business Insider*, August 12, 2020. https://www.businessinsider.com/mms-bleach-killed-7-americans-new-from-colombia-arrest-2020-8.

Powdthavee, Nattavudh, Yohanes E. Riyanto, Erwin C. L. Wong, Jonathan X. W. Yeo, and Qi Yu Chan. "When Face Masks Signal Social Identity: Explaining the Deep Face-Mask Divide During the COVID-19 Pandemic." *PLOS One* 16, no. 6 (2021): e0253195. https://doi.org/10.1371/journal.pone.0253195.

Powell, Kendall. "The Power of Diversity." *Nature* 558, no. 7708 (2018): 19–22. https://doi.org/10.1038/d41586-018-05316-5.

Preston, Stephanie, Anthony Anderson, David J. Robertson, Mark P. Shephard, and Narisong Huhe. "Detecting Fake News on Facebook: The Role of Emotional Intelligence." *PLOS One* 16, no. 3 (2021): e0246757. https://doi.org/10.1371/journal.pone.0246757.

Prior, Ryan. "Most Americans Think They Can Spot Fake News: They Can't, Study Finds." CNN, May 31, 2021. https://www.cnn.com/2021/05/31/health/fake-news-study/index.html.

Pronin, Emily. "The Introspection Illusion." In *Advances in Experimental Social Psychology*, vol. 41, ed. Mark P. Zanna, 1–67. Academic Press, 2009.

Pruitt, Sarah. "How the US Pulled Off Midterm Elections Amid the 1918 Flu Pandemic." History, April 22, 2020. https://www.history.com/news/1918-pandemic-midterm-elections.

Putnam, Robert D. *The Upswing: How America Came Together a Century Ago and How We Can Do It Again.* Simon and Schuster, 2020.

Qiu, Xiaoyan, Diego F. M. Oliveira, Alireza Sahami Shirazi, Alessandro Flammini, and Filippo Menczer. "Limited Individual Attention and Online Virality of Low-Quality Information." *Nature Human Behaviour* 1 (2017): art. 132. https://doi.org/10.1038/s41562-017-0132.

Rabbie, J. M., and M. Horwitz. "Arousal of Ingroup-Outgroup Bias by a Chance Win or Loss." *Journal of Personality and Social Psychology* 13, no. 3 (1969): 269–77. https://doi.org/10.1037/h0028284.

Rabin, Roni. "Paxlovid Cuts Covid Deaths Among Older People, Israeli Study Finds." *New York Times*, August 30, 2022. https://www.nytimes.com/2022/08/30/health/paxlovid-efficacy-seniors.html.

Rafail, Patrick, Whitney E. O'Connell, and Emma Sager. "Polarizing Feedback Loops on Twitter: Congressional Tweets During the 2022 Midterm Elections." *Socius* 10 (2024): e23780231241228924. https://doi.org/10.1177/23780231241228924.

Rainie, Lee, Cary Funk, Monica Anderson, and Alec Tyson. "3. Mixed Views About Social Media Companies Using Algorithms to Find False Information." Pew Research Center, March 17, 2022. https://www.pewresearch.org/internet/2022/03/17/mixed-views-about-social-media-companies-using-algorithms-to-find-false-information/.

Rajadesingan, Ashwin, Paul Resnick, and Ceren Budak. "Quick, Community-Specific Learning: How Distinctive Toxicity Norms Are Maintained in Political Subreddits." *Proceedings of the International AAAI Conference on Web and Social Media* 14, no. 1 (2020): 557–68. https://doi.org/10.1609/icwsm.v14i1.7323.

Raju, Manu, and Ted Barrett. "US Capitol Police Chief to Resign After Wednesday's Riots." CNN, January 7, 2021. https://www.cnn.com/2021/01/07/politics/capitol-police-reaction-details/index.html.

Rasmussen, Frederick N. "100 Years After the Titanic Disaster." *Baltimore Sun*, April 14, 2012. https://www.baltimoresun.com/2012/04/14/100-years-after-the-titanic-disaster/.

Rathje, Steve, Clara Pretus, James Kunling He, Trisha Harjani, Jon Roozenbeek, Kurt Gray, Sander van der Linden, and Jay Joseph Van Bavel. "Unfollowing Hyperpartisan Social Media Influencers Durably Reduces Out-Party Animosity." *PsyArXiv* (2024). https://doi.org/10.31234/osf.io/acbwg.

Rathje, Steve, Jon Roozenbeek, Cecilie Steenbuch Traberg, Jay Joseph Van Bavel, and Sander van der Linden. "Letter to the Editors of *Psychological Science*: Meta-Analysis Reveals That Accuracy Nudges Have Little to No Effect for U.S. Conservatives: Regarding Pennycook et al. (2020)." Preprint, PsyArXiv, April 2, 2022. https://doi.org/10.31234/osf.io/945na.

Rathje, Steve, Jay J. Van Bavel, and Sander van der Linden. "Out-Group Animosity Drives Engagement on Social Media." *Proceedings of the National Academy of Sciences* 118, no. 26 (2021): e2024292118. https://doi.org/10.1073/pnas.2024292118.

Ray, Rashawn, Jane Fran Morgan, Lydia Wileden, Samantha Elizondo, and Destiny Wiley-Yancy. *Examining and Addressing COVID-19 Racial Disparities in Detroit*. Brookings Institution, 2021. https://www.brookings.edu/wp-content/uploads/2021/02/Detroit_Covid_report_final.pdf

Reis, Gilmar, Eduardo Augusto dos Santos Moreira Silva, Daniela Carla Medeiros Silva et al. "Effect of Early Treatment with Hydroxychloroquine or Lopinavir and Ritonavir on Risk of Hospitalization Among Patients with COVID-19: The TOGETHER Randomized Clinical Trial." *JAMA Network Open* 4, no. 4 (2021): e216468. https://doi.org/10.1001/jamanetworkopen.2021.6468.

Reiss, Stefan, Johannes Klackl, Travis Proulx, and Eva Jonas. "Strength of Socio-political Attitudes Moderates Electrophysiological Responses to Perceptual Anomalies." *PLOS One* 14, no. 8 (2019): e0220732. https://doi.org/10.1371/journal.pone.0220732.

Relman, Eliza. "The Gap Between Republicans and Democrats on Flu Shots Is 20 Percentage-Points Bigger Than It Was Pre-pandemic." *Business Insider*, November 15, 2021. https://www.businessinsider.com/theres-a-25-point-gap-between-republicans-democrats-flu-shots-2021-11.

Ren, Zhiying, Eugen Dimant, and Maurice Schweitzer. "Beyond Belief: How Social Engagement Motives Influence the Spread of Conspiracy Theories." *Journal of Experimental Social Psychology* 104 (2023): art. 104421. https://doi.org/10.1016/j.jesp.2022.104421.

Resnick, Brian. "How to Talk Someone out of Bigotry." *Vox*, January 29, 2020. https://www.vox.com/2020/1/29/21065620/broockman-kalla-deep-canvassing.

Reunanen, Esa. "Finland." Reuters Institute for the Study of Journalism, June 14, 2023. https://reutersinstitute.politics.ox.ac.uk/digital-news-report/2023/finland.

Reuters. "Big 3 U.S. Drug Distributors, Johnson & Johnson Reach Landmark $26 Billion Opioid Settlement." CNBC, July 21, 2021. https://www.cnbc.com/2021/07/21/drug-distributors-jj-reach-landmark-26-billion-opioid-settlement-.html.

Reuters. "Bots Hyped Up GameStop on Major Social Media Platforms, Analysis Finds." February 26, 2021. https://www.reuters.com/article/idUSKBN2AQ2BH/.

Reuters. "Fact Check: Courts Have Dismissed Multiple Lawsuits of Alleged Electoral Fraud Presented by Trump Campaign." February 15, 2021. https://www.reuters.com/article/idUSKBN2AF1FQ/.

Reuters. "FDA Cautions Against Use Of Hydroxychloroquine Or Chloroquine For Covid-19 Outside Of Hospital Setting Due To Risk Of Heart Rhythm Problems." April 24, 2020. https://www.reuters.com/article/business/healthcare-pharmaceuticals/fda-cautions-against-use-of-hydroxychloroquine-or-chloroquine-for-covid-19-outsi-idUSFWN2CC20M/.

Reuters. "Herman Cain, Ex-Presidential Candidate Who Refused to Wear Mask, Dies After COVID-19 Diagnosis." July 31, 2020. https://www.reuters.com/article/world/herman-cain-ex-presidential-candidate-who-refused-to-wear-mask-dies-after-covi-idUSKCN24V2OH/.

Reuters Institute for the Study of Journalism. "Digital News Report 2023." Accessed July 6, 2024. https://reutersinstitute.politics.ox.ac.uk/digital-news-report/2023.

Ribeiro, Manoel Horta, Raphael Ottoni, Robert West, Virgílio A. F. Almeida, and Wagner Meira. "Auditing Radicalization Pathways on YouTube." In *FAT* 20: Proceedings of the 2020 Conference on Fairness, Accountability, and Transparency*, 131–41. Association for Computing Machinery, 2020. https://doi.org/10.1145/3351095.3372879.

Rich, Timothy S., Ian Milden, and Mallory Treece Wagner. "Research Note: Does the Public Support Fact-Checking Social Media? It Depends Who and How You Ask." *Harvard Kennedy School Misinformation Review* 1, no. 8 (2020): 1–10. https://doi.org/10.37016/mr-2020-46.

Riedel, Stefan. "Edward Jenner and the History of Smallpox and Vaccination." *Baylor University Medical Center Proceedings* 18, no. 1 (2005): 21–25. https://doi.org/10.1080/08998280.2005.11928028.

Rick, Blake M., Eric W. Mania, Samuel L. Gaertner, Stacy A. McDonald, and Marika J. Lamoreaux. "Does a Common Ingroup Identity Reduce Intergroup Threat?" *Group Processes and Intergroup Relations* 13, no. 4 (2010): 403–23. https://doi.org/10.1177/1368430209346701.

Riggio, Olivia. "Not All Media Literacy Programs Are Created Equal—and Most Have Yet to Be Created." FAIR, December 15, 2020. https://fair.org/home/not-all-media-literacy-programs-are-created-equal-and-most-have-yet-to-be-created/.

Robertson, Ronald E., Jon Green, Damian J. Ruck, Katherine Ognyanova, Christo Wilson, and David Lazer. "Users Choose to Engage with More Partisan News Than They Are Exposed to on Google Search." *Nature* 618, no. 7964 (2023): 342–48. https://doi.org/10.1038/s41586-023-06078-5.

Robert Wood Johnson Foundation and Harvard T. H. Chan School of Public Health. *The Public's Perspective on the United States Public Health System*. Harvard Opinion Research Program, 2021. https://cdn1.sph.harvard.edu/wp-content/uploads/sites/94/2021/05/RWJF-Harvard-Report_FINAL-051321.pdf.

Robison, Joshua, Thomas J. Leeper, and James N. Druckman. "Do Disagreeable Political Discussion Networks Undermine Attitude Strength?" *Political Psychology* 39, no. 2 (2018): 479–94. https://doi.org/10.1111/pops.12374.

Roccas, Sonia, and Marilynn B. Brewer. "Social Identity Complexity." *Personality and Social Psychology Review* 6, no. 2 (2002): 88–106. https://doi.org/10.1207/S15327957PSPR0602_01.

Roose, Kevin. "The Making of a YouTube Radical." *New York Times*, June 8, 2019. https://www.nytimes.com/interactive/2019/06/08/technology/youtube-radical.html.

Root, Tik. "Inside the Long War to Protect Plastic." Center for Public Integrity, May 16, 2019. https://publicintegrity.org/environment/pollution/pushing-plastic/inside-the-long-war-to-protect-plastic/.

Roozenbeek, Jon, Eileen Culloty, and Jane Suiter. "Countering Misinformation: Evidence, Knowledge Gaps, and Implications of Current Interventions." *European Psychologist* 28, no. 3 (2023): 189–205. https://doi.org/10.1027/1016-9040/a000492.

Roozenbeek, Jon, Rakoen Maertens, Stefan M. Herzog et al. "Susceptibility to Misinformation Is Consistent Across Question Framings and Response Modes and Better Explained by Myside Bias and Partisanship than Analytical Thinking." *Judgment and Decision Making* 17, no. 3 (2022): 547–73. https://doi.org/10.1017/S1930297500003570.

Roozenbeek, Jon, Cecilie S. Traberg, and Sander van der Linden. "Technique-Based Inoculation Against Real-World Misinformation." *Royal Society Open Science* 9, no. 5 (2022): art. 211719. https://doi.org/10.1098/rsos.211719.

Roozenbeek, Jon, and Sander van der Linden. "Breaking *Harmony Square*: A Game That 'Inoculates' Against Political Misinformation." *Harvard Kennedy School Misinformation Review* 1, no. 8 (2020): 1–26. https://doi.org/10.37016/mr-2020-47.

Roozenbeek, Jon, and Sander van der Linden. "Fake News Game Confers Psychological Resistance Against Online Misinformation." *Palgrave Communications* 5 (2019): art. 65. https://doi.org/10.1057/s41599-019-0279-9.

Roozenbeek, Jon, Sander van der Linden, Beth Goldberg, Steve Rathje, and Stephan Lewandowsky. "Psychological Inoculation Improves Resilience Against Misinformation on Social Media." *Science Advances* 8, no. 34 (2022): eabo6254. https://doi.org/10.1126/sciadv.abo6254.

Roozenbeek, Jon, Sander van der Linden, and Thomas Nygren. "Prebunking Interventions Based on the Psychological Theory of 'Inoculation' Can Reduce Susceptibility to Misinformation Across Cultures." *Harvard Kennedy School Misinformation Review* 1, no. 2 (2020): 1–23. https://doi.org/10.37016//mr-2020-008.

Rose, Joel. "Even If It's 'Bonkers,' Poll Finds Many Believe QAnon and Other Conspiracy Theories." NPR, December 30, 2020. https://www.npr.org/2020/12/30/951095644/even-if-its-bonkers-poll-finds -many-believe-qanon-and-other-conspiracy-theories.

Rosenberg, David, Natalia Szura, and The Conversation. "Teens Are Spending the Equivalent of a 40-Hour Work Week on Their Devices: Here's How to Help Them Find the Right Balance." *Fortune*, October 24, 2023. https://fortune.com/well/2023/10/24/teens-too-much-screen-time-find-balance/.

Rosenblum, Nancy L., and Russell Muirhead. *A Lot of People Are Saying: The New Conspiracism and the Assault on Democracy*. Princeton University Press, 2019.

Rosendaal, Frits R. "Review of: 'Hydroxychloroquine and Azithromycin as a Treatment of COVID-19: Results of an Open-Label Non-randomized Clinical Trial Gautret et al 2010, DOI:10.1016/j.ijantimicag .2020.105949.' " *International Journal of Antimicrobial Agents* 56, no. 1 (2020): art. 106063. https://doi.org /10.1016%2Fj.ijantimicag.2020.106063.

Ross, Robert M., David G. Rand, and Gordon Pennycook. "Beyond 'Fake News': Analytic Thinking and the Detection of False and Hyperpartisan News Headlines." *Judgment and Decision Making* 16, no. 2 (2021): 484–504. https://doi.org/10.1017/S1930297500008640.

Rothwell, Jonathan, and Sonal Desai. "How Misinformation Is Distorting COVID Policies and Behaviors." *Brookings*, December 22, 2020. https://www.brookings.edu/research/how-misinformation-is-distorting -covid-policies-and-behaviors/.

Rozado, David, and Musa al-Gharbi. "Using Word Embeddings to Probe Sentiment Associations of Politically Loaded Terms in News and Opinion Articles from News Media Outlets." *Journal of Computational Social Science* 5, no. 1 (2022): 427–48. https://doi.org/10.1007/s42001-021-00130-y.

Rozado, David, Ruth Hughes, and Jamin Halberstadt. "Longitudinal Analysis of Sentiment and Emotion in News Media Headlines Using Automated Labelling with Transformer Language Models." *PLOS One* 17, no. 10 (2022): e0276367. https://doi.org/10.1371/journal.pone.0276367.

Ruggeri, Kai, Bojana Većkalov, Lana Bojanić et al. "The General Fault in Our Fault Lines." *Nature Human Behaviour* 5, no. 10 (2021): 1369–80. https://doi.org/10.1038/s41562-021-01092-x.

Saad, Lydia. "Gallup Vault: A Country Unified After Pearl Harbor." Gallup, December 5, 2016. https://news .gallup.com/vault/199049/gallup-vault-country-unified-pearl-harbor.aspx.

Saag, Michael S. "Misguided Use of Hydroxychloroquine for COVID-19: The Infusion of Politics Into Science." *JAMA* 324, no. 21 (2020): 2161–62. https://doi.org/10.1001/jama.2020.22389.

Sagan, Carl. *The Demon-Haunted World: Science as a Candle in the Dark*. Random House, 1995.

Sager, Mike. "The Fabulist Who Changed Journalism." *Columbia Journalism Review* 54 (2016): 52–60. https://www.cjr.org/the_feature/the_fabulist_who_changed_journalism.php.

Saint Laurent, Constance de, Gillian Murphy, Karen Hegarty, and Ciara M. Greene. "Measuring the Effects of Misinformation Exposure and Beliefs on Behavioural Intentions: A COVID-19 Vaccination Study." *Cognitive Research: Principles and Implications* 7, no. 1 (2022): 87. https://doi.org/10.1186/s41235-022-00437-y.

Salah, Omnia. "First Egyptian, Arab Woman to Go to Space Recounts Her Journey." *Al-Monitor*, November 25, 2022. https://www.al-monitor.com/originals/2022/11/first-egyptian-arab-woman-go-space-recounts-her -journey.

Sanders, Linley, and Kathy Frankovic. "Two-Thirds of Americans Believe That the COVID-19 Virus Originated from a Lab in China." YouGov, March 10, 2023. https://today.yougov.com/topics/politics/articles -reports/2023/03/10/americans-believe-covid-origin-lab.

Sands, John. "Local News Is More Trusted than National News—but That Could Change." Knight Foundation, October 29, 2019. https://knightfoundation.org/articles/local-news-is-more-trusted-than-national -news-but-that-could-change/.

Sanford, Claire. "Facebook Whistleblower Frances Haugen Testifies Before UK Parliament Transcript." *Rev*, October 26, 2021. https://www.rev.com/blog/transcripts/facebook-whistleblower-frances-haugen-testifies -before-uk-parliament-transcript.

Santini, Rose Marie, Débora Salles, and Giulia Tucci. "Comparative Approaches to Mis/Disinformation: When Machine Behavior Targets Future Voters—The Use of Social Bots to Test Narratives for Political Campaigns in Brazil." *International Journal of Communication* 15 (2021): 1220–43. https://doi.org/10.1111/isj.12427.

Santoro, Erik, and David E. Broockman. "The Promise and Pitfalls of Cross-Partisan Conversations for Reducing Affective Polarization: Evidence from Randomized Experiments." *Science Advances* 8, no. 25 (2022): eabn5515. https://doi.org/10.1126/sciadv.abn5515.

Satariano, Adam, and Paul Mozur. "The People Onscreen Are Fake: The Disinformation Is Real." *New York Times*, February 7, 2023. https://www.nytimes.com/2023/02/07/technology/artificial-intelligence-training-deepfake.html.

Satija, Neena, and Lena H. Sun. "A Major Funder of the Anti-Vaccine Movement Has Made Millions Selling Natural Health Products." *Washington Post*, October 15, 2019. https://www.washingtonpost.com/investigations/2019/10/15/fdc01078-c29c-11e9-b5e4-54aa56d5b7ce_story.html.

Schaeffer, Katherine. "Those on Ideological Right Favor Fewer COVID-19 Restrictions in Most Advanced Economies." Pew Research Center, July 30, 2021. https://www.pewresearch.org/short-reads/2021/07/30/those-on-ideological-right-favor-fewer-covid-19-restrictions-in-most-advanced-economies/.

Scheffer, Marten, Ingrid van de Leemput, Els Weinans, and Johan Bollen. "The Rise and Fall of Rationality in Language." *Proceedings of the National Academy of Sciences* 118, no. 51 (2021): e2107848118. https://doi.org/10.1073/pnas.2107848118.

Schipani, Vanessa. "False Claims About Flint Water." FactCheck.org, April 27, 2016. https://www.factcheck.org/2016/04/false-claims-about-flint-water/.

Schmaltz, Rodney, and Scott O. Lilienfeld. "Hauntings, Homeopathy, and the Hopkinsville Goblins: Using Pseudoscience to Teach Scientific Thinking." *Frontiers in Psychology* 5 (2014): art. 336. https://doi.org/10.3389/fpsyg.2014.00336.

Schmidt, Ana Lucía, Fabiana Zollo, Antonio Scala, Cornelia Betsch, and Walter Quattrociocchi. "Polarization of the Vaccination Debate on Facebook." *Vaccine* 36, no. 25 (2018): 3606–12. https://doi.org/10.1016/j.vaccine.2018.05.040.

Schmuck, Desirée, Miriam Tribastone, Jörg Matthes, Franziska Marquart, and Eva Maria Bergel. "Avoiding the Other Side? An Eye-Tracking Study of Selective Exposure and Selective Avoidance Effects in Response to Political Advertising." *Journal of Media Psychology* 32, no. 3 (2020): 158–64. https://doi.org/10.1027/1864-1105/a000265.

Schrader, Adam. "Gallup: Chief Justice John Roberts Earns Top Approval Rating Among Federal Leaders." UPI, December 27, 2021. https://www.upi.com/Top_News/US/2021/12/27/chief-justice-john-roberts-highets-approval-rating-federal-leaders/9461640629282/.

Schraer, Rachel, and Jack Goodman. "Ivermectin: How False Science Created a Covid 'Miracle' Drug." BBC, October 6, 2021. https://www.bbc.com/news/health-58170809.

Schulman, Jeremy. "59 Percent of Republicans Say It's Important to Believe Trump Won the Election." *Mother Jones*, September 12, 2021. https://www.motherjones.com/mojo-wire/2021/09/59-percent-of-republicans-say-its-important-to-believe-trump-won-the-election/.

Schwalbe, Michael C., Katie Joseff, Samuel Woolley, and Geoffrey L. Cohen. "When Politics Trumps Truth: Political Concordance versus Veracity as a Determinant of Believing, Sharing, and Recalling the News." *Journal of Experimental Psychology. General* 153, no. 10 (2024): 2524–51.

Schwartz, Dana. "Director of *Behind the Curve* Shares How to Argue with People Who Believe the Earth Is Flat." *Entertainment Weekly*, March 1, 2019. https://ew.com/movies/2019/03/01/behind-the-curve-netflix-interview/.

Scientific American. "Prime Time Fox News and WSJ Editorial Climate Coverage Mostly Wrong." September 21, 2012. https://www.scientificamerican.com/podcast/episode/primetime-fox-news-and-wsj-editoria-12-09-21/.

Scott, Dylan. "Why People Who Don't Trust Vaccines Are Embracing Unproven Drugs." *Vox*, October 1, 2021. https://www.vox.com/coronavirus-covid19/22686147/covid-19-vaccine-betadine-hydroxychloroquine-ivermectin-trump-conspiracy.

Scurich, Nicholas, and Adam Shniderman. "The Selective Allure of Neuroscientific Explanations." *PLOS One* 9, no. 9 (2014): e107529. https://doi.org/10.1371/journal.pone.0107529.

Senapathy, Kavin. "The Anti-Vaccine and Anti-GMO Movements Are Inextricably Linked and Cause Preventable Suffering." *Forbes*, May 18, 2017. https://www.forbes.com/sites/kavinsenapathy/2017/05/18/the-anti-vaccine-and-anti-gmo-movements-are-inextricably-linked-and-cause-preventable-suffering/.

Senger, Amy R., and Ho P. Huynh. "Intellectual Humility's Association with Vaccine Attitudes and Intentions." *Psychology, Health and Medicine* 26, no. 9 (2021): 1053–62. https://doi.org/10.1080/13548506.2020.1778753.

Serpe, Richard T., Fritz Long-Yarrison, Jan E. Stets, and Sheldon Stryker. "Multiple Identities and Self-Esteem." In *Identities in Everyday Life*, ed. Jan E. Stets and Richard T. Serpe, 53–72. Oxford University Press, 2019. https://doi.org/10.1093/oso/9780190873066.003.0004.

Settle, Jaime E. *Frenemies: How Social Media Polarizes America*. Cambridge University Press, 2018.

Settle, Jaime E., and Taylor N. Carlson. "Opting Out of Political Discussions." *Political Communication* 36, no. 3 (2019): 476–96. https://doi.org/10.1080/10584609.2018.1561563.

Severns, Maggie. "From Distraction to Disaster: How Coronavirus Crept Up on Washington." *Politico*, March 30, 2020. https://www.politico.com/news/2020/03/30/how-coronavirus-shook-congress-complacency-155058.

Sewall, Craig J. R. "Flawed Data Led to Findings of a Connection Between Time Spent on Devices and Mental Health Problems—New Research." *The Conversation*, June 23, 2021. http://theconversation.com/flawed-data-led-to-findings-of-a-connection-between-time-spent-on-devices-and-mental-health-problems-new-research-162585.

Shah, Khushbu. "When Your Family Spreads Misinformation." *The Atlantic*, June 16, 2020. https://www.theatlantic.com/family/archive/2020/06/when-family-members-spread-coronavirus-misinformation/613129/.

Shao, Chengcheng, Giovanni Luca Ciampaglia, Onur Varol, Kai-Cheng Yang, Alessandro Flammini, and Filippo Menczer. "The Spread of Low-Credibility Content by Social Bots." *Nature Communications* 9, no. 1 (November 20, 2018): 4787. https://doi.org/10.1038/s41467-018-06930-7.

Shao, Chengcheng, Pik-Mai Hui, Lei Wang et al. "Anatomy of an Online Misinformation Network." *PLOS One* 13, no. 4 (2018): e0196087. https://doi.org/10.1371/journal.pone.0196087.

Sharma, Nikhil, Q. Vera Liao, and Ziang Xiao. "Generative Echo Chamber? Effects of LLM-Powered Search Systems on Diverse Information Seeking." Preprint, arXiv:2402.05880v2 [cs.CL], February 8, 2024. https://doi.org/10.48550/arXiv.2402.05880.

Sharot, Tali. "To Quell Misinformation, Use Carrots—Not Just Sticks." *Nature*, March 17, 2021. https://doi.org/10.1038/d41586-021-00657-0.

Shaughnessy, Brittany, Myiah J. Hutchens, and Eliana DuBosar. "That Is So Mainstream: The Impact of Hyper-Partisan Media Use and Right-, Left-Wing Alternative Media Repertoires on Consumers' Belief in Political Misperceptions in the United States." *International Journal of Communication* 18 (2024): 1561–81. 1932-8036/20240005.

Shearer, Elisa. "Americans Are Wary of the Role Social Media Sites Play in Delivering the News." Pew Research Center, October 2, 2019. https://www.pewresearch.org/journalism/2019/10/02/americans-are-wary-of-the-role-social-media-sites-play-in-delivering-the-news/.

Sheldon, Robert, Stacey Peterson, and Sarah Wilson. "Gigabyte (GB)." TechTarget, October 21, 2021. https://www.techtarget.com/searchstorage/definition/gigabyte.

Shen, Chun, Edmund T. Rolls, Wei Cheng et al. "Associations of Social Isolation and Loneliness with Later Dementia." *Neurology* 99, no. 2 (2022): e164–75. https://doi.org/10.1212/WNL.0000000000200583.

Sherman, Amy. "Biden Said People Vaccinated for COVID-19 'Do Not Spread the Disease to Anyone Else': That Contradicts a CDC Presentation in Dec That Said It's Likely That Vaccinated People 'Can Spread the Virus to Others.'" PolitiFact, December 22, 2021. https://www.politifact.com/factchecks/2021/dec/22/joe-biden/biden-says-vaccinated-people-cant-spread-covid-19-/.

Sherman, David K. "Self-Affirmation: Understanding the Effects." *Social and Personality Psychology Compass* 7, no. 11 (2013): 834–45. https://doi.org/10.1111/spc3.12072.

Sherman, David K., and Geoffrey L. Cohen. "The Psychology of Self-Defense: Self-Affirmation Theory." In *Advances in Experimental Social Psychology*, vol. 38, ed. Mark P. Zanna, 183–242. Academic Press, 2006.

Sherman, Jenna. "Gendered Health Misinformation." Meedan, October 2022. https://assets-global.website-files.com/615e270f23c94c3fc683f12c/6360182ce09baba276f9d96d_Gendered%20Health%20Misinformation%20-%20Meedan.pdf.

Sherman, Lauren E., Ashley A. Payton, Leanna M. Hernandez, Patricia M. Greenfield, and Mirella Dapretto. "The Power of the *Like* in Adolescence: Effects of Peer Influence on Neural and Behavioral Responses to Social Media." *Psychological Science* 27, no. 7 (2016): 1027–35. https://doi.org/10.1177/0956797616645673.

Shewale, Rohit. "Facebook Users Statistics 2024 (Worldwide Data)." DemandSage, April 5, 2024. https://www.demandsage.com/facebook-statistics/.

Shi, Feng, Misha Teplitskiy, Eamon Duede, and James A. Evans. "The Wisdom of Polarized Crowds." *Nature Human Behaviour* 3, no. 4 (2019): 329–36. https://doi.org/10.1038/s41562-019-0541-6.

Shierholz, Heidi, and Ben Zipperer. "Here Is What's at Stake with the Conflict of Interest ('Fiduciary') Rule." Economic Policy Institute, May 30, 2017. https://www.epi.org/publication/here-is-whats-at-stake-with-the-conflict-of-interest-fiduciary-rule/.

Shinkareva, Svetlana V., Jing Wang, Jongwan Kim, Matthew J. Facciani, Laura B. Baucom, and Douglas H. Wedell. "Representations of Modality-Specific Affective Processing for Visual and Auditory Stimuli Derived from Functional Magnetic Resonance Imaging Data." *Human Brain Mapping* 35, no. 7 (2014): 3558–68. https://doi.org/10.1002/hbm.22421.

Siani, Alessandro, and Amy Tranter. "Is Vaccine Confidence an Unexpected Victim of the COVID-19 Pandemic?" *Vaccine* 40, no. 50 (2022): 7262–69. https://doi.org/10.1016/j.vaccine.2022.10.061.

Silverman, Craig. "This Analysis Shows How Viral Fake Election News Stories Outperformed Real News on Facebook." *BuzzFeed News*, November 16, 2016. https://www.buzzfeednews.com/article/craigsilverman/viral-fake-election-news-outperformed-real-news-on-facebook.

Simione, Luca, Monia Vagni, Camilla Gnagnarella, Giuseppe Bersani, and Daniela Pajardi. "Mistrust and Beliefs in Conspiracy Theories Differently Mediate the Effects of Psychological Factors on Propensity for COVID-19 Vaccine." *Frontiers in Psychology* 12 (2021): art. 683684. https://doi.org/10.3389/fpsyg.2021.683684.

Simon, Felix M., Sacha Altay, and Hugo Mercier. "Misinformation Reloaded? Fears About the Impact of Generative AI on Misinformation Are Overblown." *Harvard Kennedy School Misinformation Review* 4, no. 5 (2023): 1–11. https://doi.org/10.37016/mr-2020-127.

Simpson, Michael. "Anti-Vaccine Activists Support Big Pharma Profits—My Irony Meter Dies." Skeptical Raptor, March 8, 2022. https://www.skepticalraptor.com/skepticalraptorblog.php/anti-vaccine-activists-support-big-pharma-profits-my-irony-meter-dies/

Sinclair, David. *The Land That Never Was: Sir Gregor MacGregor and the Most Audacious Fraud in History*. Da Capo Press, 2004.

Sinclair, Samantha, and Jens Agerström. "Do Social Norms Influence Young People's Willingness to Take the COVID-19 Vaccine?" *Health Communication* 38, no. 1 (2023): 152–59. https://doi.org/10.1080/10410236.2021.1937832.

Singh, Shubham. "How Many People Use Instagram 2024 [Global Data]." DemandSage, May 6, 2024. https://www.demandsage.com/instagram-statistics/.

Singh, Shubham. "TikTok User Statistics 2024 [Global Data]." DemandSage, April 29, 2024. https://www.demandsage.com/tiktok-user-statistics/.

Sirlin, Nathaniel, Ziv Epstein, Antonio A. Arechar, and David G. Rand. "Digital Literacy Is Associated with More Discerning Accuracy Judgments but Not Sharing Intentions." *Harvard Kennedy School Misinformation Review* 2, no. 6 (2021): 1–13. https://doi.org/10.37016/mr-2020-83.

60 Minutes. "Social Media and Political Polarization in America." YouTube, November 6, 2022,. Video, 13 min., 38 sec. https://www.youtube.com/watch?v=WLfr7sU5W2E.

Skelley, Geoffrey, and Amelia Thomson-DeVeaux. "How Americans Are Reacting to Trump's COVID-19 Diagnosis." FiveThirtyEight, October 5, 2020. https://fivethirtyeight.com/features/will-trumps-diagnosis -change-the-way-republicans-think-about-covid-19/.

Slater, Michael D. "Reinforcing Spirals Model: Conceptualizing the Relationship Between Media Content Exposure and the Development and Maintenance of Attitudes." *Media Psychology* 18, no. 3 (2015): 370–95. https://doi.org/10.1080/15213269.2014.897236.

Slater, Michael D. "Reinforcing Spirals: The Mutual Influence of Media Selectivity and Media Effects and Their Impact on Individual Behavior and Social Identity." *Communication Theory* 17, no. 3 (2007): 281–303. https://doi.org/10.1111/j.1468-2885.2007.00296.x.

Slawson, Nicola. " 'Women Have Been Woefully Neglected': Does Medical Science Have a Gender Problem?" *The Guardian*, December 18, 2019. https://www.theguardian.com/education/2019/dec/18/women -have-been-woefully-neglected-does-medical-science-have-a-gender-problem.

Sloman, Steven A. "How Do We Believe?" *Topics in Cognitive Science* 14, no. 1 (2022): 31–44. https://doi.org /10.1111/tops.12580.

Sloman, Steven, and Philip Fernbach. *The Knowledge Illusion: Why We Never Think Alone.* Penguin, 2018.

Smith, Christian, Michael Emerson, Sally Gallagher, Paul Kennedy, and David Sikkink. *American Evangelicalism: Embattled and Thriving.* University of Chicago Press, 1998.

Smith-Schoenwalder, Cecelia. "Fauci Disagrees with Trump's Claim About Rounding the 'Final Turn' on the COVID-19 Outbreak." *U.S. News & World Report*, September 11, 2020. https://www.usnews .com/news/national-news/articles/2020-09-11/fauci-disagrees-with-trumps-claim-about-rounding-the -corner-on-the-coronavirus-outbreak.

Smolkin, Doran. "Puzzles About Trust." *Southern Journal of Philosophy* 46, no. 3 (2008): 431–49. https:// doi.org/10.1111/j.2041-6962.2008.tb00127.x.

Sokolow, Amy. "With Science and Scripture, a Baltimore Pastor Is Fighting Covid-19 Vaccine Skepticism." *STAT*, August 31, 2020. https://www.statnews.com/2020/08/31/with-science-and-scripture-a-baltimore -pastor-is-fighting-covid-19-vaccine-skepticism/.

Solender, Andrew. "Trump Said U.S. Was 'Rounding the Final Turn' on Aug. 31—and on 39 of the 57 Days Since." *Forbes*, October 27, 2020. https://www.forbes.com/sites/andrewsolender/2020/10/27 /trump-said-us-was-rounding-the-final-turn-on-aug-31-and-on-39-of-the-57-days-since/.

Sommer, Will. "QAnon Conspiracy Theorists' Magic Cure for Coronavirus Is Drinking Lethal Bleach." *Daily Beast*, January 28, 2020. https://www.thedailybeast.com/qanon-conspiracy-theorists-magic-cure -for-coronavirus-is-drinking-lethal-bleach.

Sommer, Will. "QAnon Star Cirsten Weldon Who Said Only 'Idiots' Get Vaccinated Dies of COVID." *Daily Beast*, January 7, 2022. https://www.thedailybeast.com/qanon-star-cirsten-weldon-who-said-only -idiots-get-vaccinated-dies-of-covid.

Spangler, Todd. "X/twitter Verified Blue Check-Mark Users Are 'Superspreaders' of Disinformation About Israel-Hamas War, Study Says." *Variety*, October 20, 2023. https://variety.com/2023/digital /news/x-twitter-blue-check-mark-users-superspreaders-disinformation-israel-hamas-war-1235763100/.

Sparkman, Gregg, Nathan Geiger, and Elke U. Weber. "Americans Experience a False Social Reality by Underestimating Popular Climate Policy Support by Nearly Half." *Nature Communications* 13, no. 1 (2022): art. 4779. https://doi.org/10.1038/s41467-022-32412-y.

Spencer, Saranac Hale. "Grant Wahl Died from Aortic Aneurysm, No Link to COVID-19 Vaccine." FactCheck.org, December 16, 2022. https://www.factcheck.org/2022/12/scicheck-grant-wahl-died-from -aortic-aneurysm-no-link-to-covid-19-vaccine/.

Spencer, Saranac Hale. "Tucker Carlson Misrepresents Vaccine Safety Reporting Data." FactCheck.org, May 14, 2021. https://www.factcheck.org/2021/05/scicheck-tucker-carlson-misrepresents-vaccine-safety -reporting-data/.

Stanford University. "Strengthening Democracy Challenge: Winning Interventions." Accessed June 27, 2024. https://www.strengtheningdemocracychallenge.org/winning-interventions.

Stanley, Matthew L., Alyssa H. Sinclair, and Paul Seli. "Intellectual Humility and Perceptions of Political Opponents." *Journal of Personality* 88, no. 6 (2020): 1196–1216. https://doi.org/10.1111/jopy.12566.

Stantcheva, Stefanie, Alberto Alesina, and Armando Miano. "The Polarization of Reality." Discussion Paper No. 14348. Centre for Economic Policy Research, 2020.

Starbird, Kate. "How a Crisis Researcher Makes Sense of Covid-19 Misinformation." *Medium*, March 9, 2020. https://onezero.medium.com/reflecting-on-the-covid-19-infodemic-as-a-crisis-informatics-researcher-ce0656fa4d0a.

Starbird, Kate, Renée DiResta, and Matt DeButts. "Influence and Improvisation: Participatory Disinformation Suring the 2020 US Election." *Social Media + Society* 9, no. 2 (2023). https://doi.org/10.1177/20563051231177943.

Statista. "Data Growth Worldwide 2010–2025." Accessed June 30, 2024. https://www.statista.com/statistics/871513/worldwide-data-created/.

Statista. "Internet and Social Media Users in the World 2024." Accessed July 6, 2024. https://www.statista.com/statistics/617136/digital-population-worldwide/.

Statista. "Number of Internet Users Worldwide from 2005 to 2023." May 22, 2024. https://www.statista.com/statistics/273018/number-of-internet-users-worldwide/.

Statista. "WhatsApp Market Share Among Messaging App Users Worldwide 2022, by Country." Accessed July 6, 2024. https://www.statista.com/statistics/1311229/whatsapp-usage-messaging-app-users-by-country/.

Stecula, Dominik A., Ozan Kuru, Dolores Albarracin, and Kathleen Hall Jamieson. "Policy Views and Negative Beliefs About Vaccines in the United States, 2019." *American Journal of Public Health* 110, no. 10 (2020): 1561–63. https://doi.org/10.2105/AJPH.2020.305828.

Stecuła, Dominik A., and Matt Motta. "Unverified Reports of Vaccine Side Effects in VAERS Aren't the Smoking Guns Portrayed by Right-Wing Media Outlets—They Can Offer Insight Into Vaccine Hesitancy." *The Conversation*, August 25, 2021. http://theconversation.com/unverified-reports-of-vaccine-side-effects-in-vaers-arent-the-smoking-guns-portrayed-by-right-wing-media-outlets-they-can-offer-insight-into-vaccine-hesitancy-166401.

Steele, Molly K., Alexia Couture, Carrie Reed et al. "Estimated Number of COVID-19 Infections, Hospitalizations, and Deaths Prevented Among Vaccinated Persons in the US, December 2020 to September 2021." *JAMA Network Open* 5, no. 7 (2022): e2220385. https://doi.org/10.1001/jamanetworkopen.2022.20385.

Steenbuch Traberg, Cecilie. "Misinformation: Broaden Definition to Curb Its Societal Influence." *Nature* 606, no. 7915 (2022): 653. https://doi.org/10.1038/d41586-022-01700-4.

Steenbuch Traberg, Cecilie, and Sander van der Linden. "Birds of a Feather Are Persuaded Together: Perceived Source Credibility Mediates the Effect of Political Bias on Misinformation Susceptibility." *Personality and Individual Differences* 185 (2022): art. 111269. https://doi.org/10.1016/j.paid.2021.111269.

Steffens, Maryke S., Adam G. Dunn, Kerrie E. Wiley, and Julie Leask. "How Organisations Promoting Vaccination Respond to Misinformation on Social Media: A Qualitative Investigation." *BMC Public Health* 19, no. 1 (2019): art. 1348. https://doi.org/10.1186/s12889-019-7659-3.

Stein, Jonas, Marc Keuschnigg, and Arnout van de Rijt. "Network Segregation and the Propagation of Misinformation." *Scientific Reports* 13, no. 1 (2023): 917. https://doi.org/10.1038/s41598-022-26913-5.

Stern, Joanna. "Social-Media Algorithms Rule How We See the World: Good Luck Trying to Stop Them." *Wall Street Journal*, January 17, 2021. https://www.wsj.com/articles/social-media-algorithms-rule-how-we-see-the-world-good-luck-trying-to-stop-them-11610884800.

Stets, Jan E., John Aldecoa, Quinn Bloom, and Joel Winegar. "Using Identity Theory to Understand Homophily in Groups." In *Identities in Action*, ed. Philip S. Brenner, Jan E. Stets, and Richard T. Serpe, 285–302. Springer International, 2021. https://doi.org/10.1007/978-3-030-76966-6_14.

Stets, Jan E., and Peter J. Burke. "Identity Theory and Social Identity Theory." *Social Psychology Quarterly* 63, no. 3 (2000): 224–37. https://doi.org/10.2307/2695870.

Stets, Jan E. and Peter J. Burke. "Self-Esteem and Identities." *Sociological Perspectives* 57, no. 4 (2014): 409–33. https://doi.org/10.1177/0731121414536141.

Stewart, Emily. "Anti-Maskers Explain Themselves." *Vox*, August 7, 2020. https://www.vox.com/the-goods /2020/8/7/21357400/anti-mask-protest-rallies-donald-trump-covid-19.

Stieger, Stefan, and Ulf-Dietrich Reips. "A Limitation of the Cognitive Reflection Test: Familiarity." *PeerJ* 4 (2016): e2395. https://doi.org/10.7717/peerj.2395.

Stillman, Tyler F., and Roy F. Baumeister. "Uncertainty, Belongingness, and Four Needs for Meaning." *Psychological Inquiry* 20, no. 4 (2009): 249–51. https://doi.org/10.1080/10478400903333544.

Stokel-Walker, Chris. "ChatGPT Listed as Author on Research Papers: Many Scientists Disapprove." *Nature* 613, no. 7945 (2023): 620–21. https://doi.org/10.1038/d41586-023-00107-z.

Strandberg, Kim, Staffan Himmelroos, and Kimmo Grönlund. "Do Discussions in Like-Minded Groups Necessarily Lead to More Extreme Opinions? Deliberative Democracy and Group Polarization." *International Political Science Review* 40, no. 1 (2019): 41–57. https://doi.org/10.1177/0192512117692136.

Strandberg, Thomas, Jay A. Olson, Lars Hall, Andy Woods, and Petter Johansson. "Depolarizing American Voters: Democrats and Republicans Are Equally Susceptible to False Attitude Feedback." *PLOS One* 15, no. 2 (2020): e0226799. https://doi.org/10.1371/journal.pone.0226799.

Street, Richard L., Kimberly J. O'Malley, Lisa A. Cooper, and Paul Haidet. "Understanding Concordance in Patient-Physician Relationships: Personal and Ethnic Dimensions of Shared Identity." *Annals of Family Medicine* 6, no. 3 (2008): 198–205. https://doi.org/10.1370/afm.821.

Stroud, Natalie Jomini. "Polarization and Partisan Selective Exposure." *Journal of Communication* 60, no. 3 (2010): 556–76. https://doi.org/10.1111/j.1460-2466.2010.01497.x.

Stryker, Sheldon, and Richard T. Serpe. "Commitment, Identity Salience, and Role Behavior: Theory and Research Example." In *Personality, Roles, and Social Behavior*, ed. William Ickes and Eric S. Knowles, 199–218. Springer New York, 1982.

Stubenvoll, Marlis, and Jörg Matthes. "Why Retractions of Numerical Misinformation Fail: The Anchoring Effect of Inaccurate Numbers in the News." *Journalism and Mass Communication Quarterly* 99, no. 2 (2022): 368–89. https://doi.org/10.1177/10776990211021800.

Suhay, Elizabeth, Emily Bello-Pardo, and Brianna Maurer. "The Polarizing Effects of Online Partisan Criticism: Evidence from Two Experiments." *International Journal of Press/Politics* 23, no. 1 (2018): 95–115. https://doi.org/10.1177/1940161217740697.

Sullivan, Kate, and Jennifer Agiesta. "Biden's Popular Vote Margin Over Trump Tops 7 Million." CNN, December 4, 2020. https://www.cnn.com/2020/12/04/politics/biden-popular-vote-margin-7-million /index.html.

Sunstein, Cass R. "The Law of Group Polarization." *Journal of Political Philosophy* 10 (2002): 175–95.

Sunstein, Cass R. "Sunstein on the Internet and Political Polarization." University of Chicago Law School, December 14, 2007. https://www.law.uchicago.edu/news/sunstein-internet-and-political-polarization.

Susmann, Mark W., Mengran Xu, Jason K. Clark et al. "Persuasion Amidst a Pandemic: Insights from the Elaboration Likelihood Model." *European Review of Social Psychology* 33, no. 2 (2022): 323–59. https:// doi.org/10.1080/10463283.2021.1964744.

Swann, William B., Jr., and Michael D. Buhrmester. "Identity Fusion." *Current Directions in Psychological Science* 24, no. 1 (2015): 52–57. https://doi.org/10.1177/0963721414551363.

Tajfel, Henri, ed. *Differentiation Between Social Groups: Studies in the Social Psychology of Intergroup Relations*. Academic Press, 1978. https://psycnet.apa.org/fulltext/1980-50696-000.pdf.

Tajfel, Henri, M. G. Billig, R. P. Bundy, and Claude Flament. "Social Categorization and Intergroup Behaviour." *European Journal of Social Psychology* 1, no. 2 (1971): 149–78. https://doi.org/10.1002/ejsp.2420010202.

Tajfel, Henri, John C. Turner, William G. Austin, and Stephen Worchel. "An Integrative Theory of Intergroup Conflict." *Organizational Identity: A Reader* 56, no. 65 (1979).

Tambuscio, Marcella, Giancarlo Ruffo, Alessandro Flammini, and Filippo Menczer. "Fact-Checking Effect on Viral Hoaxes: A Model of Misinformation Spread in Social Networks." In *WWW '15 Companion:*

Proceedings of the 24th International Conference on World Wide Web, 977–82. Association for Computing Machinery, 2015. https://doi.org/10.1145/2740908.2742572.

Tamerius, Karin. "Resources," Smart Politics. Accessed July 15, 2024. https://www.joinsmart.org/resources/.

Tamerius, Karin. "The 10 Types of Trust You Need to Persuade a Republican." *Medium*, February 17, 2022. https://medium.com/progressively-speaking/trust-is-everything-when-talking-politics-5cd84140a11f.

Tang, Lu, Kayo Fujimoto, Muhammad (Tuan) Amith et al. "'Down the Rabbit Hole' of Vaccine Misinformation on YouTube: Network Exposure Study." *Journal of Medical Internet Research* 23, no. 1 (2021): e23262. https://doi.org/10.2196/23262.

Tanis, Martin, and Tom Postmes. "A Social Identity Approach to Trust: Interpersonal Perception, Group Membership and Trusting Behaviour." *European Journal of Social Psychology* 35, no. 3 (2005): 413–24. https://doi.org/10.1002/ejsp.256.

Tappin, Ben M., Chloe Wittenberg, Luke B. Hewitt, Adam J. Berinsky, and David G. Rand. "Quantifying the Potential Persuasive Returns to Political Microtargeting." *Proceedings of the National Academy of Sciences* 120, no. 25 (2023): e2216261120. https://doi.org/10.1073/pnas.2216261120.

Tardelli, Serena, Marco Avvenuti, Maurizio Tesconi, and Stefano Cresci. "Characterizing Social Bots Spreading Financial Disinformation." In *Social Computing and Social Media: Design, Ethics, User Behavior, and Social Network Analysis*, ed. Gabriele Meiselwitz:376–92. Springer International, 2020. https://doi.org/10.1007/978-3-030-49570-1_26.

Tavris, Carol. "Episode 1: Carol Tavris on Mistakes, Justification, and Cognitive Dissonance." Interview by Sean Carroll. *Sean Carroll's Mindscape*, podcast, July 9, 2018. https://www.preposterousuniverse.com/podcast/2018/07/09/episode-1-carol-tavris-on-mistakes-justification-and-cognitive-dissonance/.

Taylor, Luke. "Covid-19 Misinformation Sparks Threats and Violence Against Doctors in Latin America." *BMJ* 370 (2020). https://doi.org/10.1136/bmj.m3088.

Taylor, Steven, and Gordon J. G. Asmundson. "Negative Attitudes About Facemasks During the COVID-19 Pandemic: The Dual Importance of Perceived Ineffectiveness and Psychological Reactance." *PLOS One* 16, no. 2 (2021): e0246317. https://doi.org/10.1371/journal.pone.0246317.

Teale, Chris. "7 in 10 Voters Support Teaching Social Media Literacy in Schools." Morning Consult Pro, December 8, 2021. https://morningconsult.com/2021/12/08/social-media-literacy-polling/.

Ternovski, John, Joshua Kalla, and Peter Michael Aronow. "Deepfake Warnings for Political Videos Increase Disbelief but Do Not Improve Discernment: Evidence from Two Experiments." OSF Preprints, January 14, 2021. https://doi.org/10.31219/osf.io/dta97.

Tesser, Abraham, Leonard Martin, and Marilyn Mendolia. "The Impact of Thought on Attitude Extremity and Attitude-Behavior Consistency." In *Attitude Strength*, ed. Richard E. Petty and Jon A. Krosnick, 73–92. Psychology Press, 2014.

Thompson, Derek. "Deep Cleaning Isn't a Victimless Crime." *The Atlantic*, April 13, 2021. https://www.theatlantic.com/ideas/archive/2021/04/end-hygiene-theater/618576/.

Thompson, Derek. "Hygiene Theater Is a Huge Waste of Time." *The Atlantic*, July 27, 2020. https://www.theatlantic.com/ideas/archive/2020/07/scourge-hygiene-theater/614599/.

Thompson, Derek. "The Pandemic's Wrongest Man." *The Atlantic*, April 1, 2021. https://www.theatlantic.com/ideas/archive/2021/04/pandemics-wrongest-man/618475/.

Thompson, Stuart A. "To Fight Election Falsehoods, Social Media Companies Ready a Familiar Playbook." *New York Times*, August 23, 2022. https://www.nytimes.com/2022/08/23/technology/midterms-misinformation-tiktok-facebook.html.

3M. "3M State of Science Index: Connecting the 2023 Survey to 3M Forward." Accessed August 14, 2024. https://www.3m.com/3M/en_US/3m-forward-us/about-the-survey/.

Tilley, James. "Why So Many People Believe Conspiracy Theories." BBC, February 12, 2019. https://www.bbc.com/news/world-47144738.

Toor, Jaspreet, Susy Echeverria-Londono, Xiang Li et al. "Lives Saved with Vaccination for 10 Pathogens Across 112 Countries in a Pre-COVID-19 World." *eLife* 10 (2021): e67635. https://doi.org/10.7554/eLife.67635.

Torres, Lesley. "Unvaccinated Adults Are More Likely to Be Uninsured, Study Finds." *Bloomberg Law*, June 11, 2021. https://news.bloomberglaw.com/pharma-and-life-sciences/unvaccinated-adults-are-more-likely-to-be-uninsured-study-finds.

Tran, Lucky. "Don't Believe Those Who Claim Science Proves Masks Don't Work." *The Guardian*, February 27, 2023. https://www.theguardian.com/commentisfree/2023/feb/27/dont-believe-those-who-claim-science-proves-masks-dont-work.

Trecek-King, Melanie. "Is What You Believe True? Use These 6 Questions to Find Out." Thinking Is Power, December 17, 2020. https://thinkingispower.com/the-power-of-questioning-our-beliefs/.

Trecek-King, Melanie. "Teach Skills, Not Facts." Thinking Is Power, accessed February 23, 2021. https://thinkingispower.com/from-non-majors-biology-to-critical-thinking-an-educators-journey/.

Treen, Kathie M. d'I., Hywel T. P. Williams, and Saffron J. O'Neill. "Online Misinformation About Climate Change." *Wiley Interdisciplinary Reviews: Climate Change* 11, no. 5 (2020). https://doi.org/10.1002/wcc.665.

Tripodi, Francesca Bolla. *The Propagandists' Playbook: How Conservative Elites Manipulate Search and Threaten Democracy*. Yale University Press, 2022.

Tucker, Joshua A., and Nathaniel Persily. "How to Fix Social Media? Start with Independent Research." *Brookings*, December 1, 2021. https://www.brookings.edu/articles/how-to-fix-social-media-start-with-independent-research/.

Tufekci, Zeynep. "5 Pandemic Mistakes We Keep Repeating." *The Atlantic*, February 26, 2021. https://www.theatlantic.com/ideas/archive/2021/02/how-public-health-messaging-backfired/618147/.

Tufekci, Zeynep. "Keep the Parks Open." *The Atlantic*, April 7, 2020. https://www.theatlantic.com/health/archive/2020/04/closing-parks-ineffective-pandemic-theater/609580/.

Tufekci, Zeynep. "Scolding Beachgoers Isn't Helping." *The Atlantic*, July 4, 2020. https://www.theatlantic.com/health/archive/2020/07/it-okay-go-beach/613849/.

Turrentine, Jeff. "Climate Misinformation on Social Media Is Undermining Climate Action." Natural Resources Defense Council, April 19, 2022. https://www.nrdc.org/stories/climate-misinformation-social-media-undermining-climate-action.

Tversky, Amos, and Daniel Kahneman. "Loss Aversion in Riskless Choice: A Reference-Dependent Model." *Quarterly Journal of Economics* 106, no. 4 (1991): 1039–61. https://doi.org/10.2307/2937956.

Tyler, Matthew, Justin Grimmer, and Shanto Iyengar. "Partisan Enclaves and Information Bazaars: Mapping Selective Exposure to Online News." *Journal of Politics* 84, no. 2 (2022): 1057–73. https://doi.org/10.1086/716950.

Tyson, Alec, Cary Funk, Brian Kennedy, and Courtney Johnson. "Majority in U.S. Says Public Health Benefits of COVID-19 Restrictions Worth the Costs, Even as Large Shares Also See Downsides." Pew Research Center, September 15, 2021. https://www.pewresearch.org/science/2021/09/15/majority-in-u-s-says-public-health-benefits-of-covid-19-restrictions-worth-the-costs-even-as-large-shares-also-see-downsides/.

Ueki, Hiroshi, Yuri Furusawa, Kiyoko Iwatsuki-Horimoto et al. "Effectiveness of Face Masks in Preventing Airborne Transmission of SARS-CoV-2." *mSphere* 5, no. 5 (2020): e00637-20. https://doi.org/10.1128/mSphere.00637-20.

Understanding Science. "The Scientific Community: Diversity Makes the Difference." April 14, 2022. https://undsci.berkeley.edu/article/0_0_0/socialsideofscience_02.

Unsworth, Kerrie L., and Kelly S. Fielding. "It's Political: How the Salience of One's Political Identity Changes Climate Change Beliefs and Policy Support." *Global Environmental Change* 27 (2014): 131–37. https://doi.org/10.1016/j.gloenvcha.2014.05.002.

U.S. Department of Justice. "Leaders of 'Genesis II Church of Health and Healing,' Who Sold Toxic Bleach as Fake 'Miracle' Cure for COVID-19 and Other Serious Diseases, Sentenced to More than 12 Years in Federal Prison." October 6, 2023. https://www.justice.gov/usao-sdfl/pr/leaders-genesis-ii-church-health-and-healing-who-sold-toxic-bleach-fake-miracle-cure.

U.S. Food and Drug Administration. "FDA Warns Consumers About the Dangerous and Potentially Life-Threatening Side Effects of Miracle Mineral Solution." News release, August 12, 2019. https://

www.fda.gov/news-events/press-announcements/fda-warns-consumers-about-dangerous-and-potentially-life-threatening-side-effects-miracle-mineral.

U.S. Holocaust Memorial Museum. "How Did Public Opinion About Entering World War II Change Between 1939 and 1941?" Accessed June 20, 2024. https://exhibitions.ushmm.org/americans-and-the-holocaust/us-public-opinion-world-war-II-1939-1941.

Vaccari, Cristian, and Andrew Chadwick. "Deepfakes and Disinformation: Exploring the Impact of Synthetic Political Video on Deception, Uncertainty, and Trust in News." *Social Media + Society* 6, no. 1 (2020). https://doi.org/10.1177/2056305120903408.

Vaisey, Stephen, and Omar Lizardo. "Can Cultural Worldviews Influence Network Composition?" *Social Forces* 88, no. 4 (2010): 1595–1618. https://doi.org/10.1353/sof.2010.0009.

Valdes, Isabelle, Ashley Kirzinger, Shannon Schumacher, and Liz Hamel. "KFF Misinformation Poll Snapshot: Public Views Misinformation as a Major Problem, Feels Uncertain About Accuracy of Information on Current Events." *KFF*, December 15, 2023. https://www.kff.org/health-misinformation-and-trust/poll-finding/kff-misinformation-poll-snapshot-public-views-misinformation-as-a-major-problem-feels-uncertain-about-accuracy-of-information/.

Valika, Taher S., Sarah E. Maurrasse, and Lara Reichert. "A Second Pandemic? Perspective on Information Overload in the COVID-19 Era." *Otolaryngology–Head and Neck Surgery* 163, no. 5 (2020): 931–33. https://doi.org/10.1177/0194599820935850.

van Baar, Jeroen M., David J. Halpern, and Oriel FeldmanHall. "Intolerance of Uncertainty Modulates Brain-to-Brain Synchrony During Politically Polarized Perception." *Proceedings of the National Academy of Sciences* 118, no. 20 (2021). https://doi.org/10.1073/pnas.2022491118.

Van Bavel, Jay J., Aleksandra Cichocka, Valerio Capraro et al. "National Identity Predicts Public Health Support During a Global Pandemic." *Nature Communications* 13, no. 1 (2022): 517. https://doi.org/10.1038/s41467-021-27668-9.

Van Bavel, Jay J., Elizabeth A. Harris, Philip Pärnamets, Steve Rathje, Kimberly C. Doell, and Joshua A. Tucker. "Political Psychology in the Digital (Mis)Information Age: A Model of News Belief and Sharing." *Social Issues and Policy Review* 15, no. 1 (2021): 84–113. https://doi.org/10.1111/sipr.12077.

Van Bavel, Jay J., and Andrea Pereira. "The Partisan Brain: An Identity-Based Model of Political Belief." *Trends in Cognitive Sciences* 22, no. 3 (2018): 213–24. https://doi.org/10.1016/j.tics.2018.01.004.

Van Bavel, Jay J., Steve Rathje, Madalina Vlasceanu, and Clara Pretus. "Updating the Identity-Based Model of Belief: From False Belief to the Spread of Misinformation." *Current Opinion in Psychology* 56 (April 2024): art. 101787. https://doi.org/10.1016/j.copsyc.2023.101787.

Van Beest, Ilja, and Kipling D. Williams. "When Inclusion Costs and Ostracism Pays, Ostracism Still Hurts." *Journal of Personality and Social Psychology* 91, no. 5 (2006): 918–28. https://doi.org/10.1037/0022-3514.91.5.918.

van der Linden, Sander, Anthony Leiserowitz, Seth Rosenthal, and Edward Maibach. "Inoculating the Public Against Misinformation About Climate Change." *Global Challenges* 1, no. 2 (2017): 1600008. https://doi.org/10.1002/gch2.201600008.

van Prooijen, Jan-Willem. "Why Education Predicts Decreased Belief in Conspiracy Theories." *Applied Cognitive Psychology* 31, no. 1 (2017): 50–58. https://doi.org/10.1002/acp.3301.

Vanderbilt Institute for Infection, Immunology and Inflammation, Vanderbilt University Medical Center. "How Does a mRNA Vaccine Compare to a Traditional Vaccine?" November 16, 2020. https://www.vumc.org/viiii/infographics/how-does-mrna-vaccine-compare-traditional-vaccine.

Vargas, Edward D., and Gabriel R. Sanchez. "American Individualism Is an Obstacle to Wider Mask Wearing in the US." *Brookings*, August 31, 2020. https://www.brookings.edu/blog/up-front/2020/08/31/american-individualism-is-an-obstacle-to-wider-mask-wearing-in-the-us/.

Varol, Onur, Emilio Ferrar, Clayton Davis, Filippo Menczer, and Alessandro Flammini. "Online Human-Bot Interactions: Detection, Estimation, and Characterization." *Eleventh International AAAI Conference on Web and Social Media* 11, no. 1 (2017): 280–89. https://aaai.org/ocs/index.php/ICWSM/ICWSM17/paper/view/15587.

Velásquez, N., R. Leahy, N. Johnson Restrepo et al. "Online Hate Network Spreads Malicious COVID-19 Content Outside the Control of Individual Social Media Platforms." *Scientific Reports* 11, no. 1 (2021): art. 11549. https://doi.org/10.1038/s41598-021-89467-y.

Verplanken, Bas, and Jie Sui. "Habit and Identity: Behavioral, Cognitive, Affective, and Motivational Facets of an Integrated Self." *Frontiers in Psychology* 10 (2019): art. 1504. https://doi.org/10.3389/fpsyg.2019.01504.

Vetterkind, Riley. "Sen. Ron Johnson Doubles Down on Unproven Early COVID-19 Treatments Including Ivermectin." *Wisconsin State Journal*, September 2, ""?2021. https://madison.com/wsj/news/local/govt-and-politics/sen-ron-johnson-doubles-down-on-unproven-early-covid-19-treatments-including-ivermectin/article_22d17b96-3b26-5a6b-a1e7-a778dde40893.html.

Voelkel, Jan G., Michael Stagnaro, James Chu et al., "Megastudy Identifying Successful Interventions to Strengthen Americans' Democratic Attitudes." Working Paper No. 22-38. Institute for Policy Research, Northwestern University, 2022.

Volz, Kirsten G., Thomas Kessler, and D. Yves von Cramon. "In-Group as Part of the Self: In-Group Favoritism Is Mediated by Medial Prefrontal Cortex Activation." *Social Neuroscience* 4, no. 3 (2009): 244–60. https://doi.org/10.1080/17470910802553565.

Vosoughi, Soroush, Deb Roy, and Sinan Aral. "The Spread of True and False News Online." *Science* 359, no. 6380 (2018): 1146–51. https://doi.org/10.1126/science.aap9559.

Vraga, Emily K., and Leticia Bode. "Correction as a Solution for Health Misinformation on Social Media." *American Journal of Public Health* 110, no. S3 (2020): S278–80. https://doi.org/10.2105/AJPH.2020.305916.

Vraga, Emily K., and Leticia Bode. "Defining Misinformation and Understanding Its Bounded Nature: Using Expertise and Evidence for Describing Misinformation." *Political Communication* 37, no. 1 (2020): 136–44. https://doi.org/10.1080/10584609.2020.1716500.

Vranic, Andrea, Ivana Hromatko, and Mirjana Tonković. "'I Did My Own Research': Overconfidence, (Dis)trust in Science, and Endorsement of Conspiracy Theories." *Frontiers in Psychology* 13 (2022): 1–9. https://doi.org/10.3389/fpsyg.2022.931865.

Wagner, Kurt. "Here's How Facebook Allowed Cambridge Analytica to Get Data for 50 Million Users." *Vox*, March 17, 2018. https://www.vox.com/2018/3/17/17134072/facebook-cambridge-analytica-trump-explained-user-data.

Wallace, Jacob, Paul Goldsmith-Pinkham, and Jason L. Schwartz. "Excess Death Rates for Republican and Democratic Registered Voters in Florida and Ohio During the COVID-19 Pandemic." *JAMA Internal Medicine* 183, no. 9 (2023): 916–23.

Wallace, Jacob, Paul Goldsmith-Pinkham, and Jason L. Schwartz. "Excess Death Rates for Republicans and Democrats During the COVID-19 Pandemic." Working Paper No. 30512. National Bureau of Economic Research, 2022.

Walls, Margaret. "Extended Producer Responsibility and Product Design: Economic Theory and Selected Case Studies." Discussion Paper No. 06–08. Resources for the Future, March 1, 2006. https://doi.org/10.2139/ssrn.901661.

Walter, Nathan, Jonathan Cohen, R. Lance Holbert, and Yasmin Morag. "Fact-Checking: A Meta-Analysis of What Works and for Whom." *Political Communication* 37, no. 3 (2020): 350–75. https://doi.org/10.1080/10584609.2019.1668894.

Wang, Chris, Michael J. Platow, and Eryn J. Newman. "There Is an 'I' in Truth: How Salient Identities Shape Dynamic Perceptions of Truth." *European Journal of Social Psychology* 53, no. 2 (2023). https://doi.org/10.1002/ejsp.2909.

Wang, Luxuan. "Many Americans Find Value in Getting News on Social Media, but Concerns About Inaccuracy Have Risen." Pew Research Center, February 7, 2024. https://www.pewresearch.org/short-reads/2024/02/07/many-americans-find-value-in-getting-news-on-social-media-but-concerns-about-inaccuracy-have-risen/.

Wang, Xiaowen, Enrico G. Ferro, Guohai Zhou, Dean Hashimoto, and Deepak L. Bhatt. "Association Between Universal Masking in a Health Care System and SARS-CoV-2 Positivity Among Health Care Workers." *JAMA* 324, no. 7 (2020): 703–4. https://doi.org/10.1001/jama.2020.12897.

Ward, Adrian F., Jianqing (Frank) Zheng, and Susan M. Broniarczyk. "I Share, Therefore I Know? Sharing Online Content—Even Without Reading It—Inflates Subjective Knowledge." *Journal of Consumer Psychology* 33, no. 3 (2022): 469–88. https://doi.org/10.1002/jcpy.1321.

Wardle, Claire. "Misunderstanding Misinformation." *Issues in Science and Technology* 39, no. 3 (2023): 38–40. https://issues.org/misunderstanding-misinformation-wardle/.

Wardle, Claire, and Hossein Derakhshan. *Information Disorder: Toward an Interdisciplinary Framework for Research and Policymaking.* Council of Europe, 2017.

Warwick Business School. "Research Reveals Why People Refuse to Wear Face Masks." August 10, 2021. https://www.wbs.ac.uk/news/research-reveals-why-people-refuse-to-wear-face-masks/.

Washington Post. "As a Conservative Twitter User Sleeps, His Account Is Hard at Work." February 5, 2017. https://www.washingtonpost.com/business/economy/as-a-conservative-twitter-user-sleeps-his-account-is-hard-at-work/2017/02/05/18d5a532-df31-11e6-918c-99ede3c8cafa_story.html.

Washington Post. "Booster Shots in U.S. Have Strongly Protected Against Severe Disease from Omicron Variant, CDC Studies Show." January 21, 2022. https://www.washingtonpost.com/health/2022/01/21/cdc-studies-booster-shots-omicron/.

Washington Post. "CDC, Under Fire, Lays Out Plan to Become More Nimble and Accountable." August 17, 2022. https://www.washingtonpost.com/health/2022/08/17/walensky-revamp-cdc-culture-covid/.

Washington Post. "Fox News and Trump Are Still Pushing Hydroxychloroquine: Here's What the Data Actually Shows." June 21, 2021. https://www.washingtonpost.com/politics/2021/06/21/hydroxycholoroquine-coronavirus-treatment-trump-allies-cant-quit/.

Washington Post. "A Group of Moms on Facebook Built an Island of Good-Faith Vaccine Debate in a Sea of Misinformation." August 23, 2021. https://www.washingtonpost.com/technology/2021/08/23/facebook-vaccine-talk-group/.

Washington Post. "Jon Stewart, Again in the Crossfire." October 19, 2004. https://www.washingtonpost.com/archive/lifestyle/2004/10/19/jon-stewart-again-in-the-crossfire/cd6ffdbb-6f06-42cd-9479-21af28ac5b81/.

Washington Post. "New Research Explores How Conservative Media Misinformation May Have Intensified the Severity of the Pandemic." June 25, 2020. https://www.washingtonpost.com/business/2020/06/25/fox-news-hannity-coronavirus-misinformation/.

Washington Post. "This Former Surgeon General Says There's a 'Loneliness Epidemic' and Work Is Partly to Blame." October 4, 2017. https://www.washingtonpost.com/news/on-leadership/wp/2017/10/04/this-former-surgeon-general-says-theres-a-loneliness-epidemic-and-work-is-partly-to-blame/.

Washington Post. "Washington Post-ABC News Poll March 22–25, 2020." March 29, 2020. https://www.washingtonpost.com/context/washington-post-abc-news-poll-march-22-25-2020/974c3312-5a40-4764-afb1-4bb6b86f1cf4/.

Washington Post. "A Year Into the Pandemic, It's Even More Clear That It's Safer to Be Outside." April 13, 2021. https://www.washingtonpost.com/health/2021/04/13/covid-outside-safety/.

Washington Post. "'You've Got Bad Blood': The Horror of the Tuskegee Syphilis Experiment." May 16, 2017. https://www.washingtonpost.com/news/retropolis/wp/2017/05/16/youve-got-bad-blood-the-horror-of-the-tuskegee-syphilis-experiment/.

Waszkiewicz, Paweł, Piotr Lewulis, Michał Górski, Adam Czarnecki, and Wojciech Feleszko. "Public Vaccination Reluctance: What Makes Us Change Our Minds? Results of a Longitudinal Cohort Survey." *Vaccines* 10, no. 7 (2022): art. 1081. https://doi.org/10.3390/vaccines10071081.

Weatherbed, Jess. "Twitter Replaces Its Free API with a Paid Tier in Quest to Make More Money." *The Verge*, February 2, 2023. https://www.theverge.com/2023/2/2/23582615/twitter-removing-free-api-developer-apps-price-announcement.

Weger, Harry, Jr., Gina Castle Bell, Elizabeth M. Minei, and Melissa C. Robinson. "The Relative Effectiveness of Active Listening in Initial Interactions." *International Journal of Listening* 28, no. 1 (2014): 13–31. https://doi.org/10.1080/10904018.2013.813234.

Weintraub, Rebecca, Julia Raifman, and Benjy Renton. "We Must 'Boost' COVID Vaccinations to Prevent a Winter Surge." *The Hill*, October 12, 2021. https://thehill.com/opinion/healthcare/575879-we-must-boost-covid-vaccinations-to-prevent-a-winter-surge/.

Wellcome. "Wellcome Global Monitor 2018." September 18, 2020. https://wellcome.org/reports/wellcome-global-monitor/2018.

Weng, L., Alessandro Flammini, Alessandro Vespignani, and Filippo Menczer. "Competition Among Memes in a World with Limited Attention." *Scientific Reports* 2 (2012): art. 335. https://doi.org/10.1038/srep00335.

West, Jevin D., and Carl T. Bergstrom. "Misinformation in and About Science." *Proceedings of the National Academy of Sciences* 118, no. 15 (2021): e1912444117. https://doi.org/10.1073/pnas.1912444117.

Westen, Drew, Pavel S. Blagov, Keith Harenski, Clint Kilts, and Stephan Hamann. "An fMRI Study of Motivated Reasoning: Partisan Political Reasoning in the US Presidential Election." Unpublished manuscript, Emory University, Psychology Department, 2006.

Wetsman, Nicole. "Masks May Be Good, but the Messaging Around Them Has Been Very Bad." *The Verge*, April 3, 2020. https://www.theverge.com/2020/4/3/21206728/cloth-face-masks-white-house-coronavirus-covid-cdc-messaging.

Whalen, Andrew. " 'Behind the Curve' Ending: Flat Earthers Disprove Themselves with Own Experiments in Netflix Documentary." *Newsweek*, February 25, 2019. https://www.newsweek.com/behind-curve-netflix-ending-light-experiment-mark-sargent-documentary-movie-1343362.

Wilholt, Torsten. "Epistemic Trust in Science." *British Journal for the Philosophy of Science* 64, no. 2 (2013): 233–53. http://www.jstor.org/stable/24563046.

Wilkins, Matt. "More Recycling Won't Solve Plastic Pollution." *Scientific American*, July 6, 2018. https://blogs.scientificamerican.com/observations/more-recycling-wont-solve-plastic-pollution/.

Williams, Kip. "Ostracism: The Impact of Being Rendered Meaningless." In *Meaning, Mortality, and Choice: The Social Psychology of Existential Concerns*, ed. P. R. Shaver and M. Mikulincer, 309–23. American Psychological Association, 2012.

Wineburg, Sam, Joel Breakstone, Sarah McGrew, Mark D. Smith, and Teresa Ortega. "Lateral Reading on the Open Internet: A District-Wide Field Study in High School Government Classes." *Journal of Educational Psychology* 114, no. 5 (2022): 893–909. https://doi.org/10.1037/edu0000740.

Wineburg, Sam, and Sarah McGrew. "Lateral Reading and the Nature of Expertise: Reading Less and Learning More When Evaluating Digital Information." *Teachers College Record* 121, no. 11 (2019): 1–40. https://doi.org/10.1177/016146811912101102.

Winter, Kevin, Lotte Pummerer, Matthew J. Hornsey, and Kai Sassenberg. "Pro-Vaccination Subjective Norms Moderate the Relationship Between Conspiracy Mentality and Vaccination Intentions." *British Journal of Health Psychology* 27, no. 2 (2022): 390–405. https://doi.org/10.1111/bjhp.12550.

Witynski, Max. "False Balance in News Coverage of Climate Change Makes It Harder to Address Crisis." Phys.org, July 22, 2022. https://phys.org/news/2022-07-false-news-coverage-climate-harder.html.

Wojcieszak, Magdalena, Andreu Casas, Xudong Yu, Jonathan Nagler, and Joshua A. Tucker. "Most Users Do Not Follow Political Elites on Twitter; Those Who Do Show Overwhelming Preferences for Ideological Congruity." *Science Advances* 8, no. 39 (2022): eabn9418. https://doi.org/10.1126/sciadv.abn9418.

Wolak, Jennifer. "The Social Foundations of Public Support for Political Compromise." *The Forum* 20, no. 1 (2022): 185–207. https://doi.org/10.1515/for-2022-2050.

Wong, Norman C. H. " 'Vaccinations Are Safe and Effective': Inoculating Positive HPV Vaccine Attitudes Against Antivaccination Attack Messages." *Communication Reports* 29, no. 3 (2016): 127–38. https://doi.org/10.1080/08934215.2015.1083599.

Wood, Michelle L. M. "Rethinking the Inoculation Analogy: Effects on Subjects with Differing Preexisting Attitudes." *Human Communication Research* 33, no. 3 (2007): 357–78. https://doi.org/10.1111/j.1468-2958.2007.00303.x.

Woodward, Alex. "'Fake News': A Guide to Trump's Favourite Phrase—and the Dangers It Obscures." *The Independent*, October 2, 2020. https://www.independent.co.uk/news/world/americas/us-election/trump-fake-news-counter-history-b732873.html.

World Health Organization. "COVID-19 Cases." Accessed June 20, 2024. https://covid19.who.int/region/amro/country/us.

World Health Organization. "Ten Threats to Global Health in 2019." Accessed July 4, 2024. https://www.who.int/news-room/spotlight/ten-threats-to-global-health-in-2019.

World Population Review. "Facebook Users by Country 2024." Accessed July 6, 2024. https://worldpopulationreview.com/country-rankings/facebook-users-by-country.

World Population Review. "400 [U.S.] Cities." Accessed June 20, 2024. https://worldpopulationreview.com/countries/cities/united-states.

World War I Document Archive. Propaganda leaflets. Brigham Young University Library, last edited June 30, 2009. https://wwi.lib.byu.edu/index.php/Propaganda_Leaflets.

Worobey, Michael, Joshua I. Levy, Lorena Malpica Serrano et al. "The Huanan Seafood Wholesale Market in Wuhan Was the Early Epicenter of the COVID-19 Pandemic." *Science* 377, no. 6609 (2022): 951–59. https://doi.org/10.1126/science.abp8715.

Wright, Turner. "Fake News: Litecoin Price Surges 35 Percent Following Walmart Adoption Hoax." Cointelegraph, September 13, 2021. https://cointelegraph.com/news/fake-news-litecoin-price-surges-35-following-walmart-adoption-hoax.

Xin, Sufei, Ziqiang Xin, and Chongde Lin. "Effects of Trustors' Social Identity Complexity on Interpersonal and Intergroup Trust." *European Journal of Social Psychology* 46, no. 4 (2016): 428–40. https://doi.org/10.1002/ejsp.2156.

Yaden, David B., Jonathan Iwry, Kelley J. Slack et al. "The Overview Effect: Awe and Self-Transcendent Experience in Space Flight." *Psychology of Consciousness: Theory, Research, and Practice* 3, no. 1 (2016): 1–11. https://doi.org/10.1037/cns0000092.

Yale Medicine. "Why Doctors Wear Masks." September 1, 2020. https://www.yalemedicine.org/news/why-doctors-wear-masks.

YaleNews. "Yale Experts Join Campaign to Boost Vaccinations in Communities of Color." December 17, 2021. https://news.yale.edu/2021/12/17/yale-experts-join-campaign-boost-vaccinations-communities-color.

Yan, Harry Yaojun, Kai-Cheng Yang, Filippo Menczer, and James Shanahan. "Asymmetrical Perceptions of Partisan Political Bots." *New Media and Society* 23, no. 10 (2021): 3016–37. https://doi.org/10.1177/1461444820942744.

Yang, JungHwan, Hernando Rojas, Magdalena Wojcieszak et al. "Why Are 'Others' So Polarized? Perceived Political Polarization and Media Use in 10 Countries." *Journal of Computer-Mediated Communication* 21, no. 5 (2016): 349–67. https://doi.org/10.1111/jcc4.12166.

Yang, Yang, Tanya Y. Tian, Teresa K. Woodruff, Benjamin F. Jones, and Brian Uzzi. "Gender-Diverse Teams Produce More Novel and Higher-Impact Scientific Ideas." *Proceedings of the National Academy of Sciences* 119, no. 36 (2022): e2200841119. https://doi.org/10.1073/pnas.2200841119.

Yankoski, Michael, Tim Weninger, and Walter Scheirer. "An AI Early Warning System to Monitor Online Disinformation, Stop Violence, and Protect Elections." *Bulletin of the Atomic Scientists* 76, no. 2 (2020): 85–90. https://doi.org/10.1080/00963402.2020.1728976.

Yarrow, David. "From Fact-Checking to Value-Checking: Normative Reasoning in the New Public Sphere." *Political Quarterly* 92, no. 4 (2021): 621–28. https://doi.org/10.1111/1467-923X.12999.

Yasmin, Seema. *Viral BS: Medical Myths and Why We Fall for Them.* Johns Hopkins University Press, 2021.

Yeager, Ashley. "Government's Mixed Messages on Coronavirus Are Dangerous: Experts." *The Scientist*, February 28, 2020. https://www.the-scientist.com/news-opinion/governments-mixed-messages-on-coronavirus-are-dangerous-experts-67202.

Yong, Ed. "America Is Getting Unvaccinated People All Wrong." *The Atlantic*, July 22, 2021. https://www.theatlantic.com/health/archive/2021/07/unvaccinated-different-anti-vax/619523/.

Yong, Ed. "It's a Terrible Idea to Deny Medical Care to Unvaccinated People." *The Atlantic*, January 20, 2022. https://www.theatlantic.com/health/archive/2022/01/unvaccinated-medical-care-hospitals-omicron /621299/.

Young, Dannagal G., and Amy Bleakley. "Ideological Health Spirals: An Integrated Political and Health Communication Approach to COVID Interventions." *International Journal of Communication Systems* 14, no. 14 (2020): 3508–24.

Yousefinaghani, Samira, Rozita Dara, Samira Mubareka, Andrew Papadopoulos, and Shayan Sharif. "An Analysis of COVID-19 Vaccine Sentiments and Opinions on Twitter." *International Journal of Infectious Diseases* 108 (2021): 256–62. https://doi.org/10.1016/j.ijid.2021.05.059.

Yudkin, Daniel, Stephen Hawkins, and Tim Dixon. *The Perception Gap: How False Impressions Are Pulling Americans Apart*. More in Common, 2019.

Zakrzewski, Cat, and Cristiano Lima-Strong. "Former Facebook Employee Frances Haugen Revealed as 'Whistleblower' Behind Leaked Documents That Plunged the Company Into Scandal." *Washington Post*, October 3, 2021. https://www.washingtonpost.com/technology/2021/10/03/facebook-whistleblower -frances-haugen-revealed/.

Zerback, Thomas, Florian Töpfl, and Maria Knöpfle. "The Disconcerting Potential of Online Disinformation: Persuasive Effects of Astroturfing Comments and Three Strategies for Inoculation Against Them." *New Media and Society* 23, no. 5 (2021): 1080–98. https://doi.org/10.1177/1461444820908530.

Zhang, Yini, Fan Chen, and Josephine Lukito. "Network Amplification of Politicized Information and Misinformation About COVID-19 by Conservative Media and Partisan Influencers on Twitter." *Political Communication* 40, no. 1 (2023): 24–47. https://doi.org/10.1080/10584609.2022.2113844.

Zhao, Nan, Song Yao, Raphael Thomadsen, and Chong Bo Wang. "The Impact of Government Interventions on COVID-19 Spread and Consumer Spending." *Management Science* 70, no. 5 (2024): 3302–18. https://doi.org/10.1287/mnsc.2023.4853.

Zhu, Xun, and Youllee Kim. "Mitigating Identity Threat in Health Messaging: A Social Identity Complexity Perspective." *Health Communication*, May 22, 2024, 1–12. https://doi.org/10.1080/10410236.2024 .2358275.

Zinn, Anna K., Aureliu Lavric, Mark Levine, and Miriam Koschate. "Social Identity Switching: How Effective Is It?" *Journal of Experimental Social Psychology* 101 (2022): 104309. https://doi.org/10.1016/j .jesp.2022.104309.

Zion Market Research. "Insights on Global Homeopathy Products Market Size & Share Projected to Hit at USD 50,203.3 Million and Rise at a CAGR of 18.7 Percent by 2028: Industry Trends, Demand, Value, Analysis & Forecast Report." PR Newswire, May 17, 2022. https://www.prnewswire.com/news-releases /insights-on-global-homeopathy-products-market-size--share-projected-to-hit-at-usd-50-203-3-million -and-rise-at-a-cagr-of-18-7-by-2028-industry-trends-demand-value-analysis--forecast-report--zion-market -research-301549050.html.

Zmigrod, Leor, and Manos Tsakiris. "Computational and Neurocognitive Approaches to the Political Brain: Key Insights and Future Avenues for Political Neuroscience." *Philosophical Transactions of the Royal Society of London, Series B, Biological Sciences* 376, no. 1822 (2021): 20200130. https://doi.org/10.1098/rstb .2020.0130.

Zubrow, Keith. "Facebook Whistleblower Says Company Incentivizes 'Angry, Polarizing, Divisive Content.'" *60 Minutes Overtime*, CBS News, October 4, 2021. https://www.cbsnews.com/news/facebook-whistleblower -frances-haugen-60-minutes-polarizing-divisive-content/.

Zuckerman, Ethan. *Mistrust: Why Losing Faith in Institutions Provides the Tools to Transform Them*. Norton, 2021.

INDEX

GPSR Authorized Representative: Easy Access System Europe, Mustamäe tee
50, 10621 Tallinn, Estonia, gpsr.requests@easproject.com

www.ingramcontent.com/pod-product-compliance
Lightning Source LLC
Chambersburg PA
CBHW022133020426
42334CB00015B/876